# Diagnosis and Treatment for Hepatocellular Tumors

# Diagnosis and Treatment for Hepatocellular Tumors

Editor

**Georgios Germanidis**

MDPI

Basel • Beijing • Wuhan • Barcelona • Belgrade • Novi Sad • Cluj • Manchester

*Editor*
Georgios Germanidis
AHEPA University Hospital,
Aristotle University of
Thessaloniki
Thessaloniki, Greece

*Editorial Office*
MDPI
St. Alban-Anlage 66
4052 Basel, Switzerland

This is a reprint of articles from the Special Issue published online in the open access journal *Cancers* (ISSN 2072-6694) (available at: https://www.mdpi.com/journal/cancers/special_issues/Diagnosis_Treatment_Hepatocellular).

For citation purposes, cite each article independently as indicated on the article page online and as indicated below:

Lastname, A.A.; Lastname, B.B. Article Title. *Journal Name* **Year**, *Volume Number*, Page Range.

**ISBN 978-3-0365-9320-3 (Hbk)**
**ISBN 978-3-0365-9321-0 (PDF)**
**doi.org/10.3390/books978-3-0365-9321-0**

# Contents

# About the Editor

**Georgios Germanidis**

Georgios Germanidis was born on the 21st of February, 1963, and is a Full Professor of Gastroenterology. He graduated from the Medical School of the University of Athens in 1987 and from 25-1-1987 to 4-4-1988 was a postgraduate student at the 2nd Department of Internal Medicine, Hippokrateion Hospital, University of Athens. From 19/11/1987 to 19/11/1989, he served in the Greek Military and was a postgraduate student at the Immunology Laboratory of the University of Ioannina. Moreover, from 13-3-1991 to 12-1-1995, he was a Resident of Internal Medicine at the General Hospital of Thessaloniki G.Papanikolaou, and from 15-1-1995 to 15-2-1999, he was a Resident in Gastroenterology, Service de Gastroenterologie et. Hepatologie, at the Hopital Henri Mondor, Universite Paris XII, with Prof. Daniel Dhumeaux, Prof. Jean Charles Delchier., and Prof. Jean Michel Pawlotsky. In 3-8-1999, he received his Specialty in Gastroenterology, and from 9-2-2000 to 29-3-2007, he was a Staff Gastroenterologist at the First Department of Medicine in Papageorgiou General Hospital of Thessaloniki. In 2005, he presented his Doctorate Thesis at the Medical School, Aristotle University of Thessaloniki, entitled 'Hepatitis B virus kinetics under antiviral therapy'. From 30-3-2007 to 17-6-2010, he was Staff Gastroenterologist at the First Department of Internal Medicine, in Thessaloniki, Greece. From 17-6-2010 to 26-9-2013, he was a Lecturer in Gastroenterology at the Medical School, Aristotle University of Thessaloniki, while from 26-9-2013 to 30-8-2018, he was an Assistant Professor in Gastroenterology, and from August 2018 until June 2022, he was an Associate Professor in Gastroenterology, at the Medical School, Aristotle University of Thessaloniki. Since 1st July 2022, he has been a Full Professor of Gastroenterology and Head of the Hepatology and Gastroenterology Unit of the 1st Department of Internal Medicine, Aristotle University of Thessaloniki, St. Kiriakidi 1, 54636, Greece.

*cancers*

*Editorial*

# The Liver Cancer Immune Microenvironment: Emerging Concepts for Myeloid Cell Profiling with Diagnostic and Therapeutic Implications

**Konstantinos Arvanitakis [1,2], Ioannis Mitroulis [3], Antonios Chatzigeorgiou [4], Ioannis Elefsiniotis [5] and Georgios Germanidis [1,2,*]**

1 First Department of Internal Medicine, AHEPA University Hospital, Aristotle University of Thessaloniki, 54636 Thessaloniki, Greece
2 Basic and Translational Research Unit (BTRU) of Special Unit for Biomedical Research and Education (BRESU), Faculty of Health Sciences, School of Medicine, Aristotle University of Thessaloniki, 54636 Thessaloniki, Greece
3 First Department of Internal Medicine, University Hospital of Alexandroupolis, Democritus University of Thrace, 68100 Alexandroupolis, Greece
4 Department of Physiology, Medical School, National and Kapodistrian University of Athens, 11527 Athens, Greece
5 University Department of Internal Medicine, General and Oncology Hospital of Kifisia Agioi Anargyroi, 14564 Athens, Greece
* Correspondence: geogerm@auth.gr; Tel.: +30-2313303156; Fax: +30-2310994638

**Citation:** Arvanitakis, K.; Mitroulis, I.; Chatzigeorgiou, A.; Elefsiniotis, I.; Germanidis, G. The Liver Cancer Immune Microenvironment: Emerging Concepts for Myeloid Cell Profiling with Diagnostic and Therapeutic Implications. *Cancers* **2023**, *15*, 1522. https://doi.org/10.3390/cancers15051522

Received: 12 February 2023
Revised: 22 February 2023
Accepted: 23 February 2023
Published: 28 February 2023



## 1. Introduction

Hepatocellular carcinoma (HCC) is one of the leading causes of cancer-related deaths worldwide. To date, systemic treatment for patients with unresectable or advanced disease was composed of sorafenib and other multikinase inhibitors, with limited efficacy and high toxicity. Nevertheless, immune checkpoint inhibitors (ICIs) have greatly broadened the treatment landscape of unresectable HCC [1]. Specifically, the IMbrave150 trial demonstrated that in patients with unresectable HCC, combined treatment with the programmed-death ligand-1 (PD-L1) inhibitor, atezolizumab, alongside the vascular endothelial growth factor (VEGF) inhibitor, bevacizumab, prolonged the median overall survival (OS) to 19.2 months as compared to sorafenib treatment alone [2]. However, around 20–25% of patients exhibit complete primary resistance to atezolizumab, plus bevacizumab, indicating that the identification of patients who might gain the most from this therapy is crucial.

To date, there are no definite biomarkers in HCC that can accurately predict response or resistance to ICIs, while as the HCC treatment regimens have shifted towards immunotherapy, the identification of potent predictive and prognostic biomarkers has attracted attention. Although HCC formation has been attributed to certain viral or non-viral causes, nonalcoholic steatohepatitis (NASH), is a major driver of HCC as well [3]. Recent data suggest that NASH-related HCC might have decreased sensitivity to immunotherapy, due to the presence of CD8+ T cells and, especially, of the hepatic steatosis-induced CXCR6+ subset that was correlated with hepatocyte injury and potentiated the pathogenesis of NASH-related HCC via the secretion of pro-inflammatory cytokines and direct hepatocyte killing, mediated by the tumor necrosis factor (TNF) [4]. Moreover, local and systemic inflammation are considered hallmarks of cancer, and they have a pivotal role in HCC pathogenesis and progression [5]. An increased peripheral blood absolute neutrophil count and an elevated neutrophil to lymphocyte ratio (NLR $\geq$ 5) are considered markers of advanced disease, poor prognosis, and poor response to treatment with hepatic resection, transplantation, locoregional therapy, and tyrosine kinase inhibitors in patients with HCC. Indeed, systemic inflammation measured by NLR is independently a negative prognostic factor for patients with HCC under ICI therapy [6]. The measurement of the NLR across various time points could provide insight into how different values of this inflammatory marker could accurately predict patient response to systemic therapy, patient outcomes, or the development of adverse events (AEs).

## 2. TANs and TAMs in the Immune Microenvironment of HCC

The tumor immune microenvironment (TIME), being heterogeneous and comprised of a multitude of immune and stromal cells, is an essential factor that leads to tumor metastasis and relapse, as well as resistance to therapy, whereas the way in which different TIME cell subtypes are connected with the clinical relevance in liver cancer remains unclear [7]. Indeed, cells of innate and adaptive immunity coexist and interact within the liver microenvironment during HCC, especially in the case of NASH, during which chronic hepatic inflammation preexists the emergence of HCC [8]. As far as the innate immune cells are concerned, in a mouse model of NASH-related HCC, neutrophils were shown to predominantly increase in the course of the disease, in comparison to other immune cells [9]. Indeed, in humans, high numbers of tumor-associated neutrophils (TANs) are a biomarker of poor prognosis in HCC and various other cancers [10]. Nevertheless, whether TANs, either within the HCC TIME or in a peritumoral hepatic location, account for this association remains unclear. However, these results should be carefully assessed, since they derive from HCC patients undergoing liver resection or liver transplantation, and it is typical for such patients to present with early, localized disease with preserved performance status and liver function. In countries with a low prevalence of viral hepatitis and a high prevalence of nonalcoholic fatty liver disease (NAFLD), approximately 15% of patients with HCC present with early disease and are considered candidates for curative resection. Consequently, the results of the aforementioned studies regarding TANs, might not be similar and comparable for individuals with more advanced HCC that are usually candidates for systemic treatment and account for the majority of patients.

Neutrophils have considerable phenotypic plasticity and can exist in both tumor-promoting (TAN2) and tumor-suppressing (TAN1) states. Neutrophils may also have the ability to influence ICIs therapy. Recent data report that CXCR2+ neutrophils were found in human NASH and within the tumor of both human and mouse models of NASH-related HCC. The resistance of NASH-related HCC to anti-PD1 therapy is being overcome by co-treatment with a CXCR2 small molecule inhibitor, with evidence of reduced tumor burden and extended survival [11]. Anti-PD1 and CXCR2 inhibitors combine to selectively reprogram TANs from a protumor to an antitumor phenotype, which unlocks their potential for cancer therapy. The ability of CXCR2 antagonism to combine with ICI therapy in order to lead to enhanced therapeutic benefit in NASH-related HCC (and potentially in HCC related to other aetiologies) warrants further clinical investigation. Along the same line, a recent animal study added that ferroptosis, caused by a tumor-suppressive immune response, is characterised by a CXCL10-dependent infiltration of cytotoxic CD8+ T cells, which at the same time was counterbalanced by a PD-L1 upregulation on tumor cells, as well as by a marked myeloid-derived suppressor cell (MDSC) infiltration. A triple combination of a ferroptosis-inducing agent, a CXCR2 inhibitor, and an anti-PD1, greatly improved the survival of wild-type mice with liver tumors [12].

Furthermore, another important study [13], by integrated analyses on molecular correlates of clinical response and resistance to atezolizumab in combination with bevacizumab in advanced patient samples of HCC collected within the phase Ib GO30140 and the phase III IMbrave150 trials, identified key molecular correlates of the combination therapy and highlighted that anti-VEGF might synergize with anti-PD-L1 by targeting angiogenesis, regulatory T-cells (Tregs) proliferation and myeloid cell inflammation. The presence of pre-existing T-cell immunity is a key phenotypic characteristic that correlates with the response to atezolizumab plus bevacizumab. TIME potentiating the enrichment of the effector T-cell response over immunosuppressive Tregs identified patients that achieved significantly improved overall survival from the aforementioned combination. These findings were further validated by analyses of paired pre- and post-treatment biopsies, in situ analyses, and in vivo mouse models. Recent studies have also revealed the critical role of antigen non-specific auto-aggressive CD8+ T cells in instigating liver damage and promoting liver cancer in human NASH [14]. In addition, hepatic CD8+ PD1+ CXCR6+ T cells of humans with NASH, as well as neutrophil extracellular traps (NETs), contributed to the development of NASH-related HCC by promoting Treg differentiation, thus suppressing HCC immune surveillance [9,15] (Figure 1).

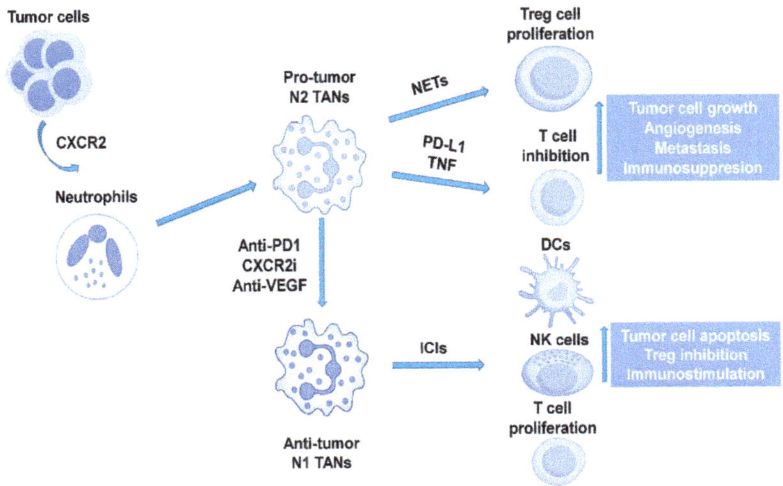


**Figure 1.** The role of neutrophils in the liver cancer immune microenvironment. CXCR2: C-X-C motif chemokine receptor 2; TAN: tumor-associated neutrophil; PD1: programmed cell death protein 1; CXCR2i: C-X-C motif chemokine receptor 2 inhibitor; NET: neutrophil extracellular trap; PD-L1: programmed death-ligand 1; ICI: immune checkpoint inhibitor; Treg: regulatory T; DC: dentritic cell; NK: natural killer.

Through in vitro induction of TANs and ex vivo analyses of human TANs, a recent study also showed that CCL4+ TANs can recruit macrophages and that PD-L1 + TANs can suppress T cell cytotoxicity [16]. Monocytes are recruited into the tumor site by the release of tumoral and stromal chemokines, such as CCL2 and CCL15. Monocytes can be polarized into different subtypes such as CD14+, CCR1+, and CD14+ [7], while macrophages are in the epicentre of the molecular pathways regulating NASH-related HCC pathogenesis [17]. All of these subtypes promote a strong immunosuppressive environment with the expression of ICIs (PD-L1/2, B7-H3, and TIM3) and cytokines (IL-10, CXCL2, and CXCL8), inhibiting natural killer (NK) cytotoxicity, inducing Tregs. They also interact with neutrophils to promote tumor invasiveness through the oncostatin M pathway. A way to control tumorigenesis through monocytes would be through the prevention of their recruitment to the tumor site by inhibiting the CCL15 pathway, via blockade of their polarization by the inhibition of the p38 pathway, or via repressing the IL-6 pathway in order to prevent the formation of Tregs. The CD68 marker is commonly used for liver tumor-associated macrophage (TAM) localization and distribution, while the expression levels of CD86 (M1), CD163 (M2), and CD206 (M2) are used to distinguish between M1-like (inflammatory) and M2-like (anti-inflammatory) macrophages in vitro [18]. Collectively, these data show that non-viral HCC, and particularly NASH-related HCC, might be less responsive to immunotherapy, at least partially due to the presence of TANs and TAMs in the TIME of HCC.

## 3. Myeloid Cells and HCC Profiling

However, an important unmet clinical need is demonstrated by the lack of accurate biomarkers that can influence therapeutic choices. This unfulfilled need is being met by very few novel studies that incorporate multiregional single cell-dissection landscape of tumor and immune cells in HCC with the sole purpose of shedding light on the biological tumor characteristics and identifying potential tumor and blood biomarkers, in order to categorize specific groups of TIME and identify patients who might have a benefit from a specific treatment option. In brief, the combination of two single-cell RNA sequencing technologies [19], produced transcriptomes of CD45+ immune cells for HCC patients from five immune-relevant sites, and demonstrated an aggregate of LAMP3+ dendritic cells (DCs) that could modulate different subtypes of lymphocytes. Moreover, TAMs were correlated with poor prognosis, and the authors provided evidence of the inflammatory role of SLC40A1 and GPNMB in those cells.

Additionally, Sun et al. [20], by performing single-cell profiling in relapsed HCC, remarkably found that CD8+ T cells in recurrent tumors overexpressed KLRB1 (CD161) and displayed an innate-like low cytotoxic state, with low clonal expansion, unlike the classical exhausted state observed in primary HCC. The enrichment of these cells was associated with a worse prognosis. In addition, by performing multiregional single-cell RNA sequencing (scRNA-seq) analysis, Ma et al. [21], identified and further validated the cellular dynamics of malignant cells and their communication networks with tumor-associated immune cells in terms of ligand-receptor interaction pairs associated with unique transcriptome. These molecular networks of malignant ecosystems, may open a path for therapeutic exploration. Very recently, too, the first proteogenomic characterization of hepatitis B virus (HBV)-related HCC using paired tumor and adjacent liver tissues from 159 patients was performed by Gao Q et al. [22], and two prognostic biomarkers, PYCR2 and ADH1A, which were related to proteomic subgrouping and were involved in HCC metabolic reprogramming, were identified. CTNNB1 and TP53 mutation-associated signaling and metabolic profiles were revealed, among which, mutated CTNNB1-associated ALDOA phosphorylation was demonstrated to promote glycolysis and cell proliferation.

In a molecular study of HCC in patients with NASH, NASH-related HCCs were characterized by bile and fatty acid signaling, oxidative stress, and inflammation, and demonstrated an increased fraction of Wnt/TGF-β subclass of tumors and a decreased fraction of the CTNNB1 subclass. In comparison to other etiologies, NASH-related HCC had a considerably higher prevalence of an immunosuppressive cancer field [23]. Moreover, it was also demonstrated that the prognostic liver signature (PLS)-NAFLD predicted incident HCC over up to 15 years of longitudinal observation, while high-risk PLS-NAFLD was associated with IDO1+ dendritic cells and dysfunctional CD8+ T cells in fibrotic portal tracts, with impaired metabolic regulators. PLS-NAFLD was affected by bariatric surgery, lipophilic statins, and the use of IDO1 inhibitors, implicating that it could be utilized in pharmacotherapy and HCC chemoprevention [24].

Interestingly, treatment modalities aiming towards specific genomic alterations form the basis of personalized medicine and constitute the epitome of systemic treatment for many malignancies, but are still not available in HCC. Tools such as liquid biopsy and circulating tumor DNA (ctDNA), even though most studies have not analyzed HCC tissue concomitantly, may be of aid in identifying biomarkers of early diagnosis, response, or resistance to treatment, and their role in HCC represents an ongoing research field [25]. In addition, extracellular vesicles (EVs) or exosomes provide a critical mechanistic way of bidirectional intercellular communication in the TIME of various cancers and it would be very interesting to characterize tumor-derived versus immune-cell-TIME-derived EVs for HCC, in terms of their functionally important genomic, lipidomic, and proteomic cargo [26]. Finally, an integrative analysis of RNA and whole exome sequencing, T-cell receptor (TCR) sequencing, multiplex immunofluorescence, and immunohistochemistry was performed in a cohort of 240 patients with HCC and was validated in other cohorts of 660 patients in total [27]. A 20-gene signature, characterized by high interferon signalling and type I antigen-presenting genes, defined the inflamed class of HCC and was able to capture ~90% of these tumors and was associated with response to immunotherapy. Proteins identified in liquid biopsies recapitulated the inflamed class with an area under the ROC curve (AUC) of 0.91.

## 4. Critical Analysis of Data and Future Perspectives

Taking into account the aforementioned complex molecular omics alongside the heterogeneous cellular landscape of the TIME of HCC, future studies are expected to highlight in a simple way that is practical for clinical use, the hepatic and peripheral blood inflammatory and immunosuppressive tumorigenic function and composition of TANs and TAMs in NASH-related HCC in comparison with other aetiologies and, more importantly, to shed further light on potential prognostic or predictive molecular markers for future immunotherapies targeting TANs and TAMs, acting in synergy with ICIs, in order to overcome resistance and eventually improve the percentage of patients with HCC that respond to treatment. In addition, we underline the significance of predictive biomarkers of response to ICIs in order to (i) enhance the overall survival of patients that are likely to respond to therapy, (ii) reduce the risk of treatment-related adverse effects conveyed through the combination of drugs such as bevacizumab, (iii) maximize efficacious application, and therefore the cost-effectiveness of different treatment modalities, and (iv) characterize the molecular landscape of patients with advanced HCC responding to anti-PD1 therapy and define a novel tool for patient selection in future clinical trials.

All of these data render HCC an oncological diagnosis in which spontaneous immunogenicity is critical for the efficacy of immunotherapy. Although the predictive value of histopathologic assessment is unparalleled, TIME immunogenicity is influenced by density, functional polarization, and distribution of the infiltrate. The diversity of the HCC TIME and the demand for readily applicable biomarkers rather than complex transcriptomics is still challenging, while the ultimate goal of expanding the reach of effective cancer immunotherapy to a wider proportion of patients via clinical stratification of trial participants or targeted testing of novel combinations prognostically modulating adverse traits, such as TANs and TAMs infiltration, is of the utmost importance.

**Author Contributions:** K.A., I.M., A.C., I.E. and G.G. performed the writing and review of this paper. All authors have read and agreed to the published version of the manuscript.

**Funding:** This research received no external funding.

**Conflicts of Interest:** All of the authors declare no conflict of interest.

## References

1. Foerster, F.; Gairing, S.J.; Ilyas, S.I.; Galle, P.R. Emerging immunotherapy for HCC: A guide for hepatologists. *Hepatology* **2022**, *75*, 1604–1626. [CrossRef]
2. Cheng, A.L.; Qin, S.; Ikeda, M.; Galle, P.R.; Ducreux, M.; Kim, T.Y.; Lim, H.Y.; Kudo, M.; Breder, V.; Merle, P.; et al. Updated efficacy and safety data from IMbrave150: Atezolizumab plus bevacizumab vs. sorafenib for unresectable hepatocellular carcinoma. *J. Hepatol.* **2022**, *76*, 862–873. [CrossRef]
3. Anstee, Q.M.; Reeves, H.L.; Kotsiliti, E.; Govaere, O.; Heikenwalder, M. From NASH to HCC: Current concepts and future challenges. *Nat. Rev. Gastroenterol. Hepatol.* **2019**, *16*, 411–428. [CrossRef]
4. Pfister, D.; Nunez, N.G.; Pinyol, R.; Govaere, O.; Pinter, M.; Szydlowska, M.; Gupta, R.; Qiu, M.; Deczkowska, A.; Weiner, A.; et al. NASH limits anti-tumour surveillance in immunotherapy-treated HCC. *Nature* **2021**, *592*, 450–456. [CrossRef]
5. Muhammed, A.; Fulgenzi, C.A.M.; Dharmapuri, S.; Pinter, M.; Balcar, L.; Scheiner, B.; Marron, T.U.; Jun, T.; Saeed, A.; Hildebrand, H.; et al. The Systemic Inflammatory Response Identifies Patients with Adverse Clinical Outcome from Immunotherapy in Hepatocellular Carcinoma. *Cancers* **2021**, *14*, 186. [CrossRef]
6. Wu, Y.L.; Fulgenzi, C.A.M.; D'Alessio, A.; Cheon, J.; Nishida, N.; Saeed, A.; Wietharn, B.; Cammarota, A.; Pressiani, T.; Personeni, N.; et al. Neutrophil-to-Lymphocyte and Platelet-to-Lymphocyte Ratios as Prognostic Biomarkers in Unresectable Hepatocellular Carcinoma Treated with Atezolizumab plus Bevacizumab. *Cancers* **2022**, *14*, 5834. [CrossRef]
7. Donne, R.; Lujambio, A. The liver cancer immune microenvironment: Therapeutic implications for hepatocellular carcinoma. *Hepatology* **2022**. *early view*. [CrossRef]

8. Pinter, M.; Pinato, D.J.; Ramadori, P.; Heikenwalder, M. NASH and Hepatocellular Carcinoma: Immunology and Immunotherapy. *Clin. Cancer Res.* **2023**, *29*, 513–520. [CrossRef]
9. Wang, H.; Zhang, H.; Wang, Y.; Brown, Z.J.; Xia, Y.; Huang, Z.; Shen, C.; Hu, Z.; Beane, J.; Ansa-Addo, E.A.; et al. Regulatory T-cell and neutrophil extracellular trap interaction contributes to carcinogenesis in non-alcoholic steatohepatitis. *J. Hepatol.* **2021**, *75*, 1271–1283. [CrossRef]
10. Arvanitakis, K.; Mitroulis, I.; Germanidis, G. Tumor-Associated Neutrophils in Hepatocellular Carcinoma Pathogenesis, Prognosis, and Therapy. *Cancers* **2021**, *13*, 2899. [CrossRef]
11. Leslie, J.; Mackey, J.B.G.; Jamieson, T.; Ramon-Gil, E.; Drake, T.M.; Fercoq, F.; Clark, W.; Gilroy, K.; Hedley, A.; Nixon, C.; et al. CXCR2 inhibition enables NASH-HCC immunotherapy. *Gut* **2022**, *71*, 2093–2106. [CrossRef]
12. Conche, C.; Finkelmeier, F.; Pesic, M.; Nicolas, A.M.; Bottger, T.W.; Kennel, K.B.; Denk, D.; Ceteci, F.; Mohs, K.; Engel, E.; et al. Combining ferroptosis induction with MDSC blockade renders primary tumours and metastases in liver sensitive to immune checkpoint blockade. *Gut* **2023**, 1–9. [CrossRef]
13. Zhu, A.X.; Abbas, A.R.; de Galarreta, M.R.; Guan, Y.; Lu, S.; Koeppen, H.; Zhang, W.; Hsu, C.H.; He, A.R.; Ryoo, B.Y.; et al. Molecular correlates of clinical response and resistance to atezolizumab in combination with bevacizumab in advanced hepatocellular carcinoma. *Nat. Med.* **2022**, *28*, 1599–1611. [CrossRef]
14. Dudek, M.; Pfister, D.; Donakonda, S.; Filpe, P.; Schneider, A.; Laschinger, M.; Hartmann, D.; Huser, N.; Meiser, P.; Bayerl, F.; et al. Auto-aggressive CXCR6(+) CD8 T cells cause liver immune pathology in NASH. *Nature* **2021**, *592*, 444–449. [CrossRef]
15. Velliou, R.I.; Mitroulis, I.; Chatzigeorgiou, A. Neutrophil extracellular traps contribute to the development of hepatocellular carcinoma in NASH by promoting Treg differentiation. *Hepatobiliary Surg. Nutr.* **2022**, *11*, 415–418. [CrossRef]
16. Xue, R.; Zhang, Q.; Cao, Q.; Kong, R.; Xiang, X.; Liu, H.; Feng, M.; Wang, F.; Cheng, J.; Li, Z.; et al. Liver tumour immune microenvironment subtypes and neutrophil heterogeneity. *Nature* **2022**, *612*, 141–147. [CrossRef]
17. Kohlhepp, M.S.; Liu, H.; Tacke, F.; Guillot, A. The contradictory roles of macrophages in non-alcoholic fatty liver disease and primary liver cancer—Challenges and opportunities. *Front. Mol. Biosci.* **2023**, *10*, 1129831. [CrossRef]
18. Arvanitakis, K.; Koletsa, T.; Mitroulis, I.; Germanidis, G. Tumor-Associated Macrophages in Hepatocellular Carcinoma Pathogenesis, Prognosis and Therapy. *Cancers* **2022**, *14*, 226. [CrossRef]
19. Zhang, Q.; He, Y.; Luo, N.; Patel, S.J.; Han, Y.; Gao, R.; Modak, M.; Carotta, S.; Haslinger, C.; Kind, D.; et al. Landscape and Dynamics of Single Immune Cells in Hepatocellular Carcinoma. *Cell* **2019**, *179*, 829–845.e20. [CrossRef]
20. Sun, Y.; Wu, L.; Zhong, Y.; Zhou, K.; Hou, Y.; Wang, Z.; Zhang, Z.; Xie, J.; Wang, C.; Chen, D.; et al. Single-cell landscape of the ecosystem in early-relapse hepatocellular carcinoma. *Cell* **2021**, *184*, 404–421.e416. [CrossRef]
21. Ma, L.; Heinrich, S.; Wang, L.; Keggenhoff, F.L.; Khatib, S.; Forgues, M.; Kelly, M.; Hewitt, S.M.; Saif, A.; Hernandez, J.M.; et al. Multiregional single-cell dissection of tumor and immune cells reveals stable lock-and-key features in liver cancer. *Nat. Commun.* **2022**, *13*, 7533. [CrossRef]
22. Gao, Q.; Zhu, H.; Dong, L.; Shi, W.; Chen, R.; Song, Z.; Huang, C.; Li, J.; Dong, X.; Zhou, Y.; et al. Integrated Proteogenomic Characterization of HBV-Related Hepatocellular Carcinoma. *Cell* **2019**, *179*, 561–577.e22. [CrossRef] [PubMed]
23. Pinyol, R.; Torrecilla, S.; Wang, H.; Montironi, C.; Pique-Gili, M.; Torres-Martin, M.; Wei-Qiang, L.; Willoughby, C.E.; Ramadori, P.; Andreu-Oller, C.; et al. Molecular characterisation of hepatocellular carcinoma in patients with non-alcoholic steatohepatitis. *J. Hepatol.* **2021**, *75*, 865–878. [CrossRef]
24. Fujiwara, N.; Kubota, N.; Crouchet, E.; Koneru, B.; Marquez, C.A.; Jajoriya, A.K.; Panda, G.; Qian, T.; Zhu, S.; Goossens, N.; et al. Molecular signatures of long-term hepatocellular carcinoma risk in nonalcoholic fatty liver disease. *Sci. Transl. Med.* **2022**, *14*, eabo4474. [CrossRef]
25. Campani, C.; Zucman-Rossi, J.; Nault, J.-C. Genetics of Hepatocellular Carcinoma: From Tumor to Circulating DNA. *Cancers* **2023**, *15*, 817. [CrossRef]
26. Simon, T.; Jackson, E.; Giamas, G. Breaking through the glioblastoma micro-environment via extracellular vesicles. *Oncogene* **2020**, *39*, 4477–4490. [CrossRef]
27. Montironi, C.; Castet, F.; Haber, P.K.; Pinyol, R.; Torres-Martin, M.; Torrens, L.; Mesropian, A.; Wang, H.; Puigvehi, M.; Maeda, M.; et al. Inflamed and non-inflamed classes of HCC: A revised immunogenomic classification. *Gut* **2023**, *72*, 129–140. [CrossRef]

cancers

MDPI

*Review*

# Etiology, Pathogenesis, Diagnosis, and Practical Implications of Hepatocellular Neoplasms

**Prodromos Hytiroglou [1,*], Paulette Bioulac-Sage [2], Neil D. Theise [3] and Christine Sempoux [4]**

1 Department of Pathology, Aristotle University School of Medicine, 54124 Thessaloniki, Greece
2 INSERM, BRIC, U1312, University Bordeaux, F-33000 Bordeaux, France; paulette.bioulac@u-bordeaux.fr
3 Department of Pathology, New York University Grossman School of Medicine, New York, NY 10016, USA; neil.theise@nyulangone.org
4 Service of Clinical Pathology, Institute of Pathology, Lausanne University Hospital, University of Lausanne, CH-1007 Lausanne, Switzerland; christine.sempoux@chuv.ch
* Correspondence: pchytiro@auth.gr; Tel.: +30-2310-999-218

**Simple Summary:** In recent years, significant progress has been made in elucidating the mechanisms via which hepatocellular neoplasms, i.e., hepatocellular adenoma and hepatocellular carcinoma, arise. Hepatocellular carcinoma usually occurs in livers with chronic disease, due to deregulation of important intracellular pathways of signal transmission. Recent studies suggest that subclassification of hepatocellular carcinoma is practically useful. On the other hand, subclassification of hepatocellular adenomas has been well established through correlation of molecular alterations with morphology and protein expression. Advances in hepatic imaging have resulted in a new approach for diagnostic assessment of lesions arising in advanced chronic liver disease. Histologic examination, aided by immunohistochemistry, is the gold standard for the diagnosis and subclassification of hepatocellular neoplasms, while clinicopathologic correlation is essential for best patient management. We summarize the etiology and pathogenesis of hepatocellular neoplasms, provide practical information for their histologic diagnosis, and address various frequently asked questions regarding their diagnosis and practical implications.

**Abstract:** Hepatocellular carcinoma (HCC), a major global contributor of cancer death, usually arises in a background of chronic liver disease, as a result of molecular changes that deregulate important signal transduction pathways. Recent studies have shown that certain molecular changes of hepatocarcinogenesis are associated with clinicopathologic features and prognosis, suggesting that subclassification of HCC is practically useful. On the other hand, subclassification of hepatocellular adenomas (HCAs), a heterogenous group of neoplasms, has been well established on the basis of genotype–phenotype correlations. Histologic examination, aided by immunohistochemistry, is the gold standard for the diagnosis and subclassification of HCA and HCC, while clinicopathologic correlation is essential for best patient management. Advances in clinico-radio-pathologic correlation have introduced a new approach for the diagnostic assessment of lesions arising in advanced chronic liver disease by imaging (LI-RADS). The rapid expansion of knowledge concerning the molecular pathogenesis of HCC is now starting to produce new therapeutic approaches through precision oncology. This review summarizes the etiology and pathogenesis of HCA and HCC, provides practical information for their histologic diagnosis (including an algorithmic approach), and addresses a variety of frequently asked questions regarding the diagnosis and practical implications of these neoplasms.

**Keywords:** hepatocellular adenoma; hepatocellular carcinoma; molecular pathology; histologic diagnosis; diagnostic algorithm; LI-RADS; frequently asked questions

**Citation:** Hytiroglou, P.; Bioulac-Sage, P.; Theise, N.D.; Sempoux, C. Etiology, Pathogenesis, Diagnosis, and Practical Implications of Hepatocellular Neoplasms. *Cancers* **2022**, *14*, 3670. https://doi.org/10.3390/cancers14153670

Academic Editor: Alessandro Vitale

Received: 10 June 2022
Accepted: 25 July 2022
Published: 28 July 2022



## 1. Introduction

In the past decade, application of novel methodologies of molecular medicine in hepatocellular neoplasms has significantly improved our understanding of the pathogenesis of hepatocellular adenoma (HCA) and hepatocellular carcinoma (HCC), as well as provided useful new markers for pathologic diagnosis. At the same time, advances in clinico-radio-pathologic correlation have resulted in a new approach for the diagnostic assessment of focal hepatic lesions arising in advanced stage chronic liver disease by imaging, termed LI-RADS. In addition to providing new diagnostic and prognostic markers, elucidation of the molecular pathways of these neoplasms also has significant implications for treatment. This is particularly important for patients with HCC, for whom precision oncology strategies are finally starting to emerge, following many years of intensive research. This article briefly reviews the etiology and pathogenesis of HCA and HCC, provides practical information for their histologic diagnosis, and addresses a variety of frequently asked questions regarding the diagnosis and practical implications of these neoplasms.

## 2. Etiology and Pathogenesis of Hepatocellular Adenomas

It is now well recognized that hepatocellular adenoma (HCA), occurring mainly in young women taking oral contraception (OC), is a heterogeneous entity comprising different morpho-molecular subtypes, with various clinical and etiological backgrounds, risk for complications (bleeding and malignant transformation), and pathogenesis [1]. While most HCAs appear in normal liver, several clinical conditions and genetic syndromes have also been found to be linked to the development of HCAs [1].

The first well-recognized subtype is related to *HNF1A*-inactivating mutations (H-HCA). These tumors may be solitary or multiple, or they may occur in the context of liver adenomatosis. H-HCA is usually characterized by steatosis within the lesion and has a low risk of complications.

The second subtype is the inflammatory hepatocellular adenoma (IHCA), often developing on a background of NAFLD or in the context of alcohol consumption, predominantly but not exclusively in obese women. These lesions are often multiple. Typically characterized by sinusoidal dilatation and inflammation, IHCAs are related to different mutations leading to IL6/JAK/STAT inflammatory pathway activation.

A third subtype is the HCA with β-catenin-activating mutations (b-HCA). A proportion of these mutations occur in IHCA, thus giving rise to b-IHCA. By contrast with the other subtypes, b-(I)HCAs are overrepresented in men and have a higher risk of malignant transformation. This risk depends on the level of activation of the β-catenin pathway, which is linked to the type of *CTNNB1* mutation that results also in different immunohistochemical features [2].

A recently identified fourth HCA subtype is related to activation of the sonic hedgehog pathway (shHCA). These tumors are prone to bleeding, even when small, and can be recognized by argininosuccinate synthase 1 (ASS1) overexpression on immunohistochemistry [3,4]. This subtype has been described so far only in women, often overweight, and in the context of the metabolic syndrome.

Figure 1 illustrates the different subtypes of HCA with their principal immunohistochemical characteristics: H-HCA and liver fatty-acid-binding protein (LFABP), IHCA and C-reactive protein (CRP), b-HCA/b-IHCA and glutamine synthetase (GS), and shHCA and ASS1.

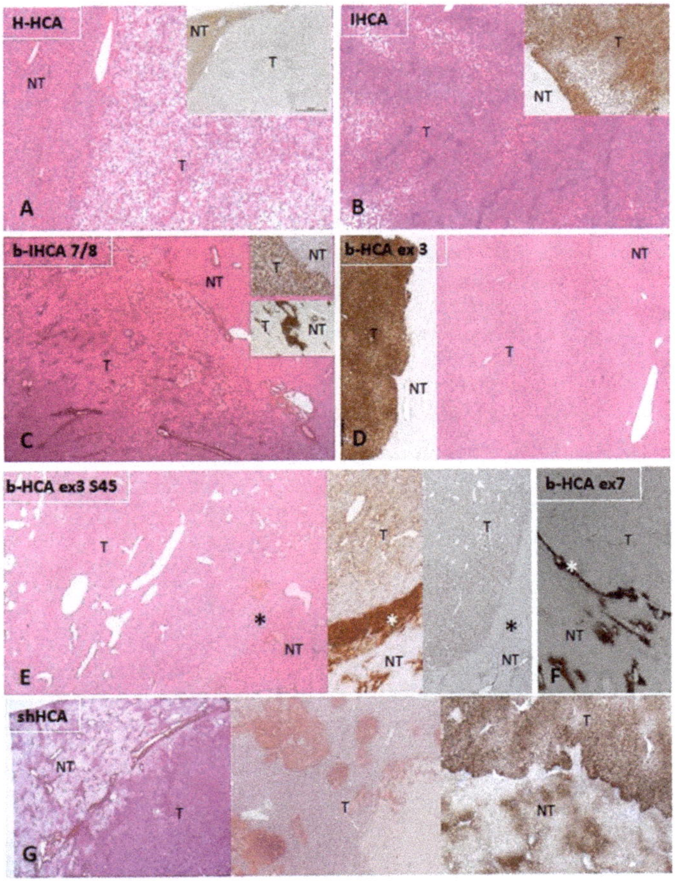


**Figure 1.** Characteristic histologic and immunohistochemical features of HCAs. (**A**) H-HCA: The tumor (T) appears highly steatotic on H&E with a complete lack of LFABP by immunohistochemistry (insert), contrasting with the normal expression in the nontumorous liver (NT). (**B**) IHCA: The tumor (T) exhibits sinusoidal dilatation on H&E with a strong CRP expression by immunohistochemistry (insert), sharply demarcated from the nontumorous liver (NT). (**C**) *CTNNB1* exon 7/8 mutated b-IHCA: This tumor exhibits a classical appearance of IHCA (sinusoidal dilatation, numerous thick arteries, and strong expression of CRP (top insert); in addition, GS is very faint in the tumor but with a strong GS rim between tumor (T) and nontumorous liver (NT) (bottom insert); molecular analysis identified a mutation on *CTNNB1* exon 7/8 (see [2]). (**D**) *CTNNB1* exon 3 mutated b-HCA: This tumor (T), which is not well delimited from the nontumorous liver (NT) on H&E, exhibits a strong and diffuse GS expression (left insert), identifying a high level of activation of the β-catenin pathway (large deletion on exon 3). (**E**) Exon 3 S45 mutated b-HCA: This tumor (T) exhibits numerous irregular vessels below the rim (asterisk) that separates T from nontumorous liver (NT); heterogeneous expression of GS is seen in T, whereas a strong GS expression characterizes the rim (middle insert); a corresponding diffuse CD34 immunostaining is seen in the endothelial cells of T, with no CD34 expression in the rim (asterisk) (right insert) (see [2]). (**F**) Exon 7 mutated b-HCA: GS is very faint in the tumor (T), and a thin GS rim (asterisk) separates T from the nontumorous liver (NT); molecular methods identified a β-catenin exon 7 mutation. (**G**) shHCA: This tumor developed in a highly steatotic nontumorous

liver (NT, left picture) and exhibits focally large hemorrhagic foci (middle picture); ASS1 immunohistochemistry shows an overexpression in the tumor (T), in comparison with the nontumorous liver (NT), in which its expression is restricted to the periportal/septal zones (right picture). *Abbreviations:* H-HCA, *HNF1A*-mutated hepatocellular adenoma; IHCA, inflammatory HCA; b-IHCA, β-catenin-mutated inflammatory HCA; b-HCA, β-catenin-mutated HCA; shHCA, sonic hedgehog-activated HCA; LFABP, liver fatty-acid-binding protein; CRP, C reactive protein; GS, glutamine synthetase.

## 3. Diagnosis and Subtyping of Hepatocellular Adenomas

The histologic diagnosis of HCA requires careful assessment of representative hematoxylin and eosin (H&E)-stained sections. HCAs are characterized by a benign hepatocellular proliferation, devoid of portal tracts. "Unpaired" arteries (i.e., arteries unaccompanied by veins or bile ducts) are present among the neoplastic cells. Other characteristic features include steatosis, inflammation, sinusoidal dilatation, and/or areas of hemorrhage. After H&E assessment, immunohistochemical evaluation follows with specific antibodies recognizing the targets identified by the genotype–phenotype studies [1,2]. An algorithm for the diagnosis is proposed in Figure 2.

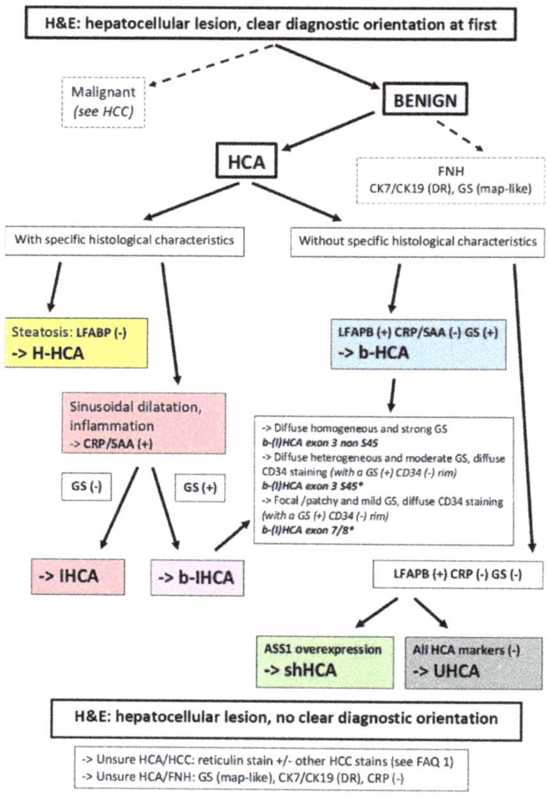


**Figure 2.** Diagnostic algorithm for HCAs. From a practical point of view, most of the cases are easily recognized as benign or malignant, but some are not. In the situation of an obvious HCA, if there is steatosis, with LFABP (−) and GS (−), there is no need to perform further IHC staining; it can be concluded that the tumor is an H-HCA. If an HCA shows sinusoidal dilatation and inflammation, with LFABP (+) and GS (−), it is mandatory to perform CRP and/or SAA immunostaining in order to diagnose an IHCA. GS immunostaining is mandatory in all IHCAs in order to diagnose a b-IHCA.

Different patterns of GS staining exist, linked to the type of underlying mutations (see [2]). If LFABP is positive and all other markers are negative, then an overexpression of ASS1 will lead to the identification of a shHCA, whereas, if it is not overexpressed, it is an UHCA. * Importantly, the GS(+)/CD34(−) rim can be irregular or discontinuous and is usually better represented in b-HCA than in b-IHCA. Its recognition on biopsies can be challenging (see [2]). In case of an uncertain diagnosis, HCA versus HCC or HCA versus FNH, additional histochemical and immunohistochemical stains are needed. The differential diagnosis of HCA versus HCC is discussed in FAQ 1. Reticulin stain might help to recognize alterations of the framework, although it is not a strict feature. Cytokeratin 7 and cytokeratin 19 stains help to recognize ductular reaction, and GS has a specific map-like pattern in FNH. *Abbreviations:* HCA, hepatocellular adenoma; H-HCA, *HNF1A*-mutated HCA; IHCA, inflammatory HCA; b-HCA, β-catenin-activated HCA; b-IHCA, β-catenin-activated and inflammatory HCA; shHCA, sonic hedgehog-activated HCA; UHCA, unclassified HCA; HCC, hepatocellular carcinoma; FNH, focal nodular hyperplasia; LFABP, liver fatty-acid-binding protein; CRP, C reactive protein; SAA, serum amyloid A; GS, glutamine synthetase; ASS1, argininosuccinate synthase; CK7, cytokeratin 7; CK19, cytokeratin 19; DR, ductular reaction.

## 4. Etiology and Pathogenesis of Hepatocellular Carcinoma

Hepatocellular carcinoma (HCC) usually arises in livers with chronic disease, and it is most often discovered when disease has reached an advanced stage, traditionally known as cirrhosis. The most common chronic diseases that are associated with HCC are chronic hepatitis B, chronic hepatitis C, and alcoholic liver disease, accounting together for 84% of the cases occurring globally in 2015 [5]. In the meanwhile, nonalcoholic steatohepatitis (NASH), associated with the metabolic syndrome, is emerging as a major risk factor for HCC [6]. Other risk factors include hereditary metabolic disorders (such as hemochromatosis, α1-antitrypsin deficiency, and tyrosinemia), aflatoxin B1 exposure (in individuals chronically infected with HBV), and tobacco smoking. Chronic liver diseases other than those mentioned above (e.g., autoimmune hepatitis, primary biliary cholangitis, primary sclerosing cholangitis, and Wilson disease) are uncommonly associated with development of HCC.

HCC arising in noncirrhotic livers is often caused by HBV, which is a virus with known carcinogenic effects. HBV DNA insertion in the host genome can deregulate genes involved in cell signaling and replication (such as *TERT, PDGFR, MLL4,* and *CCNE1*), while the HBV X protein transactivates genes involved in signal transduction pathways and inhibits *TP53* expression [7–9]. NASH and hereditary hemochromatosis are also increasingly recognized as causes of HCC arising in noncirrhotic livers [6,10]. However, HCC can also arise in apparently normal liver. Some of these cases may represent evolution of HCA (mostly b-HCA and b-IHCA) to HCC (discussed in the previous sections), while others, usually occurring in older individuals, remain unexplained. A special HCC subtype arising in normal livers of young individuals is fibrolamellar carcinoma, which is associated with a characteristic somatic gene fusion, *DNAJB1–PRKACA*, resulting from deletions in chromosome 19 and activating protein kinase A [11].

In chronic liver diseases, continuous cell loss results in cell proliferation occurring in a noxious microenvironment, characterized by oxidative stress due to chronic inflammation, overexpression of growth factors, and epigenetic changes due to derangements of DNA methyltransferases [12–14]. Thus, the possibility of mutations that initiate or promote carcinogenesis is increased, while mutations providing survival benefits to hepatocytes favor clonal expansion. This process is accelerated in the advanced stages of chronic liver diseases when vascular changes, including intrahepatic vein thrombosis and vascular reorganization, result in extensive cell loss. In that setting, hepatic regeneration largely depends on progenitor cell proliferation due to senescence of hepatocytes. Therefore, critical mutations in progenitor cells have the potential to produce large numbers of clonally expanding hepatocytes with increased likelihood to progress to precancerous lesions and then to HCC.

The diverse molecular changes that are associated with HCC have been recently reviewed [15]. Whole-exome and whole-genome sequencing studies have revealed 40–60 somatic coding mutations per HCC, including 4–6 driver mutations [16]. The most frequent mutations in HCC are those involving the promoter of telomerase reverse transcriptase (*TERT*), occurring in 60% of cases [17]. In an additional 30% of HCCs, *TERT* is deregulated by other molecular mechanisms, such as viral insertion [18]. *TERT* promoter mutations have also been detected in precancerous nodules and are considered an early event in hepatocarcinogenesis [19]. Other frequently mutated genes in HCC include *CTNNB1, TP53, RB1, ARID1A, ARID2, AXIN1*, albumin, and apolipoprotein B [20–22]. The mutations occurring in hepatocarcinogenesis can disrupt various signal transduction pathways, such as telomere maintenance (*TERT*), cell-cycle control (*TP53, CDKN2A*), Wnt/β-catenin (*CTNNB1, AXIN1*), epigenetic (*ARID1A, ARID2, MLL2*), and oxidative stress (*NFE2L2, KEAP1*) [17,23,24]. "Druggable" genetic alterations are under intense investigation because, at the present time, targeted therapeutic agents for HCC are limited to a small number of multikinase inhibitors. On the other hand, understanding the interaction between neoplastic cells and their microenvironment will be crucial for identifying biomarkers and developing new therapies based on immune checkpoint inhibition [25]

Recent studies have shown that certain molecular changes in HCC are associated with specific clinicopathologic features and prognosis, suggesting the possibility of a molecular classification for the future [26–29]. This active research has resulted in the recognition of several HCC subtypes (also called "variants") that hold promise for a more personalized treatment of HCC patients. Eight HCC subtypes, considered to represent distinct clinicopathological/molecular entities and accounting together for up to 35% of HCCs, have been included in the latest edition of the WHO classification of liver tumors [30]. The characteristic features of these subtypes are briefly presented in Section 5. It should be kept in mind that subclassification of HCC is a work in progress that will achieve significantly more importance if it becomes useful from a therapeutic point of view.

## 5. Diagnosis of Hepatocellular Carcinoma

Diagnosis of HCC is traditionally made by histologic examination of biopsy, surgical, or autopsy specimens, and it is based on the recognition of two basic attributes in the histologic material: (i) hepatocellular differentiation, and (ii) malignancy. Features suggesting hepatocellular differentiation include resemblance of neoplastic cells to hepatocytes, bile production by neoplastic cells, positive immunostaining of neoplastic cells for "hepatocytic" markers, such as arginase-1 and carbamoyl phosphate synthetase-1 (recognized by the antibody HepPar1), and detection of albumin mRNA by in situ hybridization. Except for bile production by neoplastic cells, none of the other features mentioned above is entirely specific for HCC. On the other hand, features indicating malignancy include stromal invasion, vascular invasion, metastatic spread, trabeculae thicker than three cells, and immunopositivity of neoplastic cells for oncofetal antigens α-fetoprotein and/or glypican-3.

In addition to the most common trabecular growth pattern, HCCs often display solid (compact), pseudoglandular, and macrotrabecular patterns of growth, including combinations thereof. Similar to hepatocytes, the neoplastic cells may contain fat, glycogen (resulting in clear cell change), hyaline bodies, Mallory–Denk bodies, or pale bodies. Scattered arteries unaccompanied by veins or bile ducts (i.e., "unpaired" arteries) are a characteristic histologic finding. Portal tracts are not a feature of classic HCC, except in the invasive front of some tumors. Similar to other carcinomas, HCC is also histologically classified as well, moderately and poorly differentiated [30] (Figure 3). Histologic diagnosis of poorly differentiated HCC is often difficult and requires immunohistochemical stains in support of the diagnosis (arginase-1, HepPar1, α-fetoprotein, and glypican-3), as well as appropriate markers for other tumors that are included in the differential diagnosis, on a case-per-case basis.

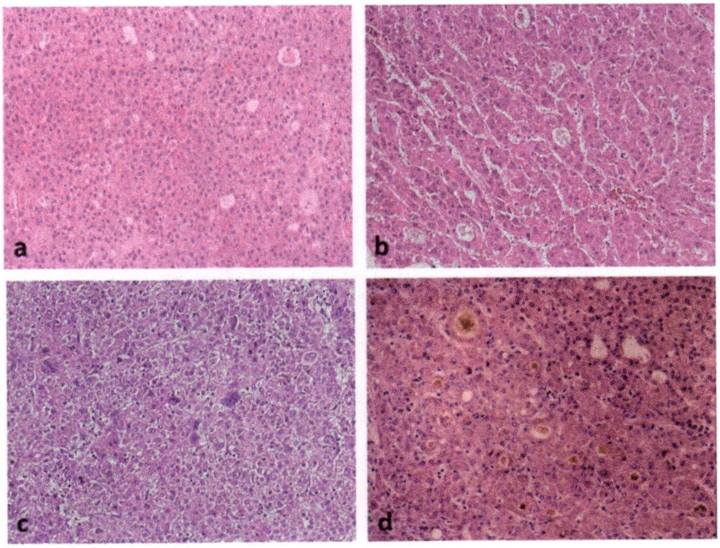

**Figure 3.** Degrees of differentiation in HCC: (**a**) This well-differentiated HCC consists of neoplastic cells resembling hepatocytes, which are arranged in trabeculae and pseudoglandular structures. (**b**) As compared to (**a**), this moderately differentiated HCC displays an increased nuclear–cytoplasmic ratio, larger nuclei with prominent nucleoli, and increased cytoplasmic basophilia. (**c**) This poorly differentiated HCC is characterized by marked tumor cell pleomorphism, including multinucleated cells; the architecture is trabecular and compact. (**d**) Bile production by neoplastic cells, often in pseudoglandular structures, as illustrated here, is a diagnostic feature of HCC.

On the other hand, some HCCs are difficult to recognize histologically, especially in biopsy material, because of well-differentiated features. Absence of portal tracts and presence of unpaired arteries in the biopsy material are features suggesting hepatocellular neoplasm, but do not allow distinction between HCA and well-differentiated HCC, while thin cell plates (<3 cells) do not exclude HCC. This difficult differential diagnosis is discussed below (see FAQ 1). It is emphasized that correlation of clinical, radiologic, and pathologic findings is essential for correct classification of difficult cases. This is particularly true in the interpretation of biopsy material from small (<2 cm) nodular lesions in cirrhotic livers, where the differential diagnosis includes large regenerative nodule, dysplastic nodule (low or high grade), early HCC, and classic HCC (see Section 6). This interpretation is facilitated when biopsy material from the hepatic parenchyma away from the lesion is available for comparison.

Early HCC (eHCC) has recently been recognized as a distinct step in hepatocarcinogenesis, characterized by ability for stromal invasion, but not for vascular invasion or metastatic spread [31]. By definition, eHCC is a well-differentiated, early-stage tumor that measures less than 2 cm in diameter. On gross examination, eHCC often appears vaguely nodular, without distinct pushing boundaries or pseudocapsule, whereas small HCC of the classic (also called "progressed") type usually has distinct boundaries marked by a pseudocapsule comprising compressed portal tracts or disease-associated scars [32]. Small classic HCCs tend to be better differentiated than larger ones, but have similar histologic features. On the other hand, many histologic features of eHCCs are reminiscent of those seen in high-grade dysplastic nodules. Early HCCs are usually composed of crowded, relatively small neoplastic cells, arranged in thin trabeculae and occasional small pseudoglandular structures. High cellularity (more than twice that of the surrounding parenchyma) and indistinct borders are characteristic features on low-power microscopic examination. Unpaired arteries are usually sparse and small, as compared to those of classic HCC. "Entrapped" portal tracts may be present in eHCC, especially in peripheral regions of the lesion. Steatosis is also often seen in eHCC, and it has been attributed to reduced oxygen supply compared to surrounding parenchyma [32]. On occasion, histologic examination of hepatic nodules may reveal classic HCC arising within eHCC (Figure 4). The vascular supply of eHCC (portal tract vessels and poorly developed unpaired arteries) significantly overlaps with that of dysplastic nodules; therefore, distinction between these lesions with imaging methods is difficult to impossible. The histologic features distinguishing eHCC from high-grade dysplastic nodules are discussed below (see Section 6).

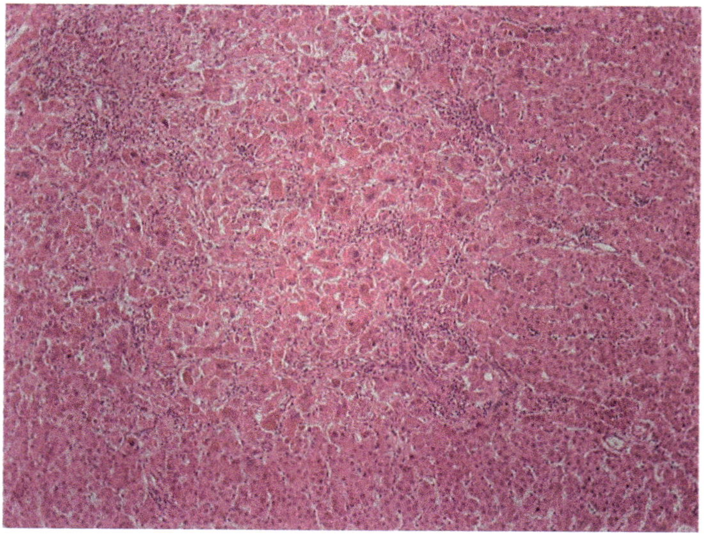

**Figure 4.** Classic HCC arising within early HCC (right and lower parts of the picture). Note the small unpaired arteries (right middle and lower part of the picture).

Table 1 provides a comparison of the etiology, pathogenesis, and diagnostically useful histopathologic features of HCA and HCC.

*HCC Subtypes*

The *steatohepatitic subtype* of HCC occurs usually, but not exclusively, in patients with metabolic syndrome or alcohol use and is characterized by histologic features similar to those of steatohepatitis occurring in nontumorous liver, i.e., macrovesicular steatosis, inflammation, ballooned cells, Mallory–Denk bodies, and pericellular fibrosis [33,34] (Figure 5a). This subtype was found to be associated with frequent IL6/JAK/STAT pathway activation, without *CTNNB1*, *TERT*, and *TP53* alterations [29]. At this point in time, steatohepatitic HCC does not seem to prognostically differ from average classic HCC.

The *clear cell subtype* owes its appearance to glycogen accumulation in tumor cells, thus simulating clear-cell carcinoma of the kidney and other organs (Figure 5b). No characteristic molecular alterations have been found in this subtype, which appears to be associated with a better-than-average prognosis [35]. Distinction from metastatic renal cell carcinoma may require immunohistochemical stains for hepatocytic markers (arginase-1, HepPar1) and renal transcription factor PAX-8.

The *macrotrabecular massive* subtype is histologically characterized by thick trabeculae, although the exact thickness (>6 cells vs. ≥10 cells thick) differs among authors [36] (Figure 5c). This subtype is associated with high serum α-fetoprotein and poor prognosis [29]. *TP53* mutations and *FGF19* amplifications are common in these tumors.

The *scirrhous subtype* is characterized by diffuse fibrosis, and it has been associated with *TSC1/TSC2* mutations [29] (Figure 5d). The prognosis of this subtype does not appear to differ from the average classic HCC. On histologic examination, this subtype should be distinguished from cholangiocarcinoma. Immunohistochemistry for hepatocytic markers arginase-1 and HepaPar1 is useful in this regard, whereas cytokeratin 7 is positive in most scirrhous HCCs and almost all cholangiocarcinomas.

The *chromophobe subtype* is characterized by light staining cytoplasm of the neoplastic cells, mostly bland nuclei, as well as scattered cells with large atypical nuclei. Another characteristic feature is the presence of scattered cystic spaces, filled with serum-like material. On a molecular basis, this subtype is characterized by alternative lengthening of telomeres, a mechanism for telomere preservation without *TERT* promoter mutation [37]. The prognosis of this subtype does not appear to differ from the average classic HCC.

**Table 1.** Comparison of etiology, pathogenesis, and diagnostically useful histopathologic features of hepatocellular neoplasms.

| | Hepatocellular Adenoma | Hepatocellular Carcinoma |
|---|---|---|
| **Etiology and Pathogenesis** | | |
| Chronic liver disease | Usually absent | Usually present |
| Molecular changes | Four specific morpho-molecular subtypes, including the following:<br>- H-HCA: *HNF1A*-inactivating mutations<br>- IHCA: mutations activating IL6/JAK/STAT<br>- b-HCA, b-IHCA: *CTNNB1*-activating mutations<br>- shHCA: *INHBE–GLI1* gene fusion | Large variety of mutations affecting a number of signal transduction pathways; most frequent mutations Involve TERT promoter |
| **Tumor architecture** | | |
| Thickness of cell plates | 1–2 cells | Variable |
| Pseudoglandular structures | Absent or few | Absent or present |
| Reticulin fibers | Preserved or focally disorganized | Decreased, disorganized |
| Invasive growth in stroma or vessels | Absent | Present |
| **Cytologic features** | | |
| Small cell size | Uncommon | Sometimes present |
| Nuclear hyperchromasia | Uncommon | Commonly present |
| Nuclear contour irregularities | Uncommon | Commonly present |
| Nuclear pleomorphism | Uncommon | Commonly present |
| Nuclear–cytoplasmic ratio | Usually normal | Often increased |
| Cytoplasmic basophilia | Usually absent | Commonly present |
| Mitotic figures | Absent or rare | Often present |
| **Nonlesional hepatic parenchyma** | | |
| Evidence of cirrhosis | Absent (rarely present in IHCA) | Present or absent |
| **Positive immunohistochemical staining** | | |
| Alpha-fetoprotein | Absent | Present or absent |
| Glypican-3 | Absent | Present or absent |

The *fibrolamellar subtype* has long been considered a distinctive HCC variant occurring in young individuals (median age: 25 years) without liver disease. These tumors are well differentiated and consist of groups and trabeculae of large polygonal cells, separated by bands of lamellar fibrosis. The neoplastic cells have abundant eosinophilic cytoplasm, often displaying pale bodies, as well as large nuclei with prominent nucleoli (Figure 5e). In contrast to most other HCCs, those of the fibrolamellar subtype are positive for cytokeratin 7 and CD68. Almost all fibrolamellar HCCs have the characteristic somatic gene fusion *DNAJB1–PRKACA*, the detection of which can aid diagnosis [11]. The prognosis of fibrolamellar HCC is similar to that of classic well-differentiated HCC occurring in noncirrhotic liver.

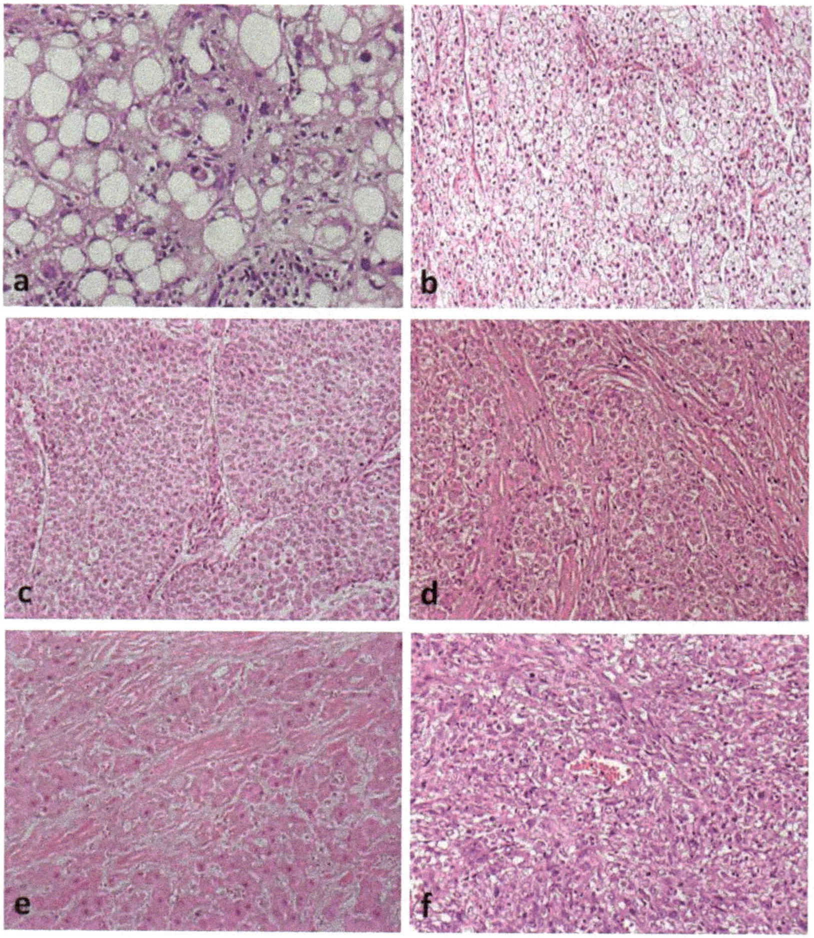


**Figure 5.** Examples of HCC subtypes: (**a**) steatohepatitic; (**b**) clear cell; (**c**) macrotrabecular; (**d**) scirrhous; (**e**) fibrolamellar; (**f**) sarcomatoid.

The *neutrophil-rich subtype* is characterized by abundant intratumoral neutrophils, due to granulocyte colony-stimulating factor (G-CSF) produced by neoplastic cells. Most tumors are poorly differentiated and may have sarcomatoid areas. The patients have elevated peripheral white blood cell counts, serum IL-6 levels, and often serum C-reactive protein. The prognosis of this subtype is worse than the average classic HCC [30].

The *lymphocyte-rich subtype* is characterized by abundant intratumoral lymphocytes. Cases tested for Epstein–Barr virus (EBV) were found to be negative. No prognostic significance has been attributed to this subtype. The lymphocyte-rich subtype should be distinguished from *lymphoepithelioma-like HCC*, a rare, poorly differentiated carcinoma, composed of tumor cells growing in poorly defined groups within a dense lymphoplasmacytic infiltrate [36,38]. Most cases of this neoplasm, which has similar histologic features to nasopharyngeal carcinoma and lymphoepithelioma-like carcinomas arising in other organs, have also been found to be negative for EBV.

In addition to lymphoepithelioma-like HCC, *sarcomatoid HCC* is another poorly differentiated variant that has not been recognized as a separate subtype in the latest edition of the WHO classification of liver tumors [30]. However, sarcomatoid HCC merits specific mention because it has a poor prognosis, as well as a spindle cell morphology that mimics various sarcomas [39] (Figure 5f). Extensive sampling may be required to reveal areas of typical HCC in these tumors, while immunohistochemical stains demonstrating expression of epithelial and hepatocytic markers can be useful, especially in cases with limited histologic material. Heterologous differentiation may be found in these rare tumors, in which case the term *carcinosarcoma* is appropriately used. It should be kept in mind that sarcomatoid change may develop in HCC following chemotherapy or transarterial chemoembolization [40].

The characteristic histologic and molecular findings of hepatocellular carcinoma subtypes are summarized in Table 2.

**Table 2.** Characteristic histologic and molecular findings of hepatocellular carcinoma subtypes.

| Subtype | Characteristic Histologic Findings | Characteristic Molecular Findings |
|---|---|---|
| Steatohepatitic | Features simulating steatohepatitis (macrovesicular steatosis, inflammation, ballooned cells, Mallory–Denk bodies, and pericellular fibrosis) | IL6/JAK/STAT pathway activation |
| Clear cell | Glycogen accumulation in tumor cells | None to date |
| Macrotrabecular massive | Thick trabeculae (>6 cells thick) | *TP53* mutations, *FGF19* amplifications |
| Scirrhous | Diffuse fibrosis | *TSC1/TSC2* mutations |
| Chromophobe | Light staining cytoplasm, mostly bland nuclei, occasional large atypical nuclei; cystic spaces with serum-like material | Alternative lengthening of telomeres |
| Fibrolamellar | Large polygonal cells with abundant eosinophilic cytoplasm, large nuclei and prominent nucleoli; pale bodies; lamellar fibrosis; immunopositivity for cytokeratin 7 and CD68 | *DNAJB1–PRKACA* gene fusion |
| Neutrophil-rich | Abundant intratumoral neutrophils | G-CSF production by neoplastic cells |
| Lymphocyte-rich | Abundant intratumoral lymphocytes | None to date |
| Sarcomatoid | Spindle cell morphology | None to date |

## 6. Precancerous Lesions in Hepatocarcinogenesis

Clonal populations of hepatocytes bearing molecular alterations of the early steps of carcinogenesis may be morphologically recognized in chronically diseased livers as precancerous lesions. These include the following [41]:

(i) dysplastic foci (DFs), which are incidentally detected on microscopic examination and measure less than 1 mm in diameter;
(ii) dysplastic nodules (DNs), which are larger than dysplastic foci, occasionally measuring over 1 cm in diameter, and may be detected on imaging studies and gross examination

The diagnosis of both DFs and DNs is made by histologic examination. Detection of such lesions is associated with an increased risk of HCC.

DFs are most commonly composed of hepatocytes with small cell change forming a roundish area with increased proliferative activity, as compared to the surrounding parenchyma. Small cell change is characterized by small cell size, increased nuclear–cytoplasmic ratio, mild nuclear pleomorphism and hyperchromasia, and cytoplasmic basophilia [42]. Small cell change of hepatocytes cytologically resembles early HCC. In livers with hereditary hemochromatosis, DFs are characterized by resistance to iron accumulation ("iron-free foci") [43].

DNs are grossly defined on the basis of comparisons to surrounding liver tissue as "distinctive nodules". They are most typically distinctive in terms of size, being larger than surrounding cirrhotic nodules [31,44]. However, they may also differ in terms of color (yellow if steatotic, tan-white if fibrotic, dark brown or black if iron-retentive, and green if cholestatic). These lesions are not distinguishable from small HCCs on gross examination. Confirmation that a distinctive nodule is a DN rather than HCC depends on histologic examination. DNs may display cytologic and architectural atypia, but to a degree that is insufficient for a diagnosis of HCC. Most consistently, DNs contain portal tracts, sometimes in a virtually normal distribution, while small, classic HCCs will have destroyed these or pushed them out of the way as they expand. Small classic HCCs will also often display all the histologic features of larger HCCs, such as overt cytologic atypia and thick trabeculae. Distinction between DNs and eHCC is more difficult; this is why eHCC was internationally recognized as an entity only in 2009 [31]. The histologic and immunohistochemical features that are useful for this distinction are discussed below. Sometimes, there are subnodules with features histologically suggestive of HCC within a DN; this is evidence of the DN's premalignant nature and is also further discussed below.

### 6.1. Low-Grade vs. High-Grade Dysplastic Nodules

DNs are subclassified in two categories, low-grade (LGDNs) and high-grade (HGDNs) [41]. LGDNs are lacking cellular atypia or architectural atypia that would be suspicious for HCC, although they may have large cell change. HGDNs are defined as having cytologic atypia (increased nuclear–cytoplasmic ratio, mild nuclear contour irregularities and hyperchromasia, cytoplasmic basophilia, and small cell change), or architectural atypia (thickened—but less than three cells thick—trabeculae, occasional pseudoglandular structures), which are reminiscent of an emerging HCC but insufficiently extensive to confidently denote a fully progressed HCC. HGDNs may display nodule-in-nodule type of growth, with a distinctive subnodule showing more atypical features. Sometimes the subnodule will merely be more expansile than the surrounding DN parenchyma with increased proliferation producing a "pushing border" at its edges. On occasion, the subnodule will be an overt HCC, displaying stromal invasion into portal tracts or fibrous septa contained within the surrounding DN (Figure 6) [45].

DNs are now understood to represent clonal neoplastic expansions of cells that often develop long before advanced stage liver disease is established [44,46]. They are generally lesions with *low* proliferation compared to surrounding, hyperplastic cirrhotic nodules [47]. (Figure 7). DNs are able to spread, however, because they are also resistant to apoptosis. This resistance gives them a slight survival advantage compared to non-neoplastic hepatocytes in adjacent parenchyma which, in response to the underlying chronic liver disease, have increased turnover [44]. The measure of how slight this advantage must be is that they may take many years to achieve sizes of up to 1.5 cm. DNs' resistance to the disease affecting the liver as a whole is also evidenced by diminished activation of hepatic stellate cells (HSCs) leading to an absence of scar within the DN or at least diminished scarring compared to the rest of the liver (Figure 7) [48].

### 6.2. Low-Grade Dysplastic Nodules vs. Large Regenerative Nodules

In early studies of DNs in sequential cirrhotic explants, the primary criterion for identifying DNs was a size cutoff (either 0.8 or 1.0 cm, depending on the study). The majority of livers containing DNs have a small number, rarely over 10; however, a subset of liver explants in patients with "macronodular cirrhosis" following either autoimmune hepatitis or hepatitis B had "uncountable" numbers of DNs by this criterion [49]. None of these were HGDN and none of the livers had HCC. Thus, it was clear that sometimes large regenerative nodules (LRNs) can mimic LGDNs.

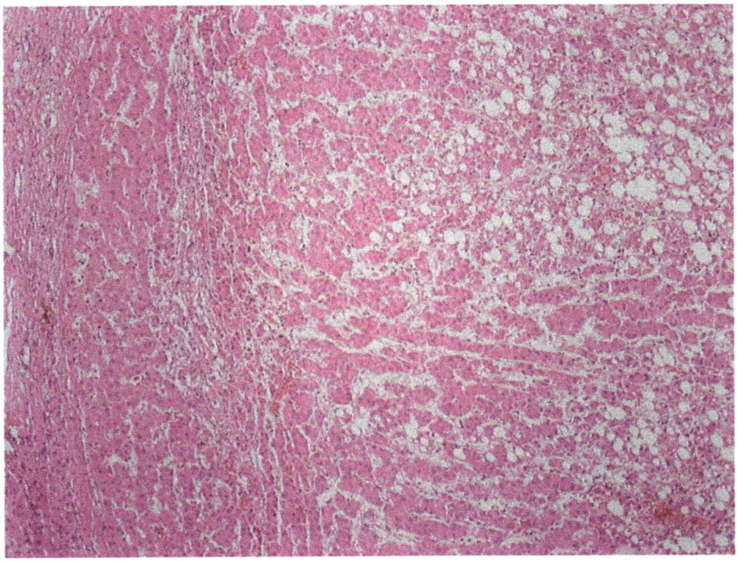

**Figure 6.** HCC (central and right part of the picture) arising in dysplastic nodule (**left part**). The tumor has features of early HCC (**central part**) and classic HCC with steatosis (**right part**).

In resection specimens, histologic distinctions between LGDN and LRN can be counterintuitive. LGDNs are more likely to show relatively preserved, even "normal appearing" parenchymal architecture, while LRNs may show significant disturbances of organization and function, such as variably regenerative or atrophic hepatocytes, large cell change, and hepatocyte injury such as ballooning or cholestasis. Thus, paradoxically, the neoplastic lesions, LGDNs, will appear more like normal liver, while the hyperplastic LRNs will appear reactive and, therefore, abnormal.

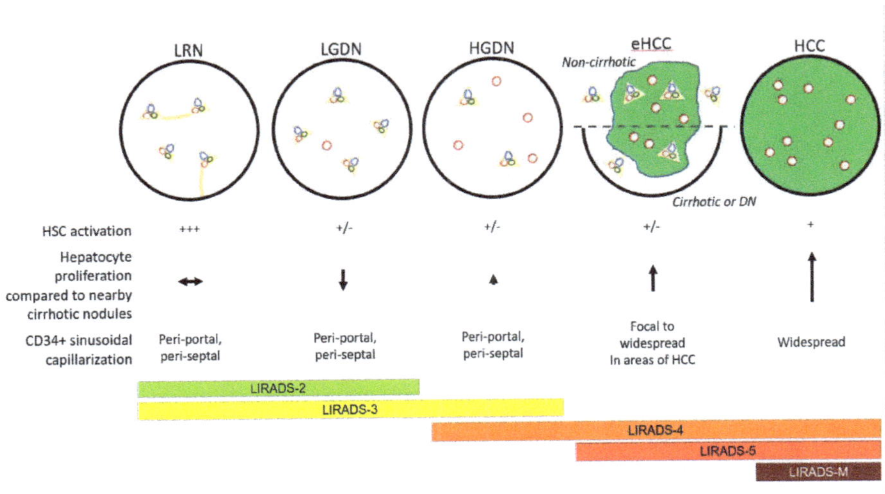


**Figure 7.** Important histologic features in hepatocellular nodules emerging in chronically diseased livers, and their LIRADS correlation. *Abbreviations:* LRN, large regenerative nodule; LGDN, low-grade dysplastic nodule; HGDN, high-grade dysplastic nodule; eHCC, early hepatocellular carcinoma; HCC, classic (progressed) hepatocellular carcinoma; HSC, hepatic stellate cell.

If the nodule has some distinctive features that might suggest clonality, this would support a diagnosis of LGDN over LRN. Such changes include diffuse iron or copper accumulation not seen in the surrounding liver or diffuse steatosis, with or without steatohepatitis, in the absence of background fatty liver disease. These findings favor the nodule being a true neoplasm. If one wishes to be more certain, one could do further studies to examine hepatocyte proliferation rates and HSC activation (both low in LGDN and high in LRN) (Figure 7) [47,48]. Moreover, LRNs lack unpaired arteries indicating neoplasia-associated angiogenesis, while LGDNs often have many such vessels (Figure 7) [50,51]. However, in many instances, the distinction between LGDN and LRN may be impossible, particularly in biopsy samples, but even when the whole nodule is present in a resection or autopsy specimen [31].

### *6.3. High-Grade Dysplastic Nodules vs. Hepatocellular Carcinoma*

Distinguishing HGDN from well-differentiated HCC can be challenging, especially on needle biopsy material. Recognition of invasive properties, in the stroma or vessels, a hallmark of malignancy (Figure 8), is obviously of paramount importance, but is often difficult to detect. Stromal invasion is the feature distinguishing eHCC from HGDN, and it is suspected when hepatocytes, even some without significant atypia, are present within the stroma of a portal tract or a septum in a large nodule. In such cases, absence of a ductular reaction, confirmed by immunohistochemical stains for cytokeratins 7 or 19, will support the presence of stromal invasion and, therefore, the diagnosis of HCC [45]. Immunohistochemistry can also be useful in biopsy material from nodules where HCC is suspected despite the lack of any evidence of invasion. Immunopositivity of lesional cells for two out of three markers, including glypican-3, glutamine synthetase, and HSP70, is considered diagnostic for HCC (either early or classic), whereas positivity for one or no marker does not resolve the issue of differential diagnosis between HGDN and HCC [52,53].

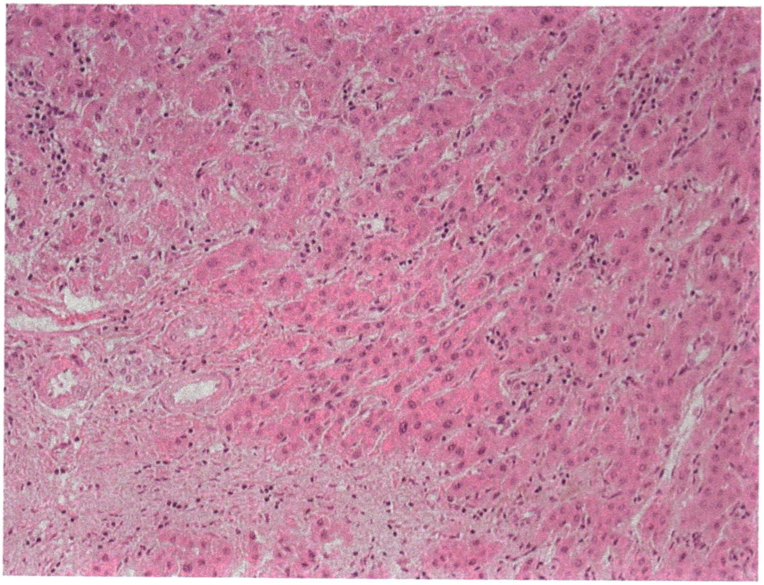

**Figure 8.** Well-differentiated HCC invading portal tract and fibrous septum in liver with advanced stage chronic hepatitis C. Note the absence of ductular reaction.

## 7. Frequently Asked Questions

***FAQ 1—Can all hepatocellular neoplasms be definitely classified as either benign or malignant?***

Recognizing a hepatocellular proliferation as benign is usually relatively easy, but can be difficult or even impossible in some cases. In livers with advanced chronic disease, the differential diagnosis is basically between high-grade dysplastic nodule and well-differentiated HCC (early or classic). An algorithmic approach to this differential diagnosis has recently been proposed [54]. In livers without chronic disease the difficulties in distinguishing HCA from well-differentiated HCC have long been recognized by experienced liver pathologists and are variably termed in the literature as "atypical hepatocellular adenoma/neoplasm", "HCA with borderline features", and "hepatocellular neoplasm with uncertain malignant potential" [1,55]. The worrisome features for the pathologist include architectural abnormalities, such as thickening of liver cell plates, presence of more than occasional pseudoglandular structures, and reticulin disorganization or disappearance, as well as cytological atypia, including presence of small cells, nuclear hyperchromasia, nuclear contour irregularities, nuclear pleomorphism, increased nuclear–cytoplasmic ratio, cytoplasmic basophilia, and presence of more than rare mitotic figures (see Table 1). In such cases, a careful search for features that allow a definite diagnosis of HCC (such as stromal or vascular invasion, trabeculae thicker than three cells, or immunopositivity for the oncofetal proteins α-fetoprotein and glypican-3) is warranted. However, despite careful histopathologic assessment, this differential diagnosis may occasionally remain unresolved. Detection of *TERT* promoter mutation, a marker of approximately 60% of HCCs [17], would be an argument for malignancy in such borderline lesions and holds promise as a diagnostic tool for the future. From a practical point of view, it is currently recommended to indicate this diagnostic difficulty in the report, especially when dealing with a biopsy specimen, in order to trigger appropriate clinical management and/or surveillance.

***FAQ 2—Some HCCs arise in completely normal liver. Do these HCCs arise from HCAs?***

HCAs are monoclonal neoplasms carrying a risk of malignant transformation reported to be in the range of 4–10%, depending on the series [1]. This percentage is obviously biased because (a) some lesions do not get a biopsy, and (b) many HCAs measuring more than 5 cm are surgically resected or ablated before expressing any potential to evolve to HCC. Since the majority of HCAs arise in normal livers, HCCs arising from and replacing HCAs will also be surrounded by normal hepatic parenchyma, except when an adenomatous rim will still be present at the periphery of the HCC. On the other hand, a minority of HCCs are discovered in normal livers, raising the possibility of a preexisting HCA that cannot be morphologically recognized. None of the immunohistochemical or molecular tools used to diagnose the different subtypes of HCA are useful at this point, because their expression can be modified in malignant lesions; LFABP can be decreased in HCC [56], CRP can be expressed by some HCCs [57], and *CTNNB1* mutations are commonly found in HCC. Therefore, none of these markers can be used for an argument to prove that an HCC arose from an HCA [1]. It is important for the pathologist to check the past medical and imaging history in order to identify clues of a preexisting HCA.

***FAQ 3—Do HCAs arise in cirrhotic livers?***

Theoretically, the definition of cirrhosis (i.e., a stage in the evolution of chronic liver diseases characterized by scarring and diffuse development of nodules) should not exclude the possibility of HCA of any subtype occurring in cirrhotic livers. However, the clinical context of HCA development is different from chronic liver disease, and pathologists are hesitant to make a diagnosis of HCA in cirrhotic livers. To date, the only HCA subtype that has been reported in livers with cirrhosis is IHCA. Rare IHCAs have been well documented in advanced-stage fatty liver disease, associated with alcohol or metabolic syndrome, with characteristic pathologic, immunohistochemical (overexpression of SAA/CRP), and molecular (different somatic mutations leading to IL6/JAK/STAT pathway activation) features [58,59]. In this context, one must be very cautious and not assert the diagnosis of IHCA only on the basis of immunohistochemical features, since cirrhotic nodules, large regenerative nodules, and dysplastic nodules can overexpress SAA or CRP [58]. Therefore, it is necessary to confirm the presence of a specific IHCA mutation by molecular analysis before reaching a diagnosis of IHCA developing in cirrhotic liver. A fortiori, it is not advisable to affirm this diagnosis on a needle biopsy. As mentioned above, HCC can express CRP, independently from the development in a preexisting IHCA [29].

***FAQ 4—Are there any minimum requirements for the use of immunohistochemistry in the diagnosis of HCA?***

After confirming that a tumor is an HCA on the basis of H&E-stained sections, it is important to define the subtype, which will determine further patient management. The choice of immunohistochemical stains depends on the pathological features, as demonstrated in Figure 2. If the tumor is highly steatotic, LFABP is mandatory to assert the diagnosis of H-HCA, provided nontumoral liver with normal expression of LFABP is available for comparison. If the tumor exhibits inflammatory features, sinusoidal dilatation, thick arteries, and pseudoportal tracts, CRP and/or SAA is first requested and will lead to the diagnosis of IHCA, if overexpressed. Of note, some H-HCA can be devoid of steatosis and some IHCA can show very little inflammation or show steatosis, which makes both immunostains (LFABP and CRP) useful for the right diagnosis in such cases. On the other hand, in case of a completely characteristic H-HCA, with steatosis and loss of LFABP expression, one can easily conclude that this is the diagnosis. However, in routine practice, even if a step-by-step approach seems to be logical, most of the time, LFABP, CRP, and glutamine synthetase (GS) are determined from the beginning in order to save time and materials. GS is mandatory for three reasons: (1) this marker is very useful to recognize and differentiate the tumoral area from the non-tumoral liver, something not always easy, particularly on biopsy specimens (GS in nontumoral liver is expressed only in a few rows of hepatocytes around the central veins); (2) GS helps to rule out focal nodular hyperplasia (FNH) in case of doubt (absence of classical map-like staining pattern in HCA); (3) GS is the major tool to diagnose *CTNNB1*-mutated HCA with or without associated inflammation allowing the diagnosis of b-HCA and b-IHCA. If GS is strong and diffuse, it means that there is a high level of activation of the β-catenin pathway (most likely due to exon 3 non-S45 mutation). Lower levels of activation of this pathway exist [60], and the pattern of GS expression is a good reflection of this phenomenon, with different immunohistochemical features suggesting different underlying molecular abnormalities, such as at the hotspot S45 of exon 3 or in exon 7/8, resulting in a moderate or low level of β-catenin pathway activation, respectively; in these latter cases, the diffuse CD34 staining in the tumor endothelial cells, except at the peripheral rim, is a good additional argument for the diagnosis [1,2].

It is emphasized that GS is mandatory in all IHCAs in order to reach a diagnosis of b-IHCA, which has the same risk of developing malignant transformation as b-HCA in the case of high-level β-catenin pathway activation. GS immunohistochemistry is much more reliable than β-catenin immunohistochemistry, which is not sensitive enough to identify *CTNNB1*-mutated HCAs. Indeed, this is positive only when GS is strongly expressed and, most of the times, positivity is focal, in a few nuclei. Therefore, there is no need to perform β-catenin immunostaining in HCA subtypes other than b-HCA or b-IHCA.

If LFABP is normally expressed, and stains for CRP and GS are negative, ASS1 is a useful new marker allowing to diagnose shHCA [3,4]. While ASS1 is normally expressed in nontumor liver with a periportal/periseptal pattern ("honeycomb pattern"), overexpression in tumor cells, as compared to nontumor is a requirement in order to make the diagnosis of shHCA. It is important to recognize shHCAs because of their high risk of bleeding. The algorithm (Figure 2) summarizes how to proceed in daily practice.

***FAQ 5—Do molecular studies provide any benefit in terms of diagnosis or prognosis of HCA, as compared to standard immunohistochemical stains?***

In routine diagnosis, standard immunohistochemical stains (i.e., LFABP, CRP, and GS) are sufficient, most of the time, for the diagnosis of H-HCA, IHCA, b-HCA, and b-IHCA, together representing more than 90% of HCA cases. There is no further benefit to identify inactivation of the *HNF1A* gene by molecular analysis or to search which mutation leads to IL6/JAK/STAT pathway activation, in order to reach a diagnosis of H-HCA or IHCA, respectively.

Concerning the β-catenin pathway, if GS immunostaining is strong and diffuse, it represents evidence that the activation level is high, which means a probable mutation in exon 3, not at the S45 hotspot. In this situation, there is no added value to search which hotspot of exon 3 is mutated for patient management decisions. Indeed, it is well known that these b-HCA/b-IHCA cases have to be resected since they have a high risk of malignant transformation. When the GS immunostaining is heterogeneous or very faint, when the GS-positive peripheral rim is not obvious, particularly in biopsy specimens, or when there are technical problems with immunohistochemistry, molecular methods are useful to search for mutations in exon 3 S45 or exon 7/8, the latter having a very low potential of malignant transformation but a high risk of bleeding, which makes recognition on biopsy material important for further patient management.

Many molecular analyses, such as those concerning *CTNNB1* mutations, can be performed today on formalin-fixed, paraffin-embedded tissue (FFPET), which is easier to obtain than frozen tissue. However, DNA of FFPET may be degraded and, therefore, without value for molecular analysis.

Regarding prognosis of HCA, it has been proposed to search for *TERT* promoter mutations (this is feasible on FFPET) as evidence of malignancy. This would be particularly useful in cases of b-HCA and b-IHCA, when atypical features are present.

In summary, apart from research protocols in referral centers, molecular studies in daily practice add value in terms of subtype diagnosis in b-HCA and b-IHCA, but are not necessary to determine prognosis when resection is mandatory (i.e., men and malignant transformation).

***FAQ 6—Should there be different guidelines for the treatment of different types of HCA?***

So far, the literature and the existing guidelines [61] indicate that (1) *CTNNB1*-mutated HCAs must be surgically resected or ablated, even if they measure less than 5 cm, (2) HCAs occurring in men also have to be resected or ablated, (3) any HCA measuring more than 5 cm should be resected or ablated, and (4) any HCA that is causing symptoms should be resected or ablated. Emerging evidence from the recent literature suggests that management should be adapted to the subtype more than to the size of the tumors [62]. In cases of adenomatosis, most residual HCAs after resection stabilize or regress, if steatohepatitis and obesity are corrected and/or the OC is discontinued; however, this evolution can take some time [63].

H-HCAs are usually indolent, even if they are large, and they can remain for years without regression and without giving rise to complications, except if they occur in specific clinical contexts, such as vascular liver diseases [64]. Not all b-HCAs and b-IHCAs are at risk of malignant transformation; the risk depends on the type of mutation, with those of exon 3 having the highest risk. On the other hand, shHCAs have a high risk of bleeding, which is clinically significant, even if they are smaller than 5 cm. It is probable that these specificities will guide the establishment of the future guidelines for the management for HCAs. With the aim of building guidelines in mind, it is important to collect standardized clinical and imaging data that led to the clinical management decision in each case [65].

***FAQ 7—When should we conclude that an HCA is "unclassified"?***

An HCA should be considered unclassified (UHCA) when all other HCA subtypes have been ruled out by currently recommended immunomarkers. Therefore, UHCA should be LFABP-positive, CRP-negative, and SAA-negative, with no abnormal staining of GS and with no abnormal expression of ASS1 (in comparison with the nontumoral liver; see above).

It is recommended, particularly for biopsy specimens, to be cautious with the interpretation because (1) some cases with very light GS staining could be a b-HCA with exon 7/8 mutation and not UHCA, and (2) ASS1 overexpression may be difficult to appreciate in comparison with nontumor liver. In both such situations, it might be advisable to repeat and interpret immunohistochemical stains at referral centers.

***FAQ 8—Can we recognize an IHCA when the nontumorous liver is positive for CRP on immunohistochemistry?***

It is not rare that nontumorous liver surrounding an IHCA or b-IHCA is CRP-positive, for instance, after portal or arterial embolization, or when there is a severe general inflammatory syndrome with a high level of blood CRP. In such cases, before concluding that a tumor is an IHCA, it is important to be sure that CRP immunopositivity is stronger in the tumor than in nontumorous liver, and to also perform SAA staining for comparison; otherwise, the staining may not be interpretable.

***FAQ 9—Is a specialized liver center needed for the management of HCAs?***

The clinical management of HCA relies on hepatologists, surgeons, radiologists, and pathologists, sharing their expertise in tumor board meetings. Imaging techniques are reliable to identify most cases of the H-HCA and IHCA subtypes, provided the radiologist has some experience and uses specific techniques. By contrast, specific recognition of b-HCA, b-IHCA, and shHCA by imaging is still under investigation. In their routine practice, with the help of immunohistochemistry, pathologists can also provide a diagnosis of H-HCA, IHCA, and b-(I)HCA (in case of a strong diffuse GS staining). b-(I)HCA with other patterns of GS staining and shHCA are less well known, and interpretation of immunohistochemistry can be difficult requiring confirmation by molecular methods. In clinical centers where HCAs are rare, referring patients to specialized centers will improve diagnosis and decision making and, for the rarer subtypes, will also help to foster cutting-edge guidelines for patient management.

***FAQ 10—Should the HCC grade (i.e., degree of differentiation) be reported in biopsy and in surgical specimens? What is the best grading system?***

Similar to other carcinomas, HCCs are graded as well, moderately, or poorly differentiated. The grade is a marker of prognosis, and has been found to predict patient survival and disease-free survival after both surgical resection and liver transplantation [66–68]. Therefore, HCC grade represents useful information that should be included in pathology reports. However, HCC often displays variable differentiation in different parts of the tumor. While prognosis would be expected to be primarily related to the least differentiated component of the neoplasm, knowledge of the existence of other components may be useful for the assessment of additional specimens, such as those obtained at later dates from metastatic sites. On the other hand, biopsy specimens may not be representative of the entire range of differentiation present in any given HCC, due to sampling error. Nevertheless, a study has found significant correlation between grade assessment of the biopsy and subsequent surgical specimen of HCC arising in cirrhotic patients [69]. Therefore, HCC grade should be reported in biopsy specimens, with the understanding that it may not always be entirely representative.

Various grading systems for HCC have been devised over the years. What is important for practical purposes is reproducibility of grading and clinical usefulness of the system. While the four-tiered Edmondson–Steiner system has been widely used in clinical studies, a three-tiered system is currently favored for daily practice [30]. It is hoped that adequate description of the characteristic features of each grade will result in high reproducibility among pathologists.

***FAQ 11—Which is the best immunohistochemical panel to assure hepatocellular origin of a malignancy?***

The most commonly used immunohistochemical markers of HCC are those mentioned in Section 5, i.e., arginase-1, HepPar1, glypican-3, and α-fetoprotein. Arginase-1 is the most sensitive and specific marker for HCC, staining over 90% of cases, while nonhepatocellular tumors are rarely positive for this marker [70]. HepPar1 stains most well differentiated HCCs, but less than 50% of poorly differentiated ones. Furthermore, various adenocarcinomas may occasionally be positive for HepPar1, particularly those from the small intestine, the normal enterocytes of which also uniformly express the antigen [71]. Glypican-3 is positive in 65–80% of HCCs, more often in poorly differentiated than well differentiated tumors; however, glypican-3 may also be positive in a variety of other malignancies, such as carcinomas from other sites, melanoma, and germ cell tumors [70]. α-Fetoprotein has low sensitivity for HCC (<50%) and may be expressed in germ cell tumors and rare other malignant neoplasms. In addition to these markers, HCCs arising in patients with chronic hepatitis B may occasionally be positive for HBsAg, an uncommon but most specific finding. Lastly, in situ hybridization for albumin mRNA can be very useful in distinguishing HCC from other malignancies but is available in a limited number of institutions.

Immunohistochemical stains for carcinoembryonic antigen utilizing polyclonal antibodies (pCEA) often provide a canalicular pattern of staining that is useful for HCC diagnosis. However, poorly differentiated HCCs often lack this pattern and may, instead, display membranous or even cytoplasmic staining, similar to that of adenocarcinomas. Immunohistochemical stains for CD10 often demonstrate in HCC a similar canalicular pattern of staining as that seen with pCEA. Again, this is usually absent in poorly differentiated HCC. An example of establishing the diagnosis of HCC with the aid of immunohistochemical stains is shown in Figure 9.

However, demonstrating the hepatocellular nature of a poorly differentiated carcinoma may be difficult in some cases. This is especially true in biopsy specimens with limited material, taking into account that immunopositivity of tumor cells for the markers mentioned above may be focal, resulting in false-negative findings. Clinicopathologic correlation taking into account all the clinical, imaging, and pathologic findings will be essential in such cases. Appropriate additional markers for other tumors should also be included, as per the differential diagnosis in each particular case.

On the other hand, if a well-differentiated carcinoma consists of what appear to be hepatocytes, then an appropriate panel would include arginase-1, HepPar1, pCEA (or CD10), and glypican-3, as α-fetoprotein has very low yield in well-differentiated HCC. If the lesion is truly HCC, any one marker may be positive, or a pair or more may show staining. However, stains for mimics of well differentiated HCC—particularly renal cell carcinoma, adrenal cortical carcinoma, neuroendocrine tumors, and follicular thyroid carcinomas—should also be considered [72]. Again, clinicopathologic correlation is essential.

***FAQ 12—Are there any HCC subtypes that need to be specified on histologic diagnosis?***

The importance of HCC subtyping is related to differences in clinical correlations, prognosis and treatment among subtypes. A subtype that is important to identify is the fibrolamellar HCC, which is characterized by young patient age, lack of underlying liver disease and tendency to metastasize to hilar lymph nodes. Therefore, the mainstay of therapy is surgery, including regional lymphadenectomy. The recent discovery of a characteristic somatic gene fusion in fibrolamellar HCC (see Section 4) allows optimism for future development of targeted therapies for nonresectable tumors. Other subtypes are worth identifying because of prognostic differences, as compared to average; for example, the macrotrabecular massive subtype and the neutrophil-rich subtype are associated with worse prognosis, while the clear cell subtype with better prognosis.

It is emphasized that HCC subtyping is a work in progress. With the exception of the fibrolamellar subtype, HCC had been regarded until recently as a tumor of bleak prognosis; therefore, there was little interest in subtyping. This view is now changing because of the availability of surveillance programs and the hopes for new therapies based on molecular profiling. Therefore, correlation of the clinical, pathologic, and molecular features in large series of patients with the various subtypes of HCC may allow a more individualized approach for treatment. As more data become available, the importance of subtyping will likely increase. Histologic examination is the basis for HCC subtyping; therefore, criteria of each subtype should be clear and reproducible.

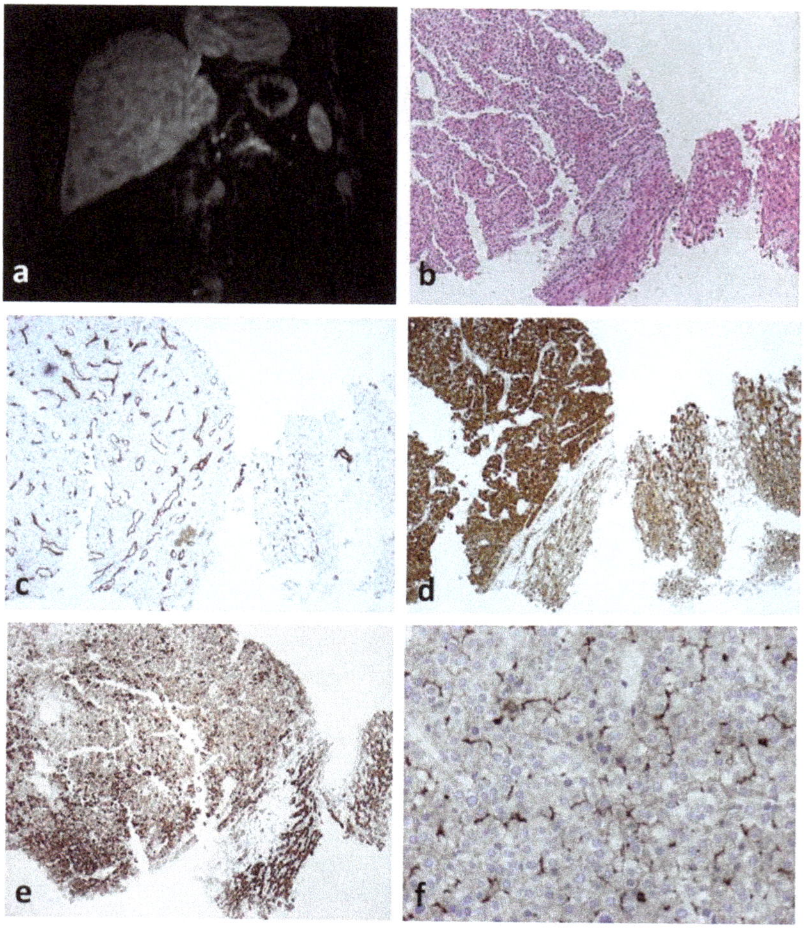


**Figure 9.** Abdominal MRI (**a**) and guided liver biopsy specimen (**b–f**) from a 53 year old man with breathing difficulty and history of metabolic syndrome, including obesity, type 2 diabetes mellitus, and hyperlipidemia. (**a**) There was marked hepatomegaly with innumerable, scattered nodules, measuring up to 3 cm, suggesting metastatic disease. However, needle biopsy revealed HCC. In this limited biopsy material, polygonal tumor cells appeared to be arranged in a compact sheet (**b, left side**); nevertheless, immunohistochemical stain for CD34 (**c**), highlighting the endothelial cells, demonstrated trabecular architecture. Further stains showed positivity of tumor cells for arginase-1 (**d**) and HepPar1 (**e**). Adjacent hepatocytes (**d,e, right side**) are also positive for these markers, serving as "internal controls". Tumor cells also displayed a canalicular pattern of staining with pCEA (**f**), as well as positivity for glypican-3 (not shown).

***FAQ 13—Do immunohistochemical stains or molecular studies provide any actionable items for HCC? Do molecular studies provide any added value in terms of diagnosis or prognosis of HCC, as compared to standard immunohistochemical stains?***

The main use of immunohistochemical stains in cases of suspected HCC is to confirm the diagnosis and rule out other neoplasms. However, there is also a stain that has been found to be of prognostic significance in HCC; cytokeratin 19-positive HCCs have higher recurrence rates than usual, as well as higher resistance to locoregional therapies [73–75]. It should be kept in mind that cytokeratin 19 is positive in a variety of adenocarcinomas, including cholangiocarcinoma; therefore, it is used as a prognostic, but not as a diagnostic marker for HCC.

Molecular methods have been used extensively in recent years to identify the molecular changes occurring in hepatocarcinogenesis, including potential targets for treatment. Substantial molecular data have been accumulated, and several molecular classifications have been proposed on the basis of a correlation of clinical, pathologic, and molecular data [22,26–29]. However, these classifications have not yet found their way to clinical practice. As a result of these studies, most HCCs can now be grouped into two classes [25,76]: (i) the proliferation class, which is etiologically related to HBV infection, and displays molecular and histologic features associated with aggressive clinical behavior; (ii) the nonproliferation class, which is etiologically related to HCV infection or alcohol, and displays features associated with better clinical outcome. HCCs of the proliferation class are characterized by *TP53* mutations, chromosomal instability, and activation of various oncogenic pathways, tend to be poorly differentiated, and are associated with high serum α-fetoprotein. On the other hand, HCCs of the nonproliferation class often have *CTNNB1* mutations and a gene expression profile resembling that of normal hepatocytes. These tumors tend to be better differentiated and with lower incidence of vascular invasion than those of the proliferation class.

The characteristic molecular changes of the various HCC subtypes are summarized in the section on diagnosis of hepatocellular carcinoma (see Table 2). Although not routinely used in daily diagnosis, detection of these changes can be used in support of the diagnosis. For instance, detection of the gene fusion *DNAJB1–PRKACA* can confirm the diagnosis of fibrolamellar HCC.

***FAQ 14—What should pathologists know and do about combined hepatocellular-cholangiocarcinomas (cHCC–CCA)?***

*FAQ14.1. Tissue Diagnosis of cHCC–CCA*

A tissue diagnosis of a primary cHCC–CCA is straightforwardly made by routine hematoxylin–eosin stains; immunostains for markers of hepatocyte or cholangiocyte differentiation are merely confirmatory [30,77]. The presence of stainable hepatocyte markers in glandular epithelium (e.g., arginase-1, HepPar1, α-fetoprotein, glypican-3, and albumin mRNA) or, conversely, of cholangiocyte markers in HCC (e.g., keratins 7 and 19, and EpCAM) are not proof of cHCC–CCA given the possibilities of aberrant gene expression in any malignancy [77]. Differentiated components of cHCC–CCA may be located in distinct areas of a tumor, or they may be intimately intermingled throughout the lesion. Boundaries between the components may be sharply defined or indistinct. There are, as yet, no definitive cutoffs for a percentage requirement for the presence of the component. A minute component of intrahepatic CCA (iCCA) within an otherwise clear HCC is sufficient to call it cHCC–CCA and vice versa.

A common pitfall of diagnosis is when cHCC–CCA is suspected on radiographic grounds, but only one element is present in the biopsy specimen. In this case, the pathologist must be careful to comment on the limitations of biopsy. Small biopsy specimens may sample only one component of such a heterogeneous tumor; the absence of the other component does not exclude cHCC–CCA and a formal statement to that effect in the pathology report is important. On the other hand, metastatic lesions associated with a primary cHCC–CCA may comprise either component alone or mimic the cHCC–CCA appearance of the primary tumor. Biopsy specimens from metastatic lesions must be cautiously interpreted in this light [78].

Another pitfall for diagnosis can occur when there appear to be two separate mass lesions in the liver that are merging together. An HCC and a separate, but simultaneous CCA may grow into each other forming a "collision tumor", particularly in chronic liver diseases that predispose to both malignancies. Each of these tumors should be assessed pathologically as independent entities.

Lastly, a very rare variant of "intermediate cell type" of cHCC–CCA notably breaks all the rules; its tumor cells appear morphologically intermediate between hepatocytes and cholangiocytes and do not show typical growth patterns of either HCC (e.g., trabeculae and pseudoglandular structures) or iCCA (e.g., mucin-producing glands, tubules, and signet ring cells), often appearing homogeneous throughout. Dual differentiation in these tumors is, indeed, at the cellular level, with each cell showing combined hepatocyte and cholangiocyte marker expression [30].

It should also be noted that, while cHCC–CCA may present, de novo, as a primary hepatic malignancy, it has also been seen to emerge from HCCs that have undergone loco-regional treatments [79]. It has been suggested that hypoxia of surviving tumor cells after transarterial chemoembolization leads to expression of proteins, such as EpCAM and cytokeratin 19 [80]. Such adaptive changes might explain emergence of cHCC-CCA from a treated HCC, although the possibility that there was a previously undetected minor component of CCA originally is difficult to exclude. In any case, while this occurrence seems uncommon, it should be considered when tumor recurs post treatment, particularly if imaging features no longer show classic features of HCC, alone.

Subpopulations of tumor cells in cHCC–CCA may have what has been described as a "stem-cell appearance", i.e., small cells with high nuclear–cytoplasmic ratio, sometimes arrayed with larger hepatobiliary cells in what appear to be lineage relationships like those seen in ductular reactions in diseased or injured liver. While subclasses of "stem-cell tumors" were characterized in the 2010 edition of the WHO "Blue Book" [81], the more recent edition [82] has eliminated the term as a diagnostic category in all primary liver cancers. Nonetheless, the question of its importance remains uncertain, and it has been recommended that the presence of "stem-cell features" be noted in the pathology report of tumors that contain them [83].

*FAQ14.2. Pathology–Radiology Collaboration for cHCC–CCA*

A biphasic radiographic appearance of a lesion may help a pathologist avoid missing the opportunity for including cHCC–CCA in the differential diagnosis when only one component is sampled in a biopsy specimen [84]. However, in multidisciplinary conferences for liver malignancies, clinicians and radiologists may miss clues to cHCC–CCA given their rarity, but the attentive pathologist can help guide radiologists toward how to best sample a lesion for successful complete diagnosis.

In early-stage liver disease or in sporadic tumors in which there is no predisposing hepatic disease, cHCC–CCA may appear biphasic, with separate areas showing typical features of HCC or iCCA, or they may merely be atypical, without imaging characteristics specific for either, thus being more suggestive of metastasis [84]. In advanced-stage liver disease, in which the Liver Imaging Reporting and Data System (LI-RADS) classification is applicable, radiologists may label a lesion LIRADS-M because unusual features suggest a metastasis, but the lesion may actually be a cHCC–CCA [85,86]. Alternatively, they may recognize that one or more parts of the lesion have an LIRADS-5 (diagnostic for HCC) appearance and, thus, label the whole lesion with that designation, even though some areas appear distinct [84,85]. In both these settings, a pathologist who is attentive to radiographic descriptions that might hint at cHCC–CCA may save the day.

*FAQ14.3. Molecular Pathology and Treatment Implications of the Diagnosis of cHCC–CCA*

Molecular studies support that these tumors may sometimes derive from a malignantly transformed hepatobiliary stem/progenitor cell or from de-/redifferentiation of malignantly transformed hepatocytes or cholangiocytes, and that they may be more like iCCA, more like HCC, or intermediate between them. All of these data confirm that they are certainly, at least to some degree, heterogeneous in origin and in behavior [87]. On the other hand, comparison of clinicopathological characteristics of cHCC–CCA with regard to the newest WHO classification [82] supports its relevance and that cHCC–CCA has intermediate survival between HCC and iCCA, if not actually tilting toward the dire outcomes for iCCA [87,88]. Given the propensity for early and distant spread of CCA components along lymphatic and perineural pathways that are typical of iCCA itself, it is no surprise that clinical outcomes after resection are, overall, worse than for HCC. On the other hand, if transplanted within the Milan criteria, cHCC–CCA showed similar overall survival to HCC after transplantation [89].

Unfortunately, the rarity of these tumors has interfered with the performance of randomized clinical trials. There is a paucity of data regarding immunotherapies [90], although studies suggest that at least some cHCC–CCA should be responsive to these types of treatment [91,92]. Broad genomic profiling of malignancies for actionable mutations specific to each case is currently the most likely path to any possible clinical benefit [93].

***FAQ 15—I do not work in a transplant center and am so unlikely to ever see a dysplastic nodule specimen. Do I need to know about them? If so, why?***

Yes! One needs to know about them! Currently, screening for emergence of malignancy in chronic liver disease depends largely on radiographic criteria defined by the Liver Imaging Reporting and Data System (LI-RADS) classification [94]. Pathologists may find themselves involved in liver tumor multidisciplinary conferences in which radiologists will report distinctive nodules that are subclassified into LIRADS-1 through 5, with a higher score indicating the higher confidence that a lesion is an actual HCC (Figure 7) [95]. The repetition of the phrase "distinctive nodule" is not a coincidence; LI-RADS was formulated to reflect the pathologic understanding of DNs as neoplastic and often premalignant lesions.

A classification of LIRADS-5 is so specific for HCC that, in most medical centers, it is sufficient for diagnosis without confirmatory biopsy. LIRADS-1 lesions are considered likely to be benign, probably merely large regenerative nodules. LIRADS-2 through 4 probably reflect LGDN through HGDN (although direct pathology–radiology correlations for these have not been reported) [94,95].

It is not necessary for the pathologist to know the full and subtle criteria for the LIRADS classifications, but it is vital to know what lesions each designation may reflect, thus enabling the pathologist to carefully guide the clinicians and radiologists in terms of follow-up screening or treatment of the patient. It is worth knowing, however, that the increasing stages of the LI-RADS classification probably reflect the changes in vascular supply. Regenerative nodules and LGDNs have mostly intact portal vein blood flow with little increase in arterialization. However, the increasing ratio between enlarging arterial blood flow (angiogenesis inside truly neoplastic LGDN and HGDN) and the diminishing blood supply (as portal tracts are degraded or pushed to the sides) result in the characteristic LIRADS features (Figure 7).

Examples are provided below.

- **LIRADS-1**: The pathologist may advise the clinical team that, while this lesion is probably just a regenerative nodule, the possibility that it is a DN, possibly even an HGDN, is not excluded. Repeat screening at a shorter time interval may be warranted. If the lesion disappears, it was probably large regenerative nodule that underwent involution or further scarring that eliminated its distinctive appearance on imaging. If it does indeed disappear, return to normal surveillance screening is reasonable.
- **LIRADS-2 or -3**: These lesions are more likely to be DNs, either LGDN or HGDN. Repeat imaging should be performed more frequently. If the nodule disappears, it was probably regenerative. If it persists, then it may be LGDN or HGDN and the patient is considered at higher risk for HCC and should return more frequently for imaging. If the lesion progresses upward in LIRADS score, it is probably a DN giving rise to an HCC. Continued imaging or ablation may be considered depending on the clinical circumstances.
- **LIRADS-4**: These lesions are probably an HGDN, possibly with an emerging focus of HCC, or possibly a small HCC. Continued imaging or ablation may be considered depending on clinical circumstances.
- **LIRADS-5:** This feature is diagnostic for HCC. While a biopsy is not necessary for diagnosis, oncologists are increasingly requesting a pre-ablation biopsy for molecular studies to inform future treatments if the lesion is resistant to ablation or if there is post-treatment (i.e., ablation, resection, or transplant) recurrence.

- **LIRADS-M:** These are lesions without typical imaging features of HCC, but highly suspicious for malignancy. The pathologists may aver that the lesion could be metastatic (M), but they could also be iCCA or cHCC–CCA. Such lesions probably require biopsy for diagnosis.

With regard to cHCC–CCA, it behooves the pathologist to watch for lesions described by the radiologist as "complex" or as having an isolated part with typical LIRADS-5 features while the other parts of the lesion are atypical for HCC [84–86]. Because cHCC–CCA may have distinct regions of the tumor that are either HCC or CCA, in such tumors, some regions will show LIRADS-5 changes, while others will not. This situation is one in which the pathologist can make a decisive difference, recommending targeted biopsies of *both* the LIRADS-5 and the atypical areas. iCCA's worse prognosis and different treatment implications make it important to diagnose as early as possible in the treatment course. The pathologist may be the only person in the room sensitive to this uncommon cancer. Alerting the clinical team to make sure that cHCC–CCA has been completely evaluated may be crucial for saving the life of this patient or preventing an inappropriate transplant for an incurable malignancy.

***FAQ 16—When should tumors with imaging features of HCC be biopsied?***

It is currently believed that the findings which define LIRADS-5, i.e., those lesions with all imaging features of HCC by imaging, do not require confirmatory biopsy. However, ablative treatment of these lesions is then likely, and tissue for molecular analysis for determining possible targeted therapies will not be available. For this reason, there may be a shift in clinical practice toward biopsy of LIRADS-5 HCCs in the near future, not for diagnosis or prognosis, but for determination of suitable targeted therapies in the event of recurrence. Subtyping of HCC might also be found to be relevant in the near future to guide treatment decisions, also with the support of artificial intelligence [96].

As noted above, if part of a lesion displays typical HCC imaging features, but other parts do not, the possibility of a cHCC–CCA cannot be excluded. In addition, even rarer HCC variants could appear in combination with classic HCC, such as the sarcomatoid one, which might have nontypical imaging features. In such cases, biopsy of both the classic HCC component and of the nontypical component is warranted.

**Author Contributions:** Conceptualization, P.H., P.B.-S., N.D.T. and C.S.; methodology, P.H., P.B.-S., N.D.T. and C.S.; validation, P.H., P.B.-S., N.D.T. and C.S.; formal analysis, P.H., P.B.-S., N.D.T. and C.S.; investigation, P.H., P.B.-S., N.D.T. and C.S.; resources, P.H., P.B.-S., N.D.T. and C.S.; data curation, P.H., P.B.-S., N.D.T. and C.S.; writing—original draft preparation, P.H., P.B.-S., N.D.T. and C.S.; writing—review and editing, P.H., P.B.-S., N.D.T. and C.S.; visualization, P.H., P.B.-S., N.D.T. and C.S.; project administration, P.H. All authors have read and agreed to the published version of the manuscript.

**Funding:** This research received no external funding.

**Acknowledgments:** The authors wish to acknowledge the members of the International Liver Pathology Study Group (the "Elves"), past and present, whose research and clinical wisdom inform all of the concepts in this manuscript. In particular, the authors are grateful to Venancio Alves, Alberto Quaglia, and Charles Balabaud, for their suggested FAQs.

**Conflicts of Interest:** The authors declare no conflict of interest.

## References

1. Bioulac-Sage, P.; Gouw, A.S.; Balabaud, C.; Sempoux, C. Hepatocellular adenoma: What we know, what we do not know, and why it matters. *Histopathology* **2022**, *80*, 878–897. [CrossRef] [PubMed]
2. Sempoux, C.; Gouw, A.S.; Dunet, V.; Paradis, V.; Balabaud, C.; Bioulac-Sage, P. Predictive Patterns of Glutamine Synthetase Immunohistochemical Staining in CTNNB1-mutated Hepatocellular Adenomas. *Am. J. Surg. Pathol.* **2021**, *45*, 477–487. [CrossRef] [PubMed]
3. Henriet, E.; Hammoud, A.A.; Dupuy, J.-W.; Dartigues, B.; Ezzoukry, Z.; Dugot-Senant, N.; Leste-Lasserre, T.; Pallares-Lupon, N.; Nikolski, M.; Le Bail, B.; et al. Argininosuccinate synthase 1 (ASS1): A marker of unclassified hepatocellular adenoma and high bleeding risk. *Hepatology* **2017**, *66*, 2016–2028. [CrossRef]

4. Sala, M.; Gonzales, D.; Leste-Lasserre, T.; Dugot-Senant, N.; Paradis, V.; Di Tommaso, S.; Dupuy, J.; Pitard, V.; Dourthe, C.; Sciarra, A.; et al. ASS1 Overexpression: A Hallmark of Sonic Hedgehog Hepatocellular Adenomas; Recommendations for Clinical Practice. *Hepatol. Commun.* **2020**, *4*, 809–824. [CrossRef] [PubMed]
5. Global Burden of Disease Liver Cancer Collaboration. The Burden of Primary Liver Cancer and Underlying Etiologies From 1990 to 2015 at the Global, Regional, and National Level: Results from the Global Burden of Disease Study 2015. *JAMA Oncol.* **2017**, *3*, 1683–1691. [CrossRef]
6. Ioannou, G.N. Epidemiology and risk-stratification of NAFLD-associated HCC. *J. Hepatol.* **2021**, *75*, 1476–1484. [CrossRef] [PubMed]
7. Paterlini-Bréchot, P.; Saigo, K.; Murakami, Y.; Chami, M.; Gozuacik, D.; Mugnier, C.; Lagorce, D.; Bréchot, C. Hepatitis B virus-related insertional mutagenesis occurs frequently in human liver cancers and recurrently targets human telomerase gene. *Oncogene* **2003**, *22*, 3911–3916. [CrossRef]
8. Murakami, Y.; Saigo, K.; Takashima, H.; Minami, M.; Okanoue, T.; Bréchot, C.; Brechot, P.P. Large scaled analysis of hepatitis B virus (HBV) DNA integration in HBV related hepatocellular carcinomas. *Gut* **2005**, *54*, 1162–1168. [CrossRef]
9. Sung, W.-K.; Zheng, H.; Li, S.; Chen, R.; Liu, X.; Li, Y.; Lee, N.P.; Lee, W.H.; Ariyaratne, P.N.; Tennakoon, C.; et al. Genome-wide survey of recurrent HBV integration in hepatocellular carcinoma. *Nat. Genet.* **2012**, *44*, 765–769. [CrossRef]
10. Goh, J.; Callagy, G.; McEntee, G.; O'Keane, J.C.; Bomford, A.; Crowe, J. Hepatocellular carcinoma arising in the absence of cirrhosis in genetic haemochromatosis: Three case reports and review of literature. *Eur. J. Gastroenterol. Hepatol.* **1999**, *11*, 915–919. [CrossRef] [PubMed]
11. Honeyman, J.N.; Simon, E.P.; Robine, N.; Chiaroni-Clarke, R.; Darcy, D.G.; Lim, I.I.P.; Gleason, C.E.; Murphy, J.M.; Rosenberg, B.R.; Teegan, L.; et al. Detection of a Recurrent *DNAJB1-PRKACA* Chimeric Transcript in Fibrolamellar Hepatocellular Carcinoma. *Science* **2014**, *343*, 1010–1014. [CrossRef] [PubMed]
12. Thorgeirsson, S.S.; Grisham, J.W. Molecular pathogenesis of human hepatocellular carcinoma. *Nat. Genet.* **2002**, *31*, 339–346. [CrossRef]
13. Hatziapostolou, M.; Polytarchou, C.; Aggelidou, E.; Drakaki, A.; Poultsides, G.A.; Jaeger, S.A.; Ogata, H.; Karin, M.; Struhl, K.; Hadzopoulou-Cladaras, M.; et al. An HNF4α-miRNA inflammatory feedback circuit regulates hepatocellular oncogenesis. *Cell* **2011**, *147*, 1233–1247. [CrossRef]
14. Brody, R.I.; Theise, N.D. An inflammatory proposal for hepatocarcinogenesis. *Hepatology* **2012**, *56*, 382–384. [CrossRef]
15. Hytiroglou, P.; Bioulac-Sage, P. Molecular Pathogenesis and Diagnostics of Hepatocellular Tumors. In *Odze and Goldblum Surgical Pathology of the GI Tract, Liver, Biliary Tract, and Pancreas*, 4th ed.; Odze, R.D., Goldblum, J.R., Eds.; Elsevier: Philadelphia, PA, USA, 2023; pp. 1369–1383.
16. Gerbes, A.; Zoulim, F.; Tilg, H.; Dufour, J.-F.; Bruix, J.; Paradis, V.; Salem, R.; Peck-Radosavljevic, M.; Galle, P.R.; Greten, T.F.; et al. Gut roundtable meeting paper: Selected recent advances in hepatocellular carcinoma. *Gut* **2018**, *67*, 380–388. [CrossRef]
17. Nault, J.C.; Mallet, M.; Pilati, C.; Calderaro, J.; Bioulac-Sage, P.; Laurent, C.; Laurent, A.; Cherqui, D.; Balabaud, C.; Zucman-Rossi, J. High frequency of telomerase reverse-transcriptase promoter somatic mutations in hepatocellular carcinoma and preneoplastic lesions. *Nat. Commun.* **2013**, *4*, 2218. [CrossRef]
18. Nault, J.-C.; Ningarhari, M.; Rebouissou, S.; Zucman-Rossi, J. The role of telomeres and telomerase in cirrhosis and liver cancer. *Nat. Rev. Gastroenterol. Hepatol.* **2019**, *16*, 544–558. [CrossRef] [PubMed]
19. Nault, J.C.; Calderaro, J.; Di Tommaso, L.; Balabaud, C.; Zafrani, E.S.; Bioulac-Sage, P.; Roncalli, M.; Zucman-Rossi, J. Telomerase reverse transcriptase promoter mutation is an early somatic genetic alteration in the transformation of premalignant nodules in hepatocellular carcinoma on cirrhosis. *Hepatology* **2014**, *60*, 1983–1992. [CrossRef] [PubMed]
20. The Cancer Genome Atlas Research Network. Comprehensive and integrative genomic characterization of hepatocellular carcinoma. *Cell* **2017**, *169*, 1327–1341. [CrossRef]
21. Dhanasekaran, R.; Nault, J.-C.; Roberts, L.R.; Zucman-Rossi, J. Genomic Medicine and Implications for Hepatocellular Carcinoma Prevention and Therapy. *Gastroenterology* **2019**, *156*, 492–509. [CrossRef]
22. Nault, J.; Martin, Y.; Caruso, S.; Hirsch, T.; Bayard, Q.; Calderaro, J.; Charpy, C.; Copie-Bergman, C.; Ziol, M.; Bioulac-Sage, P.; et al. Clinical Impact of Genomic Diversity From Early to Advanced Hepatocellular Carcinoma. *Hepatology* **2020**, *71*, 164–182. [CrossRef]
23. Totoki, Y.; Tatsuno, K.; Covington, K.R.; Ueda, H.; Creighton, C.J.; Kato, M.; Tsuji, S.; Donehower, L.A.; Slagle, B.L.; Nakamura, H.; et al. Trans-ancestry mutational landscape of hepatocellular carcinoma genomes. *Nat. Genet.* **2014**, *46*, 1267–1273. [CrossRef] [PubMed]
24. Schulze, K.; Imbeaud, S.; Letouzé, E.; Alexandrov, L.B.; Calderaro, J.; Rebouissou, S.; Couchy, G.; Meiller, C.; Shinde, J.; Soysouvanh, F.; et al. Exome sequencing of hepatocellular carcinomas identifies new mutational signatures and potential therapeutic targets. *Nat. Genet.* **2015**, *47*, 505–511. [CrossRef] [PubMed]
25. Villanueva, A. Hepatocellular carcinoma. *N. Engl. J. Med.* **2019**, *380*, 1450–1462. [CrossRef] [PubMed]
26. Boyault, S.; Rickman, D.S.; Bioulac-Sage, P.; De Reyniès, A.; Balabaud, C.; Rebouissou, S.; Jeannot, E.; Hérault, A.; Saric, J.; Belghiti, J.; et al. Transcriptome classification of HCC is related to gene alterations and to new therapeutic targets. *Hepatology* **2007**, *45*, 42–52. [CrossRef] [PubMed]
27. Hoshida, Y.; Nijman, S.M.B.; Kobayashi, M.; Chan, J.A.; Brunet, J.-P.; Chiang, D.Y.; Villanueva, A.; Newell, P.; Ikeda, K.; Hashimoto, M.; et al. Integrative Transcriptome Analysis Reveals Common Molecular Subclasses of Human Hepatocellular Carcinoma. *Cancer Res.* **2009**, *69*, 7385–7392. [CrossRef] [PubMed]

28. Nault, J.; De Reyniès, A.; Villanueva, A.; Calderaro, J.; Rebouissou, S.; Couchy, G.; Decaens, T.; Franco, D.; Imbeaud, S.; Rousseau, F.; et al. A Hepatocellular Carcinoma 5-Gene Score Associated With Survival of Patients After Liver Resection. *Gastroenterology* **2013**, *145*, 176–187. [CrossRef]
29. Calderaro, J.; Couchy, G.; Imbeaud, S.; Amaddeo, G.; Letouzé, E.; Blanc, J.-F.; Laurent, C.; Hajji, Y.; Azoulay, D.; Bioulac-Sage, P.; et al. Histological subtypes of hepatocellular carcinoma are related to gene mutations and molecular tumour classification. *J. Hepatol.* **2017**, *67*, 727–738. [CrossRef] [PubMed]
30. Torbenson, M.S.; Ng, I.O.L.; Park, Y.N.; Roncalli, M.; Sakamoto, M. Hepatocellular carcinoma. In *Digestive System Tumours*, 5th ed.; WHO Classification of Tumours Editorial Board; WHO Classification of Tumours Series; International Agency for Research on Cancer: Lyon, France, 2019; Volume 1, pp. 229–239.
31. International Consensus Group for Hepatocellular Neoplasia. Pathologic diagnosis of early hepatocellular carcinoma: A report of the International Consensus Group for Hepatocellular Neoplasia. *Hepatology* **2009**, *49*, 658–664. [CrossRef]
32. Kojiro, M. *Pathology of Hepatocellular Carcinoma*; Blackwell: Malden, MA, USA, 2006.
33. Salomao, M.; Woojin, M.Y.; Brown, R.S., Jr.; Emond, J.C.; Lefkowitch, J.H. Steatohepatitic hepatocellular carcinoma (SH-HCC): A distinctive histological variant of HCC in hepatitis C virus-related cirrhosis with associated NAFLD/NASH. *Am. J. Surg. Pathol.* **2010**, *34*, 1630–1636. [CrossRef]
34. Yeh, M.M.; Liu, Y.; Torbenson, M. Steatohepatitic variant of hepatocellular carcinoma in the absence of metabolic syndrome or background steatosis: A clinical, pathological, and genetic study. *Hum. Pathol.* **2015**, *46*, 1769–1775. [CrossRef]
35. Li, T.; Fan, J.; Qin, L.-X.; Zhou, J.; Sun, H.-C.; Qiu, S.-J.; Ye, Q.-H.; Wang, L.; Tang, Z.-Y. Risk Factors, Prognosis, and Management of Early and Late Intrahepatic Recurrence After Resection of Primary Clear Cell Carcinoma of the Liver. *Ann. Surg. Oncol.* **2011**, *18*, 1955–1963. [CrossRef]
36. Torbenson, M.S. Hepatocellular carcinoma: Making sense of morphological heterogeneity, growth patterns, and subtypes. *Hum. Pathol.* **2021**, *112*, 86–101. [CrossRef]
37. Wood, L.D.; Heaphy, C.M.; Daniel, H.D.-J.; Naini, B.V.; Lassman, C.R.; Arroyo, M.R.; Kamel, I.R.; Cosgrove, D.P.; Boitnott, J.K.; Meeker, A.K.; et al. Chromophobe hepatocellular carcinoma with abrupt anaplasia: A proposal for a new subtype of hepatocellular carcinoma with unique morphological and molecular features. *Mod. Pathol.* **2013**, *26*, 1586–1593. [CrossRef]
38. Torbenson, M.S. Morphologic Subtypes of Hepatocellular Carcinoma. *Gastroenterol. Clin. N. Am.* **2017**, *46*, 365–391. [CrossRef] [PubMed]
39. Maeda, T.; Adachi, E.; Kajiyama, K.; Takenaka, K.; Sugimachi, K.; Tsuneyoshi, M. Spindle cell hepatocellular carcinoma: A clinicopathologic and immunohistochemical analysis of 15 cases. *Cancer* **1996**, *77*, 51–57. [CrossRef]
40. Kojiro, M.; Sugihara, S.; Kakizoe, S.; Nakashima, O.; Kiyomatsu, K. Hepatocellular carcinoma with sarcomatous change: A special reference to the relationship with anticancer therapy. *Cancer Chemother. Pharmacol.* **1989**, *23*, S4–S8. [CrossRef]
41. International Working Party. Terminology of nodular hepatocellular lesions. *Hepatology* **1995**, *22*, 983–993. [CrossRef]
42. Watanabe, S.; Okita, K.; Harada, T.; Kodama, T.; Numa, Y.; Takemoto, T.; Takahashi, T. Morphologic studies of the liver cell dysplasia. *Cancer* **1983**, *51*, 2197–2205. [CrossRef]
43. Deugnier, Y.M.; Charalambous, P.; Le Quilleuc, D.; Turlin, B.; Searle, J.; Brissot, P.; Powell, L.W.; Halliday, J.W. Preneoplastic significance of hepatic iron-free foci in genetic hemochromatosis: A study of 185 patients. *Hepatology* **1993**, *18*, 1363–1369.
44. Hytiroglou, P.; Park, Y.N.; Krinsky, G.; Theise, N.D. Hepatic Precancerous Lesions and Small Hepatocellular Carcinoma. *Gastroenterol. Clin. N. Am.* **2007**, *36*, 867–887. [CrossRef]
45. Park, Y.N.; Kojiro, M.; Di Tommaso, L.; Dhillon, A.P.; Kondo, F.; Nakano, M.; Sakamoto, M.; Theise, N.D.; Roncalli, M. Ductular reaction is helpful in defining early stromal invasion, small hepatocellular carcinomas, and dysplastic nodules. *Cancer* **2007**, *109*, 915–923. [CrossRef]
46. Hytiroglou, P.; Theise, N.D. The differential diagnosis of liver nodules. *Sem. Diag. Pathol.* **1998**, *15*, 285–299.
47. Theise, N.D.; Marcelin, K.; Goldfischer, M.; Hytiroglou, P.; Ferrell, L.; Thung, S.N. Low proliferative activity in macroregenerative nodules: Evidence for an alternate hypothesis concerning human hepatocarcinogenesis. *Liver* **1996**, *16*, 134–139. [CrossRef]
48. Park, Y.N.; Yang, C.-P.; Cubukcu, O.; Thung, S.N.; Theise, N.D. Hepatic stellate cell activation in dysplastic nodules: Evidence for an alternate hypothesis concerning human hepatocarcinogenesis. *Liver Int.* **2008**, *17*, 271–274. [CrossRef] [PubMed]
49. Hytiroglou, P.; Theise, N.D.; Schwartz, M.; Mor, E.; Miller, C.; Thung, S.N. Macroregenerative nodules in a series of adult cirrhotic liver explants: Issues of classification and nomenclature. *Hepatology* **1995**, *21*, 703–708. [PubMed]
50. Park, Y.N.; Yang, C.-P.; Fernandez, G.J.; Cubukcu, O.; Thung, S.N.; Theise, N.D. Neoangiogenesis and Sinusoidal "Capillarization" in Dysplastic Nodules of the Liver. *Am. J. Surg. Pathol.* **1998**, *22*, 656–662. [CrossRef] [PubMed]
51. Theise, N.D.; Park, Y.N.; Thung, S.N. "Vascular profiles" of regenerative and dysplastic nodules. *Hepatology* **2000**, *31*, 1380–1381. [CrossRef] [PubMed]
52. Di Tommaso, L.; Franchi, G.; Park, Y.N.; Fiamengo, B.; Destro, A.; Morenghi, E.; Montorsi, M.; Torzilli, G.; Tommasini, M.; Terracciano, L.; et al. Diagnostic value of HSP70, glypican 3, and glutamine synthetase in hepatocellular nodules in cirrhosis. *Hepatology* **2007**, *45*, 725–734.
53. Di Tommaso, L.; Destro, A.; Seok, J.Y.; Balladore, E.; Terracciano, L.; Sangiovanni, A.; Iavarone, M.; Colombo, M.G.; Jang, J.J.; Yu, E.; et al. The application of markers (HSP70 GPC3 and GS) in liver biopsies is useful for detection of hepatocellular carcinoma. *J. Hepatol.* **2009**, *50*, 746–754.

54. Hytiroglou, P. Well-differentiated hepatocellular nodule: Making a diagnosis on biopsy and resection specimens of patients with advanced stage chronic liver disease. *Semin. Diagn. Pathol.* **2017**, *34*, 138–145. [CrossRef]
55. Bioulac-Sage, P.; Kakar, S.; Nault, J.C. Hepatocellular adenoma. In *Digestive System Tumours*, 5th ed.; WHO Classification of Tumours Editorial Board; WHO Classification of Tumours Series; International Agency for Research on Cancer: Lyon, France, 2019; Volume 1, pp. 224–228.
56. Joseph, N.M.; Blank, A.; Shain, A.H.; Gill, R.M.; Umetsu, S.E.; Shafizadeh, N.; Torbenson, M.S.; Kakar, S. Hepatocellular neoplasms with loss of liver fatty acid binding protein: Clinicopathologic features and molecular profiling. *Hum. Pathol.* **2022**, *122*, 60–71. [CrossRef]
57. Julien, C.; Le Bail, B.; Chiche, L.; Balabaud, C.; Bioulac-Sage, P. Malignant transformation of hepatocellular adenoma. *JHEP Rep.* **2022**, *4*, 100430. [CrossRef] [PubMed]
58. Calderaro, J.; Nault, J.C.; Balabaud, C.; Couchy, G.; Saint-Paul, M.-C.; Azoulay, D.; Mehdaoui, D.; Luciani, A.; Zafrani, E.S.; Bioulac-Sage, P.; et al. Inflammatory hepatocellular adenomas developed in the setting of chronic liver disease and cirrhosis. *Mod. Pathol.* **2016**, *29*, 43–50. [CrossRef]
59. Sasaki, M.; Yoneda, N.; Sawai, Y.; Imai, Y.; Kondo, F.; Fukusato, T.; Yoshikawa, S.; Kobayashi, S.; Sato, Y.; Matsui, O.; et al. Clinicopathological characteristics of serum amyloid A-positive hepatocellular neoplasms/nodules arising in alcoholic cirrhosis. *Histopathology* **2015**, *66*, 836–845. [CrossRef] [PubMed]
60. Rebouissou, S.; Franconi, A.; Calderaro, J.; Letouzé, E.; Imbeaud, S.; Pilati, C.; Nault, J.-C.; Couchy, G.; Laurent, A.; Balabaud, C.; et al. Genotype-phenotype correlation of CTNNB1 mutations reveals different β-catenin activity associated with liver tumor progression. *Hepatology* **2016**, *64*, 2047–2061. [CrossRef]
61. European Association for the Study of the Liver (EASL). EASL Clinical Practice Guidelines on the Management of Benign Liver Tumours. Available online: https://www.journal-of-hepatology.eu/article/S0168-8278(16)30101-5/pdf (accessed on 31 January 2022).
62. Julien, C.; Le-Bail, B.; Touhami, K.O.; Frulio, N.; Blanc, J.; Adam, J.; Laurent, C.; Balabaud, C.; Bioulac-Sage, P.; Chiche, L. Hepatocellular adenoma risk factors of hemorrhage: Size is not the only concern! Single center retrospective experience of 261 patients. *Ann. Surg.* **2021**, *274*, 843–850. [CrossRef]
63. Klompenhouwer, A.J.; Rosmalen, B.V.; Haring, M.P.D.; Thomeer, M.G.J.; Doukas, M.; Verheij, J.; Meijer, V.E.; Gulik, T.M.; Takkenberg, R.B.; Kazemier, G.; et al. A multicentre retrospective analysis on growth of residual hepatocellular adenoma after resection. *Liver Int.* **2020**, *40*, 2272–2278. [CrossRef]
64. Putra, J.; Ferrell, L.D.; Gouw, A.S.H.; Paradis, V.; Rishi, A.; Sempoux, C.; Balabaud, C.; Thung, S.N.; Bioulac-Sage, P. Malignant transformation of liver fatty acid binding protein-deficient hepatocellular adenomas: Histopathologic spectrum of a rare phenomenon. *Mod. Pathol.* **2020**, *33*, 665–675. [CrossRef] [PubMed]
65. Haring, M.P.D.; Cuperus, F.J.C.; Duiker, E.W.; de Haas, R.J.; de Meijer, V.E. Scoping review of clinical practice guidelines on the management of benign liver tumors. *BMJ Open Gastroenterol.* **2021**, *8*, e000592.
66. Lang, H.; Sotiropoulos, G.C.; Brokalaki, E.I.; Schmitz, K.J.; Bertona, C.; Meyer, G.; Frilling, A.; Paul, A.; Malagó, M.; Broelsch, C.E. Survival and Recurrence Rates after Resection for Hepatocellular Carcinoma in Noncirrhotic Livers. *J. Am. Coll. Surg.* **2007**, *205*, 27–36. [CrossRef] [PubMed]
67. Zhou, L.; Rui, J.-A.; Wang, S.-B.; Chen, S.-G.; Qu, Q.; Chi, T.-Y.; Wei, X.; Han, K.; Zhang, N.; Zhao, H.-T. Outcomes and prognostic factors of cirrhotic patients with hepatocellular carcinoma after radical major hepatectomy. *World J. Surg.* **2007**, *31*, 1782–1787. [CrossRef] [PubMed]
68. Jonas, S.; Bechstein, W.O.; Steinmüller, T.; Herrmann, M.; Radke, C.; Berg, T.; Settmacher, U.; Neuhaus, P. Vascular invasion and histopathologic grading determine outcome after liver transplantation for hepatocellular carcinoma in cirrhosis. *Hepatology* **2001**, *33*, 1080–1086. [CrossRef] [PubMed]
69. Colecchia, A.; Scaioli, E.; Montrone, L.; Vestito, A.; Di Biase, A.R.; Pieri, M.; D'Errico-Grigioni, A.; Bacchi-Reggiani, M.L.; Ravaioli, M.; Grazi, G.L.; et al. Pre-operative liver biopsy in cirrhotic patients with early hepatocellular carcinoma represents a safe and accurate diagnostic tool for tumour grading assessment. *J. Hepatol.* **2011**, *54*, 300–305. [CrossRef]
70. Ferrell, L.D.; Kakar, S.; Terraciano, M.; Wee, A. Tumours and Tumour-like lesions of the liver. In *MacSween's Pathology of the Liver*, 7th ed.; Burt, A.D., Ferrell, L.D., Huebscher, S.G., Eds.; Elsevier: Phladelphia, PA, USA, 2018; pp. 780–879.
71. Lagana, S.; Hsiao, S.; Bao, F.; Sepulveda, A.; Moreira, R.; Lefkowitch, J.; Remotti, H. HepPar-1 and Arginase-1 Immunohistochemistry in Adenocarcinoma of the Small Intestine and Ampullary Region. *Arch. Pathol. Lab. Med.* **2015**, *139*, 791–795. [CrossRef]
72. Saxena, R.; Albores-Saavedra, J.; Bioulac-Sage, P.; Hytiroglou, P.; Sakamoto, M.; Theise, N.D.; Sui, W.M.S. Diagnostic algorithms for tumours of the liver. In *WHO Classification of Tumours of the Digestive System*, 4th ed.; Bosman, F.T., Carneiro, F., Hruban, R.H., Theise, N.D., Eds.; IARC: Lyon, France, 2010.
73. Kim, H.; Choi, G.H.; Na, D.C.; Ahn, E.Y.; Kim, G.I.; Lee, J.E.; Cho, J.Y.; Yoo, J.E.; Choi, J.S.; Park, Y.N. Human hepatocellular carcinomas with "Stemness"-related marker expression: Keratin 19 expression and a poor prognosis. *Hepatology* **2011**, *54*, 1707–1717. [CrossRef]
74. Tsuchiya, K.; Komuta, M.; Yasui, Y.; Tamaki, N.; Hosokawa, T.; Ueda, K.; Kuzuya, T.; Itakura, J.; Nakanishi, H.; Takahashi, Y.; et al. Expression of Keratin 19 Is Related to High Recurrence of Hepatocellular Carcinoma after Radiofrequency Ablation. *Oncology* **2011**, *80*, 278–288. [CrossRef]

75. Rhee, H.; Nahm, J.H.; Kim, H.; Choi, G.H.; Yoo, J.E.; Lee, H.S.; Koh, M.J.; Park, Y.N. Poor outcome of hepatocellular carcinoma with stemness marker under hypoxia: Resistance to transarterial chemoembolization. *Mod. Pathol.* **2016**, *29*, 1038–1049. [CrossRef]
76. Llovet, J.M.; Pinyol, R.; Kelley, R.K.; El-Khoueiry, A.; Reeves, H.L.; Wang, X.W.; Gores, G.J.; Villanueva, A. Molecular pathogenesis and systemic therapies for hepatocellular carcinoma. *Nat. Cancer* **2022**, *3*, 386–401. [CrossRef] [PubMed]
77. Brunt, E.; Aishima, S.; Clavien, P.-A.; Fowler, K.; Goodman, Z.; Gores, G.; Gouw, A.; Kagen, A.; Klimstra, D.; Komuta, M.; et al. cHCC-CCA: Consensus terminology for primary liver carcinomas with both hepatocytic and cholangiocytic differentation. *Hepatology* **2018**, *68*, 113–126. [CrossRef] [PubMed]
78. De Vito, C.; Sarker, D.; Ross, P.; Heaton, N.; Quaglia, A. Histological heterogeneity in primary and metastatic classic combined hepatocellular-cholangiocarcinoma: A case series. *Virchows Arch.* **2017**, *471*, 619–629. [CrossRef]
79. Zen, C.; Zen, Y.; Mitry, R.R.; Corbeil, D.; Karbanova, J.; O'Grady, J.; Karani, J.; Kane, P.; Heaton, N.; Portmann, B.C.; et al. Mixed phenotype hepatocellular carcinoma after transarterial chemoembolization and liver transplantation. *Liver Transplant.* **2011**, *17*, 943–954. [CrossRef]
80. Lai, J.-P.; Conley, A.; Knudsen, B.S.; Guindi, M. Hypoxia after transarterial chemoembolization may trigger a progenitor cell phenotype in hepatocellular carcinoma. *Histopathology* **2015**, *67*, 442–450. [CrossRef] [PubMed]
81. Theise, N.D.; Nakashima, O.; Park, Y.N.; Nakanuma, Y. Combined hepatocellular-cholangiocarcinoma. In *WHO Classification of Tumours of the Digestive System*, 4th ed.; Bosman, F.T., Carneiro, F., Hruban, R.H., Theise, N.D., Eds.; IARC: Lyon, France, 2010.
82. Sempoux, C.; Kakar, S.; Kondo, F. Combined hepatocellular-cholangiocarcinoma and undifferentiated primary liver carcinoma. In *Digestive System Tumours*, 5th ed.; WHO Classification of Tumours Editorial Board; WHO Classification of Tumours Series; International Agency for Research on Cancer: Lyon, France, 2019; Volume 1, pp. 260–262.
83. Kim, M.; Hwang, S.; Ahn, C.-S.; Kim, K.-H.; Moon, D.-B.; Song, G.-W.; Jung, D.-H.; Hong, S.-M. Post-resection prognosis of combined hepatocellular carcinoma-cholangiocarcinoma cannot be predicted by the 2019 World Health Organization classification. *Asian J. Surg.* **2021**, *44*, 1389–1395. [CrossRef]
84. Gigante, E.; Ronot, M.; Bertin, C.; Ciolina, M.; Bouattour, M.; Dondero, F.; Cauchy, F.; Soubrane, O.; Vilgrain, V.; Paradis, V. Combining imaging and tumour biopsy improves the diagnosis of combined hepatocellular-cholangiocarcinoma. *Liver Int.* **2019**, *39*, 2386–2396. [CrossRef] [PubMed]
85. Jeon, S.K.; Joo, I.; Lee, D.H.; Lee, S.M.; Kang, H.-J.; Lee, K.-B.; Lee, J.M. Combined hepatocellular cholangiocarcinoma: LI-RADS v2017 categorisation for differential diagnosis and prognostication on gadoxetic acid-enhanced MR imaging. *Eur. Radiol.* **2019**, *29*, 373–382. [CrossRef]
86. Zou, X.; Luo, Y.; Morelli, J.N.; Hu, X.; Shen, Y.; Hu, D. Differentiation of hepatocellular carcinoma from intrahepatic cholangiocarcinoma and combined hepatocellular-cholangiocarcinoma in high-risk patients matched to MR field strength: Diagnostic performance of LI-RADS version 2018. *Abdom. Radiol.* **2021**, *46*, 3168–3178. [CrossRef]
87. Beaufrère, A.; Calderaro, J.; Paradis, V. Combined hepatocellular-cholangiocarcinoma: An update. *J. Hepatol.* **2021**, *74*, 1212–1224. [CrossRef] [PubMed]
88. Yen, C.-C.; Yen, C.-J.; Shan, Y.-S.; Lin, Y.J.; Liu, I.-T.; Huang, H.-Y.; Yeh, M.M.; Chan, S.-H.; Tsai, H.-W. Comparing the clinicopathological characteristics of combined hepatocellular-cholangiocarcinoma with those of other primary liver cancers by use of the updated World Health Organization classification. *Histopathology* **2021**, *79*, 556–572. [CrossRef]
89. Dageforde, L.A.; Vachharajani, N.; Tabrizian, P.; Agopian, V.; Halazun, K.; Maynard, E.; Croome, K.; Nagorney, D.; Hong, J.C.; Lee, D.; et al. Multi-Center Analysis of Liver Transplantation for Combined Hepatocellular Carcinoma-Cholangiocarcinoma Liver Tumors. *J. Am. Coll. Surg.* **2021**, *232*, 361–371. [CrossRef]
90. Rizell, M.; Åberg, F.; Perman, M.; Ny, L.; Stén, L.; Hashimi, F.; Svanvik, J.; Lindnér, P. Checkpoint Inhibition Causing Complete Remission of Metastatic Combined Hepatocellular-Cholangiocarcinoma after Hepatic Resection. *Case Rep. Oncol.* **2020**, *13*, 478–484. [CrossRef] [PubMed]
91. Nguyen, C.T.; Caruso, S.; Maille, P.; Beaufrère, A.; Augustin, J.; Favre, L.; Pujals, A.; Boulagnon-Rombi, C.; Rhaiem, R.; Amaddeo, G.; et al. Immune Profiling of Combined Hepatocellular- Cholangiocarcinoma Reveals Distinct Subtypes and Activation of Gene Signatures Predictive of Response to Immunotherapy. *Clin. Cancer Res.* **2022**, *28*, 540–551. [CrossRef]
92. Yagi, N.; Suzuki, T.; Mizuno, S.; Kojima, M.; Kudo, M.; Sugimoto, M.; Kobayashi, S.; Gotohda, N.; Ishii, G.; Nakatsura, T. Component with abundant immune-related cells in combined hepatocellular cholangiocarcinoma identified by cluster analysis. *Cancer Sci.* **2022**, *113*, 1564–1574. [CrossRef]
93. Murugesan, K.; Sharaf, R.; Montesion, M.; Moore, J.A.; Pao, J.; Pavlick, D.C.; Frampton, G.M.; Upadhyay, V.A.; Alexander, B.M.; Miller, V.A.; et al. Genomic Profiling of Combined Hepatocellular Cholangiocarcinoma Reveals Genomics Similar to Either Hepatocellular Carcinoma or Cholangiocarcinoma. *JCO Precis. Oncol.* **2021**, *5*, 1285–1296. [CrossRef]
94. Fowler, K.J.; Burgoyne, A.; Fraum, T.J.; Hosseini, M.; Ichikawa, S.; Kim, S.; Kitao, A.; Lee, J.M.; Paradis, V.; Taouli, B.; et al. Pathologic, Molecular, and Prognostic Radiologic Features of Hepatocellular Carcinoma. *RadioGraphics* **2021**, *41*, 1611–1631. [CrossRef]
95. Fung, A.; Shambhogue, K.; Tafel, M.; Theise, N.D. Imaging of hepatocarcinogenesis with clinical and pathological correlation. *Magn. Reson. Imaging Clin. N. Am.* **2021**, *29*, 359–374. [CrossRef] [PubMed]
96. Zeng, Q.; Klein, C.; Caruso, S.; Maille, P.; Laleh, N.G.; Sommacale, D.; Laurent, A.; Amaddeo, G.; Gentien, D.; Rapinat, A.; et al. Artificial intelligence predicts immune and inflammatory gene signatures directly from hepatocellular carcinoma histology. *J. Hepatol.* **2022**, *77*, 116–127. [CrossRef] [PubMed]

cancers

MDPI

*Review*

# Current Imaging Diagnosis of Hepatocellular Carcinoma

**Evangelos Chartampilas [1,*], Vasileios Rafailidis [1], Vivian Georgopoulou [2], Georgios Kalarakis [3,4,5], Adam Hatzidakis [1] and Panos Prassopoulos [1]**

1 Radiology Department, AHEPA University Hospital, Medical School, Aristotle University of Thessaloniki, 54636 Thessaloniki, Greece
2 Radiology Department, Ippokratio General Hospital of Thessaloniki, 54642 Thessaloniki, Greece
3 Department of Diagnostic Radiology, Karolinska University Hospital, 14152 Stockholm, Sweden
4 Department of Clinical Science, Division of Radiology, Intervention and Technology (CLINTEC), Karolinska Institutet, 14152 Stockholm, Sweden
5 Department of Radiology, Medical School, University of Crete, 71500 Heraklion, Greece
* Correspondence: evaharta@yahoo.gr

**Simple Summary:** The role of imaging in the management of hepatocellular carcinoma (HCC) has significantly evolved and expanded beyond the plain radiological confirmation of the tumor based on the typical appearance in a multiphase contrast-enhanced CT or MRI examination. The introduction of hepatobiliary contrast agents has enabled the diagnosis of hepatocarcinogenesis at earlier stages, while the application of ultrasound contrast agents has drastically upgraded the role of ultrasound in the diagnostic algorithms. Newer quantitative techniques assessing blood perfusion on CT and MRI not only allow earlier diagnosis and confident differentiation from other lesions, but they also provide biomarkers for the evaluation of treatment response. As distinct HCC subtypes are identified, their correlation with specific imaging features holds great promise for estimating tumor aggressiveness and prognosis. This review presents the current role of imaging and underlines its critical role in the successful management of patients with HCC.

**Citation:** Chartampilas, E.; Rafailidis, V.; Georgopoulou, V.; Kalarakis, G.; Hatzidakis, A.; Prassopoulos, P. Current Imaging Diagnosis of Hepatocellular Carcinoma. *Cancers* **2022**, *14*, 3997. https://doi.org/10.3390/cancers14163997

Academic Editor: Georgios Germanidis

Received: 9 July 2022
Accepted: 15 August 2022
Published: 18 August 2022



**Abstract:** Hepatocellular carcinoma (HCC) is the fourth leading cause of cancer related death worldwide. Radiology has traditionally played a central role in HCC management, ranging from screening of high-risk patients to non-invasive diagnosis, as well as the evaluation of treatment response and post-treatment follow-up. From liver ultrasonography with or without contrast to dynamic multiple phased CT and dynamic MRI with diffusion protocols, great progress has been achieved in the last decade. Throughout the last few years, pathological, biological, genetic, and immune-chemical analyses have revealed several tumoral subtypes with diverse biological behavior, highlighting the need for the re-evaluation of established radiological methods. Considering these changes, novel methods that provide functional and quantitative parameters in addition to morphological information are increasingly incorporated into modern diagnostic protocols for HCC. In this way, differential diagnosis became even more challenging throughout the last few years. Use of liver specific contrast agents, as well as CT/MRI perfusion techniques, seem to not only allow earlier detection and more accurate characterization of HCC lesions, but also make it possible to predict response to treatment and survival. Nevertheless, several limitations and technical considerations still exist. This review will describe and discuss all these imaging modalities and their advances in the imaging of HCC lesions in cirrhotic and non-cirrhotic livers. Sensitivity and specificity rates, method limitations, and technical considerations will be discussed.

**Keywords:** hepatocellular carcinoma (HCC); ultrasound (US); contrast-enhanced ultrasound (CEUS); computed tomography (CT); magnetic resonance imaging (MRI); perfusion imaging; MR diffusion imaging; multiparametric imaging; diagnostic algorithms; locoregional treatment

## 1. Introduction

Hepatocellular carcinoma (HCC) is the commonest primary liver tumor, comprising 75–85% of cases. HCC ranks sixth in global incidence after breast, lung, colorectal, prostate, and gastric cancer. In terms of mortality, it ranks third for both genders, and rates of incidence and mortality are 2–3 times higher in men than in women [1]. Although the increasing prevalence of metabolic syndrome has shifted the etiology of liver cancer, hepatitis B and C account for 56% and 20% of global mortality, respectively. Variations by world region exist; for example, alcohol consumption accounts for 22% of all HCC cases in Europe and North America in 2020 [2]. The vast majority of HCCs are diagnosed in patients with cirrhosis and/or chronic hepatitis B infection, necessitating close surveillance in these groups in order to detect the tumor at an early stage.

Imaging plays a key role in surveillance, diagnosis, and staging, as well as post-treatment follow-up. Ultrasound surveillance improves survival in a cost-effective way and is endorsed by all major practice guidelines, i.e., American [3], European [4], and Asian-Pacific [5]. Dynamic contrast-enhanced Computed Tomography (CT) takes advantage of the hemodynamic changes that occur in the cirrhotic nodule as it progresses to early HCC during the multistep process of hepatocarcinogenesis. The gradual decrease in both normal arterial and portal supply and the formation of unpaired arteries are exemplarily exploited in multiphasic examinations after contrast agent administration; the appearance of the suspicious nodule on the acquired images serves as the basis for its characterization and subsequent management decisions. Magnetic resonance imaging (MRI) has the advantage of assessing additional features such as nodule cellularity and presence of fat, which are also important in nodule assessment. The use of hepatospecific contrast media has significantly augmented the diagnostic performance of MRI: as these drugs are taken up by specific transporters, whose expression decreases as carcinogenesis progresses, lesion hypointensity on the hepatobiliary phase not only is a sensitive feature—even allowing detection of high-grade dysplastic nodules—but is useful for predicting histologic differentiation too. Contrast-enhanced ultrasound may characterize tumor hemodynamics with comparative capacity; furthermore, use of a specific sonographic contrast medium can also aid tumor detection via imaging on the late Kupffer phase.

The characteristic tumor hypervascularity on the arterial phase and hypoperfusion on the portal phase constitute the hallmark of HCC diagnosis on dynamic CT or MRI. Clinically-important prognostic features can be derived from the imaging appearance of HCC. Disease staging can also be performed during the examination in order to identify satellite or multifocal lesions, portal vein invasion, or extrahepatic metastases. Additionally, as HCCs are supplied almost exclusively by the hepatic artery, response assessment after locoregional treatment or systemic therapy can be performed based on tumor enhancement on the arterial phase during follow-up CT or MRI scans [6].

In the following paragraphs of this review, a thorough discussion of the imaging characteristics of HCC on the various imaging modalities—including more advanced techniques—as well as a comparison of their diagnostic performance will be provided. A brief comment on the most widely-used diagnostic algorithms is also included—for completeness of the presentation.

## 2. Ultrasound

Liver cirrhosis (LC) is the primary risk factor for HCC, with patients requiring periodical imaging surveillance. Ultrasound (US) is very-well suited for this purpose, thanks to its wide availability, cost-effectiveness, and accuracy in detecting focal liver lesions (FLL). Once a FLL is detected, US can assist its characterization, particularly using the full spectrum of ultrasonographic techniques, including B-mode, colour, and power Doppler techniques, such as pulsed-wave Doppler, non-Doppler flow-visualization, and, currently, contrast-enhanced ultrasound (CEUS); a wide range of techniques justifies the term, multiparametric ultrasound [7,8].

The appearances of HCC on US vary depending on the size and degree of differentiation. An important distinction in terms of size is the cut-off of 2 cm [9]. On B-mode, HCC is hypo-echoic in more than 50% of cases, although it can be hyperechoic or of mixed echogenicity in approximately 25% of cases, respectively [9]. Consequently, it is important not to misdiagnose a hyperechoic nodule in a cirrhotic liver as a hemangioma and discard it without further investigation. The hyperechoic element of an HCC may represent a fatty component. HCCs smaller than 1 cm can be iso-echoic, and hence difficult to detect. As a general rule, tumor echogenicity reflects cell density. Given the gradual carcinogenesis of HCC inside a cirrhotic liver, an HCC is typically nodular in shape, except for the massive type, which appears irregular. The lesion margins are usually relatively well-circumscribed in the nodular type of HCC, but poorly defined in the massive type [9]. A peripheral hypo-echoic halo may be noted, corresponding to a thin fibrous capsule in 90% of cases [10]. The so called "mosaic pattern" and the "nodule in nodule" appearance are two characteristic demonstrations of HCC in every modality, including US and CEUS [9], which is seen with a frequency proportionally increasing with tumor size. All these B-mode characteristics have been classified into five macroscopic types, as follows: small nodular type with indistinct margins, simple nodular type, simple nodular type with extranodular growth, confluent multinodular type, and infiltrative type, with the potential for malignancy increasing accordingly [11–13].

The vascularity of a FLL is important, pointing towards a benign or malignant diagnosis. From a technical point-of-view, colour Doppler is the first-line modality to assess for intratumoral or peripheral vascularity, but it suffers from technical limitations such as Doppler-angle-dependence, low sensitivity to slow flow, and overwriting artifact. Power Doppler and modern non-Doppler flow visualization techniques are now available for improved characterization of HCC vascular architecture. HCCs less than 2 cm commonly appear avascular due to the technique's low sensitivity, while in some cases blood vessels are visualized as lines or dots inside, or surrounding, the tumor. A continuous waveform on spectral analysis can be seen in these tumors, in keeping with feeding portal flow. Once the tumor increases in size, more characteristic vascular patterns can be appreciated. Namely, the "basket" pattern of vascularity refers to the presence of a fine network of arterial branches surrounding the lesion. Using spectral analysis, both pulsatile and continuous waveforms can be recorded, which correspond to the hepatic artery and hepatic or portal vein origin of blood supply, respectively. In the massive-type HCC, an overall irregular pattern of vascularity can be appreciated [9]. As a general rule, a continuous (portal vein-like) waveform indicates a dysplastic nodule or a well-differentiated HCC; contrarily, a pulsatile arterial waveform is suggestive of advanced HCC [14]. Power Doppler tends to detect intratumoral colour signals in 19% or more of angiographically hypervascular lesions compared to colour Doppler. Tumors appearing hypovascular on angiography typically exhibit no flow signals on either colour or power Doppler. Power Doppler is less affected than colour Doppler by the small size and the deep tumor location [15]. HCCs with a higher resistive and pulsatility index were associated with early recurrence, suggesting a more malignant nature [16].

Doppler limitations have been addressed with newer non-Doppler techniques, like the superb microvascular imaging (SMI, Canon Medical Systems, Otawara, Japan) or b-Flow/high-definition color (HDC, GE Healthcare, Chalfont St. Giles, UK), which exclusively visualizes intratumoral blood vessels without artifacts with high sensitivity, high spatial resolution, and in real-time [9]. When compared to colour Doppler, SMI visualized more signals, while a hypervascular pattern on SMI was significantly more common in HCC, compared with other lesions [17]. A study using SMI on FLL concluded that HCC demonstrates a "diffuse honeycomb" or a non-specific type of vascularity, which is significantly different from hemangioma [18].

Given that patients at risk for HCC formation undergo US surveillance, an important advance was the introduction of US LI-RADS® (Liver Imaging Reporting and Data System), a classification system in accordance with CT/MRI LI-RADS® that was issued by the

American College of Radiology. Briefly, this system assesses the quality of examination and the potential of a FLL to represent HCC in three classes and suggests further management. The features taken into account include size and echogenicity [19]. In the setting of HCC screening and surveillance, US LI-RADS® yielded a sensitivity of 58–89% and a specificity >90% [20]. US in general has a reported sensitivity of 98% and specificity of 85% for overall HCC detection [21]. Tumor size is nonetheless a significant factor as the technique's sensitivity can only be 65% for lesions <2 cm. The same applies to non-alcoholic steatohepatitis, where the overall change of liver echogenicity lowers the sensitivity [22].

## 3. Contrast-Enhanced Ultrasound

A decisive turning point in the ultrasonographic diagnosis of HCC was the introduction of contrast agents in combination with modern techniques like the pulse-inversion technique. These features not only increased the sensitivity and specificity of CEUS in diagnosing HCC, but also augmented the modality's role in diagnostic algorithms. There are two contrast agents widely used for HCC: SonoVue (Bracco, Milan, Italy) and Sonazoid (GE Healthcare, Amersham, UK). Both agents work with a low mechanical index (MI) and in real-time, but the latter is thought to be phagocytosed by reticuloendothelial (Kuppfer) cells of the liver parenchyma, thus generating a late arterial phase (post-vascular), starting from 10 min, when the liver parenchyma is normally enhanced, whereas malignant lesions appear washed-out due to lack of Kuppfer cells.

In Europe, CEUS is usually performed with SonoVue®, which is not uptaken by Kupffer cells and hence produces an arterial, portal-venous, and late arterial phase lasting up to 6 min [23]. The hallmark of HCC on CEUS using SonoVue® is a homogeneous and intense arterial phase hyper-enhancement (APHE) with mild wash-out starting >60 s after injection (Figure 1). Nodules measuring >5 cm may show heterogeneous enhancement due to necrosis (Figure 2). Both the size and degree of differentiation affect the enhancement pattern of HCC [24,25]. Wash-out is less often seen in HCC nodules <2 cm, but is more frequent in HCC with poorer grades of differentiation [23]. The timing and degree of wash-out are important for the characterization of HCC, which typically shows milder hypo-enhancement compared to metastasis and cholangiocarcinoma. HCC washout should start at least 60-s post-injection, while a quarter of cases may become hypo-enhancing 3-min post-injection, justifying the need for prolonged scanning in patients with cirrhosis. Early wash-out (<60 s) has been associated with poorly-differentiated HCC and non-hepatocellular malignancies [23]. An important study published in 2020 looked at CEUS findings of the entire spectrum of carcinogenesis in the cirrhotic liver by examining regenerative nodules, dysplastic nodules, and small HCC. It was concluded that shorter contrast–arrival times in lesions compared with background liver was associated with a higher degree of malignancy [26]. A different study determined 0.5 s as the minimum difference in contrast arrival time between the nodule and liver as a criterion for HCC [27]. On the contrary, arterial phase iso- or hypo-enhancement and late arterial phase iso-enhancement were associated with a lower degree of malignancy [28]. Regenerative and dysplastic nodules generally tend to appear iso-enhancing to adjacent liver cells on all phases of the scan [29] (Figure 3). In a study analyzing HCC $\leq$3 cm, moderately- and poorly-differentiated HCCs exhibited arterial phase hyperenhancement (APHE) more often than well-differentiated HCCs [30]. Other researchers found that moderately-differentiated HCCs are more commonly hypervascular during the arterial phase, while atypical appearances (iso- or hypo-vascular) may occur more often in well- and poorly-differentiated HCCs. Wash-out time has been found to be shorter in poorly-differentiated HCCs, as compared to well-differentiated lesions [31]. Despite ongoing research, overlap in blood supply origin between different steps of hepatocellular carcinogenesis may still limit the technique's accuracy in diagnosing HCC [29,32,33]. CEUS advantages and disadvantages are summarized in Table 1. In cases followed-up with CEUS or US, an increase in size or change in echogenicity indicates malignant transformation [27]. According to the European Federation of Societies for Ultrasound in Medicine and Biology (EFSUMB) guidelines, CEUS can be used in cases with inconclusive CT or MRI or if the patient is not suitable for biopsy, as well as for monitoring changes in enhancement in nodules requiring follow-up [23].

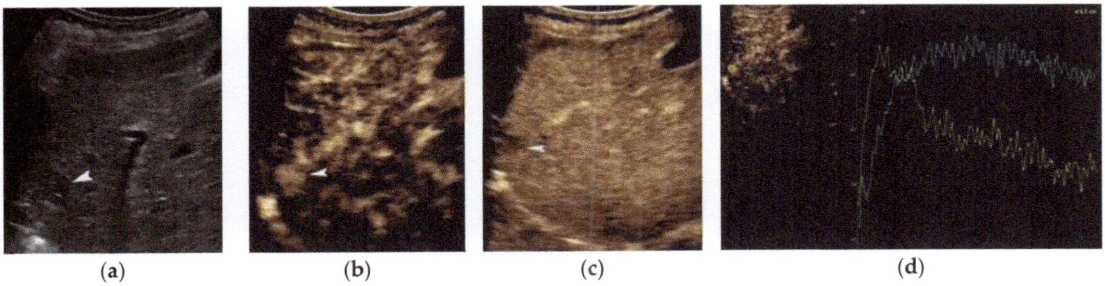


**Figure 1.** Typical CEUS findings of HCC nodule. B-mode (**a**) showing a hypo-echoic nodule (arrowhead) inside a heterogeneous cirrhotic liver. Arterial (**b**) and portal-venous phase (**c**) CEUS image showing homogeneous arterial phase hyperenhancement and mild wash-out, respectively, in keeping with HCC (arrowhead). Time-Intensity-Curve analysis (**d**) confirming the earlier enhancement of the nodule on arterial phase and wash-out. Quantitative parameters can be extracted using this type of analysis.

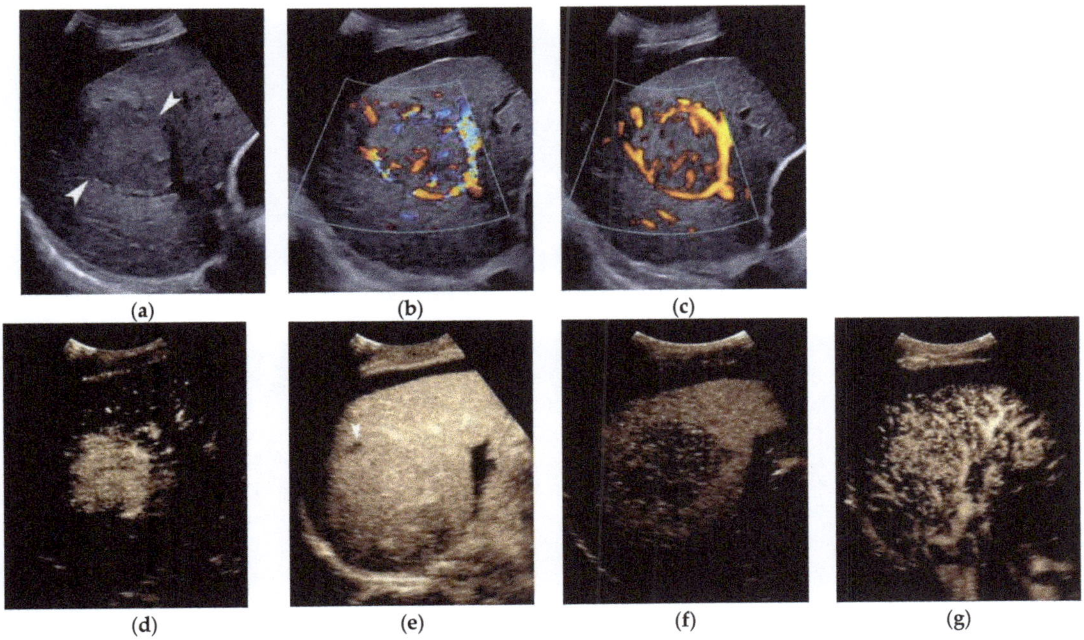


**Figure 2.** CEUS findings of a large HCC. B-mode (**a**) showing an ill-defined mildly hypo-echoic mass (outlined by arrowheads) inside the right lobe of a cirrhotic liver complicated with ascites. Colour (**b**) and power (**c**) Doppler technique visualizing the irregular internal vascular pattern of the mass. Note the increased vascularity locally. Arterial (**d**), venous (**e**), and delayed (**f**) CEUS image showing arterial phase hyperenhancement prior to adjacent hepatic parenchyma, iso-enhancement on portal-venous phase, and wash-out on the delayed phase (approximately 3 min). Note the area of necrosis appearing as non-enhancing (arrowhead on (**e**)). Temporal MIP image (**g**) showing the vascular architecture of the mass. Note the dense and irregular vascularity indicated inside the lesion.

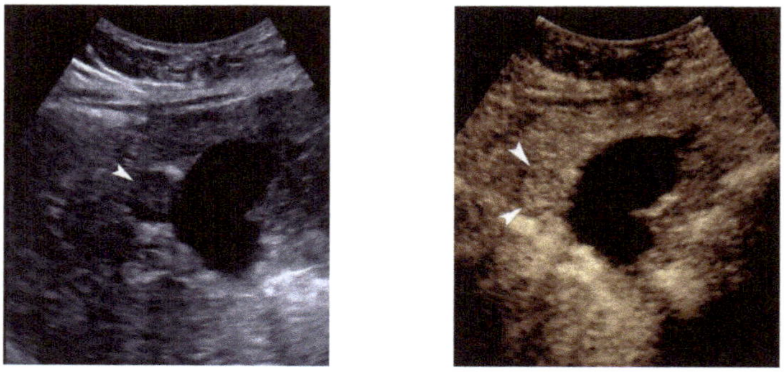

(a) (b)

**Figure 3.** A regenerative nodule in liver cirrhosis. B-mode (**a**) depicted a rounded hypo-echoic nodule (arrowhead) with smooth border. CEUS (**b**) showed that the nodule (outlined by the arrowheads) demonstrated enhancement identical to the adjacent parenchyma with no arterial phase hyperenhancement or wash-out. These findings are in keeping with a regenerative nodule.

**Table 1.** Advantages and disadvantages of CEUS for diagnosis of HCC.

| Advantages | Disadvantages |
|---|---|
| • Real-time scanning allowing for optimal detection of APHE with sensitivity<br>• Capability to the time of arrival of microbubbles (arterialization) or wash-out; potential biomarkers of malignancy<br>• Capability to be performed with free-breathing (good patient tolerability)<br>• Non-nephrotoxic agent | • Focused field-of-view, hindering staging of the entire liver<br>• Operator-dependency<br>• Limited by body-habitus |

Since both post-ablation necrotic areas and malignancies appear non-enhancing on the post-vascular phase, a way to discriminate those is a second dose of Sonazoid while targeting the non-enhancing area; arterial enhancement within the non-enhancing area thus establishes malignancy. This approach appears to outperform contrast-enhanced CT with a sensitivity of 95.4% and an accuracy of 94.7% [34]. CEUS with Sonazoid® has been found to be equivalent or even better than contrast-enhanced CT for the detection of arterial hypervascularity of HCC nodules [35–37]. The post-vascular phase can also be assessed using a high-MI Doppler pulse where the disrupted microbubbles within the normal parenchyma generate a signal, whereas the lack of microbubbles inside malignancies is accentuated, appearing in the form of "punched out" areas [9].

Similar to CT and MRI, a CEUS LI-RADS® algorithm has been introduced by the American College of Radiology to aid in the accurate characterization of nodules in liver cirrhosis patients. The major criteria are APHE, nodule size, and late-mild wash-out. A rim APHE or early (<60 s) or marked wash-out represent LR-M criteria, favoring the diagnosis of a non-hepatocellular malignancy. Ancillary features suggesting malignancy include definite growth, while the mosaic architecture and nodule-in-nodule architecture indicate HCC. On the other hand, size reduction or stability for ≥2 years indicates benignity. Classes LR1–2 can return to follow-up, LR-3 needs a second modality sooner than 6 months, while LR-4 and LR-5 require biopsy or treatment [33]. An important difference between CT/MRI and CEUS is that the latter technique does not visualize arterio-portal shunts, meaning that any arterially hyperenhancing lesion represents a true lesion and not a false finding [20]. Upon meta-analysis, CEUS is 93% sensitive and 90% specific in differentiating benign from malignant FLL [38]. In detecting HCC in patients with cirrhosis, a multi-center study showed that category LR-5 has a 98.5% positive predictive value, a 15.5 positive likelihood ratio, but only a 62% sensitivity for diagnosing HCC, making CEUS LR-5 a highly-specific tool [39]. In a study looking exclusively at HCC nodules up to 2 cm, CEUS LR-5 was 73.3% sensitive and 97.1% specific [26].

Temporal Maximum Intensity Projection (MIP) is a useful CEUS technique that generates vascular maps of FLL. A study assessing HCCs with this technique concluded that well-differentiated HCCs exhibited either normal or not-clearly visible intratumoral vasculature. Contrarily, poorly-differentiated HCCs showed tortuous, meandering, or tapering and interrupted intratumoral blood vessels. These parameters were 85% sensitive, 92.7% specific, and 90% accurate [40]. In another study, dysmorphic arteries have been found in 72% of HCCs [31]. The quantification of CEUS signal has also been used to consolidate a qualitative assessment by the performing radiologist, showing significant differences between benign and malignant lesions in terms of parameters such as rise time and late-phase ratio between lesion and liver [41]. This type of analysis, also termed "dynamic CEUS", is able to detect microvascular invasion of HCC, since various parameters such as wash-in rate and wash-out rate are significantly higher in invasive tumors [42]. Dynamic CEUS also detected higher wash-out in cholangiocarcinoma than HCC by using quantitative parameters, while the arterial enhancement profiles of these tumors were identical [43]. This analysis also has implications in tumor response to anti-angiogenetic treatment where perfusional parameter changes can be monitored [24]. Parametric maps based on the time of microbubble arrival is an alternative and simplistic quantitative approach (Figure 4).

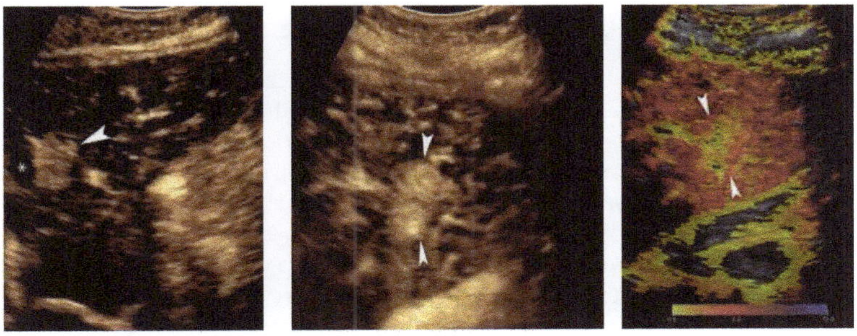

(**a**) (**b**) (**c**)

**Figure 4.** CEUS temporal MIP and parametric image in HCC. Sagittal arterial-phase CEUS image (**a**) showing an HCC nodule (arrowhead) developing next to a previous area of ablation (asterisk). Axial temporal MIP image (**b**) delineating the entire HCC nodule (outlined by arrowheads). Parametric colour map (**c**) confirming the earlier and homogeneous arrival of contrast in the HCC nodule.

HCC in the non-cirrhotic liver typically appears on CEUS with APHE and wash-out on the portal-venous or late-phase, while a chaotic vascular architecture may be seen. The fibrolamellar variant of HCC has non-specific appearances on CEUS but generally demonstrates rapid heterogeneous wash-in and quick and intense wash-out [23,44].

## 4. CT

Multidetector Computed Tomography (MDCT) plays a pivotal role in the diagnostic work-up of cirrhotic patients who are at increased risk of developing HCC. According to almost all guidelines, recognition of a nodule ≥10 mm by ultrasonography (US) during HCC surveillance should be followed by dynamic MDCT or MRI examination [45].

Nowadays, MDCT is widely-available, rapid, and robust. Most modern CT scanners have the capability to image with wide-detector arrays, typically more than 8-row detectors, allowing large z-axis coverage in a single rotation with high spatial resolution [46]. In comparison to MRI, MDCT is a well-tolerated examination, less prone to motion artifacts, even in elderly or non-cooperative patients who are unable to hold their breath. The main disadvantages include radiation exposure and relatively low contrast resolution and tissue differentiation. Moreover, studies have demonstrated a slightly higher sensitivity and specificity of MRI compared to CT, especially for lesions smaller than 20 mm.

Consistent and appropriate CT imaging protocols are absolutely requisite for optimal detection and characterization of liver lesions in a cirrhotic patient, thereby allowing for the reproducibility of LI-RADS categories [47,48]. The Technique Working Group of LI-RADS has proposed minimal technical requirements for the performance of CT in order to achieve wide acceptability and optimal imaging (Table 2).

**Table 2.** Basic technical recommendations for CT [46–48].

| Feature | Recommendation |
|---|---|
| CT Scanner Configuration | ≥8-row multidetector CT |
| Slice Thickness | 2–5 mm |
| Multiplanar Reformations | Suggested coronal and sagittal planes |
| Non-contrast Imaging | Suggested for initial diagnosis<br>Required for patients with prior locoregional therapy |
| Dynamic Contrast-Enhanced Phases | Late Arterial Phase<br>Portal-Venous<br>Delayed Phase |
| Contrast Administration | ≥300 mgI/mL for a dose of 1.5–2 mL/kg body weight (521–647 mgI/kg)<br>Injection rate ≥ 3 mL/s<br>Saline chaser bolus (30–50 mL) |

Multiphase contrast-enhanced imaging consisting of the late arterial, portal-venous, and delayed phase is invaluable for a confident imaging diagnosis of HCC.

The typical imaging hallmark diagnostic feature of HCC is the combination of non-rim arterial hyperenhancement (non-rim APHE) on the late arterial phase and non-peripheral wash-out appearance on the portal-venous and/or delayed phases on MDCT or MRI, thereby reflecting the peculiar vascular derangements that occur during hepatocarcinogenesis [49] (Figure 5). According to the latest EASL/EORTC (European Association for the Study of the Liver/European Organization for Research and Treatment of Cancer) guidelines issued in 2018, a definite diagnosis of HCC can be established in a nodule measuring ≥10 mm or based on a background of liver cirrhosis or other risk factors for HCC, if these typical imaging hallmark features are encountered [4]. Additionally, AASLD (American Association for the Study of Liver Diseases)/LI-RADS guidelines have endorsed two other major imaging features, namely, an enhancing "capsule" depicted on the venous-portal and/or delayed phase, and the threshold growth defined as ≥50% increase in size of a mass over ≤6 months [50,51] (Table 3).

**The non-contrast phase** is mainly suggested at initial diagnosis and in patients who have prior locoregional therapy, as it is essential to distinguish lipiodol staining and blood products from true arterial hyperenhancement.

The **late hepatic arterial phase** is considered the most determinant vascular phase for the assessment of HCC, as APHE is an essential feature of HCC and most current guidelines do not allow a definitive diagnosis of HCC for nodules without it [4,5,46,52].

The late arterial phase is characterized by full hepatic arterial enhancement with good portal vein enhancement, but no antegrade enhancement of the hepatic veins. The early arterial phase, which is characterized by the intense enhancement of the hepatic arteries without enhancement of the portal vein is considered inappropriate for the depiction of hypervascular lesions, as most HCCs are not conspicuous until the late arterial phase. As the late arterial phase occurs during a restricted time interval, using a fixed time delay, imaging can be inaccurate in patients with cardiovascular disease or other co-morbidities. This is the reason, most institutions recommend the use of patient individualized scan protocols such as test-bolus or bolus-tracking methods, with the latter performing better [46] (Table 4).

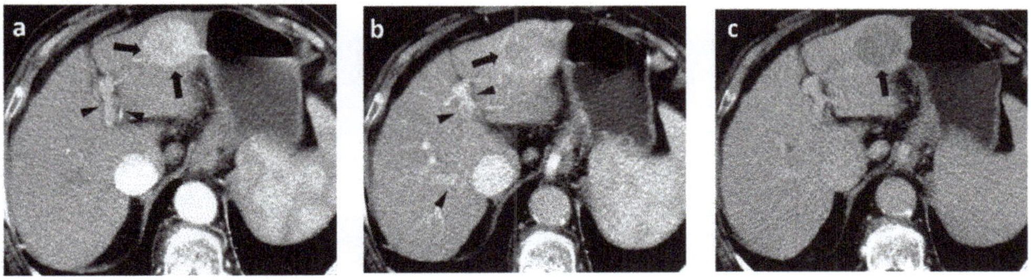


**Figure 5.** Typical hallmark imaging features of HCC in a 60-year-old patient with HBV cirrhosis. (**a**) Late arterial phase, (**b**) Portal-venous phase, and (**c**) Delayed phase. MDCT images show a 30-mm mass in the left liver lobe with global APHE (arrows) (**a**) "wash-out" on the PVP (arrow) (**b**), and capsule appearance on both the PVP and delayed phase (arrow) (**b**,**c**). Note the enhancement of both the left hepatic artery and portal vein with no enhancement of the hepatic veins on the late arterial phase (arrowheads) (**a**) and enhancement of the left hepatic artery, portal vein, and hepatic veins on PVP (arrowheads) (**b**).

**Table 3.** CT/MRI LI-RADS® v2018 Diagnostic Table by the American College of Radiology (ACR) [51]. LR-3 observations are marked in yellow, LR-4 are marked in orange and LR-5 are marked in red.

| **Arterial Phase Hyperenhancement (APHE)** | | **No APHE** | | **Nonrim APHE** | | |
|---|---|---|---|---|---|---|
| **Observation Size (mm)** | | <20 | ≥20 | <10 | 10-19 | ≥20 |
| Count additional major features:<br>•Enhancing "capsule"<br>•Non peripheral "washout"<br>•Threshold growth | None | LR-3 | LR-3 | LR-3 | LR-3 | LR-4 |
| | One | LR-3 | LR-4 | LR-4 | LR-4 / LR-5 | LR-5 |
| | ≥Two | LR-4 | LR-4 | LR-4 | LR-5 | LR-5 |

LR-4 / LR-5: Observations in this cell are categorized based on one additional major feature:

- LR-4 if enhancing "capsule"
- LR-5 if nonperipheral "washout" OR threshold growth

**Table 4.** Multiphase Liver CT Imaging and Scan Delay Parameters [46–48].

| Method | Late Arterial Phase | Portal Venous Phase | Delayed Phase |
|---|---|---|---|
| **Bolus-Tracking**<br>Individualized scan delay | Image acquisition: 10–30 s after aortic threshold density of 100–150 HU | 60–80 s after start of injection or 45–60 s after aortic threshold | 3–5 min after start of injection |
| **Test-Bolus**<br>Individualized scan delay | Image acquisition: 10–20 s after peaking aortic enhancement | 60–80 s after start of injection | 3–5 min after start of injection |
| **Fixed Scan Delay**<br>Alternatively for young patients with no comorbidities | 35–45 s after start of injection | 60–80 s after start of injection | 3–5 min after start of injection |

The **portal-venous phase (PVP)** is characterized by the maximal enhancement of the portal veins, adequate enhancement of the hepatic arteries and the hepatic veins by antegrade flow, and peak enhancement of the hepatic parenchyma. Scan timing for the portal-venous phase is not as restricted as for the late arterial phase and generally occurs 60–80 s after the start of the injection [46]. This phase is essential for visually assessing the reduction in enhancement of HCC nodules relative to the surrounding liver parenchyma, which is a subjective perception designated as "wash-out" or "wash-out appearance".

Several investigators have demonstrated that early and profound "wash-out" is related to higher tumor grade and the probability of microvascular invasion [53–55].

The **delayed phase**, also known as the equilibrium phase—due to an equilibrium state between the vascular space and the interstitial space—is obtained >120 s after the start of injection, preferably with a scan delay of 3–5 min [46]. It is characterized by decreased but persistent enhancement of the portal and hepatic veins and liver parenchyma. The use of the delayed phase increases HCC's conspicuity, particularly in tumors smaller than 2 cm, as in some cases the wash-out appearance is better depicted during that phase. Furthermore, in almost 10% of cases, "wash-out" is observed only in the delayed phase. The delayed phase is also optimal for the detection of enhanced capsular and mosaic appearance, features that are characteristic and relatively specific for progressed HCC.

An important imaging feature included in the AASLD/LI-RADS' major criteria is the **enhancing "capsule"**, which is visible as a smooth, sharp, and uniformly thick, thereby enhancing the peripheral rim during the portal and, mainly, the delayed phase [49]. The enhancing "capsule" has a specificity of 86–96% for HCC in high-risk patients, and according to the latest LI-RADS v.2018, an observation measuring >2 cm with non-rim APHE and an enhancing "capsule" can be diagnosed definitely as HCC (LR5), even in the absence of "wash-out" [50]. The tumor capsule is found in about 70% of cases and is a characteristic pathologic feature of progressed HCCs, exhibiting expansive growth [49]. It is not observed in dysplastic nodules or early HCCs and does not occur with intrahepatic cholangiocarcinomas (ICC). An intact capsule has been associated with better prognosis as it prohibits tumor cells from embolizing downstream. Several clinical studies have shown that, after adjusting for tumor size and grade, HCCs with intact capsules are related with lower recurrence rates and better overall survival [56].

Hypervascular lesions that may mimic HCC in patients with cirrhosis or chronic liver disease include perfusion alterations, hemangiomas, focal nodular hyperplasia (FNH)-like nodules, high-grade dysplastic nodules (HGDN), small ICCs, combined hepatocellular-cholangiocarcinomas (cHCC-CCAs) and, rarely, hypervascular metastases in a cirrhotic liver [54,57–61] (Table 5).

**Table 5.** Hypervascular HCC mimickers in patients with cirrhosis or chronic liver disease [54,57–61].

| Lesion. | Comments | Imaging Features |
|---|---|---|
| **Vascular Pseudolesions** | Attributable to arterioportal shunts Particular common in cirrhotic livers | Peripheral, round, or wedge shaped APHE nodules<br>Isodense on PVP |
| **Hemangiomas** | Rarely encountered<br>Most demonstrate fibrotic involution ("sclerosed" hemangiomas) | "Sclerosed" hemangiomas demonstrate rim APHE<br>Mimic non-HCC malignancies<br>(~6% of LR-M observations) |
| **FNH-like nodules** | Particularly common in alcoholic cirrhosis<br>SAA-HN-variant is potentially malignant | Nodules with APHE<br>Isodense or "washout" on PVP |
| **HGDN** | Rarely depicted on MDCT<br>Mimic early HCC | May demonstrate non-rim APHE and become isodense on PVP or depicted only as hypodense nodule on PVP |
| **ICC** | Comprise 10–15% of cancers in cirrhotic liver | Small ICCs (<3 cm) frequently demonstrate atypical enhancement pattern with global APHE and "washout" or isodensity on PVP |
| **cHCC-CCA** | Account for <5% of liver cancers | No constant imaging features<br>Commonly have targetoid appearance<br>but may also mimic HCC |
| **Hypervascular Metastases** | Very rarely encountered due to unfavorable microenvironment & altered portal venous flow | Lesions with APHE<br>Isodense or "washout" on PVP |

***Abbreviations***: **APHE**, arterial phase hyperenhancement; **PVP**, portal venous phase; **FNH**, focal nodular hyperplasia; **SAA-HN**, serum amyloid A(SAA)-positive hepatocellular neoplasm; **HCC**, hepatocellular carcinoma; **HGDN**, high grade dysplastic nodule; **ICC**, intrahepatic cholangiocarcinoma; **cHCC-CAA**, combined hepatocellular-cholangiocarcinoma.

It is worth noting that at least 40% of HCCs present with atypical imaging features and don't fulfill the appropriate vascular criteria to be diagnosed as definite HCCs. These generally include small-size HCCs (<2 cm), either early HCCs or well-differentiated HCCs, poorly-differentiated HCCs, progenitor-type HCCs, scirrhous HCCs, and the infiltrative and diffuse types of HCCs [54,62].

Kim et al. [63,64] have shown that approximately 17% of small HCCs can be isodense on the arterial phase and hypodense on the portal-venous phase, whereas about 40–60% of small HCCs, even if hypervascular on the arterial phase, don't exhibit "wash-out" on the portal phase. This atypical appearance has been linked to the multistep process of hepatocarcinogenesis, during which there may be diminished portal tracks before adequate recruitment of unpaired arteries, or there may be neoangiogenesis without significant loss of portal tracks [65]. Further characterization and differentiation of these nodules requires the application of ancillary imaging features, acquisition of specific MRI sequences, or the application of CEUS, which can provide further information concerning their cellularity, function, and vascularity [65–68]. Moreover, according to a recent large prospective study, including 296 observations in 240 patients [69], the combination of MDCT and MRI in LI-RADS yielded a better diagnostic performance for HCC than MDCT or MRI alone.

The **ancillary imaging features** that can be readily assessed with MDCT and indicate the presence of HCC mainly include the nodule in nodule architecture, the mosaic appearance, the non-enhancing capsule, and, to a lesser degree, the presence of intralesional fat or blood products [50] (Figure 6). These imaging features strongly favor neoplasia of hepatocellular origin and may be used to distinguish HCC from ICC or metastatic disease, with the exception of cHCC-CCA, which can share the same imaging features with HCC.

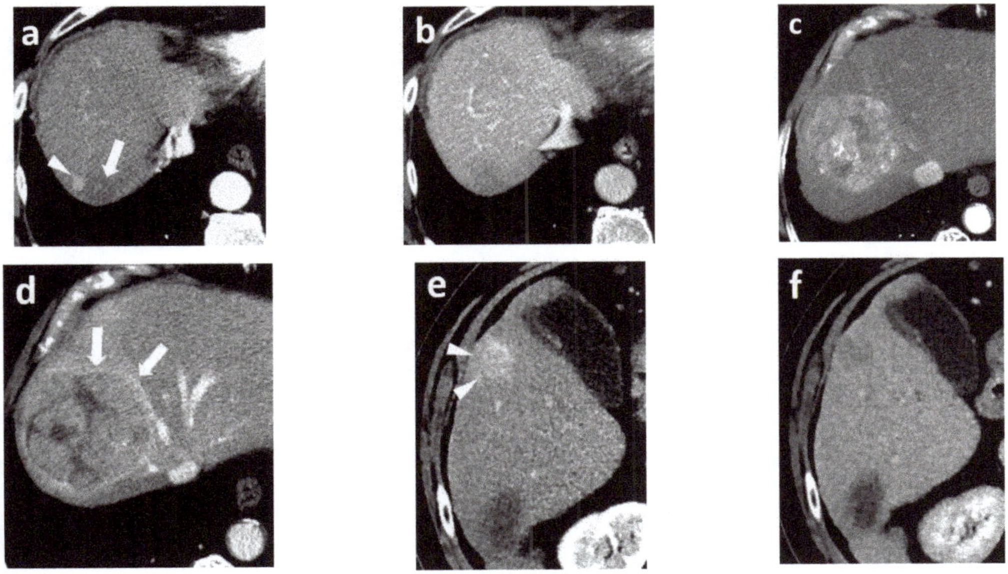


**Figure 6.** Ancillary imaging features for HCC assessment. Nodule-in-nodule architecture, mosaic architecture, and corona enhancement on the late arterial and portal-venous phase, respectively. MDCT axial images. (**a**,**b**): A small nodule (arrowhead) is located at the periphery inside a larger nodule (arrow), with APHE (**a**) and "wash-out" (**b**) on PVP. The parent nodule is not enhancing. (**c**,**d**): A heterogeneous mass, characterized by enhancing compartments and necrotic areas is seen (**c**). An enhancing capsule is depicted on PVP (**d**) (arrows). (**e**,**f**): There is a transient zone of hyperenhancement around the outer margin of the nodule (**e**) (arrowheads), which fades on PVP (**f**).

The **nodule-in-nodule architecture** reflects the emergence of a progressed HCC within a dysplastic nodule or an early HCC, which results from the clonal expansion of cells displaying less differentiation [70]. The inner nodule corresponding to progressed HCC shows arterial hyperenhancement, while the parent nodule corresponding to well-differentiated dysplastic nodule remains hypo- or iso-attenuated. The nodule-in-nodule appearance is a feature with poor prognostic value, as the inner hyperenhancing nodule has a short volume doubling time (TVDT) and grows rapidly.

Likewise, the **mosaic appearance** is the result of the clonal divergence of cells in various steps of de-differentiation inside a nodule typically larger than 3 cm. The imaging appearance is the reflection of the histology, which is comprised of randomly-distributed nodules or compartments with variable enhancements, separated by irregular enhancing septa and necrotic or hemorrhagic areas or areas with fatty metamorphosis. The mosaic pattern is observed in 28–63% nodules of HCCs [49].

**The non-enhancing capsule** refers to a capsule appearance that is hypodense and not visible as an enhancing rim. It should be unequivocally thicker and more conspicuous than fibrotic tissue around other cirrhotic nodules.

Another important ancillary imaging feature favoring malignancy in general, and not HCC in particular, is the **corona enhancement,** which refers to the enhancement of the venous drainage area of the tumor on the late arterial or early portal-venous phase. It is the result of aberrant and disorganized peritumoral drainage due to the invasion of intranodular hepatic veins, and drainage shifts to the surrounding hepatic sinusoids and, subsequently, to the portal venules [71]. It is seen in up to 80% of progressed HCCs and is strongly associated with microvascular invasion (MVI) and seeding of neoplastic cells, hence most metastatic satellite nodules originate in the peritumoral venous drainage area [71]. MVI is an important prognostic factor of overall survival and of early recurrence after resection, locoregional therapy, or transplantation. MDCT imaging features predicting MVI, apart from peritumoral enhancement, and includes the non-smooth tumor border, nodular rim, capsular disruption, prominent tortuous intratumoral arteries, and large tumor size [72,73].

Recently, Bello et al. [74] brought attention to several HCC subtypes with atypical imaging features and correlated them with their histologic and molecular features; for example, an early peripheral and progressive centripetal enhancement (a pattern similar to ICC) was indicative of the scirrhous subtype. Fowler et al. [75] have also presented distinct morphologic and pathologic subtypes of HCC with different prognostic implications. Pathologic subtypes associated with poor prognosis mainly include macrotrabecular massive HCC (MTM-HCC, presenting with necrosis on imaging), neutrophil-rich HCC (previously described as sarcomatoid subtype of HCC), scirrhous HCC (S-HCC), progenitor-type HCC (expressing stem cell markers such as CK19 or EpCAM), and diffuse- and infiltrative-type of HCC, while subtypes associated with better prognosis, apart from early and small well-differentiated HCC, include clear-cell HCC (CC-HCC), steatohepatitic (showing extensive fatty component), and b-Catenin-mutated HCC (hyperintense on the hepatobiliary phase).

This wide variability of biological behavior among progressed HCCs has been correlated to variable genetic and molecular alterations in the process of hepatocarcinogenesis and has been shown to be related with specific imaging features. Imaging features associated with aggressive biological behavior and poor survival include substantial intratumoral necrosis and targetoid appearance, as well as features suggestive of micro- or macrovascular invasion and bile duct invasion [75–78] (Figure 7). HCCs presenting with such atypical imaging features are designated according to LI-RADS as LR-M lesions and require biopsy for definitive diagnosis.

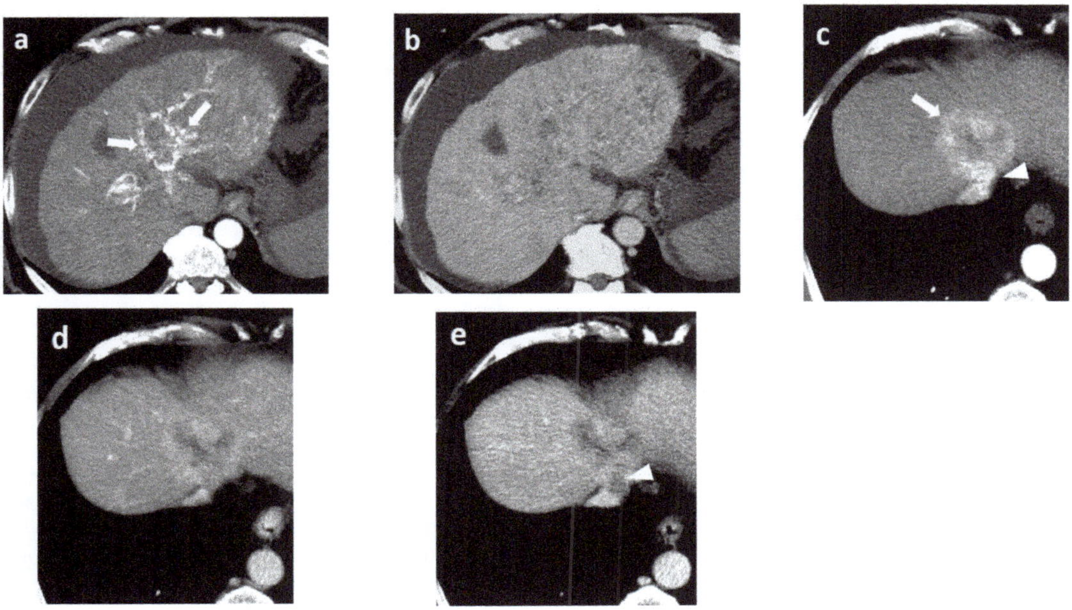


**Figure 7.** HCC with atypical imaging features; Cirrhotomimetic HCC (**a**,**b**) and Progenitor-type HCC (**c**,**d**,**e**) (late arterial, portal, and delayed phase, respectively). (**a**,**b**): Cirrhotomimetic HCC with TIV. There is marked diffuse heterogenous appearance of the left liver lobe with subtle arterial enhancement and "wash-out" accompanied with hyperenhancing portal vein tumor thrombus, displaying prominent neovascularity (thread and streak sign) (**a**) (arrows). (**c**,**d**,**e**): Progenitor-type HCC. There is a mass with irregular, non-smooth arterial enhancing rim (arrow). Note also the enhancing tissue inside the middle hepatic vein (arrowhead) (**c**) protruding into the IVC (arrowhead) (**e**). With the exception of cHCC-CCA, TIV is consider a fairly specific feature of HCC in a cirrhotic liver since ICC more frequently encases rather than invades veins.

A distinct morphologic type is the diffuse or cirrhotomimetic-type, which presents as an ill-defined permeative infiltration of the liver parenchyma with subtle or inconsistent arterial enhancement and heterogeneous "wash-out" [79]. It is frequently associated with invasion of the portal vein (68–100%) and high levels of AFP (>10,000 ng/mL). Due to its reduced conspicuity, diffuse-type HCC is often revealed only when malignant portal vein thrombosis becomes apparent. MDCT features indicating a tumor in the vein (TIV) include heterogenous thrombus enhancement, expansion of the portal vein ≥23 mm, and contiguity to the tumor. One of the most characteristic imaging features is the presence of thrombus neovascularity, which corresponds to the "thread and streak" sign [80] seen during the early arterial phase as a thin linear and chainlike opacities, reflecting the growth of the tumor into the vein (Figure 7).

Finally, as many as 20% of HCCs may involve a non-cirrhotic liver and develop without identifiable histologic precursors ("de novo hepatocarcinogenesis"). Non-cirrhotic HCC has distinct histopathologic and clinical features [81]. It usually presents as a single, large mass (>5 cm) with a mosaic architecture, intratumoral necrosis and fat, and an intact capsule, thereby reflecting the slow, expansile growth of the tumor. Extrahepatic involvement, mainly lymphadenopathy, is more frequently detected and is attributable to the delay in diagnosis. It is predominantly moderate or well differentiated, and despite the large tumor burden at the time of diagnosis, has a better overall survival rate and disease-free survival rate in comparison to HCC in cirrhotic livers.

Despite its limitations, MDCT remains a cornerstone of HCC investigation in patients at risk. Its wide availability and lower cost render MDCT as a first-line imaging modality for the evaluation of suspicious nodules in cirrhotic livers. Moreover, MDCT contributes to the prognostication of patients with HCC by identifying features associated with good or bad prognoses. Inconclusive cases require further investigation with other imaging modalities or histologic verification.

## 5. CT Perfusion

CT liver perfusion (CTLP) is a modern imaging technique that provides quantitative functional information on tissue microcirculation—in addition to morphology—and allows a more comprehensive and reproducible evaluation of focal liver lesions [82–84]. During the last decade, CTLP has been extensively studied as an imaging biomarker in hepatocellular carcinoma (HCC) and has a plethora of applications in HCC diagnosis, prognosis, treatment selection, and treatment response assessment [82].

From a technical standpoint, CTLP is based on the analysis of a dynamic CT dataset consisting of sequential images of the liver acquired over time following the injection of IV contrast medium [83,85]. Specialized software is employed to extract functional information from the image dataset by measuring the change of attenuation of liver tissue and reference input vessels over time and generating corresponding attenuation time curves. Perfusion parameters are derived, either by directly fitting the attenuation time curve of each point of liver tissue (model-free approach) or by implementing complex pharmacokinetic models (model-based approach). The results are presented as parametric maps with a color scale (Figures 8 and 9). A variety of pharmacokinetic models have been employed in the past for CTLP analysis, and can differ in the number of inputs, compartments, and fitting method. Nevertheless, most modern commercial applications implement a dual input, dual compartment model using the deconvolution method, which best approximates the perfusion characteristics of the liver. While perfusion parameter names and perfusion analysis models are vendor-specific, most manufacturers provide parameters pertaining to blood flow (BF), blood volume (BV), mean transit time (MTT), and vessel permeability (PS), as well as hepatic arterial blood flow (HaBF), portal liver perfusion (PLP), and their ratio (hepatic arterial fraction—HAF or hepatic perfusion index—HPI). Other perfusion parameters are usually reported in conjunction with MRI perfusion studies (Table 6).

**Table 6.** Glossary of commonly-reported CT and MRI perfusion parameters alongside expected changes in HCC lesions relative to surrounding liver parenchyma. Number of arrows indicates the magnitude of difference.

| Parameter (Unit) | Definition/Biological Significance | Expected Change in HCC |
|---|---|---|
| **CT Liver Perfusion/DCE-MRI**<br>*Perfusion parameters based on pharmacokinetic models (model-based approach)* | | [86–89]/[90–93] |
| Blood Flow, or Total Perfusion (mL/100 g/min) | Total flow rate of blood in liver tissue. Reflects hypervascularity. | ↑↑ |
| Blood Volume (mL/100 g) | Intravascular blood volume. Reflects hypervascularity. | ↑↑ |
| Mean Transit Time (s) | Residence time of contrast agent in tissue. Shorter MTT might suggest hypervascularity and presence of intratumoral arteriovenous shunts. | ↓ |
| Hepatic Arterial Blood Flow, or Arterial Liver Perfusion (mL/100 g/min) | Blood flow derived from hepatic artery. High in lesions with a predominant hepatic arterial supply. | ↑↑ |
| Portal Liver Perfusion (mL/100 g/min) | Blood flow derived from portal vein. High in normal liver tissue. Low in lesions with a predominant hepatic arterial supply. | ↓↓ |

**Table 6.** *Cont.*

| Parameter (Unit) | Definition/Biological Significance | Expected Change in HCC |
|---|---|---|
| Hepatic Arterial Fraction, or Hepatic Perfusion Index (%) | Percentage of blood input contributed by hepatic artery. Low in normal liver tissue. Increased in lesions with arterioportal imbalance. | ↑↑ |
| Permeability Surface area product (mL/100 g/min) | Reflects leakage rate of blood from vascular into interstitial space. Virtually zero in normal liver parenchyma where fenestrated sinusoids permit free communication between the intravascular and interstitial space. Countable in liver fibrosis and liver tumors. | ↑ |
| $K^{trans}$ ($s^{-1}$) | Transfer constant from plasma to interstitial space. Reported in studies that employ single-input dual-compartment models. Related to vessel permeability. | ↓/↑ |
| $K^{ep}$ ($s^{-1}$) | Reflux constant from interstitial space to plasma. Reported in studies that employ single-input dual-compartment models. Inverse relation to $K^{trans}$. | ↑ |
| $v_e$ (%) | Extra-vascular extra-cellular volume fraction. Related to cell density. | ↓ |
| *Descriptive perfusion parameters (model-free approach)* | | |
| Area Under the Curve (unitless) | Area under pixel density curve. High in lesions with vivid enhancement. | ↑↑ |
| Slope of Increase, or wash-in ($s^{-1}$) | Running average of the slope of the tissue density—time curve *. High in hypervascular tumors with rapid and vivid enhancement. | ↑↑ |
| Slope of Decrease, or Wash-out ($s^{-1}$) | The slope of the line connecting the point of maximum enhancement and the last point of the tissue density-time curve *. Low in lesions that display washout. | ↓↓ |
| Time to peak (s) | Time interval between onset of afferent vessel enhancement and peak of the tissue density-time curve *. Short in hypervascular lesions with rapid enhancement. | ↓↓ |
| Positive Enhancement Integral (%) | The area under the tissue density curve in each tissue voxel, divided by the area under the curve corresponding to a reference vein ROI. | ↑ |

DCE-MRI; Dynamic Contrast-Enhanced MRI, HCC; hepatocellular carcinoma ↑; increased, ↓; decreased, ↓/↑; no difference or conflicting results *; the density-time curve of CT perfusion corresponds to the signal intensity-time curve of DCE-MRI.

As previously stated, hepatocarcinogenesis is characterized by the formation of unpaired arteries that are not associated with portal vein branches, thereby leading to the presence of an arterioportal blood supply imbalance prior to the development of classic hallmarks of hypervascularity [53,65,94]. Conventional MDCT can accurately detect progressed hypervascular HCC lesions but might mischaracterize small HCC lesions that are associated with incomplete neo-angiogenesis [95]. CTLP can overcome this limitation by separating the hepatic arterial from the portal-venous component of blood flow. Studies in animals with chemically-induced liver tumors have demonstrated a significant increase in HaBF, a decrease in PLP, and a subsequent increase in HAF during the transition of pre-carcinoma lesions to early HCC [96,97]. In addition, progressed HCC lesions exhibited significant changes in perfusion parameters related to hypervascularity (increased BF and BV, decreased MTT) and permeability (increased PS) as expected. These observations have been confirmed in multiple studies in patients [86–89]. Among available CT perfusion parameters that are based on pharmacokinetic models, Fischer et al. found that a cut-off value of ≥85% HPI exhibited a sensitivity of 100%, while a cut-off value of ≥99% HPI yielded a specificity of 100% for the detection of HCCs in cirrhotic patients (Figure 8, [88]). Recently, Hatzidakis et al. highlighted that even descriptive perfusion parameters, such as the mean slope of increase, might offer high accuracy in discriminating HCC lesions from normal parenchyma (Figure 9, [89]). Other studies have demonstrated differences in CT perfusion parameters between HCC and hemangiomas [98–100], liver metastases [101–103], or arterioportal shunts [104], indicating that CTLP might be a useful tool for differential diagnosis between HCC and other focal liver lesions (Table 7).

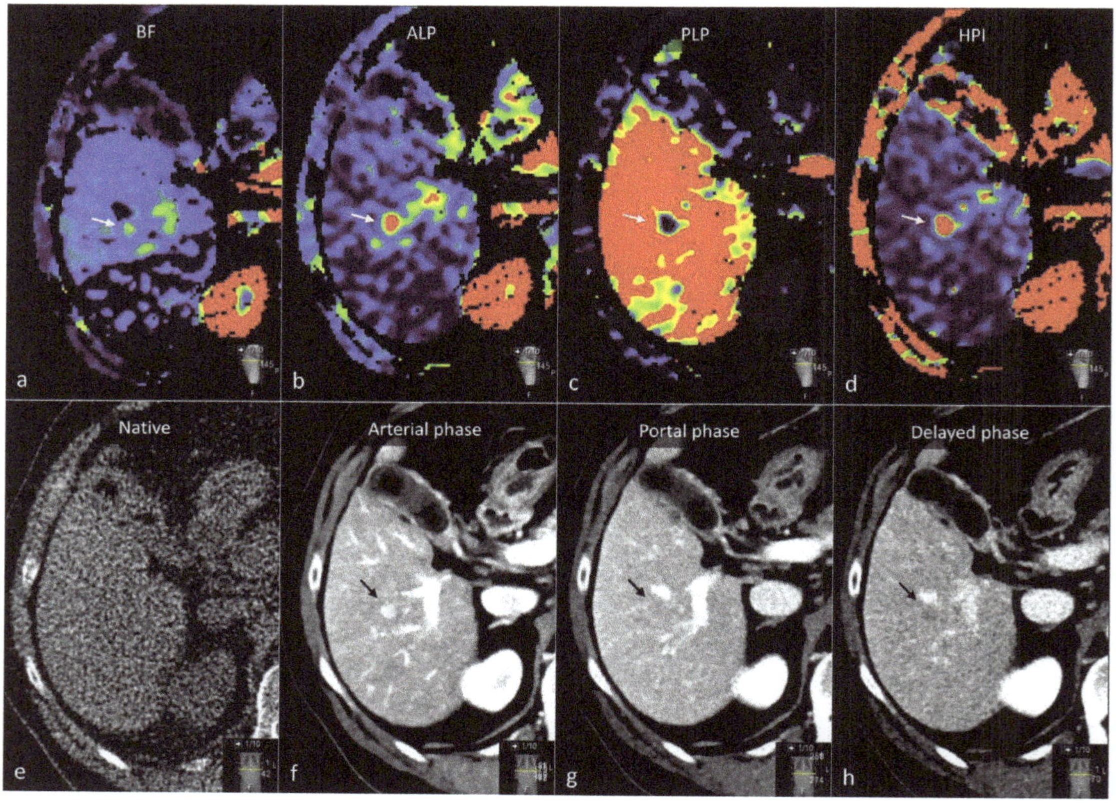


**Figure 8.** CT perfusion allows the accurate identification and characterization of HCC lesions. In this 69-year-old patient with cirrhosis, a 12-mm nodular lesion in the right liver lobe (arrow) can be easily distinguished from surrounding liver parenchyma on CT perfusion (**a**–**d**), with higher values on the Blood Flow (**a**), Arterial Liver Perfusion (**b**), and Hepatic Perfusion Index (**d**) parametric maps and a lower value on the Portal Venous Perfusion (**c**) map. The lesion is shown on conventional 4-phase CT (**e**–**h**), which was performed on the same day as CT perfusion with arterial phase hyper-enhancement (**f**) and wash-out on the portal-venous (**g**) and delayed phase (**h**), which corresponds to LI-RADS 5. BF; Blood Flow, ALP; Arterial Liver Perfusion, PLP; Portal Liver Perfusion, HPI; Hepatic Perfusion Index.

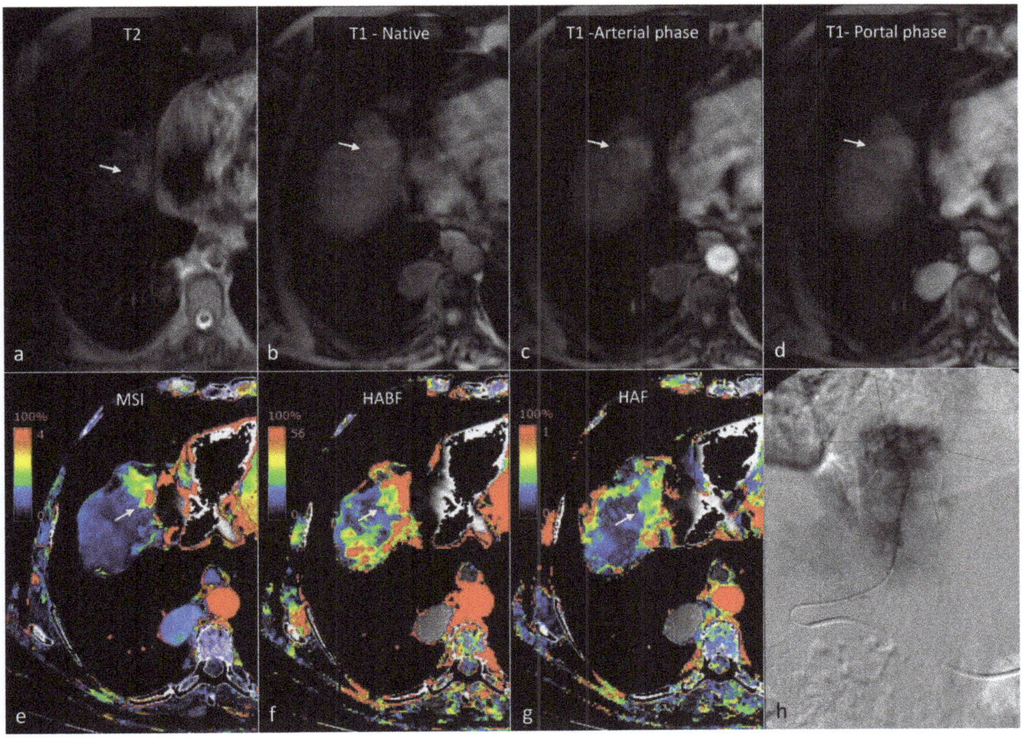


**Figure 9.** CT Liver Perfusion (CTLP) can complement other imaging modalities for establishing the diagnosis of HCC in difficult cases. This 70-year-old cirrhotic patient was previously treated for HCC with transarterial chemoembolization and presented with a 22-mm subdiaphragmatic lesion in liver segment 8 upon follow-up (arrows). The lesion had a high signal on T2 (**a**) and T1 (**b**) MRI. Contrast enhancement (**c**) and wash-out (**d**) could not be assessed on MRI due to the presence of artifacts. The MSI map of CTLP (**e**) clearly shows avid contrast enhancement in the HCC lesion, which was later confirmed with selective angiography (**h**). Although the Hepatic Arterial Blood Flow (**f**) and Hepatic Arterial Fraction (**g**) parametric maps show high values in the HCC lesion, it cannot be differentiated from surrounding parenchyma due to cirrhosis and prior chemoembolization, which alter normal liver hemodynamics. MSI; Mean Slope of Increase, HABF; Hepatic Arterial Blood Flow, HAF; Hepatic Arterial Fraction.

**Table 7.** Differential diagnosis between HCC and other common liver lesions using CT Liver Perfusion and dynamic contrast-enhanced MRI parameters. Number of upwards or downwards pointing arrows indicate magnitude of change relative to surrounding liver parenchyma and HCC lesions in comparative studies.

| CT Liver Perfusion/ DCE-MRI Parameter | Reported Behavior in Common Focal Liver Lesions | | | | |
|---|---|---|---|---|---|
| | HCC | Hemangioma | Hypovascular Metastasis | Hypervascular Metastasis | Arterioportal Shunt |
| | | [98–100]/[105,106] | [101–103]/[107,108] | [101–103] | [104] |
| Blood Flow, or Total Perfusion | ↑↑ | ↑↑↑ | ↓ | ↑↑↑ | ↑↑ |
| Blood Volume | ↑↑ | ↑↑↑ | ↑↑ | ↑↑↑ | - |

**Table 7.** *Cont.*

| CT Liver Perfusion/ DCE-MRI Parameter | Reported Behavior in Common Focal Liver Lesions | | | | |
|---|---|---|---|---|---|
| | HCC | Hemangioma | Hypovascular Metastasis | Hypervascular Metastasis | Arterioportal Shunt |
| | | [98–100]/[105,106] | [101–103]/[107,108] | [101–103] | [104] |
| Mean Transit Time | ↓ | ↓↓ | ↑↑ | ↓↓↓ | - |
| Hepatic Arterial Blood Flow, or Arterial Liver Perfusion | ↑↑ | ↑↑↑ | ↑ | ↑↑ * | ↑↑ |
| Portal Liver Perfusion | ↓↓ | ↓ | ↓ | ↓↓ * | ↓ |
| Hepatic Arterial Fraction, or Hepatic Perfusion Index | ↑↑ | ↑↑↑ | ↑ | ↑↑ * | ↑↑ |
| Permeability Surface area product | ↑ | ↑↑ | ↓/↑ | ↑ * | - |
| Slope of Increase, or wash-in | ↑↑ | ↑↑↑ | - | - | - |
| Slope of Decrease, or wash-out | ↓↓ | ↓ | - | - | - |
| Positive Enhancement Integral | ↑ | ↑↑↑ | - | - | - |

DCE-MRI; Dynamic Contrast-Enhanced MRI, HCC; hepatocellular carcinoma, -; not reported, ↑; increased, ↓; decreased, ↓/↑; no difference or conflicting results *; parameters have not been compared to those of HCC lesions.

In addition to HCC carcinoma diagnosis, CT perfusion can offer insights into tumor aggressiveness and prognosis. Sahani et al. found significant differences in BF, BV, and PS perfusion values between patients with well-differentiated HCC and other tumor grades [86]. Thaiss et al. revealed a good correlation between BF, BV, and HPI perfusion values with VEGFR-2 expression levels of HCC tumors [109]. Bai et al. showed that CLTP values in the periphery of HCC lesions correlate with microvascular density on pathology, which is an important prognostic factor for HCC [110]. Multiple studies have validated CTLP as a non-invasive tool to predict and assess the response of HCC to locoregional and systemic treatment [111–116]. However, this subject remains out of the scope of this review.

Although CTLP has many applications in HCC diagnosis and management, its use remains limited outside large reference centers. Patients may need to undergo an additional examination, which might be a burden in daily production and is associated with additional contrast media and radiation exposure. Generated parametric maps are vendor-specific, and as a result, radiologists need to acquire experience and establish intra-institutional reference values before using CTLP in HCC diagnosis. However, great effort has been made to standardize CTLP acquisition protocols [117] and reduce radiation exposure [118–120]. With the advent of comparative studies to indicate when the additional use of CT perfusion might be beneficial for the patient and further development of automated image analysis software, CTLP might reach the availability of brain perfusion, which has a central role in the management of stroke [121].

## 6. MRI

MRI is an excellent modality for lesion detection and characterization thanks to its higher contrast resolution and ability to assess more tissue properties than sole vascularization. According to a recent meta-analysis, the pooled sensitivity and specificity for HCC diagnosis were 70% and 94%, respectively, regardless of tumor size [122]. However, sensitivity is greater for lesions >2 cm (approaching 100%) but drops to 58.3–64.6% for lesions smaller than 2 cm [123–125], and it is even lower for sub-centimetre lesions. It is generally agreed, however, that MRI outperforms CT for the diagnosis of HCC smaller than 2 cm, while comparable accuracy is reported for lesions ≥2 cm [124,126,127]. It should be kept in mind that the size of an HCC is a significant prognostic factor and an important criterion in all staging systems. The use of hepatospecific contrast media, namely gadoxetate disodium and gadobenate dimeglumine, increases sensitivity by 5–10% [124,128–130].

As mentioned before, the radiological hallmark that enables a confident non-histological diagnosis of HCC is the combination of hypervascularity on the arterial phase and hypoperfusion on the portal phase; as with CT, this "wash-in/wash-out" pattern is indispensable on MRI as well. According to the LI-RADS criteria, no lesion without hyperenhancement on the arterial phase can be definitely characterized as HCC; hyperenhancement has to be "non-rim", i.e., not predominantly peripheral (in order to differentiate from metastases or cholangiocarcinoma) [51]. However, up to 40% of HCCs show no hypervascularity on the arterial phase, and these mainly represent early or poorly-differentiated HCCs [131,132]. Moreover, 40–60% of small HCCs lack wash-out during the portal phase [133,134] (Figure 10). Additional major and ancillary features are employed to help characterize the lesion and assign a LI-RADS category to it.

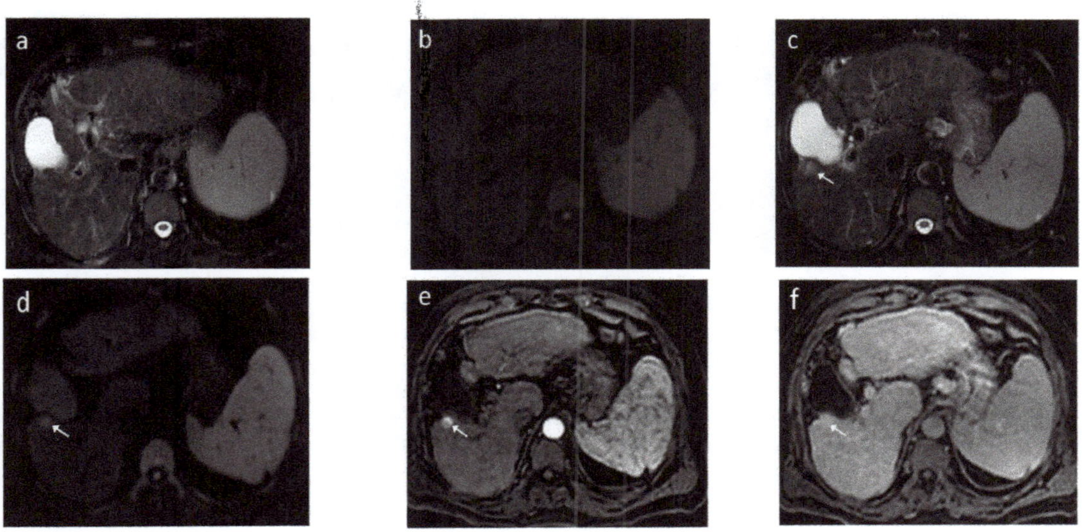


**Figure 10.** Evolution of a cirrhotic nodule into HCC. No suspicious lesions are identified on the T2 (**a**) and DWI (**b**) sequence of this 66-year-old man with cirrhosis due to hepatitis B infection. On the follow-up scan, performed 3 months later, increased T2 signal (**c**) is now observed in a nodule in segment V, which is associated with diffusion restriction (arrow) (**d**). After contrast administration, arterial enhancements (**e**) without delayed wash-out (**f**) are seen; absence of wash-out is frequent in early HCCs.

As already stated, small lesions (smaller than 1, 1.5, or 2 cm, depending on the publication) more often demonstrate atypical imaging features [135,136]. That is to say that lesions below the threshold of 1 cm cannot be characterized as HCC and follow-up is advised according to both EASL and AASLD guidelines. Small arterially-enhancing lesions may represent arterioportal shunts, perfusion disorders, or small intrahepatic cholangiocarcinomas (which may also show portal wash-out) [60]. Interval increase in size by $\geq$50% in $\leq$6 months is a major feature according to LI-RADS (Figure 11). However, it is not accepted by EASL and any lesion growth or change in enhancement pattern not typical of HCC should call for biopsy [4].

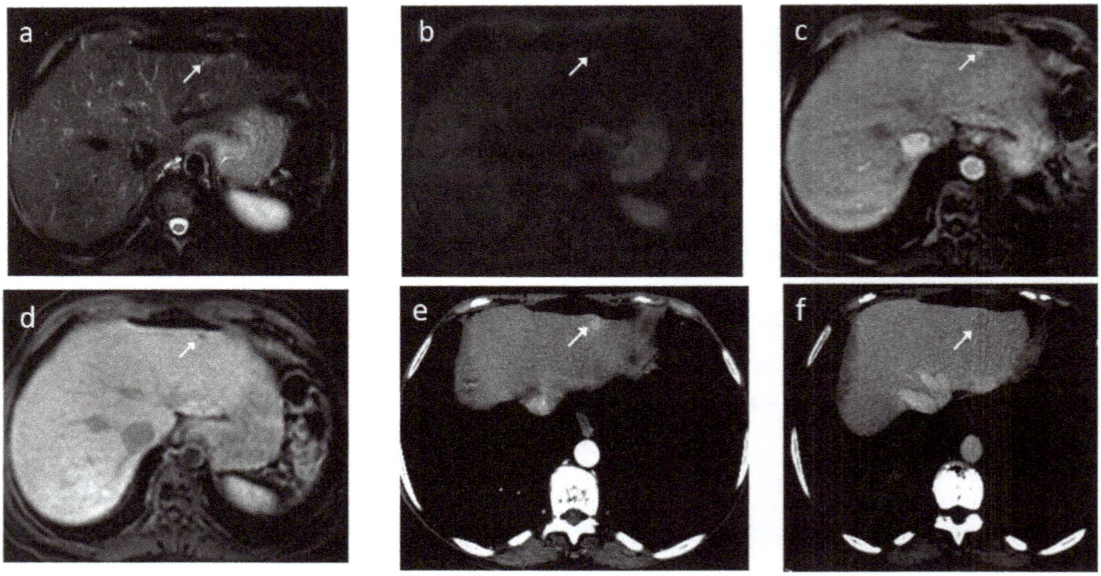


**Figure 11.** This 48-year-old woman with a history of β-thalassemia major and cirrhosis was followed-up after successful locoregional treatment of two small HCCs. In liver segment II, a 5-mm high T2 signal focus is seen in an anterior subcapsular location (**a**) with associated restricted diffusion (arrow) (**b**). The lesion shows arterial enhancement (**c**) and no uptake of the hepatospecific contrast on the hepatobiliary phase (**d**). Although findings are highly suspicious, the lesion cannot be definitely characterized as HCC, due to its small size. On the subsequent follow up CT, interval growth and typical wash-in/wash-out are now present (**e**,**f**).

The presence of a capsule (Figure 12) is a major finding according to LI-RADS, but not EASL. The capsule is a characteristic feature of progressed HCC and is absent in dysplastic nodules or early HCCs. It shows low T1 and T2 signal intensity and enhancements on the portal and delayed phase at 3 min after contrast injection (or transitional phase if hepatospecific contrast agent is used); on the contrary, corona enhancement occurs earlier on the arterial phase. A capsule should be thicker than the fibrous septa of cirrhosis, which also show delayed enhancement. An intact capsule on imaging has been associated with lower recurrence rates after treatment [137], while extracapsular tumor extension predicts poor survival [138]. It should be stressed, however, that an encapsulated progressed HCC has a worse prognosis than an unencapsulated early HCC; the presence of a capsule confers a better prognosis only when the encapsulated tumor is compared to HCCs of a similar size and grade with breached capsules or without a capsule.

On the T2 sequences, most large HCCs show mild–moderate hyperintensity; in contrast, cirrhotic and dysplastic nodules appear iso-intense or hypo-intense relative to the background liver and early HCCs, which are mostly iso-intense. T2 hyperintensity is attributed to increased arterial and decreased portal vascularity; peliotic changes may also contribute [139,140]. Mildly increased T2 signal intensity is an ancillary—but not specific—feature as it is also observed in other malignant lesions of the liver.

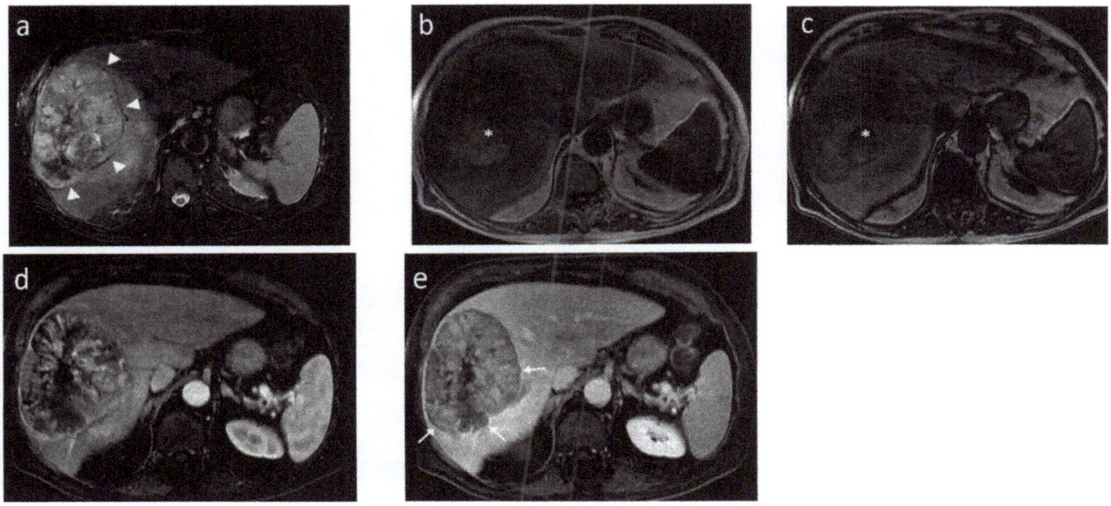


**Figure 12.** A large HCC is depicted in the right liver lobe of this 81-year-old man. The tumor is surrounded by a capsule, nicely seen as a thin, low signal line on the fat-suppressed T2 sequence (arrowheads) (**a**) and shows inhomogeneous but predominantly high T2 signal intensity. Areas of fat are clearly shown in the in/out of phase images (asterisk) (**b**,**c**). This marked heterogeneity is known as the "mosaic" pattern. After contrast administration, mottled arterial enhancement is noted (**d**); definite wash-out and capsular enhancement (arrows) are seen on the portal phase (**e**).

Regenerative nodules, dysplastic nodules, and well-differentiated HCCs may all present with T1 hyperintensity before contrast administration; if subtraction techniques are not used, an erroneous impression of enhancement could ensue or, on the contrary, subtle arterial enhancement could be missed. T1 hyperintensity may be due to the presence of fat, copper, glycogen, hemorrhage, or high protein content. Copper and glycogen tend to decrease as hepatocarcinogenesis progresses [141]; the same is also true for iron, and although siderotic nodules appear hypo-intense on all sequences—particularly the T2*—hyperintensity on the T1 sequences may also be seen. Hepatic iron overload, on the other hand, is predisposed to the development of HCC [142], and the appearance of an iron-free area in an iron-overloaded liver should be regarded as suspicious.

Fatty change is encountered in approximately 40% of early HCCs [143]. With increasing tumor size and histologic grade, fat usually regresses and the percentage drops to 6% in moderately-differentiated HCCs [144], only to increase again in highly de-differentiated tumors. This occurs along with the diminished arterial supply, suggesting a connection between reduced blood flow, hypoxia, and steatogenesis [145]. MRI is superior to CT in detecting fatty change with the use of chemical shift sequences, which show the characteristic signal drop on the opposed-phase compared to the in-phase (Figures 12 and 13). Intratumoral fat can also be used to exclude cholangiocarcinoma, which is also associated with cirrhosis. Nevertheless, the added value of fat identification in a HCC is debatable because, when detected, other more suggestive features (like the vascular pattern) are already present [123].

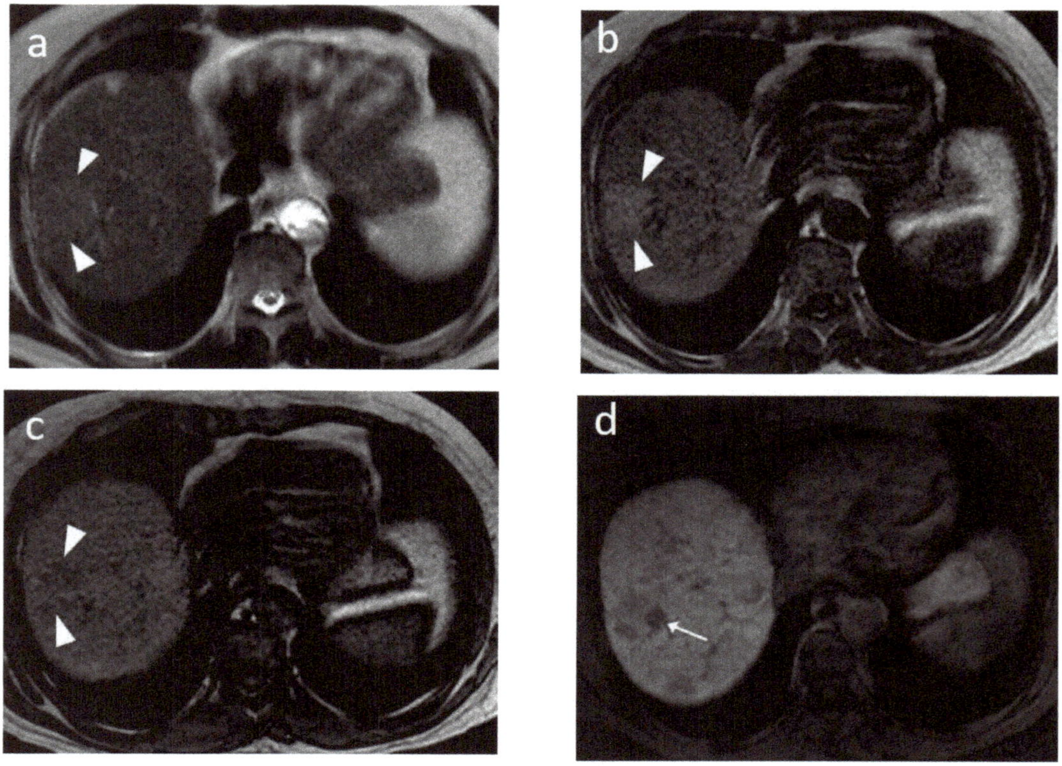


**Figure 13.** Sixty-five-year-old man with cirrhosis. On the T2 sequence, a nodular high T2 lesion is vaguely seen (arrowheads) (**a**). The lesion appears hyperintense on the in-phase image (**b**) and slightly hypo-intense on the out-of-phase image (arrowheads) (**c**), suggesting the presence of fat. On the hepatobiliary phase after gadoxetic acid administration, a small nodule with markedly decreased signal (no contrast uptake) is evident in the left aspect of the larger lesion (arrow), suggesting focal de-differentiation in a dysplastic fatty nodule and early HCC formation ("nodule in nodule" sign) (**d**).

MRI has the unique capability to assess lesion cellularity, which is translated as the reduced diffusivity of water molecules among the tightly-packed cells of a tumor. Restricted diffusion is an ancillary finding that favors malignancy in general, and not specifically HCC. DWI (Diffusion Weighted Imaging) may fail to detect early HCCs, since their cellular density and architecture do not markedly differ from the surrounding cirrhotic nodules [146]. Higher histological grades are associated with higher DWI signal intensities and corresponding lower ADC (Apparent Diffusion Coefficient) values [147,148]. However, the increased amount of fibrotic tissue in the cirrhotic parenchyma also causes restricted diffusion, reducing the conspicuity of HCC as cirrhosis advances [149]. The addition of DWI to the dynamic contrast enhanced phases or the hepatobiliary phase increases the sensitivity of HCC detection [150,151]. Even as a standalone technique, DWI appears to be an acceptable alternative for HCC diagnosis when a contrast agent is contra-indicated [152].

The introduction of hepatospecific contrast media in the 2000s has opened new perspectives in liver imaging. Following their intravenous administration, multiphasic dynamic imaging is performed—similarly to extracellular agents—and subsequently, they are taken up by functioning hepatocytes via specific transporters (organic anion-transporting polypeptide, OATP). Approximately 30% of high-grade dysplastic nodules demonstrate decreased expression of these transporters; the percentage rises to 70% in early HCC, while all poorly-differentiated tumors show decreased or absent expression of OATP transporters [153], leading to low signal intensity on the hepatobiliary phase. More importantly, the decline of OATP expression precedes the typical vascular changes of hepatocarcinogenesis, making the hypo-intensity on the hepatobiliary phase the most sensitive imaging feature for early diagnosis of HCC [125,126,130,154,155]. Typical HCC appears hypo-intense on the hepatobiliary phase and the degree of hypo-intensity has been correlated to histologic grading [156] [157,158]; however, approximately 10% of HCCs appear iso-intense or hyperintense relative to the surrounding liver on the hepatobiliary phase with gadoxetate (paradoxical uptake). The iso/hyperintensity may not be due to tumor differentiation, but rather represents a peculiar molecular/genetic subtype—probably due to genetic or epigenetic alterations—with less aggressive biological features [159,160]. Additionally, it has been shown lately that hyperintensity during the hepatobiliary phase reflects the activation of the Wnt/β-catenin pathway, which, in turn, is associated with resistance to immunotherapy, thereby suggesting that the specific imaging feature (i.e., paradoxical uptake of gadoxetate) could serve as an imaging biomarker [161,162].

Some HCCs demonstrate a transient rim of enhancement on the late arterial phase, which fades away during the subsequent phases. It corresponds to the draining pathway towards the perinodular sinusoids, a potential route for satellite metastases. It is seen in advanced tumors, which are frequently correlated with microvascular invasion [163,164], and represents an ancillary finding, although not specific for HCC.

When a smaller nodule is seen within a larger nodule, it implies de-differentiation of a cell subpopulation and progression towards hepatocarcinogenesis. The "nodule-in-nodule" sign suggests development of HCC within a dysplastic nodule (Figure 13) and the typical HCC features, such as the wash-in/wash-out pattern or diffusion restriction, are seen in the inner nodule. When numerous foci with different imaging characteristics are seen within a nodule, the appearance is known as a "mosaic" pattern and is usually encountered in large tumors (Figure 12), thereby facilitating the differentiation from cholangiocarcinoma [54].

Overall, MRI, with its superb contrast resolution, ability to assess functional parameters, and use of hepatospecific contrast agents, is the imaging modality of choice for the characterization of a suspicious nodule detected during the screening of high-risk patients. In addition to diagnosis, important prognostic features can be extracted with a direct impact on clinical decisions.

## 7. MR Perfusion

Comparable to CT perfusion, it is possible to calculate parameters related to liver microcirculation from MR images. The most common approach is called Dynamic Contrast Enhanced imaging (DCE imaging) and it is based on the quantification of positive enhancement (the 'T1 effect') of the contrast agent over time [165,166]. In a similar fashion to CT perfusion, several T1 weighted DCE-MRI source images are acquired following the injection of gadolinium and are subsequently analyzed with specialized software to extract perfusion-related parameters [167,168]. Most studies employ either in-house developed software to analyze DCE-MRI images or commercially-available solutions that do not incorporate liver-specific kinetic models. In addition, the calculation of quantitative perfusion parameters from the source images is even more challenging because the signal intensity does not have a linear relationship to contrast media concentration. As a result, a direct comparison of results between different studies is not straightforward. Nevertheless, relative comparisons of findings within the same study group are feasible.

Overall, a few DCE-MRI studies have been performed on patients with HCC, with findings in line with those of CTLP. Regarding semiquantitative (descriptive) perfusion parameters, HCCs display a higher area under the curve (AUC), higher wash-in, and positive enhancement integral (PEI), as well as lower wash-out and time to peak (TTP), compared to liver parenchyma (Table 6, [90,91,105]). Wash-in, wash-out, and PEI can be used to differentiate hemangiomas from malignant liver lesions (HCCs and metastases) (Table 7, [105]). Using a variety of dual input perfusion models, several authors have reported a higher arterial fraction (comparable to HAF or HPI) and hepatic arterial blood flow, as well as a lower portal-venous flow and distribution volume ($v_e$) in HCC lesions compared to surrounding liver parenchyma [90,92,93,107]. Permeability-related parameters like $K^{trans}$ and $K^{ep}$ seem to be more dependent on the characteristics of the employed pharmacokinetic model, and as such, might contribute less to HCC diagnosis [92]. However, Chen et al. demonstrated a significant correlation between these parameters and tumor proliferation status, histological grade, and microvascular density, which suggests that permeability parameters might offer insights into HCC prognosis [169]. In addition, arterial fraction, MTT, and BV seem to differ between HCC lesions, colorectal liver metastases, and hemangiomas and might aid in the differential diagnosis of focal liver lesions (Table 7, [107,108,169]).

Despite all the advantages related to the absence of radiation exposure, and the possibility to combine DCE-MRI with liver specific contrast agents and other MRI techniques, the application of DCE-MRI for liver imaging was, up until recently, hampered by technical limitations. Whole liver coverage was not feasible due to time constraints. As a result, researchers where obliged to select only an oblique slab of the liver containing the input vessels and the target tumor. Additionally, the lack of standardization of acquisition and analysis methods made image interpretation difficult. Recent technological advances, such as golden-angle radial sparse parallel (GRASP) MRI, have permitted the acquisition of free-breathing, whole-liver 3D images with high temporal resolution and without motion artifacts [165]. These images can be used simultaneously for morphological interpretation and for extracting DCE-MRI perfusion parameters without requiring additional contrast injection or scanning time [170]. With a growing number of modern MRI scanners that support compressed sensing sequences in tertiary centers, it is perhaps a matter of time before the development of MR perfusion surpasses that of CTLP and DCE-MRI and becomes incorporated into standard liver imaging protocols.

## 8. PET/CT

Although 18F-fluorodeoxyglucose Positron Emission Tomography (FDG-PET)/CT is an extremely useful tool in the evaluation of many cancers, it is not routinely used for HCC as it is limited by low sensitivity due to the high physiologic uptake of liver tissue and the variable expression of glucose transporters and glycolytic activity in HCC tumors [171]. FDG accumulates in poorly-differentiated HCCs but not in well-differentiated ones. However, tracers based on choline, which is an important component for the synthesis of membrane lipids, recently showed improved detection rates of well-differentiated HCCs [172]. Dual-tracer PET/CT combining choline and glucose as tracers seems to be promising in the staging of HCC patients [172–174], although significant overlap between well- and less-differentiated HCCs renders the characterization of tumors challenging based on dual-tracer PET/CT [175]. In a recent study, β-catenin-activated HCCs demonstrated increased uptake of 18F-choline but not 18F glucose, suggesting a potential imaging biomarker that may guide treatment [176]. In another promising study, a correlation between 18F glucose uptake and the expression of genes regulating metabolism was found, proposing a role of FDG-PET in selecting patients for metabolism-targeted therapy [177].

## 9. Artificial Intelligence

The research on Artificial Intelligence (AI) has greatly expanded in the last few years. The application of AI in HCC imaging has demonstrated promising results regarding differentiation from other lesions, prediction of grading and microvascular invasion, identification of specific molecular profile, prediction of response to treatment or post-operative recurrence, and guidance on treatment selection [178–180]. However, validation of these results in larger, prospective, multicenter studies is required in the years to come and before AI proves its clinical utility.

## 10. HCC Diagnostic Algorithms

It is clear that imaging has a decisive role in every step of the HCC course, starting from the screening of high-risk patients; to the non-invasive diagnosis, staging, management decisions; and finally, post-treatment follow-up. Improvement in HCC mortality is partly due to the critical contribution of imaging. As HCC can be confidently diagnosed by imaging without confirmatory biopsy, robust imaged-based guidelines are needed. Numerous diagnostic algorithms have been proposed by many scientific societies and organizations all over the world and they are continuously updated, revised, and ever evolving. Up until 2017, 14 imaging-based algorithms had been published [45]; the list has grown since then. All of the proposed imaging systems aim to set standards regarding performance and interpretation of imaging tests in order to maximize diagnostic accuracy, however, their approach differs to a greater or lesser extent.

A major difference is observed between diagnostic criteria proposed by Eastern and Western societies. Eastern societies, like the Japan Society of Hepatology (JSH) and the Asian Pacific Association for the Study of the Liver (APASL), include MRI with hepatospecific contrast media as a first-line diagnostic tool for HCC. In comparison, hypo-intensity on the hepatobiliary phase is considered an ancillary feature that is suggestive of malignancy in general (not HCC in particular) according to the LI-RADS classification system (endorsed by the AASLD) and is not included at all in the EASL guidelines. As stated earlier, hepatospecific contrast agents increase sensitivity for HCC, however some decrease in specificity is expected, even after excluding mimicking lesions with hepatobiliary phase hypo-intensity, such as hemangiomas and cholangiocarcinomas [181–183]. This difference in diagnostic strategies is explained by the differences in treatment practices; as liver transplantation is preferred as a curative therapy in Western countries, maintaining high specificity is of utmost importance for the proper allocation of the scant transplant livers. On the contrary, as locoregional treatments are broadly used in Asia, the highest achievable sensitivity is the defining parameter in imaging. For the same reason the presence of a capsule, which is a highly-specific feature of HCC, is considered a major feature in the LI-RADS categorization but is not included in Asian algorithms. Similarly, the Asian guidelines permit the non-invasive diagnosis of a non-arterially enhancing nodule if it appears hypo-intense on the hepatobiliary phase and hypoechoic on the Kuppfer phase of CEUS using Sonazoid, while arterial hyperenhancement is an absolute prerequisite for the confident diagnosis of HCC in both European and American algorithms.

Even with the strictest adherence to imaging protocols, diagnosis of HCC is not always straightforward. When the results of an examination are inconclusive, re-imaging the patient with a different modality, biopsy, or close follow-up could be selected following careful consideration of all clinical and laboratory data in the multidisciplinary team meeting.

## 11. HCC in Non-Cirrhotic Livers

Although most HCCs arise in the setting of cirrhosis through a multistep process that starts from the dysplastic focus, proceeds through dysplastic nodules, and, finally, to overt HCC, some carcinomas arise "de novo". This term describes the development of HCC bypassing the stage of cirrhosis and is not synonymous with appearance of HCC in a normal liver, as these tumors usually appear through a background of chronic hepatitis B infection or non-alcoholic steatohepatitis (NASH) or, less commonly, metabolic disorders, use of anabolic steroids, or exposure to aflatoxin B1. The relevance between metabolic syndrome, non-alcoholic fatty liver disease (NAFLD, the hepatic manifestation of metabolic syndrome), and NASH (the first stage of the inflammatory phase of NAFLD) on the one hand and hepatocarcinogenesis on the other has drawn attention worth noting recently. The metabolic syndrome deposition of free fatty acids in the liver leads to oxidative stress, inflammation, dysfunction of mitochondria, and activation of stellate cells. Activated stellate cells not only produce collagen leading to fibrosis, but also a number of cytokines, chemokines, and angiogenic factors that play an important role in HCC development by triggering specific pathways [184,185]. Incidence rates of HCC in cirrhosis range widely from 0.7 to 26 per 1000 person-years, depending on the etiology; the risk is higher in viral hepatitis compared to steatohepatitis [186,187]. The percentage of HCCs that arise in the absence of cirrhosis varies across publications and has been reported to be between 1.7–50.1% [188].

HCCs developing de novo tend to be larger than HCCs in cirrhotic patients, which is probably due to the absence of surveillance [189,190]. A large solitary mass with or without satellite nodules is commonly seen; increased tumor size is associated with inhomogeneity and necrosis (Figure 14). Imaging features do not differ and the wash-in/wash-out pattern is usually seen; however, reduced percentages of portal wash-out have been reported in the setting of NASH [191,192]. Despite the increasing contribution of non-alcoholic fatty liver disease to the burden of HCC, only the Asian-Pacific algorithm advises surveillance for NASH. Likewise, application of LI-RADS in the setting of NASH is not recommended. Although patients without underlying cirrhosis present with more advanced stages of HCC, their survival is better, which is likely due to the preserved underlying liver function [193].

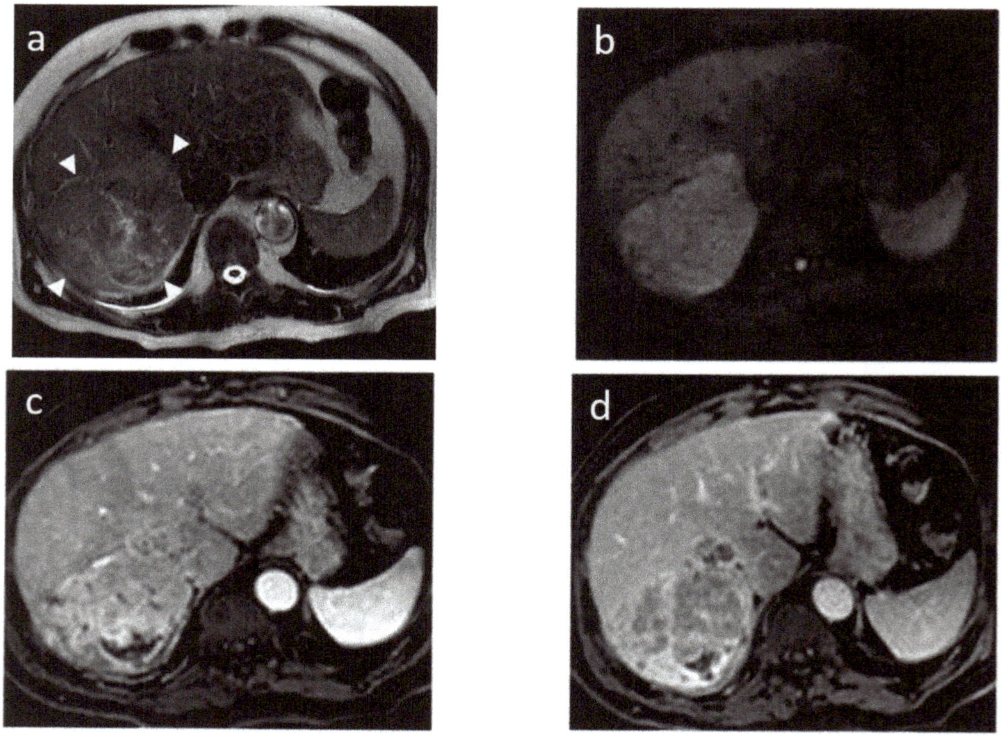


**Figure 14.** An 11-cm tumor (arrowheads) is incidentally discovered in a 70-year-old man without history of viral hepatitis, non-alcoholic fatty liver disease, or cirrhosis. Otherwise, the lesion shows typical HCC imaging features such as high T2 signal intensity (**a**), diffusion restriction (**b**), arterial hyperenhancement ©, and wash-out during the portal phase (**d**). Due to the absence of risk factors, a biopsy was performed that confirmed the radiological diagnosis.

## 12. Imaging Assessment of HCC following Percutaneous Locoregional Therapy

Percutaneous minimally-invasive locoregional treatment of HCC includes transarterial chemoembolization (TACE), transarterial radioembolization (TARE), radiofrequency (RFA), or microwave (MWA) ablation (thermal ablation), as well as cryoablation [194–200]. Among these, thermal ablation and TACE are the most widely adopted [200]. Treatment selection depends on tumor location and size, number of lesions, liver function, and patients' general condition [4,201]. Post-treatment imaging follow-up is commonly performed with multiphasic contrast-enhanced MDCT or MRI with the goal being to evaluate tumor response to treatment and to detect recurrent disease elsewhere within the liver [4,202–204]. Several HCC-specific radiological response classification systems have been developed, including the modified Response Evaluation Criteria in Solid Tumors (mRECIST) and the LI-RADS-Treatment Response (LI-RADS-TR) [205,206]. Compared to conventional tumor response evaluation criteria, LI-RADS-TR and mRECIST focus on the presence of viable contrast-enhancing HCC tissue rather than changes in total tumor size [207]. The interpretation of radiologic findings after locoregional therapy may be challenging, due to the presence of treatment-induced changes in the peritumoral liver parenchyma, which vary among different therapies [207].

Thermal ablation constitutes a first-line treatment with curative intent in patients with small, very-early and early stage HCCs, alongside surgical excision and orthotopic liver transplantation [4,208]. The principle of ablation is the application of thermal energy via an electrode or antenna placed in the tumor, with subsequent heating of the adjacent tumor cells and induction of necrosis [209]. Usually, a contrast-enhanced CT scan is performed immediately after the ablation procedure in order to confirm a "safety" necrotic margin of at least 0.5 cm around the ablated tumor (Figure 15). Coagulative tumor necrosis appears as a non-enhancing area on contrast-enhanced MDCT or MRI, while residual or recurrent tumors will often manifest as a nodular or irregular tissue with arterial phase hyperenhancement near the ablation zone margin [210,211]. A thin, smooth continuous hypervascular rim around the ablation zone might be apparent up to a few months following the procedure, as a result of the inflammation of the adjacent liver tissue due to thermal injury [207,211]. In addition, peripheral geographic areas of hyperenhancement might appear along the needle tract, which correspond to small arteriovenous shunts caused by the needle puncture [212]. A small residual tumor can escape detection in this early stage, though growth on subsequent follow-up imaging, or the presence of "wash-out", should raise suspicion for viable malignancy [213]. Although dynamic MDCT is the established method for following patients after thermal ablation, due to its wider availability, MRI might offer superior diagnostic accuracy and sensitivity for detection of recurrent HCC, especially with the use of diffusion-weighted images or liver-specific contrast media [122,214]. Few studies with small sample sizes have evaluated the role of CTLP in exploring the hemodynamic changes of HCC nodules after RFA [112,197,215]. Ippolito et al. reported 14 HCC patients treated with RFA. In the hepatic perfusion (HP), arterial perfusion (AP), and hepatic perfusion index (HPI) maps, residual tumors showed significantly higher values ($p = 0.012$, $p = 0.018$, and $p = 0.012$, respectively) compared to the ablated lesion [215]. In another study, Marquez et al. reported 10 HCC patients treated with RFA and found that HPI maps had a higher accuracy in the prediction of residual tumors [112]. CTLP can be very helpful after RFA as well as after MWA, particularly with the use of mean slope of increase (MSI) perfusion maps, in our experience (Figure 15) [89].

TACE is currently the treatment of choice for intermediate stage HCCs, including multinodular asymptomatic tumors without vascular invasion or extrahepatic spread [4,208]. The rationale of TACE treatment rests in the super-selective delivery of embolization particles (lipiodol or drug-eluting beads) together with a chemotherapeutic drug (usually Epirubicin), thereby aiming to induce complete anoxia of the malignant tissue area. This causes selective necrosis of the tumor lesion, sparing as much normal liver parenchyma as possible [216]. Even though TACE might achieve a complete therapeutic response, it is still considered a palliative method, since small numbers of malignant cells can escape necrosis. Early detection of such areas will warrant a chance for repeat treatment, usually with a second TACE session [217].

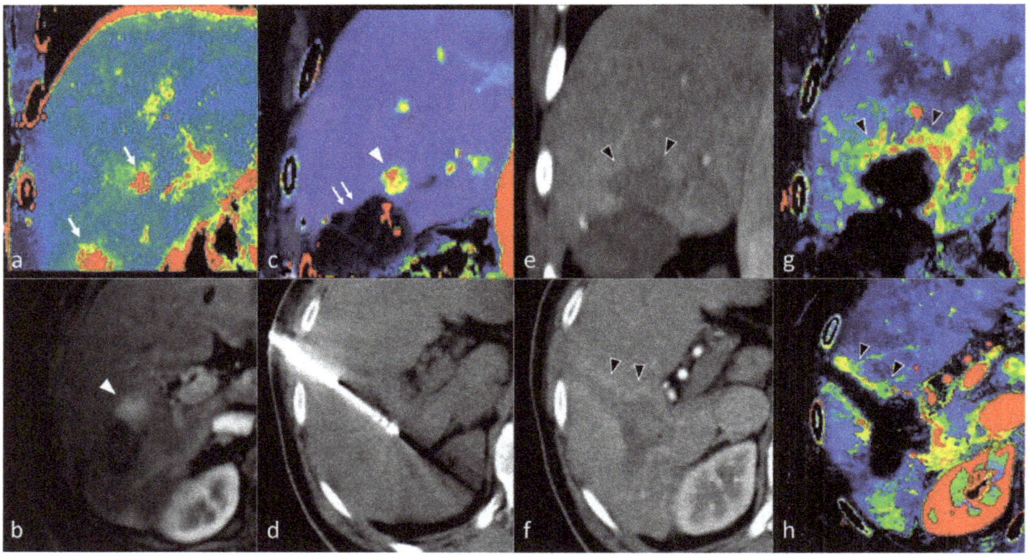


**Figure 15.** Expected post-treatment changes after microwave ablation (MWA). During US screening examination of a 70-years-old cirrhotic patient, two suspicious nodules, one 2 cm in segment VI and one 1.2 cm more centrally, were found. (**a**) Mean Slope of Increase (MSI) map of CT Liver Perfusion (CTLP) in coronal plane of the same patient shows the two hypervascular lesions in the right liver lobe, both of which proved to be HCCs (arrows). A wedge resection of the inferior tumor was performed. Post-operative arterial phase MRI in axial plane (**b**) and MSI map of CLTP in coronal plane (**c**) revealed once again the centrally-located viable tumor (arrowhead) alongside post-operative changes in the area of the resection (double arrows). A percutaneous MWA of the remaining tumor was subsequently decided. (**d**) Axial non-enhanced periprocedural CT shows the microwave antenna placed inside the tumor. Post-procedural portal phase CT in coronal (**e**) and axial (**f**) plane shows lack of contrast enhancement due to coagulative necrosis in the ablation zone and needle tract with surrounding hyperemia (black arrowheads). Note the safe ablation margin compared to reference images (**b**,**c**). MSI map of follow-up CTLP one month later in coronal (**g**) and axial plane (**h**) shows remaining hyperemia adjacent to the ablation margin, which should not be interpreted as residual or recurrent tumor.

Multiphase contrast-enhanced CT and MRI are the most commonly used imaging modalities for the follow-up of patients treated with TACE [218]. In a similar manner to thermal ablation, complete treatment response is suggested by the absence of enhancement in the tumor rather than regression in tumor size [219,220]. A thin, smooth rim of arterial phase hyperenhancement adjacent to the tumor margin corresponds to inflammatory tissue and might persist up to 1 year after TACE [207]. In addition, a geographic region of hyperenhancement might be seen in the corresponding liver segment due to occlusion of the feeding arteries and will eventually regress over time [221]. A residual or recurrent tumor usually presents as a nodular enhancing mass with or without wash-out. However, recurrent tumors can also manifest with weak or absent arterial phase hyperenhancement in previously-treated areas [207]. In cases of conventional TACE with Lipiodol or after use of special hyperdense drug-eluting beads, blooming artifacts of MDCT might hamper the recognition of small viable tumor tissue in MDCT; in such cases, MRI offers a clear diagnostic advantage [222]. Additionally, diffusion-weighted imaging can improve the depiction of residual or recurrent HCCs after TACE, and the ADC value may serve as a quantitative biomarker for treatment response [223].

CTLP has been extensively evaluated as a means for the assessment of treatment response in patients with HCC undergoing TACE [224–228]. Yang et al., in a series of 24 HCC nodules successfully treated with TACE, found a statistically significant decrease in arterial perfusion (AP) and hepatic perfusion index (HPI) values in the treated nodules [225]. Chen et al. categorized treatment response by dividing 38 HCC patients into partial response (PR), stable disease (SD), and progressive disease (PD) groups, according to RECIST criteria, and found significantly reduced AP and blood volume (BV) values in the PR group after TACE in comparison to pre-TACE. In the PD group, AP, BF, and hepatic arterial fraction (HAF) showed a significant increase after TACE [224]. Ippolito et al. reported similar results by comparing the HP, AP, HPI, BV, and TTP maps between successfully treated lesions in the site of deposited iodized oil and viable persistence tumors at CTLP performed 4 weeks after TACE; HP, AP, and HPI values were significantly higher in the residual neoplastic tissue compared to the treated lesion ($p < 0.0001$) [226]. Another two papers validated previous studies, both presenting a significant reduction of AP and HPI in successfully treated lesions [227,228]. Hypervascular findings in AP maps reflect tumor-related neo-angiogenesis of viable tissue and should be considered as the most relevant parameter to assess treatment response after TACE in our experience. Similar results can be obtained with the MSI maps with the additional advantage of increased spatial resolution (Figure 16) [89]. Finally, Su et al. demonstrated that CTLP parameters, in particular the AP, Arterial Perfusion Index (API), and PP, are useful predictors of short-term therapeutic response after TACE with a sensitivity and specificity of 87% and 95%, respectively, and an accuracy of 91% [229].

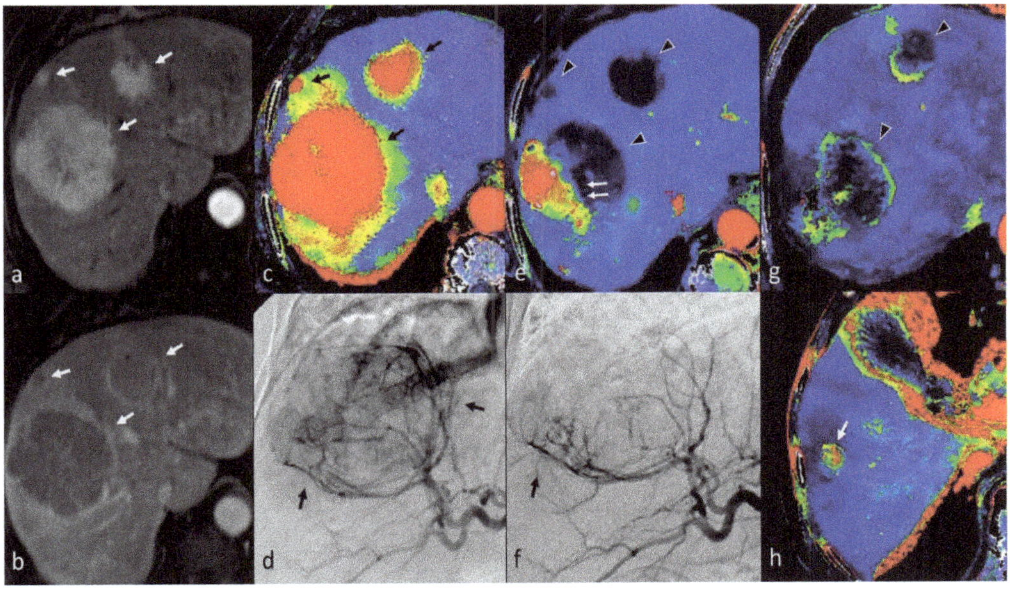


**Figure 16.** CT Liver perfusion (CTLP) imaging findings before and after transarterial chemoembolization (TACE) of multiple HCC nodules. (**a**,**b**) T1-w MRI scan of 72-year-old cirrhotic patient on arterial (**a**) and portal phase (**b**) after gadolinium injection showing three liver masses with arterial phase hyperenhancement (arrows in a), wash-out, and capsule (arrows in (**b**)) consistent with HCCs. (**c**) Mean slope of increase (MSI) after CTLP confirms MRI findings, showing all hypervascular lesions in red color (black arrows) surrounded by yellow rim, which reflects perfusion disorders.

(**d**) Angiographic view of the liver before TACE revealed the malignant hypervascularized overlapping lesions (between black arrows). The smaller subcapsular lesion could not be clearly depicted. The lesions were subsequently treated with selective TACE with Epirubicin-loaded beads. (**e**) Follow-up CTLP one month later (MSI map) revealed extensive necrosis of the chemo-embolized tumors (arrowheads), with a viable area in the dorsolateral portion of the larger mass (double arrows). (**f**) During a second TACE session, the residual malignant tumor lesions were depicted and treated with super-selective TACE. (**g**,**h**). Follow-up CTLP a month after second TACE (MSI map) revealed no signs of viability in the previously found tumors (arrowheads). However, a new, small 1.5-cm lesion had appeared.

## 13. Conclusions

The role of imaging in the multidisciplinary approach of patients with HCC is constantly evolving and expanding thanks to the application of hepatobiliary and contrast ultrasound agents, as well as newer quantitative techniques that evaluate blood perfusion on CT and MRI. As a result, earlier and more confident diagnosis of HCC has already been achieved; in addition, new imaging biomarkers are very promising in the prognosis of the biological behavior of the tumor. Hopefully, this review has persuasively demonstrated that imaging of HCC is not only an exciting field of active and intense research, but, most importantly, a critical aspect in the effective and personalized management and treatment of patients.

**Author Contributions:** All authors contributed equally to this work. All authors have read and agreed to the published version of the manuscript.

**Funding:** This review received no external funding.

**Conflicts of Interest:** The authors declare no conflict of interest.

## References

1. Global Cancer Observatory. Available online: http://globocan.iarc.fr/Default.aspx (accessed on 5 August 2022).
2. Rumgay, H.; Shield, K.; Charvat, H.; Ferrari, P.; Sornpaisarn, B.; Obot, I.; Islami, F.; Lemmens, V.E.P.P.; Rehm, J.; Soerjomataram, I. Global burden of cancer in 2020 attributable to alcohol consumption: A population-based study. *Lancet Oncol.* **2021**, *22*, 1071–1080. [CrossRef]
3. Marrero, J.A.; Kulik, L.M.; Sirlin, C.B.; Zhu, A.X.; Finn, R.S.; Abecassis, M.M.; Roberts, L.R.; Heimbach, J.K. Diagnosis, Staging, and Management of Hepatocellular Carcinoma: 2018 Practice Guidance by the American Association for the Study of Liver Diseases. *Hepatology* **2018**, *68*, 723–750. [CrossRef]
4. Galle, P.R.; Forner, A.; Llovet, J.M.; Mazzaferro, V.; Piscaglia, F.; Raoul, J.-L.; Schirmacher, P.; Vilgrain, V. EASL Clinical Practice Guidelines: Management of hepatocellular carcinoma. *J. Hepatol.* **2018**, *69*, 182–236. [CrossRef]
5. Omata, M.; Cheng, A.-L.; Kokudo, N.; Kudo, M.; Lee, J.M.; Jia, J.; Tateishi, R.; Han, K.-H.; Chawla, Y.K.; Shiina, S.; et al. Asia–Pacific clinical practice guidelines on the management of hepatocellular carcinoma: A 2017 update. *Hepatol. Int.* **2017**, *11*, 317–370. [CrossRef]
6. Masch, W.R.; Kampalath, R.; Parikh, N.; Shampain, K.A.; Aslam, A.; Chernyak, V. Imaging of treatment response during systemic therapy for hepatocellular carcinoma. *Abdom. Radiol.* **2021**, *46*, 3625–3633. [CrossRef]
7. Sidhu, P. Multiparametric Ultrasound (MPUS) Imaging: Terminology Describing the Many Aspects of Ultrasonography. *Ultraschall Med. Eur. J. Ultrasound* **2015**, *36*, 315–317. [CrossRef]
8. Sparchez, Z.; Craciun, R.; Caraiani, C.; Horhat, A.; Nenu, I.; Procopet, B.; Sparchez, M.; Stefanescu, H.; Mocan, T. Ultrasound or Sectional Imaging Techniques as Screening Tools for Hepatocellular Carcinoma: Fall Forward or Move Forward? *J. Clin. Med.* **2021**, *10*, 903. [CrossRef] [PubMed]
9. Tanaka, H. Current role of ultrasound in the diagnosis of hepatocellular carcinoma. *J. Med. Ultrason.* **2020**, *47*, 239–255. [CrossRef]
10. Makuuchi, M.; Hasegawa, H.; Yamazaki, S.; Bandai, Y.; Watanabe, G.; Ito, T. Ultrasonic characteristics of the small hepatocellular carcinoma. *Ultrasound Med. Biol.* **1983**, (Suppl. S2), 489–491.
11. Minami, Y. Hepatic malignancies: Correlation between sonographic findings and pathological features. *World J. Radiol.* **2010**, *2*, 249. [CrossRef]
12. Hui, A.-M.; Takayama, T.; Sano, K.; Kubota, K.; Akahane, M.; Ohtomo, K.; Makuuchi, M. Predictive value of gross classification of hepatocellular carcinoma on recurrence and survival after hepatectomy. *J. Hepatol.* **2000**, *33*, 975–979. [CrossRef]
13. Shimada, M.; Rikimaru, T.; Hamatsu, T.; Yamashita, Y.-i.; Terashi, T.; Taguchi, K.-i.; Tanaka, S.; Shirabe, K.; Sugimachi, K. The role of macroscopic classification in nodular-type hepatocellular carcinoma. *Am. J. Surg.* **2001**, *182*, 177–182. [CrossRef]

14. Tochio, H.; Kudo, M. Afferent and Efferent Vessels of Premalignant and Overt Hepatocellular Carcinoma: Observation by Color Doppler Imaging. *Intervirology* **2004**, *47*, 144–153. [CrossRef]
15. Lencioni, R.; Pinto, F.; Armillotta, N.; Bartolozzi, C. Assessment of tumor vascularity in hepatocellular carcinoma: Comparison of power Doppler US and color Doppler US. *Radiology* **1996**, *201*, 353–358. [CrossRef]
16. Chen, M.; Wang, D.; Zhao, Y.; Lu, D.M.; Li, H.X.; Liu, J.J.; Li, H. Preoperative color Doppler ultrasonography predicts early recurrence in AFP-positive hepatocellular carcinoma. *Oncol. Lett.* **2019**, *18*, 4703–4711. [CrossRef]
17. Yang, F.; Zhao, J.; Liu, C.; Mao, Y.; Mu, J.; Wei, X.; Jia, J.; Zhang, S.; Xin, X.; Tan, J. Superb microvascular imaging technique in depicting vascularity in focal liver lesions: More hypervascular supply patterns were depicted in hepatocellular carcinoma. *Cancer Imaging* **2019**, *19*, 92. [CrossRef]
18. He, M.-N.; Lv, K.; Jiang, Y.-X.; Jiang, T.-A. Application of superb microvascular imaging in focal liver lesions. *World J. Gastroenterol.* **2017**, *23*, 7765–7775. [CrossRef]
19. Rodgers, S.K.; Fetzer, D.T.; Gabriel, H.; Seow, J.H.; Choi, H.H.; Maturen, K.E.; Wasnik, A.P.; Morgan, T.A.; Dahiya, N.; O'Boyle, M.K.; et al. Role of US LI-RADS in the LI-RADS Algorithm. *RadioGraphics* **2019**, *39*, 690–708. [CrossRef]
20. Delli Pizzi, A.; Mastrodicasa, D.; Cianci, R.; Serafini, F.L.; Mincuzzi, E.; Di Fabio, F.; Giammarino, A.; Mannetta, G.; Basilico, R.; Caulo, M. Multimodality Imaging of Hepatocellular Carcinoma: From Diagnosis to Treatment Response Assessment in Everyday Clinical Practice. *Can. Assoc. Radiol. J.* **2021**, *72*, 714–727. [CrossRef]
21. Sangiovanni, A.; Del Ninno, E.; Fasani, P.; De Fazio, C.; Ronchi, G.; Romeo, R.; Morabito, A.; de Franchis, R.; Colombo, M. Increased survival of cirrhotic patients with a hepatocellular carcinoma detected during surveillance. *Gastroenterology* **2004**, *126*, 1005–1014. [CrossRef]
22. Samoylova, M.L.; Mehta, N.; Roberts, J.P.; Yao, F.Y. Predictors of Ultrasound Failure to Detect Hepatocellular Carcinoma. *Liver Transpl.* **2018**, *24*, 1171–1177. [CrossRef]
23. Dietrich, C.F.; Nolsøe, C.P.; Barr, R.G.; Berzigotti, A.; Burns, P.N.; Cantisani, V.; Chammas, M.C.; Chaubal, N.; Choi, B.I.; Clevert, D.-A.; et al. Guidelines and Good Clinical Practice Recommendations for Contrast-Enhanced Ultrasound (CEUS) in the Liver–Update 2020 WFUMB in Cooperation with EFSUMB, AFSUMB, AIUM, and FLAUS. *Ultrasound Med. Biol.* **2020**, *46*, 2579–2604. [CrossRef]
24. Salvatore, V.; Gianstefani, A.; Negrini, G.; Allegretti, G.; Galassi, M.; Piscaglia, F. Imaging Diagnosis of Hepatocellular Carcinoma: Recent Advances of Contrast-Enhanced Ultrasonography with SonoVue®. *Liver Cancer* **2015**, *5*, 55–66. [CrossRef]
25. Dietrich, C.; Bamber, J.; Berzigotti, A.; Bota, S.; Cantisani, V.; Castera, L.; Cosgrove, D.; Ferraioli, G.; Friedrich-Rust, M.; Gilja, O.; et al. EFSUMB Guidelines and Recommendations on the Clinical Use of Liver Ultrasound Elastography, Update 2017 (Long Version). *Ultraschall Med. Eur. J. Ultrasound* **2017**, *38*, e16–e47. [CrossRef]
26. Huang, J.-Y.; Li, J.-W.; Lu, Q.; Luo, Y.; Lin, L.; Shi, Y.-J.; Li, T.; Liu, J.-B.; Lyshchik, A. Diagnostic Accuracy of CEUS LI-RADS for the Characterization of Liver Nodules 20 mm or Smaller in Patients at Risk for Hepatocellular Carcinoma. *Radiology* **2020**, *294*, 329–339. [CrossRef]
27. Lin, M.; Lu, D.S.; Duan, Y.; Liao, P.; Sayre, J.; Xie, X.; Kuang, M. Cirrhotic Nodule Transformation to Hepatocellular Carcinoma: Natural History and Predictive Biomarkers on Contrast-Enhanced Ultrasound. *Am. J. Roentgenol.* **2020**, *214*, 96–104. [CrossRef]
28. Duan, Y.; Xie, X.; Li, Q.; Mercaldo, N.; Samir, A.E.; Kuang, M.; Lin, M. Differentiation of regenerative nodule, dysplastic nodule, and small hepatocellular carcinoma in cirrhotic patients: A contrast-enhanced ultrasound–based multivariable model analysis. *Eur. Radiol.* **2020**, *30*, 4741–4751. [CrossRef]
29. Bartolotta, T.V.; Taibbi, A.; Midiri, M.; Lagalla, R. Contrast-enhanced ultrasound of hepatocellular carcinoma: Where do we stand? *Ultrasonography* **2019**, *38*, 200–214. [CrossRef]
30. Fan, P.L.; Ding, H.; Mao, F.; Chen, L.L.; Dong, Y.; Wang, W.P. Enhancement patterns of small hepatocellular carcinoma ($\leq$30 mm) on contrast-enhanced ultrasound: Correlation with clinicopathologic characteristics. *Eur. J. Radiol.* **2020**, *132*, 109341. [CrossRef]
31. Jang, H.-J.; Kim, T.K.; Burns, P.N.; Wilson, S.R. Enhancement Patterns of Hepatocellular Carcinoma at Contrast-enhanced US: Comparison with Histologic Differentiation. *Radiology* **2007**, *244*, 898–906. [CrossRef]
32. Wilson, S.R.; Burns, P.N.; Kono, Y. Contrast-Enhanced Ultrasound of Focal Liver Masses: A Success Story. *Ultrasound Med. Biol.* **2020**, *46*, 1059–1070. [CrossRef]
33. Bartolotta, T.V.; Terranova, M.C.; Gagliardo, C.; Taibbi, A. CEUS LI-RADS: A pictorial review. *Insights Imaging* **2020**, *11*, 9. [CrossRef]
34. Hatanaka, K.; Kudo, M.; Minami, Y.; Maekawa, K. Sonazoid-Enhanced Ultrasonography for Diagnosis of Hepatic Malignancies: Comparison with Contrast-Enhanced CT. *Oncology* **2008**, *75*, 42–47. [CrossRef]
35. Mandai, M.; Koda, M.; Matono, T.; Nagahara, T.; Sugihara, T.; Ueki, M.; Ohyama, K.; Murawaki, Y. Assessment of hepatocellular carcinoma by contrast-enhanced ultrasound with perfluorobutane microbubbles: Comparison with dynamic CT. *Br. J. Radiol.* **2011**, *84*, 499–507. [CrossRef]
36. Numata, K.; Fukuda, H.; Miwa, H.; Ishii, T.; Moriya, S.; Kondo, M.; Nozaki, A.; Morimoto, M.; Okada, M.; Takebayashi, S.; et al. Contrast-enhanced ultrasonography findings using a perflubutane-based contrast agent in patients with early hepatocellular carcinoma. *Eur. J. Radiol.* **2014**, *83*, 95–102. [CrossRef]
37. Maruyama, H.; Takahashi, M.; Ishibashi, H.; Yoshikawa, M.; Yokosuka, O. Contrast-enhanced ultrasound for characterisation of hepatic lesions appearing non-hypervascular on CT in chronic liver diseases. *Br. J. Radiol.* **2012**, *85*, 351–357. [CrossRef]

38. Friedrich-Rust, M.; Klopffleisch, T.; Nierhoff, J.; Herrmann, E.; Vermehren, J.; Schneider, M.D.; Zeuzem, S.; Bojunga, J. Contrast-Enhanced Ultrasound for the differentiation of benign and malignant focal liver lesions: A meta-analysis. *Liver Int.* **2013**, *33*, 739–755. [CrossRef]
39. Terzi, E.; Iavarone, M.; Pompili, M.; Veronese, L.; Cabibbo, G.; Fraquelli, M.; Riccardi, L.; De Bonis, L.; Sangiovanni, A.; Leoni, S.; et al. Contrast ultrasound LI-RADS LR-5 identifies hepatocellular carcinoma in cirrhosis in a multicenter restropective study of 1006 nodules. *J. Hepatol.* **2018**, *68*, 485–492. [CrossRef]
40. Sugimoto, K.; Moriyasu, F.; Kamiyama, N.; Metoki, R.; Yamada, M.; Imai, Y.; Iijima, H. Analysis of morphological vascular changes of hepatocellular carcinoma by microflow imaging using contrast-enhanced sonography. *Hepatol. Res.* **2008**, *38*, 790–799. [CrossRef]
41. Schwarz, S.; Clevert, D.-A.; Ingrisch, M.; Geyer, T.; Schwarze, V.; Rübenthaler, J.; Armbruster, M. Quantitative Analysis of the Time–Intensity Curve of Contrast-Enhanced Ultrasound of the Liver: Differentiation of Benign and Malignant Liver Lesions. *Diagnostics* **2021**, *11*, 1244. [CrossRef]
42. Dong, Y.; Qiu, Y.; Yang, D.; Yu, L.; Zuo, D.; Zhang, Q.; Tian, X.; Wang, W.-P.; Jung, E.M. Potential application of dynamic contrast enhanced ultrasound in predicting microvascular invasion of hepatocellular carcinoma. *Clin. Hemorheol. Microcirc.* **2021**, *77*, 461–469. [CrossRef]
43. Wildner, D.; Pfeifer, L.; Goertz, R.; Bernatik, T.; Sturm, J.; Neurath, M.; Strobel, D. Dynamic Contrast-Enhanced Ultrasound (DCE-US) for the Characterization of Hepatocellular Carcinoma and Cholangiocellular Carcinoma. *Ultraschall Med. Eur. J. Ultrasound* **2014**, *35*, 522–527. [CrossRef]
44. Dong, Y.; Wang, W.-P.; Mao, F.; Zhang, Q.; Yang, D.; Tannapfel, A.; Meloni, M.F.; Neye, H.; Clevert, D.-A.; Dietrich, C.F. Imaging Features of Fibrolamellar Hepatocellular Carcinoma with Contrast-Enhanced Ultrasound. *Ultraschall Med. Eur. J. Ultrasound* **2021**, *42*, 306–313. [CrossRef]
45. Tang, A.; Cruite, I.; Mitchell, D.G.; Sirlin, C.B. Hepatocellular carcinoma imaging systems: Why they exist, how they have evolved, and how they differ. *Abdom. Radiol.* **2018**, *43*, 3–12. [CrossRef]
46. Kulkarni, N.M.; Fung, A.; Kambadakone, A.R.; Yeh, B.M. Computed Tomography Techniques, Protocols, Advancements, and Future Directions in Liver Diseases. *Magn. Reson. Imaging Clin. N. Am.* **2021**, *29*, 305–320. [CrossRef]
47. Kambadakone, A.R.; Fung, A.; Gupta, R.T.; Hope, T.A.; Fowler, K.J.; Lyshchik, A.; Ganesan, K.; Yaghmai, V.; Guimaraes, A.R.; Sahani, D.V.; et al. LI-RADS technical requirements for CT, MRI, and contrast-enhanced ultrasound. *Abdom. Radiol.* **2018**, *43*, 56–74. [CrossRef]
48. Bae, K.T. Intravenous Contrast Medium Administration and Scan Timing at CT: Considerations and Approaches. *Radiology* **2010**, *256*, 32–61. [CrossRef]
49. Choi, J.-Y.; Lee, J.-M.; Sirlin, C.B. CT and MR Imaging Diagnosis and Staging of Hepatocellular Carcinoma: Part I. Development, Growth, and Spread: Key Pathologic and Imaging Aspects. *Radiology* **2014**, *272*, 635–654. [CrossRef]
50. Chernyak, V.; Fowler, K.J.; Kamaya, A.; Kielar, A.Z.; Elsayes, K.M.; Bashir, M.R.; Kono, Y.; Do, R.K.; Mitchell, D.G.; Singal, A.G.; et al. Liver Imaging Reporting and Data System (LI-RADS) Version 2018: Imaging of Hepatocellular Carcinoma in At-Risk Patients. *Radiology* **2018**, *289*, 816–830. [CrossRef]
51. American College of Radiology (ACR). Liver Imaging Reporting and Data System Version 2018. Available online: https://www.acr.org/Clinical-Resources/Reporting-and-Data-Systems/LI-RADS/CT-MRI-LI-RADS-v2018 (accessed on 8 July 2022).
52. Heimbach, J.K.; Kulik, L.M.; Finn, R.S.; Sirlin, C.B.; Abecassis, M.M.; Roberts, L.R.; Zhu, A.X.; Murad, M.H.; Marrero, J.A. AASLD guidelines for the treatment of hepatocellular carcinoma. *Hepatology* **2018**, *67*, 358–380. [CrossRef]
53. Narsinh, K.H.; Cui, J.; Papadatos, D.; Sirlin, C.B.; Santillan, C.S. Hepatocarcinogenesis and LI-RADS. *Abdom. Radiol.* **2018**, *43*, 158–168. [CrossRef]
54. Choi, J.-Y.; Lee, J.-M.; Sirlin, C.B. CT and MR Imaging Diagnosis and Staging of Hepatocellular Carcinoma: Part II. Extracellular Agents, Hepatobiliary Agents, and Ancillary Imaging Features. *Radiology* **2014**, *273*, 30–50. [CrossRef]
55. Lee, J.H.; Lee, J.M.; Kim, S.J.; Baek, J.H.; Yun, S.H.; Kim, K.W.; Han, J.K.; Choi, B.I. Enhancement patterns of hepatocellular carcinomas on multiphasic multidetector row CT: Comparison with pathological differentiation. *Br. J. Radiol.* **2012**, *85*, e573–e583. [CrossRef]
56. Zhu, F.; Yang, F.; Li, J.; Chen, W.; Yang, W. Incomplete tumor capsule on preoperative imaging reveals microvascular invasion in hepatocellular carcinoma: A systematic review and meta-analysis. *Abdom. Radiol.* **2019**, *44*, 3049–3057. [CrossRef]
57. Galia, M.; Taibbi, A.; Marin, D.; Furlan, A.; Burgio, M.D.; Agnello, F.; Cabibbo, G.; Beers, B.E.V.; Bartolotta, T.V.; Midiri, M.; et al. Focal lesions in cirrhotic liver: What else beyond hepatocellular carcinoma? *Diagn. Interv. Radiol.* **2014**, *20*, 222–228. [CrossRef]
58. Vernuccio, F.; Cannella, R.; Porrello, G.; Calandra, A.; Midiri, M.; Furlan, A.; Brancatelli, G. Uncommon imaging evolutions of focal liver lesions in cirrhosis. *Abdom. Radiol.* **2019**, *44*, 3069–3077. [CrossRef]
59. Tang, M.; Li, Y.; Lin, Z.; Shen, B.; Huang, M.; Li, Z.-P.; Li, X.; Feng, S.-T. Hepatic nodules with arterial phase hyperenhancement and washout on enhanced computed tomography/magnetic resonance imaging: How to avoid pitfalls. *Abdom. Radiol.* **2020**, *45*, 3730–3742. [CrossRef]
60. Huang, B.; Wu, L.; Lu, X.-Y.; Xu, F.; Liu, C.-F.; Shen, W.-F.; Jia, N.-Y.; Cheng, H.-Y.; Yang, Y.-F.; Shen, F. Small Intrahepatic Cholangiocarcinoma and Hepatocellular Carcinoma in Cirrhotic Livers May Share Similar Enhancement Patterns at Multiphase Dynamic MR Imaging. *Radiology* **2016**, *281*, 150–157. [CrossRef]

61. Xu, J.; Igarashi, S.; Sasaki, M.; Matsubara, T.; Yoneda, N.; Kozaka, K.; Ikeda, H.; Kim, J.; Yu, E.; Matsui, O.; et al. Intrahepatic cholangiocarcinomas in cirrhosis are hypervascular in comparison with those in normal livers. *Liver Int.* **2012**, *32*, 1156–1164. [CrossRef]
62. Kim, I.; Kim, M.-J. Histologic Characteristics of Hepatocellular Carcinomas Showing Atypical Enhancement Patterns on 4-Phase MDCT Examination. *Korean J. Radiol.* **2012**, *13*, 586. [CrossRef]
63. Kim, Y.K.; Lee, W.J.; Park, M.J.; Kim, S.H.; Rhim, H.; Choi, D. Hypovascular Hypointense Nodules on Hepatobiliary Phase Gadoxetic Acid–enhanced MR Images in Patients with Cirrhosis: Potential of DW Imaging in Predicting Progression to Hypervascular HCC. *Radiology* **2012**, *265*, 104–114. [CrossRef] [PubMed]
64. Kim, Y.K.; Lee, Y.H.; Kim, C.S.; Han, Y.M. Added diagnostic value of T2-weighted MR imaging to gadolinium-enhanced three-dimensional dynamic MR imaging for the detection of small hepatocellular carcinomas. *Eur. J. Radiol.* **2008**, *67*, 304–310. [CrossRef] [PubMed]
65. Park, H.J.; Choi, B.I.; Lee, E.S.; Park, S.B.; Lee, J.B. How to Differentiate Borderline Hepatic Nodules in Hepatocarcinogenesis: Emphasis on Imaging Diagnosis. *Liver Cancer* **2017**, *6*, 189–203. [CrossRef] [PubMed]
66. Kim, T.-H.; Yoon, J.H.; Lee, J.M. Emerging Role of Hepatobiliary Magnetic Resonance Contrast Media and Contrast-Enhanced Ultrasound for Noninvasive Diagnosis of Hepatocellular Carcinoma: Emphasis on Recent Updates in Major Guidelines. *Korean J. Radiol.* **2019**, *20*, 863. [CrossRef] [PubMed]
67. Joo, I.; Kim, S.Y.; Kang, T.W.; Kim, Y.K.; Park, B.J.; Lee, Y.J.; Choi, J.-I.; Lee, C.-H.; Park, H.S.; Lee, K.; et al. Radiologic-Pathologic Correlation of Hepatobiliary Phase Hypointense Nodules without Arterial Phase Hyperenhancement at Gadoxetic Acid–enhanced MRI: A Multicenter Study. *Radiology* **2020**, *296*, 335–345. [CrossRef] [PubMed]
68. Kim, J.H.; Joo, I.; Lee, J.M. Atypical Appearance of Hepatocellular Carcinoma and Its Mimickers: How to Solve Challenging Cases Using Gadoxetic Acid-Enhanced Liver Magnetic Resonance Imaging. *Korean J. Radiol.* **2019**, *20*, 1019. [CrossRef]
69. Basha, M.A.A.; AlAzzazy, M.Z.; Ahmed, A.F.; Yousef, H.Y.; Shehata, S.M.; El Sammak, D.A.E.A.; Fathy, T.; Obaya, A.A.; Abdelbary, E.H. Does a combined CT and MRI protocol enhance the diagnostic efficacy of LI-RADS in the categorization of hepatic observations? A prospective comparative study. *Eur. Radiol.* **2018**, *28*, 2592–2603. [CrossRef]
70. Kojiro, M. 'Nodule-in-Nodule' Appearance in Hepatocellular Carcinoma: Its Significance as a Morphologic Marker of Dedifferentiation. *Intervirology* **2004**, *47*, 179–183. [CrossRef]
71. Kitao, A.; Zen, Y.; Matsui, O.; Gabata, T.; Nakanuma, Y. Hepatocarcinogenesis: Multistep Changes of Drainage Vessels at CT during Arterial Portography and Hepatic Arteriography—Radiologic-Pathologic Correlation. *Radiology* **2009**, *252*, 605–614. [CrossRef]
72. Chou, C.-T.; Chen, R.-C.; Lin, W.-C.; Ko, C.-J.; Chen, C.-B.; Chen, Y.-L. Prediction of Microvascular Invasion of Hepatocellular Carcinoma: Preoperative CT and Histopathologic Correlation. *Am. J. Roentgenol.* **2014**, *203*, W253–W259. [CrossRef]
73. An, C.; Kim, M.-J. Imaging features related with prognosis of hepatocellular carcinoma. *Abdom. Radiol.* **2019**, *44*, 509–516. [CrossRef]
74. Bello, H.R.; Mahdi, Z.K.; Lui, S.K.; Nandwana, S.B.; Harri, P.A.; Davarpanah, A.H. Hepatocellular Carcinoma with Atypical Imaging Features: Review of the Morphologic Hepatocellular Carcinoma Subtypes with Radiology-Pathology Correlation. *J. Magn. Reson. Imaging* **2021**, *55*, 681–697. [CrossRef]
75. Fowler, K.J.; Burgoyne, A.; Fraum, T.J.; Hosseini, M.; Ichikawa, S.; Kim, S.; Kitao, A.; Lee, J.M.; Paradis, V.; Taouli, B.; et al. Pathologic, Molecular, and Prognostic Radiologic Features of Hepatocellular Carcinoma. *RadioGraphics* **2021**, *41*, 1611–1631. [CrossRef]
76. Mulé, S.; Galletto Pregliasco, A.; Tenenhaus, A.; Kharrat, R.; Amaddeo, G.; Baranes, L.; Laurent, A.; Regnault, H.; Sommacale, D.; Djabbari, M.; et al. Multiphase Liver MRI for Identifying the Macrotrabecular-Massive Subtype of Hepatocellular Carcinoma. *Radiology* **2020**, *295*, 562–571. [CrossRef]
77. Kim, S.H.; Lim, H.K.; Lee, W.J.; Choi, D.; Park, C.K. Scirrhous hepatocellular carcinoma: Comparison with usual hepatocellular carcinoma based on CT–pathologic features and long-term results after curative resection. *Eur. J. Radiol.* **2009**, *69*, 123–130. [CrossRef]
78. Choi, S.-Y.; Kim, S.H.; Park, C.K.; Min, J.H.; Lee, J.E.; Choi, Y.-H.; Lee, B.R. Imaging Features of Gadoxetic Acid–enhanced and Diffusion-weighted MR Imaging for Identifying Cytokeratin 19-positive Hepatocellular Carcinoma: A Retrospective Observational Study. *Radiology* **2018**, *286*, 897–908. [CrossRef]
79. Reynolds, A.R.; Furlan, A.; Fetzer, D.T.; Sasatomi, E.; Borhani, A.A.; Heller, M.T.; Tublin, M.E. Infiltrative Hepatocellular Carcinoma: What Radiologists Need to Know. *RadioGraphics* **2015**, *35*, 371–386. [CrossRef]
80. Raab, B.-W. The Thread and Streak Sign. *Radiology* **2005**, *236*, 284–285. [CrossRef]
81. Lafitte, M.; Laurent, V.; Soyer, P.; Ayav, A.; Balaj, C.; Petit, I.; Hossu, G. MDCT features of hepatocellular carcinoma (HCC) in non-cirrhotic liver. *Diagn. Interv. Imaging* **2016**, *97*, 355–360. [CrossRef]
82. Ippolito, D.; Pecorelli, A.; Querques, G.; Drago, S.G.; Maino, C.; Franzesi, C.T.; Hatzidakis, A.; Sironi, S. Dynamic Computed Tomography Perfusion Imaging: Complementary Diagnostic Tool in Hepatocellular Carcinoma Assessment from Diagnosis to Treatment Follow-up. *Acad. Radiol.* **2019**, *26*, 1675–1685. [CrossRef]
83. Kim, S.H.; Kamaya, A.; Willmann, J.K. CT Perfusion of the Liver: Principles and Applications in Oncology. *Radiology* **2014**, *272*, 322–344. [CrossRef] [PubMed]
84. Ronot, M.; Clift, A.K.; Vilgrain, V.; Frilling, A. Functional imaging in liver tumours. *J. Hepatol.* **2016**, *65*, 1017–1030. [CrossRef]

85. Ronot, M.; Leporq, B.; Van Beers, B.E.; Vilgrain, V. CT and MR perfusion techniques to assess diffuse liver disease. *Abdom. Radiol.* **2020**, *45*, 3496–3506. [CrossRef]
86. Sahani, D.V.; Holalkere, N.-S.; Mueller, P.R.; Zhu, A.X. Advanced Hepatocellular Carcinoma: CT Perfusion of Liver and Tumor Tissue—Initial Experience. *Radiology* **2007**, *243*, 736–743. [CrossRef]
87. Ippolito, D.; Sironi, S.; Pozzi, M.; Antolini, L.; Invernizzi, F.; Ratti, L.; Leone, E.B.; Fazio, F. Perfusion CT in cirrhotic patients with early stage hepatocellular carcinoma: Assessment of tumor-related vascularization. *Eur. J. Radiol.* **2010**, *73*, 148–152. [CrossRef]
88. Fischer, M.A.; Kartalis, N.; Grigoriadis, A.; Loizou, L.; Stål, P.; Leidner, B.; Aspelin, P.; Brismar, T.B. Perfusion computed tomography for detection of hepatocellular carcinoma in patients with liver cirrhosis. *Eur. Radiol.* **2015**, *25*, 3123–3132. [CrossRef]
89. Hatzidakis, A.; Perisinakis, K.; Kalarakis, G.; Papadakis, A.; Savva, E.; Ippolito, D.; Karantanas, A. Perfusion-CT analysis for assessment of hepatocellular carcinoma lesions: Diagnostic value of different perfusion maps. *Acta Radiol.* **2019**, *60*, 561–568. [CrossRef]
90. Chen, B.-B.; Hsu, C.-Y.; Yu, C.-W.; Liang, P.-C.; Hsu, C.; Hsu, C.-H.; Cheng, A.-L.; Shih, T.T.-F. Dynamic Contrast-enhanced MR Imaging of Advanced Hepatocellular Carcinoma: Comparison with the Liver Parenchyma and Correlation with the Survival of Patients Receiving Systemic Therapy. *Radiology* **2016**, *281*, 454–464. [CrossRef]
91. Chaturvedi, A.; Bhargava, P.; Kolokythas, O.; Mitsumori, L.M.; Maki, J.H. Computer-Assisted Evaluation of Contrast Kinetics for Detection of Hepatocellular Carcinoma on Magnetic Resonance Imaging. *Curr. Probl. Diagn. Radiol.* **2015**, *44*, 8–14. [CrossRef]
92. Jajamovich, G.H.; Huang, W.; Besa, C.; Li, X.; Afzal, A.; Dyvorne, H.A.; Taouli, B. DCE-MRI of hepatocellular carcinoma: Perfusion quantification with Tofts model versus shutter-speed model—initial experience. *Magn. Reson. Mater. Phys. Biol. Med.* **2016**, *29*, 49–58. [CrossRef] [PubMed]
93. Taouli, B.; Johnson, R.S.; Hajdu, C.H.; Oei, M.T.H.; Merad, M.; Yee, H.; Rusinek, H. Hepatocellular Carcinoma: Perfusion Quantification with Dynamic Contrast-Enhanced MRI. *Am. J. Roentgenol.* **2013**, *201*, 795–800. [CrossRef] [PubMed]
94. Nascimento, C.; Bottino, A.; Nogueira, C.; Pannain, V. Analysis of morphological variables and arterialization in the differential diagnosis of hepatic nodules in explanted cirrhotic livers. *Diagn. Pathol.* **2007**, *2*, 51. [CrossRef] [PubMed]
95. Ueda, K.; Matsui, O.; Kitao, A.; Kobayashi, S.; Nakayama, J.; Miyagawa, S.; Kadoya, M. Tumor Hemodynamics and Hepatocarcinogenesis: Radio-Pathological Correlations and Outcomes of Carcinogenic Hepatocyte Nodules. *ISRN Hepatol.* **2014**, *2014*, 607628. [CrossRef]
96. Ma, G.-L.; Bai, R.-J.; Jiang, H.-J.; Hao, X.-J.; Dong, X.-P.; Li, D.-Q.; Liu, X.-D.; Wei, L. Early changes of hepatic hemodynamics measured by functional CT perfusion in a rabbit model of liver tumor. *Hepatobiliary Pancreat. Dis. Int.* **2012**, *11*, 407–411. [CrossRef]
97. Li, J.-P.; Feng, G.-L.; Li, D.-Q.; Wang, H.-B.; Zhao, D.-L.; Wan, Y.; Jiang, H.-J. Detection and differentiation of early hepatocellular carcinoma from cirrhosis using CT perfusion in a rat liver model. *Hepatobiliary Pancreat. Dis. Int.* **2016**, *15*, 612–618. [CrossRef]
98. Singh, J.; Sharma, S.; Aggarwal, N.; Sood, R.G.; Sood, S.; Sidhu, R. Role of Perfusion CT Differentiating Hemangiomas from Malignant Hepatic Lesions. *J. Clin. Imaging Sci.* **2014**, *4*, 10. [CrossRef]
99. Guo, M.; Yu, Y. Application of 128 Slice 4D CT Whole Liver Perfusion Imaging in Hepatic Tumor. *Cell Biochem. Biophys.* **2014**, *70*, 173–178. [CrossRef]
100. Li, M.; Chen, Y.; Gao, Z.; Zhu, K.; Yin, X. Evaluation of the blood flow in common hepatic tumors by multi-slice spiral CT whole-liver perfusion imaging. *Zhonghua Zhong Liu Za Zhi* **2015**, *37*, 904–908.
101. Hayano, K.; Desai, G.S.; Kambadakone, A.R.; Fuentes, J.M.; Tanabe, K.K.; Sahani, D.V. Quantitative characterization of hepatocellular carcinoma and metastatic liver tumor by CT perfusion. *Cancer Imaging* **2013**, *13*, 512–519. [CrossRef]
102. Guyennon, A. Perfusion characterization of liver metastases from endocrine tumors: Computed tomography perfusion. *World J. Radiol.* **2010**, *2*, 449. [CrossRef]
103. Reiner, C.S.; Goetti, R.; Burger, I.A.; Fischer, M.A.; Frauenfelder, T.; Knuth, A.; Pfammatter, T.; Schaefer, N.; Alkadhi, H. Liver Perfusion Imaging in Patients with Primary and Metastatic Liver Malignancy. *Acad. Radiol.* **2012**, *19*, 613–621. [CrossRef] [PubMed]
104. Fischer, M.A.; Marquez, H.P.; Gordic, S.; Leidner, B.; Klotz, E.; Aspelin, P.; Alkadhi, H.; Brismar, T.B. Arterio-portal shunts in the cirrhotic liver: Perfusion computed tomography for distinction of arterialized pseudolesions from hepatocellular carcinoma. *Eur. Radiol.* **2017**, *27*, 1074–1080. [CrossRef]
105. Alıcıoglu, B.; Guler, O.; Bulakbası, N.; Akpınar, S.; Tosun, O.; Comunoglu, C. Utility of semiquantitative parameters to differentiate benign and malignant focal hepatic lesions. *Clin. Imaging* **2013**, *37*, 692–696. [CrossRef] [PubMed]
106. Chen, J.; Si, Y.; Zhao, K.; Shi, X.; Bi, W.; Liu, S.-e.; Hua, H. Evaluation of quantitative parameters of dynamic contrast-enhanced magnetic resonance imaging in qualitative diagnosis of hepatic masses. *BMC Med. Imaging* **2018**, *18*, 56. [CrossRef] [PubMed]
107. Ghodasara, S.; Pahwa, S.; Dastmalchian, S.; Gulani, V.; Chen, Y. Free-Breathing 3D Liver Perfusion Quantification Using a Dual-Input Two-Compartment Model. *Sci. Rep.* **2017**, *7*, 17502. [CrossRef] [PubMed]
108. Abdullah, S.S.; Pialat, J.B.; Wiart, M.; Duboeuf, F.; Mabrut, J.-Y.; Bancel, B.; Rode, A.; Ducerf, C.; Baulieux, J.; Berthezene, Y. Characterization of hepatocellular carcinoma and colorectal liver metastasis by means of perfusion MRI. *J. Magn. Reson. Imaging* **2008**, *28*, 390–395. [CrossRef] [PubMed]
109. Thaiss, W.M.; Kaufmann, S.; Kloth, C.; Nikolaou, K.; Bösmüller, H.; Horger, M. VEGFR-2 expression in HCC, dysplastic and regenerative liver nodules, and correlation with pre-biopsy Dynamic Contrast Enhanced CT. *Eur. J. Radiol.* **2016**, *85*, 2036–2041. [CrossRef]

110. Bai, R.-J.; Li, J.-P.; Ren, S.-H.; Jiang, H.-J.; Liu, X.-D.; Ling, Z.-S.; Huang, Q.; Feng, G.-L. A correlation of computed tomography perfusion and histopathology in tumor edges of hepatocellular carcinoma. *Hepatobiliary Pancreat. Dis. Int.* **2014**, *13*, 612–617. [CrossRef]
111. Borgheresi, A.; Gonzalez-Aguirre, A.; Brown, K.T.; Getrajdman, G.I.; Erinjeri, J.P.; Covey, A.; Yarmohammadi, H.; Ziv, E.; Sofocleous, C.T.; Boas, F.E. Does Enhancement or Perfusion on Preprocedure CT Predict Outcomes After Embolization of Hepatocellular Carcinoma? *Acad. Radiol.* **2018**, *25*, 1588–1594. [CrossRef] [PubMed]
112. Marquez, H.P.; Puippe, G.; Mathew, R.P.; Alkadhi, H.; Pfammatter, T.; Fischer, M.A. CT Perfusion for Early Response Evaluation of Radiofrequency Ablation of Focal Liver Lesions: First Experience. *Cardiovasc. Interv. Radiol.* **2017**, *40*, 90–98. [CrossRef]
113. Lv, Y.; Jin, Y.; Yan, Q.; Yuan, D.; Wang, Y.; Li, X.; Shen, Y. The value of 64-slice spiral CT perfusion imaging in the treatment of liver cancer with argon-helium cryoablation. *Oncol. Lett.* **2016**, *12*, 4584–4588. [CrossRef] [PubMed]
114. Ippolito, D.; Querques, G.; Pecorelli, A.; Talei Franzesi, C.; Okolicsanyi, S.; Strazzabosco, M.; Sironi, S. Diagnostic Value of Quantitative Perfusion Computed Tomography Technique in the Assessment of Tumor Response to Sorafenib in Patients with Advanced Hepatocellular Carcinoma. *J. Comput. Assist. Tomogr.* **2019**, *43*, 206–213. [CrossRef] [PubMed]
115. Popovic, P.; Leban, A.; Kregar, K.; Garbajs, M.; Dezman, R.; Bunc, M. Computed tomographic perfusion imaging for the prediction of response and survival to transarterial chemoembolization of hepatocellular carcinoma. *Radiol. Oncol.* **2017**, *52*, 14–22. [CrossRef] [PubMed]
116. Reiner, C.S.; Morsbach, F.; Sah, B.-R.; Puippe, G.; Schaefer, N.; Pfammatter, T.; Alkadhi, H. Early Treatment Response Evaluation after Yttrium-90 Radioembolization of Liver Malignancy with CT Perfusion. *J. Vasc. Interv. Radiol.* **2014**, *25*, 747–759. [CrossRef] [PubMed]
117. Klotz, E.; Haberland, U.; Glatting, G.; Schoenberg, S.O.; Fink, C.; Attenberger, U.; Henzler, T. Technical prerequisites and imaging protocols for CT perfusion imaging in oncology. *Eur. J. Radiol.* **2015**, *84*, 2359–2367. [CrossRef]
118. Kalarakis, G.; Perisinakis, K.; Akoumianakis, E.; Karageorgiou, I.; Hatzidakis, A. CT liver perfusion in patients with hepatocellular carcinoma: Can we modify acquisition protocol to reduce patient exposure? *Eur. Radiol.* **2021**, *31*, 1410–1419. [CrossRef]
119. Bevilacqua, A.; Malavasi, S.; Vilgrain, V. Liver CT perfusion: Which is the relevant delay that reduces radiation dose and maintains diagnostic accuracy? *Eur. Radiol.* **2019**, *29*, 6550–6558. [CrossRef]
120. Topcuoglu, O.M.; Karcaaltincaba, M.; Akata, D.; Ozmen, M.N. Reproducibility and variability of very low dose hepatic perfusion CT in metastatic liver disease. *Diagn. Interv. Radiol.* **2016**, *22*, 495–500. [CrossRef]
121. Chung, C.Y.; Hu, R.; Peterson, R.B.; Allen, J.W. Automated Processing of Head CT Perfusion Imaging for Ischemic Stroke Triage: A Practical Guide to Quality Assurance and Interpretation. *Am. J. Roentgenol.* **2021**, *217*, 1401–1416. [CrossRef]
122. Zhao, C.; Dai, H.; Shao, J.; He, Q.; Su, W.; Wang, P.; Tang, Q.; Zeng, J.; Xu, S.; Zhao, J.; et al. Accuracy of Various Forms of Contrast-Enhanced MRI for Diagnosing Hepatocellular Carcinoma: A Systematic Review and Meta-Analysis. *Front. Oncol.* **2021**, *11*, 680691. [CrossRef]
123. Rimola, J.; Forner, A.; Tremosini, S.; Reig, M.; Vilana, R.; Bianchi, L.; Rodríguez-Lope, C.; Solé, M.; Ayuso, C.; Bruix, J. Non-invasive diagnosis of hepatocellular carcinoma ≤2cm in cirrhosis. Diagnostic accuracy assessing fat, capsule and signal intensity at dynamic MRI. *J. Hepatol.* **2012**, *56*, 1317–1323. [CrossRef] [PubMed]
124. Semaan, S.; Vietti Violi, N.; Lewis, S.; Chatterji, M.; Song, C.; Besa, C.; Babb, J.S.; Fiel, M.I.; Schwartz, M.; Thung, S.; et al. Hepatocellular carcinoma detection in liver cirrhosis: Diagnostic performance of contrast-enhanced CT vs. MRI with extracellular contrast vs. gadoxetic acid. *Eur. Radiol.* **2020**, *30*, 1020–1030. [CrossRef] [PubMed]
125. Hanna, R.F.; Miloushev, V.Z.; Tang, A.; Finklestone, L.A.; Brejt, S.Z.; Sandhu, R.S.; Santillan, C.S.; Wolfson, T.; Gamst, A.; Sirlin, C.B. Comparative 13-year meta-analysis of the sensitivity and positive predictive value of ultrasound, CT, and MRI for detecting hepatocellular carcinoma. *Abdom. Radiol.* **2016**, *41*, 71–90. [CrossRef] [PubMed]
126. Roberts, L.R.; Sirlin, C.B.; Zaiem, F.; Almasri, J.; Prokop, L.J.; Heimbach, J.K.; Murad, M.H.; Mohammed, K. Imaging for the diagnosis of hepatocellular carcinoma: A systematic review and meta-analysis. *Hepatology* **2018**, *67*, 401–421. [CrossRef] [PubMed]
127. Aubé, C.; Oberti, F.; Lonjon, J.; Pageaux, G.; Seror, O.; N'Kontchou, G.; Rode, A.; Radenne, S.; Cassinotto, C.; Vergniol, J.; et al. EASL and AASLD recommendations for the diagnosis of HCC to the test of daily practice. *Liver Int.* **2017**, *37*, 1515–1525. [CrossRef]
128. Golfieri, R.; Renzulli, M.; Lucidi, V.; Corcioni, B.; Trevisani, F.; Bolondi, L. Contribution of the hepatobiliary phase of Gd-EOB-DTPA-enhanced MRI to Dynamic MRI in the detection of hypovascular small (≤2 cm) HCC in cirrhosis. *Eur. Radiol.* **2011**, *21*, 1233–1242. [CrossRef]
129. Marin, D.; Di Martino, M.; Guerrisi, A.; De Filippis, G.; Rossi, M.; Ginanni Corradini, S.; Masciangelo, R.; Catalano, C.; Passariello, R. Hepatocellular Carcinoma in Patients with Cirrhosis: Qualitative Comparison of Gadobenate Dimeglumine–enhanced MR Imaging and Multiphasic 64-Section CT. *Radiology* **2009**, *251*, 85–95. [CrossRef]
130. Lee, Y.J.; Lee, J.M.; Lee, J.S.; Lee, H.Y.; Park, B.H.; Kim, Y.H.; Han, J.K.; Choi, B.I. Hepatocellular Carcinoma: Diagnostic Performance of Multidetector CT and MR Imaging—A Systematic Review and Meta-Analysis. *Radiology* **2015**, *275*, 97–109. [CrossRef]
131. Sano, K.; Ichikawa, T.; Motosugi, U.; Sou, H.; Muhi, A.M.; Matsuda, M.; Nakano, M.; Sakamoto, M.; Nakazawa, T.; Asakawa, M.; et al. Imaging Study of Early Hepatocellular Carcinoma: Usefulness of Gadoxetic Acid–enhanced MR Imaging. *Radiology* **2011**, *261*, 834–844. [CrossRef]

132. Yoon, S.H.; Lee, J.M.; So, Y.H.; Hong, S.H.; Kim, S.J.; Han, J.K.; Choi, B.I. Multiphasic MDCT Enhancement Pattern of Hepatocellular Carcinoma Smaller Than 3 cm in Diameter: Tumor Size and Cellular Differentiation. *Am. J. Roentgenol.* **2009**, *193*, W482–W489. [CrossRef]
133. Luca, A.; Caruso, S.; Milazzo, M.; Mamone, G.; Marrone, G.; Miraglia, R.; Maruzzelli, L.; Carollo, V.; Minervini, M.I.; Vizzini, G.; et al. Multidetector-row computed tomography (MDCT) for the diagnosis of hepatocellular carcinoma in cirrhotic candidates for liver transplantation: Prevalence of radiological vascular patterns and histological correlation with liver explants. *Eur. Radiol.* **2010**, *20*, 898–907. [CrossRef] [PubMed]
134. Khan, A.S.; Hussain, H.K.; Johnson, T.D.; Weadock, W.J.; Pelletier, S.J.; Marrero, J.A. Value of delayed hypointensity and delayed enhancing rim in magnetic resonance imaging diagnosis of small hepatocellular carcinoma in the cirrhotic liver. *J. Magn. Reson. Imaging* **2010**, *32*, 360–366. [CrossRef] [PubMed]
135. Choi, M.H.; Choi, J.-I.; Lee, Y.J.; Park, M.Y.; Rha, S.E.; Lall, C. MRI of Small Hepatocellular Carcinoma: Typical Features Are Less Frequent Below a Size Cutoff of 1.5 cm. *Am. J. Roentgenol.* **2017**, *208*, 544–551. [CrossRef] [PubMed]
136. Fowler, K.J.; Karimova, E.J.; Arauz, A.R.; Saad, N.E.; Brunt, E.M.; Chapman, W.C.; Heiken, J.P. Validation of Organ Procurement and Transplant Network (OPTN)/United Network for Organ Sharing (UNOS) Criteria for Imaging Diagnosis of Hepatocellular Carcinoma. *Transplantation* **2013**, *95*, 1506–1511. [CrossRef]
137. Lim, J.H.; Choi, D.; Park, C.K.; Lee, W.J.; Lim, H.K. Encapsulated hepatocellular carcinoma: CT-pathologic correlations. *Eur. Radiol.* **2006**, *16*, 2326–2333. [CrossRef]
138. Iguchi, T.; Aishima, S.; Sanefuji, K.; Fujita, N.; Sugimachi, K.; Gion, T.; Taketomi, A.; Shirabe, K.; Maehara, Y.; Tsuneyoshi, M. Both Fibrous Capsule Formation and Extracapsular Penetration Are Powerful Predictors of Poor Survival in Human Hepatocellular Carcinoma: A Histological Assessment of 365 Patients in Japan. *Ann. Surg. Oncol.* **2009**, *16*, 2539–2546. [CrossRef]
139. Honda, H.; Kaneko, K.; Maeda, T.; Kuroiwa, T.; Fukuya, T.; Yoshimitsu, K.; Irie, H.; Aibe, H.; Takenaka, K.; Masuda, K. Small Hepatocellular Carcinoma on Magnetic Resonance Imaging: Relation of Signal Intensity to Angiographic and Clinicopathologic Findings. *Investig. Radiol.* **1997**, *32*, 161–168. [CrossRef]
140. Shinmura, R.; Matsui, O.; Kobayashi, S.; Terayama, N.; Sanada, J.; Ueda, K.; Gabata, T.; Kadoya, M.; Miyayama, S. Cirrhotic Nodules: Association between MR Imaging Signal Intensity and Intranodular Blood Supply. *Radiology* **2005**, *237*, 512–519. [CrossRef]
141. Honda, H.; Kaneko, K.; Kanazawa, Y.; Hayashi, T.; Fukuya, T.; Matsumata, T.; Maeda, T.; Masuda, K. MR imaging of hepatocellular carcinomas: Effect of Cu and Fe contents on signal intensity. *Abdom. Imaging* **1997**, *22*, 60–66. [CrossRef]
142. Kew, M.C. Hepatic Iron Overload and Hepatocellular Carcinoma. *Liver Cancer* **2014**, *3*, 31–40. [CrossRef]
143. International Consensus Group for Hepatocellular, Neoplasia Pathologic Diagnosis of Early Hepatocellular Carcinoma: A Report of the International Consensus Group for Hepatocellular Neoplasia. *Hepatology* **2009**, *49*, 658–664. [CrossRef] [PubMed]
144. Kutami, R.; Nakashima, Y.; Nakashima, O.; Shiota, K.; Kojiro, M. Pathomorphologic study on the mechanism of fatty change in small hepatocellular carcinoma of humans. *J. Hepatol.* **2000**, *33*, 282–289. [CrossRef]
145. Asayama, Y.; Nishie, A.; Ishigami, K.; Ushijima, Y.; Takayama, Y.; Okamoto, D.; Fujita, N.; Kubo, Y.; Aishima, S.; Yoshizumi, T.; et al. Fatty change in moderately and poorly differentiated hepatocellular carcinoma on MRI: A possible mechanism related to decreased arterial flow. *Clin. Radiol.* **2016**, *71*, 1277–1283. [CrossRef]
146. Kim, Y.K.; Kim, C.S.; Han, Y.M.; Lee, Y.H. Detection of liver malignancy with gadoxetic acid-enhanced MRI: Is addition of diffusion-weighted MRI beneficial? *Clin. Radiol.* **2011**, *66*, 489–496. [CrossRef] [PubMed]
147. De Gaetano, A.M.; Catalano, M.; Pompili, M.; Marini, M.G.; Rodríguez Carnero, P.; Gullì, C.; Infante, A.; Iezzi, R.; Ponziani, F.R.; Cerrito, L.; et al. Critical analysis of major and ancillary features of LI-RADS v2018 in the differentiation of small ($\leq$ 2 cm) hepatocellular carcinoma from dysplastic nodules with gadobenate dimeglumine-enhanced magnetic resonance imaging. *Eur. Rev. Med. Pharmacol. Sci.* **2019**, *23*, 7786–7801. [CrossRef] [PubMed]
148. Nakanishi, M.; Chuma, M.; Hige, S.; Omatsu, T.; Yokoo, H.; Nakanishi, K.; Kamiyama, T.; Kubota, K.; Haga, H.; Matsuno, Y.; et al. Relationship Between Diffusion-Weighted Magnetic Resonance Imaging and Histological Tumor Grading of Hepatocellular Carcinoma. *Ann. Surg. Oncol.* **2012**, *19*, 1302–1309. [CrossRef]
149. Lim, K.S. Diffusion-weighted MRI of hepatocellular carcinoma in cirrhosis. *Clin. Radiol.* **2014**, *69*, 1–10. [CrossRef]
150. Piana, G.; Trinquart, L.; Meskine, N.; Barrau, V.; Beers, B.V.; Vilgrain, V. New MR imaging criteria with a diffusion-weighted sequence for the diagnosis of hepatocellular carcinoma in chronic liver diseases. *J. Hepatol.* **2011**, *55*, 126–132. [CrossRef]
151. Park, M.J.; Kim, Y.K.; Lee, M.W.; Lee, W.J.; Kim, Y.-S.; Kim, S.H.; Choi, D.; Rhim, H. Small Hepatocellular Carcinomas: Improved Sensitivity by Combining Gadoxetic Acid–enhanced and Diffusion-weighted MR Imaging Patterns. *Radiology* **2012**, *264*, 761–770. [CrossRef]
152. Park, M.-S.; Kim, S.; Patel, J.; Hajdu, C.H.; Do, R.K.G.; Mannelli, L.; Babb, J.; Taouli, B. Hepatocellular carcinoma: Detection with diffusion-weighted versus contrast-enhanced magnetic resonance imaging in pretransplant patients. *Hepatology* **2012**, *56*, 140–148. [CrossRef]
153. Kitao, A.; Matsui, O.; Yoneda, N.; Kozaka, K.; Shinmura, R.; Koda, W.; Kobayashi, S.; Gabata, T.; Zen, Y.; Yamashita, T.; et al. The uptake transporter OATP8 expression decreases during multistep hepatocarcinogenesis: Correlation with gadoxetic acid enhanced MR imaging. *Eur. Radiol.* **2011**, *21*, 2056–2066. [CrossRef] [PubMed]

154. Kudo, M.; Izumi, N.; Kokudo, N.; Matsui, O.; Sakamoto, M.; Nakashima, O.; Kojiro, M.; Makuuchi, M. Management of Hepatocellular Carcinoma in Japan: Consensus-Based Clinical Practice Guidelines Proposed by the Japan Society of Hepatology (JSH) 2010 Updated Version. *Dig. Dis.* **2011**, *29*, 339–364. [CrossRef]
155. Zech, C.J.; Ba-Ssalamah, A.; Berg, T.; Chandarana, H.; Chau, G.-Y.; Grazioli, L.; Kim, M.-J.; Lee, J.M.; Merkle, E.M.; Murakami, T.; et al. Consensus report from the 8th International Forum for Liver Magnetic Resonance Imaging. *Eur. Radiol.* **2020**, *30*, 370–382. [CrossRef] [PubMed]
156. Qin, X.; Yang, T.; Huang, Z.; Long, L.; Zhou, Z.; Li, W.; Gao, Y.; Wang, M.; Zhang, X. Hepatocellular carcinoma grading and recurrence prediction using T1 mapping on gadolinium-ethoxybenzyl diethylenetriamine pentaacetic acid-enhanced magnetic resonance imaging. *Oncol. Lett.* **2019**, *18*, 2322–2329. [CrossRef] [PubMed]
157. Mulé, S.; Chalaye, J.; Legou, F.; Tenenhaus, A.; Calderaro, J.; Galletto Pregliasco, A.; Laurent, A.; Kharrat, R.; Amaddeo, G.; Regnault, H.; et al. Hepatobiliary MR contrast agent uptake as a predictive biomarker of aggressive features on pathology and reduced recurrence-free survival in resectable hepatocellular carcinoma: Comparison with dual-tracer 18F-FDG and 18F-FCH PET/CT. *Eur. Radiol.* **2020**, *30*, 5348–5357. [CrossRef]
158. Huang, X.; Xiao, Z.; Zhang, Y.; Lin, N.; Xiong, M.; Huang, X.; Chen, Q.; Cao, D. Hepatocellular Carcinoma: Retrospective Evaluation of the Correlation Between Gadobenate Dimeglumine–Enhanced Magnetic Resonance Imaging and Pathologic Grade. *J. Comput. Assist. Tomogr.* **2018**, *42*, 365–372. [CrossRef]
159. Kitao, A.; Matsui, O.; Yoneda, N.; Kozaka, K.; Kobayashi, S.; Koda, W.; Inoue, D.; Ogi, T.; Yoshida, K.; Gabata, T. Gadoxetic acid-enhanced MR imaging for hepatocellular carcinoma: Molecular and genetic background. *Eur. Radiol.* **2020**, *30*, 3438–3447. [CrossRef]
160. Yoneda, N.; Matsui, O.; Kitao, A.; Kita, R.; Kozaka, K.; Koda, W.; Kobayashi, S.; Gabata, T.; Ikeda, H.; Nakanuma, Y. Hypervascular hepatocellular carcinomas showing hyperintensity on hepatobiliary phase of gadoxetic acid-enhanced magnetic resonance imaging: A possible subtype with mature hepatocyte nature. *Jpn. J. Radiol.* **2013**, *31*, 480–490. [CrossRef]
161. Aoki, T.; Nishida, N.; Ueshima, K.; Morita, M.; Chishina, H.; Takita, M.; Hagiwara, S.; Ida, H.; Minami, Y.; Yamada, A.; et al. Higher Enhancement Intrahepatic Nodules on the Hepatobiliary Phase of Gd-EOB-DTPA-Enhanced MRI as a Poor Responsive Marker of Anti-PD-1/PD-L1 Monotherapy for Unresectable Hepatocellular Carcinoma. *Liver Cancer* **2021**, *10*, 615–628. [CrossRef]
162. Kudo, M. Gd-EOB-DTPA-MRI Could Predict WNT/β-Catenin Mutation and Resistance to Immune Checkpoint Inhibitor Therapy in Hepatocellular Carcinoma. *Liver Cancer* **2020**, *9*, 479–490. [CrossRef]
163. Huang, M.; Liao, B.; Xu, P.; Cai, H.; Huang, K.; Dong, Z.; Xu, L.; Peng, Z.; Luo, Y.; Zheng, K.; et al. Prediction of Microvascular Invasion in Hepatocellular Carcinoma: Preoperative Gd-EOB-DTPA-Dynamic Enhanced MRI and Histopathological Correlation. *Contrast Media Mol. Imaging* **2018**, *2018*, 9674565. [CrossRef] [PubMed]
164. Lee, S.; Kim, S.H.; Lee, J.E.; Sinn, D.H.; Park, C.K. Preoperative gadoxetic acid–enhanced MRI for predicting microvascular invasion in patients with single hepatocellular carcinoma. *J. Hepatol.* **2017**, *67*, 526–534. [CrossRef] [PubMed]
165. Cannella, R.; Sartoris, R.; Grégory, J.; Garzelli, L.; Vilgrain, V.; Ronot, M.; Dioguardi Burgio, M. Quantitative magnetic resonance imaging for focal liver lesions: Bridging the gap between research and clinical practice. *Br. J. Radiol.* **2021**, *94*, 20210220. [CrossRef]
166. Ippolito, D.; Inchingolo, R.; Grazioli, L.; Drago, S.G.; Nardella, M.; Gatti, M.; Faletti, R. Recent advances in non-invasive magnetic resonance imaging assessment of hepatocellular carcinoma. *World J. Gastroenterol.* **2018**, *24*, 2413–2426. [CrossRef]
167. Khalifa, F.; Soliman, A.; El-Baz, A.; Abou El-Ghar, M.; El-Diasty, T.; Gimel'farb, G.; Ouseph, R.; Dwyer, A.C. Models and methods for analyzing DCE-MRI: A review: Models and methods for analyzing DCE-MRI. *Med. Phys.* **2014**, *41*, 124301. [CrossRef] [PubMed]
168. Shukla-Dave, A.; Obuchowski, N.A.; Chenevert, T.L.; Jambawalikar, S.; Schwartz, L.H.; Malyarenko, D.; Huang, W.; Noworolski, S.M.; Young, R.J.; Shiroishi, M.S.; et al. Quantitative imaging biomarkers alliance (QIBA) recommendations for improved precision of DWI and DCE-MRI derived biomarkers in multicenter oncology trials. *J. Magn. Reson. Imaging* **2019**, *49*, e101–e121. [CrossRef]
169. Chen, J.; Chen, C.; Xia, C.; Huang, Z.; Zuo, P.; Stemmer, A.; Song, B. Quantitative free-breathing dynamic contrast-enhanced MRI in hepatocellular carcinoma using gadoxetic acid: Correlations with Ki67 proliferation status, histological grades, and microvascular density. *Abdom. Radiol.* **2018**, *43*, 1393–1403. [CrossRef]
170. Weiss, J.; Ruff, C.; Grosse, U.; Grözinger, G.; Horger, M.; Nikolaou, K.; Gatidis, S. Assessment of Hepatic Perfusion Using GRASP MRI: Bringing Liver MRI on a New Level. *Investig. Radiol.* **2019**, *54*, 737–743. [CrossRef]
171. Izuishi, K.; Yamamoto, Y.; Mori, H.; Kameyama, R.; Fujihara, S.; Masaki, T.; Suzuki, Y. Molecular mechanisms of [18F]fluorodeoxyglucose accumulation in liver cancer. *Oncol. Rep.* **2014**, *31*, 701–706. [CrossRef]
172. Signore, G.; Nicod-Lalonde, M.; Prior, J.O.; Bertagna, F.; Muoio, B.; Giovanella, L.; Furlan, C.; Treglia, G. Detection rate of radiolabelled choline PET or PET/CT in hepatocellular carcinoma: An updated systematic review and meta-analysis. *Clin. Transl. Imaging* **2019**, *7*, 237–253. [CrossRef]
173. Chalaye, J.; Costentin, C.E.; Luciani, A.; Amaddeo, G.; Ganne-Carrié, N.; Baranes, L.; Allaire, M.; Calderaro, J.; Azoulay, D.; Nahon, P.; et al. Positron emission tomography/computed tomography with 18F-fluorocholine improve tumor staging and treatment allocation in patients with hepatocellular carcinoma. *J. Hepatol.* **2018**, *69*, 336–344. [CrossRef] [PubMed]
174. Castilla-Lièvre, M.-A.; Franco, D.; Gervais, P.; Kuhnast, B.; Agostini, H.; Marthey, L.; Désarnaud, S.; Helal, B.-O. Diagnostic value of combining 11C-choline and 18F-FDG PET/CT in hepatocellular carcinoma. *Eur. J. Nucl. Med. Mol. Imaging* **2016**, *43*, 852–859. [CrossRef]

175. Ghidaglia, J.; Golse, N.; Pascale, A.; Sebagh, M.; Besson, F.L. 18F-FDG /18F-Choline Dual-Tracer PET Behavior and Tumor Differentiation in HepatoCellular Carcinoma. A Systematic Review. *Front. Med.* **2022**, *9*, 924824. [CrossRef]
176. Gougelet, A.; Sartor, C.; Senni, N.; Calderaro, J.; Fartoux, L.; Lequoy, M.; Wendum, D.; Talbot, J.-N.; Prignon, A.; Chalaye, J.; et al. Hepatocellular Carcinomas with Mutational Activation of Beta-Catenin Require Choline and Can Be Detected by Positron Emission Tomography. *Gastroenterology* **2019**, *157*, 807–822. [CrossRef] [PubMed]
177. Lee, H.; Choi, J.Y.; Joung, J.-G.; Joh, J.-W.; Kim, J.M.; Hyun, S.H. Metabolism-Associated Gene Signatures for FDG Avidity on PET/CT and Prognostic Validation in Hepatocellular Carcinoma. *Front. Oncol.* **2022**, *12*, 845900. [CrossRef] [PubMed]
178. Castaldo, A.; De Lucia, D.R.; Pontillo, G.; Gatti, M.; Cocozza, S.; Ugga, L.; Cuocolo, R. State of the Art in Artificial Intelligence and Radiomics in Hepatocellular Carcinoma. *Diagnostics* **2021**, *11*, 1194. [CrossRef]
179. Feng, B.; Ma, X.-H.; Wang, S.; Cai, W.; Liu, X.-B.; Zhao, X.-M. Application of artificial intelligence in preoperative imaging of hepatocellular carcinoma: Current status and future perspectives. *World J. Gastroenterol.* **2021**, *27*, 5341–5350. [CrossRef]
180. Yao, S.; Ye, Z.; Wei, Y.; Jiang, H.-Y.; Song, B. Radiomics in hepatocellular carcinoma: A state-of-the-art review. *World J. Gastrointest. Oncol.* **2021**, *13*, 1599–1615. [CrossRef]
181. Kim, D.H.; Choi, S.H.; Kim, S.Y.; Kim, M.-J.; Lee, S.S.; Byun, J.H. Gadoxetic Acid–enhanced MRI of Hepatocellular Carcinoma: Value of Washout in Transitional and Hepatobiliary Phases. *Radiology* **2019**, *291*, 651–657. [CrossRef]
182. Joo, I.; Lee, J.M.; Lee, D.H.; Jeon, J.H.; Han, J.K. Retrospective validation of a new diagnostic criterion for hepatocellular carcinoma on gadoxetic acid-enhanced MRI: Can hypointensity on the hepatobiliary phase be used as an alternative to washout with the aid of ancillary features? *Eur. Radiol.* **2019**, *29*, 1724–1732. [CrossRef]
183. Vernuccio, F.; Cannella, R.; Meyer, M.; Choudhoury, K.R.; Gonzáles, F.; Schwartz, F.R.; Gupta, R.T.; Bashir, M.R.; Furlan, A.; Marin, D. LI-RADS: Diagnostic Performance of Hepatobiliary Phase Hypointensity and Major Imaging Features of LR-3 and LR-4 Lesions Measuring 10–19 mm With Arterial Phase Hyperenhancement. *Am. J. Roentgenol.* **2019**, *213*, W57–W65. [CrossRef] [PubMed]
184. Dhar, D.; Baglieri, J.; Kisseleva, T.; Brenner, D.A. Mechanisms of liver fibrosis and its role in liver cancer. *Exp. Biol. Med.* **2020**, *245*, 96–108. [CrossRef] [PubMed]
185. Baglieri, J.; Brenner, D.; Kisseleva, T. The Role of Fibrosis and Liver-Associated Fibroblasts in the Pathogenesis of Hepatocellular Carcinoma. *Int. J. Mol. Sci.* **2019**, *20*, 1723. [CrossRef]
186. Bengtsson, B.; Widman, L.; Wahlin, S.; Stål, P.; Björkström, N.K.; Hagström, H. The risk of hepatocellular carcinoma in cirrhosis differs by etiology, age and sex: A Swedish nationwide population-based cohort study. *UEG J.* **2022**, *10*, 465–476. [CrossRef]
187. Sharma, S.A.; Kowgier, M.; Hansen, B.E.; Brouwer, W.P.; Maan, R.; Wong, D.; Shah, H.; Khalili, K.; Yim, C.; Heathcote, E.J.; et al. Toronto HCC risk index: A validated scoring system to predict 10-year risk of HCC in patients with cirrhosis. *J. Hepatol.* **2018**, *68*, 92–99. [CrossRef] [PubMed]
188. Madhoun, M.F.; Fazili, J.; Bader, T.; Roberts, D.N.; Bright, B.C.; Bronze, M.S. Hepatitis C Prevalence in Patients with Hepatocellular Carcinoma Without Cirrhosis. *Am. J. Med. Sci.* **2010**, *339*, 169–173. [CrossRef] [PubMed]
189. Di Martino, M.; Saba, L.; Bosco, S.; Rossi, M.; Miles, K.A.; Di Miscio, R.; Lombardo, C.V.; Tamponi, E.; Piga, M.; Catalano, C. Hepatocellular carcinoma (HCC) in non-cirrhotic liver: Clinical, radiological and pathological findings. *Eur. Radiol.* **2014**, *24*, 1446–1454. [CrossRef]
190. Jamwal, R.; Krishnan, V.; Kushwaha, D.S.; Khurana, R. Hepatocellular carcinoma in non-cirrhotic versus cirrhotic liver: A clinico-radiological comparative analysis. *Abdom. Radiol.* **2020**, *45*, 2378–2387. [CrossRef]
191. Al-Sharhan, F.; Dohan, A.; Barat, M.; Feddal, A.; Terris, B.; Pol, S.; Mallet, V.; Soyer, P. MRI presentation of hepatocellular carcinoma in non-alcoholic steatohepatitis (NASH). *Eur. J. Radiol.* **2019**, *119*, 108648. [CrossRef]
192. Thompson, S.M.; Garg, I.; Ehman, E.C.; Sheedy, S.P.; Bookwalter, C.A.; Carter, R.E.; Roberts, L.R.; Venkatesh, S.K. Non-alcoholic fatty liver disease-associated hepatocellular carcinoma: Effect of hepatic steatosis on major hepatocellular carcinoma features at MRI. *Br. J. Radiol.* **2018**, *91*, 20180345. [CrossRef]
193. Gawrieh, S.; Dakhoul, L.; Miller, E.; Scanga, A.; deLemos, A.; Kettler, C.; Burney, H.; Liu, H.; Abu-Sbeih, H.; Chalasani, N.; et al. Characteristics, aetiologies and trends of hepatocellular carcinoma in patients without cirrhosis: A United States multicentre study. *Aliment Pharm.* **2019**, *50*, 809–821. [CrossRef] [PubMed]
194. Bouda, D.; Barrau, V.; Raynaud, L.; Dioguardi Burgio, M.; Paulatto, L.; Roche, V.; Sibert, A.; Moussa, N.; Vilgrain, V.; Ronot, M. Factors Associated with Tumor Progression After Percutaneous Ablation of Hepatocellular Carcinoma: Comparison Between Monopolar Radiofrequency and Microwaves. Results of a Propensity Score Matching Analysis. *Cardiovasc Interv. Radiol.* **2020**, *43*, 1608–1618. [CrossRef] [PubMed]
195. Liu, C.; Li, T.; He, J.-t.; Shao, H. TACE combined with microwave ablation therapy vs. TACE alone for treatment of early- and intermediate-stage hepatocellular carcinomas larger than 5 cm: A meta-analysis. *Diagn. Interv. Radiol.* **2020**, *26*, 575–583. [CrossRef] [PubMed]
196. Ricci, A.D.; Rizzo, A.; Bonucci, C.; Tavolari, S.; Palloni, A.; Frega, G.; Mollica, V.; Tober, N.; Mazzotta, E.; Felicani, C.; et al. The (Eternal) Debate on Microwave Ablation Versus Radiofrequency Ablation in BCLC-A Hepatocellular Carcinoma. *In Vivo* **2020**, *34*, 3421–3429. [CrossRef] [PubMed]

197. Shao, G.-L.; Zheng, J.-P.; Guo, L.-W.; Chen, Y.-T.; Zeng, H.; Yao, Z. Evaluation of efficacy of transcatheter arterial chemoembolization combined with computed tomography-guided radiofrequency ablation for hepatocellular carcinoma using magnetic resonance diffusion weighted imaging and computed tomography perfusion imaging: A prospective study. *Medicine* **2017**, *96*, e5518. [CrossRef]
198. Tan, W.; Deng, Q.; Lin, S.; Wang, Y.; Xu, G. Comparison of microwave ablation and radiofrequency ablation for hepatocellular carcinoma: A systematic review and meta-analysis. *Int. J. Hyperth.* **2019**, *36*, 263–271. [CrossRef]
199. Wang, C.; Wang, H.; Yang, W.; Hu, K.; Xie, H.; Hu, K.-Q.; Bai, W.; Dong, Z.; Lu, Y.; Zeng, Z.; et al. Multicenter randomized controlled trial of percutaneous cryoablation versus radiofrequency ablation in hepatocellular carcinoma: HEPATOLOGY, Vol. 00, No. X, 2014. *Hepatology* **2015**, *61*, 1579–1590. [CrossRef]
200. Makary, M.S.; Khandpur, U.; Cloyd, J.M.; Mumtaz, K.; Dowell, J.D. Locoregional Therapy Approaches for Hepatocellular Carcinoma: Recent Advances and Management Strategies. *Cancers* **2020**, *12*, 1914. [CrossRef]
201. Llovet, J.M.; Kelley, R.K.; Villanueva, A.; Singal, A.G.; Pikarsky, E.; Roayaie, S.; Lencioni, R.; Koike, K.; Zucman-Rossi, J.; Finn, R.S. Hepatocellular carcinoma. *Nat. Rev. Dis. Primers* **2021**, *7*, 6. [CrossRef]
202. Gerena, M.; Molvar, C.; Masciocchi, M.; Nandwana, S.; Sabottke, C.; Spieler, B.; Sharma, R.; Tsai, L.; Kielar, A. LI-RADS treatment response assessment of combination locoregional therapy for HCC. *Abdom. Radiol.* **2021**, *46*, 3634–3647. [CrossRef]
203. Ahmed, O.; Pillai, A. Hepatocellular Carcinoma: A Contemporary Approach to Locoregional Therapy. *Am. J. Gastroenterol.* **2020**, *115*, 1733–1736. [CrossRef] [PubMed]
204. Hussein, R.S.; Tantawy, W.; Abbas, Y.A. MRI assessment of hepatocellular carcinoma after locoregional therapy. *Insights Imaging* **2019**, *10*, 8. [CrossRef] [PubMed]
205. Llovet, J.M.; Lencioni, R. mRECIST for HCC: Performance and novel refinements. *J. Hepatol.* **2020**, *72*, 288–306. [CrossRef]
206. Kielar, A.; Fowler, K.J.; Lewis, S.; Yaghmai, V.; Miller, F.H.; Yarmohammadi, H.; Kim, C.; Chernyak, V.; Yokoo, T.; Meyer, J.; et al. Locoregional therapies for hepatocellular carcinoma and the new LI-RADS treatment response algorithm. *Abdom. Radiol.* **2018**, *43*, 218–230. [CrossRef] [PubMed]
207. Mendiratta-Lala, M.; Masch, W.R.; Shampain, K.; Zhang, A.; Jo, A.S.; Moorman, S.; Aslam, A.; Maturen, K.E.; Davenport, M.S. MRI Assessment of Hepatocellular Carcinoma after Local-Regional Therapy: A Comprehensive Review. *Radiol. Imaging Cancer* **2020**, *2*, e190024. [CrossRef]
208. Kloeckner, R.; Galle, P.R.; Bruix, J. Local and Regional Therapies for Hepatocellular Carcinoma. *Hepatology* **2021**, *73*, 137–149. [CrossRef]
209. Huber, T.C.; Bochnakova, T.; Koethe, Y.; Park, B.; Farsad, K. Percutaneous Therapies for Hepatocellular Carcinoma: Evolution of Liver Directed Therapies. *J. Hepatocell. Carcinoma* **2021**, *8*, 1181–1193. [CrossRef]
210. Young, S.; Taylor, A.J.; Sanghvi, T. Post Locoregional Therapy Treatment Imaging in Hepatocellular Carcinoma Patients: A Literature-based Review. *J. Clin. Transl. Hepatol.* **2018**, *6*, 189–197. [CrossRef]
211. Park, M.-h.; Rhim, H.; Kim, Y.-s.; Choi, D.; Lim, H.K.; Lee, W.J. Spectrum of CT Findings after Radiofrequency Ablation of Hepatic Tumors. *RadioGraphics* **2008**, *28*, 379–390. [CrossRef]
212. Crocetti, L.; Della Pina, C.; Cioni, D.; Lencioni, R. Peri-intraprocedural imaging: US, CT, and MRI. *Abdom. Imaging* **2011**, *36*, 648–660. [CrossRef]
213. Guan, Y.-S. Hepatocellular carcinoma treated with interventional procedures: CT and MRI follow-up. *World J. Gastroenterol.* **2004**, *10*, 3543. [CrossRef] [PubMed]
214. Imai, Y.; Katayama, K.; Hori, M.; Yakushijin, T.; Fujimoto, K.; Itoh, T.; Igura, T.; Sakakibara, M.; Takamura, M.; Tsurusaki, M.; et al. Prospective Comparison of Gd-EOB-DTPA-Enhanced MRI with Dynamic CT for Detecting Recurrence of HCC after Radiofrequency Ablation. *Liver Cancer* **2017**, *6*, 349–359. [CrossRef]
215. Ippolito, D.; Bonaffini, P.A.; Capraro, C.; Leni, D.; Corso, R.; Sironi, S. Viable residual tumor tissue after radiofrequency ablation treatment in hepatocellular carcinoma: Evaluation with CT perfusion. *Abdom. Imaging* **2013**, *38*, 502–510. [CrossRef] [PubMed]
216. Ramsey, D.E.; Kernagis, L.Y.; Soulen, M.C.; Geschwind, J.-F.H. Chemoembolization of Hepatocellular Carcinoma. *J. Vasc. Interv. Radiol.* **2002**, *13*, S211–S221. [CrossRef]
217. Bolondi, L.; Burroughs, A.; Dufour, J.-F.; Galle, P.; Mazzaferro, V.; Piscaglia, F.; Raoul, J.; Sangro, B. Heterogeneity of Patients with Intermediate (BCLC B) Hepatocellular Carcinoma: Proposal for a Subclassification to Facilitate Treatment Decisions. *Semin. Liver Dis.* **2013**, *32*, 348–359. [CrossRef]
218. Willatt, J.; Ruma, J.A.; Azar, S.F.; Dasika, N.L.; Syed, F. Imaging of hepatocellular carcinoma and image guided therapies—How we do it. *Cancer Imaging* **2017**, *17*, 9. [CrossRef]
219. Vossen, J.A.; Buijs, M.; Kamel, I.R. Assessment of Tumor Response on MR Imaging After Locoregional Therapy. *Tech. Vasc. Interv. Radiol.* **2006**, *9*, 125–132. [CrossRef]
220. Lim, H.S.; Jeong, Y.Y.; Kang, H.K.; Kim, J.K.; Park, J.G. Imaging Features of Hepatocellular Carcinoma After Transcatheter Arterial Chemoembolization and Radiofrequency Ablation. *Am. J. Roentgenol.* **2006**, *187*, W341–W349. [CrossRef]
221. Agnello, F. Imaging appearance of treated hepatocellular carcinoma. *World J. Hepatol.* **2013**, *5*, 417. [CrossRef]
222. Chiu, R.Y.W.; Yap, W.W.; Patel, R.; Liu, D.; Klass, D.; Harris, A.C. Hepatocellular Carcinoma Post Embolotherapy: Imaging Appearances and Pitfalls on Computed Tomography and Magnetic Resonance Imaging. *Can. Assoc. Radiol. J.* **2016**, *67*, 158–172. [CrossRef]

223. Liu, Z.; Fan, J.-M.; He, C.; Li, Z.-F.; Xu, Y.-S.; Li, Z.; Liu, H.-F.; Lei, J.-Q. Utility of diffusion weighted imaging with the quantitative apparent diffusion coefficient in diagnosing residual or recurrent hepatocellular carcinoma after transarterial chemoembolization: A meta-analysis. *Cancer Imaging* **2020**, *20*, 3. [CrossRef] [PubMed]
224. Chen, G.; Ma, D.-Q.; He, W.; Zhang, B.-F.; Zhao, L.-Q. Computed tomography perfusion in evaluating the therapeutic effect of transarterial chemoembolization for hepatocellular carcinoma. *World J. Gastroenterol.* **2008**, *14*, 5738. [CrossRef] [PubMed]
225. Yang, L.; Zhang, X.; Tan, B.; Liu, M.; Dong, G.; Zhai, Z. Computed Tomographic Perfusion Imaging for the Therapeutic Response of Chemoembolization for Hepatocellular Carcinoma. *J. Comput. Assist. Tomogr.* **2012**, *36*, 226–230. [CrossRef] [PubMed]
226. Ippolito, D.; Fior, D.; Franzesi, C.T.; Capraro, C.; Casiraghi, A.; Leni, D.; Vacirca, F.; Corso, R.; Sironi, S. Tumour-related neoangiogenesis: Functional dynamic perfusion computed tomography for diagnosis and treatment efficacy assessment in hepatocellular carcinoma. *Dig. Liver Dis.* **2014**, *46*, 916–922. [CrossRef]
227. Syha, R.; Gatidis, S.; Grözinger, G.; Grosse, U.; Maurer, M.; Zender, L.; Horger, M.; Nikolaou, K.; Ketelsen, D. C-arm computed tomography and volume perfusion computed tomography (VPCT)-based assessment of blood volume changes in hepatocellular carcinoma in prediction of midterm tumor response to transarterial chemoembolization: A single center retrospective trial. *Cancer Imaging* **2016**, *16*, 30. [CrossRef]
228. Rathmann, N.; Kara, K.; Budjan, J.; Henzler, T.; Smakic, A.; Schoenberg, S.O.; Diehl, S.J. Parenchymal Liver Blood Volume and Dynamic Volume Perfusion CT Measurements of Hepatocellular Carcinoma in Patients Undergoing Transarterial Chemoembolization. *Anticancer Res.* **2017**, *37*, 5681–5685. [CrossRef]
229. Su, T.-H.; He, W.; Jin, L.; Chen, G.; Xiao, G.-W. Early Response of Hepatocellular Carcinoma to Chemoembolization: Volume Computed Tomography Liver Perfusion Imaging as a Short-Term Response Predictor. *J. Comput. Assist. Tomogr.* **2017**, *41*, 315–320. [CrossRef]

cancers

MDPI

*Review*

# Tumor-Associated Macrophages in Hepatocellular Carcinoma Pathogenesis, Prognosis and Therapy

**Konstantinos Arvanitakis [1,2], Triantafyllia Koletsa [3], Ioannis Mitroulis [4,*,†] and Georgios Germanidis [1,2,*,†]**

1 First Department of Internal Medicine, AHEPA University Hospital, Aristotle University of Thessaloniki, 54636 Thessaloniki, Greece; kostarvanit@gmail.com
2 Basic and Translational Research Unit, Special Unit for Biomedical Research and Education, School of Medicine, Faculty of Health Sciences, Aristotle University of Thessaloniki, 54636 Thessaloniki, Greece
3 Department of Pathology, School of Medicine, Aristotle University of Thessaloniki, 54124 Thessaloniki, Greece; tkoletsa@auth.gr
4 First Department of Internal Medicine, University Hospital of Alexandroupolis, Democritus University of Thrace, 68100 Alexandroupolis, Greece
* Correspondence: imitroul@med.duth.gr (I.M.); geogerm@auth.gr (G.G.); Tel.: +30-2313-30-3156 (G.G.)
† I.M. and G.G. are senior co-authors.

**Simple Summary:** Hepatocellular carcinoma (HCC) constitutes a major health burden, accounting for >80% of primary liver cancers globally. Inflammation has come into the spotlight as a hallmark of cancer, and it is evident that tumor-associated inflammation drives the involvement of monocytes in tumor growth and metastasis. Tumor-associated macrophages (TAMs) actively participate in tumor-related inflammation, representing the main type of inflammatory cells in the tumor microenvironment, setting the crosstalk between tumor and stromal cells. Infiltrating TAMs exert either anti-tumorigenic (M1) or pro-tumorigenic (M2) functions. In most solid human tumors, increased TAM infiltration has been associated with enhanced tumor growth and metastasis, while other studies showcase that under certain conditions, TAMs exhibit cytotoxic and tumoricidal activity, inhibiting the progression of cancer. In this review, we summarize the current evidence on the role of macrophages in the pathogenesis and progression of HCC and we highlight their potential utilization in HCC prognosis and therapy.

**Abstract:** Hepatocellular carcinoma (HCC) constitutes a major health burden globally, and it is caused by intrinsic genetic mutations acting in concert with a multitude of epigenetic and extrinsic risk factors. Cancer induces myelopoiesis in the bone marrow, as well as the mobilization of hematopoietic stem and progenitor cells, which reside in the spleen. Monocytes produced in the bone marrow and the spleen further infiltrate tumors, where they differentiate into tumor-associated macrophages (TAMs). The relationship between chronic inflammation and hepatocarcinogenesis has been thoroughly investigated over the past decade; however, several aspects of the role of TAMs in HCC development are yet to be determined. In response to certain stimuli and signaling, monocytes differentiate into macrophages with antitumor properties, which are classified as M1-like. On the other hand, under different stimuli and signaling, the polarization of macrophages shifts towards an M2-like phenotype with a tumor promoting capacity. M2-like macrophages drive tumor growth both directly and indirectly, via the suppression of cytotoxic cell populations, including CD8+ T cells and NK cells. The tumor microenvironment affects the response to immunotherapies. Therefore, an enhanced understanding of its immunobiology is essential for the development of next-generation immunotherapies. The utilization of various monocyte-centered anticancer treatment modalities has been under clinical investigation, selectively targeting and modulating the processes of monocyte recruitment, activation and migration. This review summarizes the current evidence on the role of TAMs in HCC pathogenesis and progression, as well as in their potential involvement in tumor therapy, shedding light on emerging anticancer treatment methods targeting monocytes.

**Citation:** Arvanitakis, K.; Koletsa, T.; Mitroulis, I.; Germanidis, G. Tumor-Associated Macrophages in Hepatocellular Carcinoma Pathogenesis, Prognosis and Therapy. *Cancers* **2022**, *14*, 226. https://doi.org/10.3390/cancers14010226

Academic Editor: Melchiorre Cervello

Received: 8 December 2021
Accepted: 2 January 2022
Published: 4 January 2022

**Keywords:** tumor-associated macrophages; hepatocellular carcinoma; tumorigenesis; tumor microenvironment; treatment resistance

## 1. Introduction

Liver cancer is the sixth most commonly diagnosed cancer and the fourth leading cause of cancer-related death worldwide, while hepatocellular carcinoma (HCC) accounts for approximately 75–85% of primary liver cancers [1,2]. In the vast majority of cases, HCC arises as a result of sustained inflammatory damage, hepatocyte necrosis and regeneration in patients with liver cirrhosis [3]. The most common causes of liver cirrhosis that predispose patients to HCC are chronic hepatitis C virus (HCV), hepatitis B virus (HBV) and non-alcoholic fatty liver disease/non-alcoholic steatohepatitis (NAFLD/NASH) [4]. Despite major scientific advances, the majority of patients with HCC still face a dismal prognosis. Liver transplantation is only possible in patients with early HCC, whereas patients that undergo alternative treatment modalities—including liver resection or tumor ablation—develop recurrent disease in up to 70% of cases [5]. HCC prognosis fluctuates according to the stage at the time of diagnosis, with an overall 5-year survival rate of 20%. Patients who are not considered eligible for surgical or other curative procedures can receive palliative therapies, including tyrosine kinase inhibitors (TKI), such as sorafenib, lenvatinib, regorafenib or cabozantinib, or the VEGFR2-antibody ramucirumab [6–9]. However, these treatment options have minimal benefit in survival, increasing the necessity for novel therapeutic strategies for the treatment of HCC [10]. Recent evidence provided by Finn et al. showed that the use of the immune checkpoint inhibitor atezolizumab in combination with the antiangiogenic agent bevacizumab reduced mortality by 42% and decreased the risk of disease worsening or death by 41% compared to sorafenib alone, and it is currently accepted as the first-line systemic treatment of HCC [11]. Immunotherapies targeting the PD-1/PD-L1 axis, other than atezolizumab, have also been approved for the treatment of HCC without showing a major effect on patient survival [12]. Specifically, the efficacy of these agents is compromised in patients with NASH and this observation was linked to NASH-dependent altered immune cell function in TME [13].

Chronic liver inflammation drives a dysfunctional tissue repair process, leading to the formation of dysplastic nodules and, eventually, cancer [14,15]. The tumor microenvironment (TME) greatly contributes to the tolerogenic immune response towards HCC. It comprises myeloid-derived suppressor cells (MDSCs), tumor-associated macrophages (TAMs), tumor-associated neutrophils (TANs), cancer-associated fibroblasts (CAFs) and regulatory T cells (Tregs) [16–18]. MDSCs, either of monocytic or granulocytic origin, typically show immunosuppressive properties [19]. The liver bears the highest proportion of macrophages among all organs in the body [20], and in a healthy rodent liver, 20–40 macrophages accompany every 100 hepatocytes [21]. They are generally categorized into two distinct subsets that can be distinguished from each other based on their differential expression of cell surface markers. Kupffer cells (KCs) are the non-migratory tissue-resident macrophages of the liver, are located in the sinusoids and maintain homeostasis [22,23]. Monocyte-derived macrophages (MoMϕs) exert migratory capabilities and engraft liver tissue during inflammatory conditions or after KC depletion [24,25]. TAMs play an essential role in HCC pathogenesis, establishing a pro-inflammatory and pro-tumorigenic environment through the suppression of antitumor immune responses [26,27]. However, they can also participate in tumor immune surveillance and antitumor responses [28,29]. Given the major contribution of hepatic macrophages in normal tissue homeostasis, their pivotal role in liver inflammation and their dual promoting and inhibitory functions in tumor formation, hepatic macrophages have been at the forefront as potential therapeutic targets for various HCC treatment modalities [30].

This review summarizes the current state of knowledge regarding the involvement of TAMs in the pathogenesis and progression of HCC and the heterogeneity and plasticity that they exert in the cancer-associated microenvironment. We also highlight the principal future therapeutic options that target TAMs to treat HCC.

## 2. Monopoiesis and Tumor-Associated Monocytes

One of the main features of cancer is the tumor-associated chronic inflammation that results in the reprogramming of immune cells [31]. In addition to the immunomodulatory effect that this type of chronic inflammation elicits at the site of tumors, it has systemic effects, affecting cell populations in the bone marrow and spleen. One of these effects is the induction of emergency myelopoiesis, which results in the generation of mature cells of the myeloid lineage, including monocytes and neutrophils, that further infiltrate solid tumors and act mainly as immunosuppressive and tumor-promoting cells [32]. Hematopoietic stem and progenitor cells (HSPCs) and myeloid progenitors (MyP) can sense and are responsive to a variety of mediators that are released by tumor cells, such as granulocyte-colony stimulating factor (G-CSF), granulocyte macrophage (GM)-CSF or the chemokine CXCL-12 [33,34]. This results in their activation and differentiation towards a myeloid lineage, as well as in their mobilization and egress from the bone marrow and migration to extra medullary sites, such as the spleen [35].

Resident macrophages derive from the differentiation of progenitors in the yolk sack and fetal liver during early life, whereas bone marrow is the site of monocyte generation under steady state conditions [36]. Experiments in mice showed that monocytes derive from common myeloid progenitors, following two distinct differentiation pathways, giving rise to cells with different transcriptional programs and, probably, different functions [37]. Cancer-elicited inflammation promotes not only the generation of monocytes in the bone marrow in response to the myeloid-lineage growth factors GM-CSF, G-CSF and interleukin-6 (IL-6) [33], but also drives extramedullary monopoiesis in the spleen by mobilized HSPCs and MyP [38]. Indeed, both mice and patients with invasive cancer exhibited increased numbers of splenic granulocyte-macrophage progenitors (GMPs), which are able to generate monocytes, as well as neutrophils, that further infiltrate tumors and differentiate into TAMs and TANs, respectively [38]. The spleen is also a major site for the tumor-associated reprogramming of monocytes, which results in an accumulation of MDSCs, a monocytic cell population with potent immunosuppressive activity against CD8+ T cells [39]. For instance, a study by Jordan et al. showed that the accumulation of MDSCs of monocytic origin with T-cell-suppressive properties were observed in the spleens of patients with various types of cancer, including pancreatic and colorectal adenocarcinoma, pancreatic neuroendocrine tumors, melanoma or ovarian cancer [40]. Regarding HCC, increased numbers of hematopoietic progenitor cells, stained as CD133+, and CD11b+ myeloid cells were observed in the spleens of patients with HCC, as well as other tumors, compared to patients with cirrhosis, a condition that predisposes patients to HCC [41]. Interestingly, there was a positive correlation between the number of progenitor cells and mature myeloid cells, implying that extra medullary myelopoiesis was responsible for the generation of mature myeloid cells [41]. Taken together, enhanced myelopoiesis results in the generation of monocytic cells with immunosuppressive properties in the bone marrow and the spleen, which further migrate into tumors, where an additional step of reprogramming takes place during their differentiation to TAMs (Figure 1).

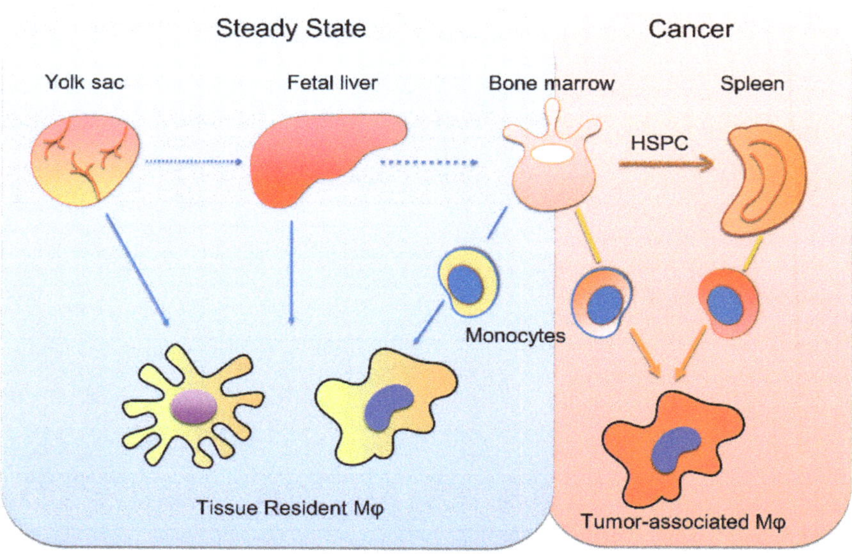


**Figure 1.** Origins of tissue-resident and tumor-associated macrophages. Tissue-resident macrophages derive from the differentiation of yolk sac and fetal liver hematopoietic progenitors and, later in life, stem from the differentiation of monocytes, generated in the bone marrow. Cancer induces myelopoiesis in the bone marrow, as well as the mobilization of hematopoietic stem cells and progenitor cells (HSPCs), which reside in the spleen. Monocytes produced in the bone marrow and the spleen further infiltrate tumors, where they differentiate into tumor-associated macrophages.

## 3. Liver Macrophages and Their Plasticity in Response to the Tumor Microenvironment

Liver macrophages (Mϕs) are a heterogeneous cell population that includes resident Kupffer cells and MoMϕs. On histological examination, Kupffer cells share common morphological characteristics with the monocyte-derived cells, MoMϕs and dendritic cells, showing morphological variability in size and shape. Kupffer cells are identified on light microscopy by their location, as they are localized adjacent to sinusoids and play a crucial role in homeostasis. MoMϕs are observed in inflammatory sites, orchestrating the immune response to tissue injury or pathogens. Mϕs play a major role in the pathogenesis of inflammatory disorders, such as NASH, promoting liver inflammation and fibrosis [42,43], as discussed in elsewhere [43,44].

The microenvironment of HCC, which consists of CAFs, hepatic stellate cells (HSCs), endothelial cells and immune cells, as well as extracellular matrix proteins [45], shapes Mϕs, altering their function. TAMs actively participate in tumor-related inflammation, setting the crosstalk between tumor and stromal cells [46,47]. TAMs are MoMϕs that are recruited into the TME, mostly by chemokine (C-C motif) ligand 2 (CCL2) and macrophage (M)-CSF, and eventually differentiate into mature macrophages. The plasticity of TAMs enables them to exert either anti- or pro-tumor activity, depending on the distinct micro-environmental signals originating from tumor cells, fibroblasts, stroma and immunocompetent cells [48]. TAMs show similar morphological features to their normal counterparts, despite their structural and functional diversity, which is controlled by the tumor microenvironment [49]. Similar to liver Mϕs, TAMs are functionally heterogeneous. However, in most human studies, TAMs are characterized based on the expression of polarization markers as classically activated (M1) and alternatively activated (M2). Classically activated M1 macrophages have pro-inflammatory activity and macrophages polarize to this direction in response to treatment with lipopolysaccharides (LPS) and interferon-gamma (IFN-$\gamma$). M1 macrophages produce proinflammatory cytokines, such as interleukin (IL)-12, and have the potential to stimulate effector T-cell proliferation and function. They also exhibit strong microbicidal and tumoricidal activity by the production of reactive oxygen species (ROS) and nitric oxide synthase (iNOS; NOS2) that promotes arginine metabolism into nitric oxide (NO) and citrulline [50,51]. Alternative activation by IL4, IL-10 and IL-13 in vitro results in the generation of macrophages with immunosuppressive properties that produce IL-10, transforming growth factor beta (TGF-β) and chemokine (C-C motif) ligand (CCL) family members, such as CCL17, CCL18, CCL22 and CCL24, and express high levels of PD-L1 [27,52]. M2 macrophages initiate the Th2 immune response, promoting angiogenesis, tissue remodeling and repair [53,54].

Histologically, the distinction between M1 and M2 macrophages is based on immunohistochemistry. CD68 is a common monocytic marker expressed by TAMs, of both M1 and M2 phenotypes, and dendritic cells. M1 macrophages are positive for several markers, such as iNOS, CD80 and CD86, whereas the M2 polarization markers are CD163, CD204 and CD206 [55,56]. CD163, in addition to being a marker for M2 macrophages, is also expressed by Kupffer cells. Moreover, a population of tolerogenic dendritic cells has been recently identified in peripheral blood, namely DC-10, which expresses CD163, releases IL-10 and induces type-1 T regulatory cells [57]. CD163 expression is upregulated by IL-10, and CD163-positive cells secrete IL-10.

The TME is the driving force behind macrophage polarization in HCC, leading to the generation of macrophages with immunosuppressive properties. For instance, the TME is characterized by an acidic pH, triggering regulatory macrophages and enhancing immune evasion [58]. Yang et al. demonstrated that Wnt/β-catenin signaling drives macrophage differentiation towards the M2 phenotype and was highly expressed in c-Myc-driven M2-polarized macrophages. Inhibition of Wnt protein secretion in HCC hindered hepatic tumor growth by regulating the tumor immune microenvironment in mice, whereas nuclear accumulation of β-catenin was observed in M2-like TAMs in human HCC biopsies [59]. Moreover, Chen et al. demonstrated that TLR4-elicited innate monocyte inflammation was necessary for IL21+ T follicular helper (Tfh)-like cell induction in HCC, and activation of STAT1 and STAT3 was critical for TFH-like cell polarization. Importantly, the TFH-like cells operated in IL21-IFNγ-dependent pathways to induce plasma cell differentiation and, thereby, create conditions for pro-tumorigenic M2 macrophage polarization and cancer progression [60]. Similarly, intratumoral macrophages were associated with increased intratumoral FoxP3+ regulatory T cells (Tregs) and poor prognosis in patients with HCC, while in vivo depletion of tissue macrophages decreased the frequency of intratumoral immunosuppressive FoxP3+ Tregs [61]. To underline the multitude and the complementarity of the mechanisms of local HCC immunosuppression, it was recently deduced that selectively increased intrahepatic Tregs can promote an immunosuppressive environment in NASH livers. Neutrophil extracellular traps link innate and adaptive immunity by promoting Treg differentiation via the metabolic reprogramming of naïve CD4+ T-cells. This mechanism may explain, at least partly, the relative resistance of NASH-related HCC to current first-line immunotherapies and could be targeted to prevent or treat liver cancer in patients with NASH etiology [62]. On the other hand, the extracellular matrix protein SPON2 and its integrin receptors α4β1 play significant roles in the recruitment of the M1-like TAM subtype in the HCC microenvironment, functioning as an opsonin for macrophage phagocytosis, resulting in anti-tumor immune responses [63]. Additionally, expression of MiR206 by Kupffer cells drives M1 polarization and the recruitment of CD8+ T cells through CCL2 production in mice with HCC [64]. In addition to T cells, Liu et al. reported that TAMs interact with B cells, since CXCR3+ B cells drive M2 polarization in HCC through IL-17 production [65]. The IL-6/STAT3 signaling pathway was shown to regulate M1/M2 macrophage polarization, as its inhibition, mediated by anti-IL6, reduced cell viability and drug resistance, suppressed cell invasion and migration and induced apoptosis of HCC cells co-cultured with M1- or M2-type macrophages, resulting in suppressed tumor formation and lung metastases [66]. In addition, TLR4 on macrophages promoted the growth of steatohepatitis-related HCC in mice, as the number of macrophages expressing Ly6C was increased and these cells were associated with inflammation and tumor progression, generating increased amounts of IL-6 and TNFα in response to LPS [67]. On the contrary, IL-12 inhibited HCC proliferation and invasiveness in vitro by the induction of M1-like polarization of macrophages through the downregulation of Stat-3 [68]. Furthermore, a study by Wang et al. demonstrated that TGF-β recruits M2 macrophages in HCC, which are in turn polarized by connective tissue growth factor (CTGF, CCN2), a protein expressed by mesenchymal-like HCC cells. TGF-β acts as a chemoattractant, recruiting monocytes from the peripheral blood, while CTGF acts as a transformant, polarizing monocytes to M2 macrophages, stimulating tumor growth. In turn, M2 macrophages secrete CCL18, promoting HCC cell migration [69]. Finally, the transition of the macrophage phenotype from antitumorigenic to protumorigenic, which has been proven to be mediated by c-Jun N-terminal phosphorylation in the liver microenvironment, occurs before overt tumorigenesis, resulting mostly in the production of CCL17 and CCL22, thus facilitating HCC growth [70,71].

Even though the aforementioned studies suggest that TAMs can be polarized between the two extremes of macrophage phenotypes, recent studies using single cell approaches demonstrated that TAMs in HCC are characterized by vast heterogeneity. A seminal study by Zhang et al. generated a large body of information regarding immune cell populations in HCC and ascites using single cell RNAseq of CD45+ cells [72]. Among others, six clusters of macrophages with distinct gene expression modules were identified [72]. Interestingly, they identified a cluster of macrophages that simultaneously expressed genes of both M1 and M2 polarization states [72]. Song et al. also engaged a similar experimental approach in HBV/HCV-related hepatocellular carcinoma and identified eight clusters of myeloid cells, showing that there is heterogeneity within macrophage populations, with a high number of macrophages sharing both M1 and M2 characteristics [73]. Among the different macrophage clusters, a cluster of CCL18-expressing macrophages with M2 features was identified that was associated with a worse clinical outcome [73]. In the same study, a population of $XCL1^+$ $CD8^+$ T cells was identified that was capable of recruiting dendritic cells, which resulted in an enhanced anti-tumor response, suggesting an interaction between T cells and myeloid cells [73]. In addition to the heterogenicity at the single cell level, it has been shown that immune cell infiltrates are distinct in intrahepatic metastatic lesions in multifocal HCC compared to multicentric occurrence, since more M2 macrophages and less T cells are observed in metastases [74] (Figure 2).

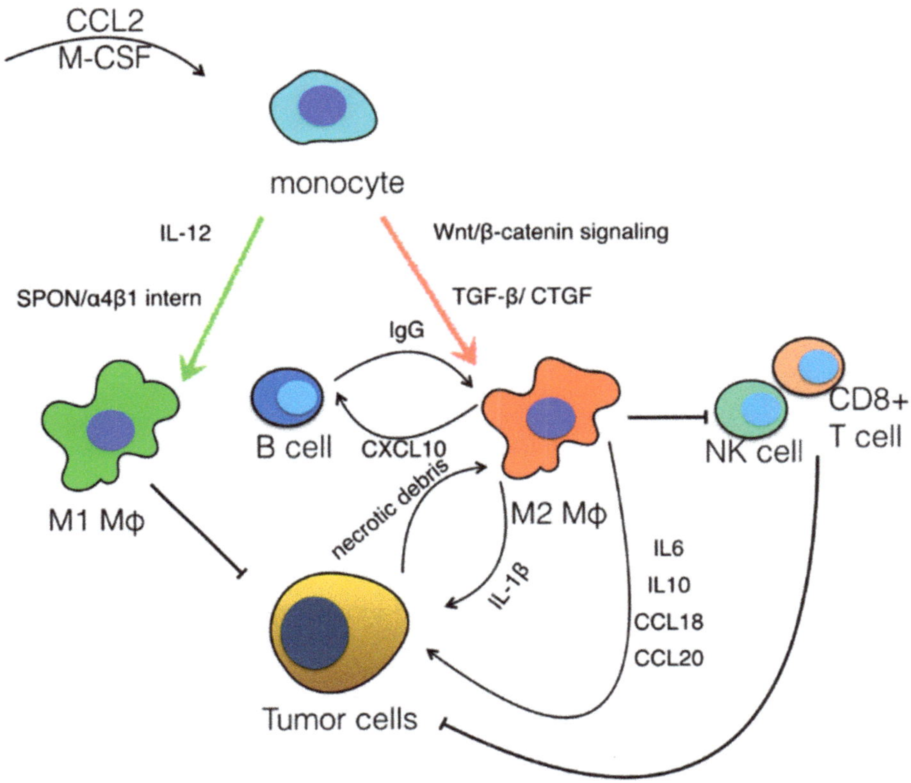


**Figure 2.** Polarization of tumor-associated macrophages. Macrophage colony-stimulating factor (M-CSF) and C–C motif chemokine ligand 2 (CCL2) drive the generation of monocytes in cancer. In response to IL-12 and SPON/α4β1 signaling, monocytes differentiate into macrophages with antitumor properties, which are classified as M1-like. On the other hand, Wnt/β-catenin signaling, the TGF-β/CTGF pathway, necrotic debris from tumor cells and immunoglobulins released by B cells facilitate the polarization of macrophages towards a M2-like phenotype with tumor promoting properties. M2-like macrophages drive tumor growth directly, through the release of IL-1β, IL-6, IL-10, CCL18 and CCL20, and indirectly, via the suppression of cytotoxic cell populations, including CD8+ T cells and NK cells.

## 4. The Role of Macrophages in HCC Pathogenesis

The involvement of macrophages in the pathogenesis and development of HCC is crucial. Zhang et al. found that M2 macrophages increased the proliferation, migration and invasion of HCC cells through a fatty acid oxidation (FAO)-dependent process. Specifically, IL-1β instigated the pro-migratory effect of M2 cells, and FAO was responsible for the upregulated secretion of IL-1β, which depended on ROS and NLRP3 inflammasome [75]. A study on the diethylnitrosamine (DEN)-induced model of carcinogenesis engaged in transcriptomic analysis and demonstrated that MoMϕs acquire a proinflammatory phenotype during carcinogenesis in this model, which is, however, distinct from the mixed pro-inflammatory and immunosuppressive phenotype of cells from the NASH-induced carcinogenesis model [76]. This observation implies that there are diverse mechanisms in the response of immune cells to liver carcinogenesis in different animal models. Schneider et al. also demonstrated that DEN-induced hepatocarcinogenesis triggers liver inflammation with an intrahepatic accumulation of macrophages and cytotoxic T cells. Interestingly, the increased macrophage accumulation in chemokine scavenger receptor D6-deficient mice did not have an impact on HCC progression [77]. TAMs have also been linked to HCC growth stimulation via STAT3 signaling, while IL-6 release by macrophages was demonstrated to enhance HCC proliferation and migration [78]. Furthermore, evidence has been provided that chemokine CCL2/chemokine receptor CCR2-dependent signaling mechanisms participate in the process of HCC development. The levels of CCL2 are increased in patients with HCC and have been associated with poor prognosis [79]. In murine liver cancer models, CCR2+ myeloid cells exhibited dual functions. Before cancer initiation, CCR2+ myeloid cells suppressed tumorigenesis by removing senescent hepatocytes. However, when HCC was established, tumor cells inhibited the differentiation of infiltrating CCR2+ immature myeloid cells, which in turn promoted tumor growth, via the inhibition of NK cells. In a model of NASH-dependent HCC, CCR2 depletion had no distinct effect on HCC tumorigenesis, suggesting that the effect of CCR2 in hepatocarcinogenesis is dependent on disease etiology [80]. Guo et al. also provided evidence that infiltrating M2-TAMs were markedly elevated in the HCC TME, producing IL-17, a pro-inflammatory cytokine, and were augmented upon oxaliplatin treatment [81]. IL-17A, secreted in concert from lymphatic endothelial cells, promotes tumorigenesis by upregulation of PD-L1 in hepatoma stem cells [82]. Regarding HCC metastasis, it was demonstrated that the exosome-mediated transfer of the functional protein CD11b/CD18 (integrin αMβ2) from TAMs to tumor cells may have the potency to boost the migratory potential of HCC cells [83].

In addition, inhibition of the CCL2/CCR2 axis resulted in the blockade of monocyte recruitment, M2 polarization and, as a result, inhibition of the CD8+ T-cell-mediated antitumor response [84]. Zhang et al. demonstrated that the hypoxia inducible factor (HIF)-1α /IL-1β feedback loop between tumor cells and TAMs in the hypoxic TME resulted in the epithelial–mesenchymal transition of cancer cells and metastasis in vitro. Specifically, they found that TAMs secreted increased amounts of IL-1β under moderate hypoxic conditions, due to the increased stability of HIF-1α, and that the necrotic debris of HCC cells increased IL-1β release by TAMs with an M2 phenotype, via TLR4/TRIF/NF-κB signaling [85]. Moreover, Zhao et al. observed a positive association between B7-H1+ monocyte/Mφ and IL-17-producing cell density in the peritumoral stroma of HCC patients and that the IL-17-exposed macrophages suppressed cytotoxic T-cell function through B7-H1/PD-1 interactions [86]. CD48/2B4 interactions mediated a high level of infiltration of peritumoral macrophages, which was correlated with the decreased activity of NK cells in HCC tissues [87]. Furthermore, IL-23 generation by liver CD14+ inflammatory macrophages in response to infected hepatocytes during chronic HBV was shown to alter macrophage function, favoring HCC growth [88] (Table 1).

**Table 1.** Summary of studies evaluating the role of TAMs in HCC pathogenesis.

| Study (Year) | Study Subjects | Primary Outcome | Secondary Outcome |
|---|---|---|---|
| Yang et al. (2018) [59] | Human/Animal | Wnt/β-catenin activation promotes M2 Mϕ polarization through c-Myc. | Nuclear accumulation of β-catenin is positively correlated with M2-like TAMs in human HCC biopsies. |
| Chen et al. (2016) [60] | Human | High level of infiltration of IL21+ TFH-like cells induces pro-tumorigenic M2b Mϕ polarization and HCC growth. | Fcγ receptor–TLR cross-talk is required for M2b Mϕ polarization and subsequent upregulation of the M2 markers IL10 and CCL1. |
| Zhang et al. (2016) [89] | Human | CD169+ Mϕs could suppress tumor progression by enhancing CD8+ T-cell activity in human HCC. | Tumor-induced autocrine TGF-β downregulates CD169 expression by Mϕs. |
| Zhang et al. (2018) [63] | Human/Animal | M1 Mϕs accumulate in the SPON2-abundant regions of HCC, exhibiting antitumor immune responses through distinct integrin-Rho GTPase-Hippo pathways. | SPON2 interactions with integrin α4β1 receptors activate RhoA and Rac1, resulting in F-Actin accumulation that promotes M1 Mϕ infiltration and migration. |

**Table 1.** *Cont.*

| Study (Year) | Study Subjects | Primary Outcome | Secondary Outcome |
|---|---|---|---|
| Zhao et al. (2012) [66] | Human | IL-6/STAT3 signaling pathway regulates Mϕ polarization in HCC, and its inhibition suppresses tumor formation and metastases. | The TME induces the formation of suppressive MΦs, leading to early T cell activation and subsequent MΦ IDO expression in HCC. |
| Zhang et al. (2018) [75] | Human/Animal | M2 Mϕs under FAO-mediated upregulated secretion of IL-1β enhance the proliferation, migration and invasion of HCC cells. | IL-1β induction is reactive oxygen species-dependent and NLRP3-dependent. |
| Schneider et al. (2012) [77] | Animal | Chemically induced hepatocarcinogenesis triggers an intrahepatic accumulation of macrophages and cytotoxic T cells. | Activation of adaptive immunity-related pathways affect survival of patients with HCC. |
| Mano et al. (2013) [78] | Human/Animal | TAMs correlate with pSTAT3 expression in HCC, expressing high levels of IL-6. | IL-6 stimulates cell proliferation and the migration of human HCC cell lines. |
| Guo et al. (2017) [81] | Human | The expression of CD68, CD163 and CD206, the M2-TAM markers, is significantly higher in HCC tissues than in normal hepatic tissues. | IL-17 expression by M2-TAMs is augmented by oxaliplatin treatment and reduces oxaliplatin-induced apoptosis in HCC cells by activating CMA. |
| Bartneck et al. (2019) [90] | Animal | Pro-inflammatory CCR2+ TAMs accumulate at the highly vascularized HCC border, whereas CD163+ immune-suppressive TAMs accrue in the HCC center. | CCR2+ M2 Mϕs express CCL6, which is involved in immune cell recruitment, and NF-κB, which is associated with many inflammatory processes. |
| Zhang et al. (2018) [85] | Human/Animal | M2 Mϕs enhance IL-1β secretion in HCC under moderate hypoxic conditions via an HIF-1α/IL-1β signaling loop, leading to increased metastasis and the poor prognosis of HCC patients. | TLR4/TRIF/NF-κB signaling mediates cell necrotic debris–induced IL-1β production by macrophages, inducing an epithelial–mesenchymal transition in HCC cells. |
| Zang et al. (2018) [88] | Human/Animal | Liver inflammatory macrophages of HBV-related HCC patients produce high amounts of IL-23, which in turn augment macrophage-induced angiogenesis in the JAK-STAT3 pathway. | Blocking IL-23 cytokine activity decreased liver cancer development in the murine model. |
| Wang et al. (2017) [69] | Human | M2 Mϕs promote HCC progression by secreting cytokine factor CCL18. | CTGF is the key factor secreted by mesenchymal-like HCC cells that leads to the polarization of Mϕs, promoting HCC progression. |

Mϕ: macrophage; TAM: tumor-associated macrophage; HCC: hepatocellular carcinoma; TME: tumor microenvironment; IL: interleukin; TFH: follicular helper T; TLR: toll-like receptor; CCL: CC chemokine ligand; CCR: CC chemokine receptor; SPON2: spondin 2; STAT: signal transducer and activator of transcription; IDO: indoleamine 2,3 dioxygenase; FAO: fatty acid oxidation; NLRP3: NOD-, LRR- and pyrin domain-containing protein 3; CMA: chaperone-mediated autophagy; NF-κB: nuclear factor kappa-light-chain-enhancer of activated B cells; HIF: hypoxia inducible factor; TRIF: toll/interleukin-1 receptor domain-containing adaptor protein inducing interferon beta; HBV: hepatitis B virus; JAK: janus kinase; CTGF: connective tissue growth factor.

## 5. Macrophages in HCC Prognosis

The prognostic role of the proportion of M1 and M2 macrophages, as well as their ratio, have been reported in several tumors [91,92], including HCC [93]. M2 macrophages are implicated in the "exhausted immune response" subclass of HCC and are correlated with adverse prognosis [93,94]. The method for their quantification differs and is not comparable between studies. Density estimation, as reflected by the number/mm$^2$, seems to be the most reliable method. Histologic distribution of TAMs and their subtypes, in the center of the tumor or the invasive front, has been proposed to have an independent prognostic value in several solid tumors [95,96]. The histological evaluation of the invasive front is preferable to be performed in surgical specimens. Further research related to macrophage phenotypes and their ratios, in conjunction with their spatial location in the tumor specimens, can facilitate a better understanding of their biological behavior in HCC.

Recent evidence suggests that miR-148b deficiency promotes HCC growth and metastasis through colony-stimulating factor-1 (M-CSF)/CSF1 receptor (CSF1R)-mediated TAM infiltration, while HCC patients with decreased miR-148b levels and increased TAM infiltration were correlated with worse prognosis [97]. Moreover, Chen et al. reported that monocytes engage in glycolysis at the peritumoral region of human HCC, inducing PD-L1 expression and attenuating cytotoxic T lymphocyte responses in cancer. Tumor-derived soluble factors upregulated PFKFB3 expression in TAMs, which in turn mediated the increased expression of PD-L1 by the activation of the NF-kB signaling pathway. Interestingly, the degree of CD68 + PFKFB3 + PD-L1 + monocyte-macrophage infiltration in peritumoral tissues was negatively associated with the overall survival of HCC patients and could serve as an independent prognostic factor for patients with HCC [98]. In addition, Li et al. demonstrated that the downregulation of SIRT4 was correlated with increased macrophage infiltration and M2-like TAMs in HCC peritumoral tissues and, consequently, with poor survival of HCC patients. SIRT4 expression was decreased in macrophages in HCC, driving M2 polarization in a FAO-PPARδ-STAT3-dependent signaling pathway, while silencing SIRT4 increased IL-6 production in TAMs. Moreover, SIRT4 silencing also resulted in M1 macrophage apoptosis due to enhanced IL-10 production in HCC peritumoral tissues [99]. Along the same line, high-mobility group protein box1 (HMGB1) expression, which is linked to increased secretion of IL-1β, IFN-γ and TNF-α, was associated with peritumoral TAM infiltration and poor prognosis in patients with HCC. High peritumoral HMGB1 expression and TAM numbers were positively correlated with tumor size and BCLC stage and acted as independent prognostic factors for the overall survival (OS) and recurrence free survival (RFS) in patients with HCC [100]. Furthermore, Zhao et al. provided evidence that the expression of macrophage migration inhibitory factor (MIF) in tumors was positively correlated with plasma MIF levels, which had a higher value for the diagnosis of HCC compared to serum a-fetoprotein (AFP). In fact, plasma MIF levels demonstrated a significant correlation with the OS and disease-free survival (DFS) of HCC patients, even in those with normal serum AFP levels and tumor-node-metastasis (TNM) stage I. In addition, the plasma MIF levels were identified as an independent factor for OS and DFS and decreased significantly within 30 days after HCC resection [101]. Another study further reported that M-CSF density and the CD163 and CD31 indices in peritumoral tissues were predictable factors for time to recurrence, DFS and OS in patients with HCC, while M-CSF was involved in the progression of hepatocellular carcinoma after curative resection [102]. Moreover, hepatocellular tumors with increased intratumoral CD204, as well as monocarboxylate transporter-4 (MCT4)-positive macrophages and MCT4-postive expressing HCC cells, were associated with an unfavorable patient outcome [103].

Another study by Zhu et al. revealed that CD206 was highly expressed in the HCC tissues compared to its peri-carcinoma tissue levels, while GdCl3 treatment suppressed the malignant potential of HCC in vitro and in vivo, mainly by downregulating the expression of CD206 in M2 macrophages, indicating the potential significance of CD206 as a biomarker for HCC prognosis [104]. Similarly, high expression of peritumoral M-CSF and the density of macrophages, which correlated with a large tumor size, presence of intrahepatic metastasis and a high TNM stage, were associated with HCC progression, disease recurrence and poor survival after curative hepatectomy [105], while the combination of tumor-derived osteopontin (OPN) and peritumoral infiltrating macrophages was associated with a high incidence of early recurrence and poor survival for early-stage HCC, after curative resection [106]. A study by Zhou et al. demonstrated that Yes-Associated Protein (YAP) activation was critical for the recruitment of TAMs towards HCC cells, as IL-6 secreted by YAP-activated HCC cells induced TAM recruitment. Together with their findings that the expression levels of IL-6 in human HCC tumors were highly associated with the prognosis of HCC patients, they highlighted the possibility of improving HCC treatment by targeting YAP-IL-6-mediated TAM recruitment [107]. In addition, evidence has been provided that macrophages contribute to the decreased expression of E-cadherin in HCC via the NF–κB/Slug pathway, leading to increased tumor invasiveness and metastasis [108] (Table 2).

**Table 2.** Summary of studies evaluating the role of TAMs in HCC prognosis.

| Study (Year) | Study Subjects | Primary Outcome | Secondary Outcome |
|---|---|---|---|
| Ke et al. (2019) [97] | Human/Animal | MiR-148b deficiency promotes HCC growth and metastasis through CSF1/CSF1R-mediated TAM infiltration. | Decreased miR-148b levels and increased TAM infiltration were correlated with worse prognoses for HCC patients. |
| Chen et al. (2019) [98] | Human/Animal | The levels of PFKFB3 + CD68+ cell infiltration in peritumoral tissues were negatively correlated with the overall survival and could serve as an independent prognostic factor for survival in patients with HCC. | Tumor-derived soluble factors upregulated PFKFB3 in TAMs, which in turn mediated the increased expression of PD-L1 by the activation of the NF-kB signaling pathway. |
| Li et al. (2019) [99] | Human/Animal | SIRT4 is downregulated in CD68+ M2-like TAMs and correlates with the poor survival of HCC patients. | Downregulation of SIRT4 in TAMs modulates the alternative activation of macrophages and promotes HCC development via the FAO-PPARδ-STAT3 axis. |
| Zhang et al. (2016) [100] | Human | High peritumoral HMGB1 expression and TAM count, which correlated positively with tumor size and the BCLC stage of HCC, are independent prognostic factors for OS and RFS. | The degree of TAM infiltration is higher in peritumoral tissues with high HMGB1 expression than in peritumoral tissues with low HMGB1 expression. |
| Kono et al. (2016) [102] | Human | M-CSF density, CD163 index and CD31 index in peritumoral tissues are independent prognostic factors HCC patients. | M-CSF, M2 Mφs and angiogenesis in the peritumoral liver tissue are correlated with DFS after surgery. |
| Ohno et al. (2014) [103] | Human/Animal | Increased intratumoral infiltration of CD204-positive or MCT4-positive macrophages suggested shorter OS in patients with HCC. | MCT4+ HCC cases correlated with higher intratumoral M2-Mφ and higher intratumoral MCT4-positive Mφ. |
| Zhu et al. (2008) [105] | Human/Animal | High peritumoral M-CSF and Mφs are associated with HCC progression, disease recurrence and poor survival after hepatectomy. | High peritumoral M-CSF and Mφ density correlate with large tumor size, presence of intrahepatic metastasis and advanced stage. |
| Zhu et al. (2014) [106] | Human | OPN, combined with PTMs, is an independent prognostic factor for both OS and TTR of early-stage HCC after curative resection. | PTM expression is closely associated with tumor recurrence and survival in HCCs with higher OPN levels, but is not significant in those with lower OPN expression. |

HCC: hepatocellular carcinoma; TAM: tumor-associated macrophage; Mφ: macrophage; CSF: colony stimulating factor; CSF1R: colony stimulating factor-1 receptor; NF-Kb: nuclear factor kappa-light-chain-enhancer of activated B cells; PD-L1: programmed death-ligand 1; SIRT4: sirtuin 4; FAO: fatty acid oxidation; STAT3: signal transducer and activator of transcription 3; PPARδ: peroxisome proliferator-activated receptor delta; HMGB1: high mobility group box 1; BCLC: Barcelona clinic liver cancer; OS: overall survival; RFS: recurrence-free survival; M-CSF: macrophage colony-stimulating factor; MCT4: monocarboxylate transporter-4; OPN: osteopontin; PTM: peritumoral macrophage; TTR: time to response.

## 6. The Potential Role of Macrophages in HCC Treatment

The implication of macrophages in various HCC treatment modalities has been at the forefront of clinical investigation, as various preclinical studies in animal models suggest that macrophages can play a pivotal role in HCC therapy. For instance, macrophage depletion by clodrolip or zoledronic acid, in combination with sorafenib, significantly inhibited HCC progression, tumor angiogenesis and lung metastasis in mice [109]. In addition, sorafenib administration at a subpharmacologic dose, augmented the antitumor effects of mouse chimeric antigen receptor CAR-T cells, partly by promoting IL12 secretion from

TAMs [110]. Along the same line, there is evidence that sorafenib triggers the proinflammatory activity of TAMs, reverts their alternative polarization, enhancing IL12 secretion, and, as a result, induces antitumor NK cell responses in a cytokine- and NF-kB-dependent fashion [111]. This ability of sorafenib to partially inhibit M2-cell activation in vivo was shown by a study by Sprinzl et al. [112]. Another study by Yao et al. indicated that a natural CCR2 antagonist potentiated TAM-mediated tumor immunosuppression and enhanced the therapeutic effect of sorafenib, indicating that the combination of an immunomodulator with a chemotherapeutic drug could be a therapeutic approach for HCC [113]. Macrophage modulation could also potentiate the anti-cancer activity of sorafenib. In addition, it has been shown that sorafenib inhibits the macrophage-mediated epithelial–mesenchymal transition in HCC via the HGF-Met signaling pathway in vitro, while sorafenib therapy reduced plasma HGF and alpha-fetoprotein concentrations in patients with HCC [114]. Finally, it inhibited miR-101 expression and enhanced DUSP1 expression and downregulated the release of TGF-β and the expression of CD206 in M2 cells, suppressing the macrophage-mediated growth of HCC [115].

A study by Yang et al. identified that 17β-estradiol (E2) could suppress HCC growth via regulation of macrophage polarization, as E2 re-administration reduced tumor growth in orthotopic and ectopic mouse HCC models, functioning as an inhibitor of macrophage alternative activation and tumor progression, by keeping estrogen receptor-β (ERβ) away from interacting with ATP5J and hindering the JAK1-STAT6 signaling pathway [116]. Moreover, delivery of recombinant adenovirus vector (rAd) expressing monocyte chemoattractant protein-1 (MCP-1) was demonstrated to potentiate the antitumor effects of suicide gene therapy against HCC by M1 macrophage activation, suggesting its potential use as a method of cancer gene therapy against HCC progression and recurrence [117]. Guerra et al. also demonstrated that, in response to HCC cells, hydrogel-embedded M1 macrophages upregulated nitrite and TNF-α, activating caspase-3-induced apoptosis in the tumor cells, leading to tumor regression in vivo [118]. It is also worth mentioning that small interfering RNA (siRNA)-mediated knockdown of MIF suppressed cyclin D1 expression and HCC cell proliferation, inducing tumor-cell apoptosis [119], while antibody mediated therapy targeting CD47 inhibited HCC progression, promoting the migration of macrophages into the tumor mass and the subsequent phagocytosis of HCC cells [120]. Tan et al. provided evidence that IRE1α mediated the inhibition of TAM activation by genipin in HCC, suppressing its growth, while the reduced association of IRE1α-TRAF2-IKK might have been responsible for a genipin-regulated inactivation of NF-κB [121]. Furthermore, co-administration of glycyrrhizin and doxorubicin by alginate nanogel particles was demonstrated to diminish the activation of macrophages through the regulation of the apoptosis pathway, via altering the Bax/Bcl-2 ratio and caspase-3 activity, enhancing the therapeutic efficacy for HCC [122]. Moreover, nanoliposome C6-ceramide (LipC6) was demonstrated to enhance the anti-tumor immune response and hinder HCC growth in mice, reducing the number of TAMs and their ability to suppress the anti-tumor immune response, allowing LipC6, a potential chemotherapeutic agent, to increase the efficacy of immune therapy in patients with HCC [123]. An additional in vivo study provided evidence that the strategy of low doses and multiple treatments of nsPEF was superior to a high dose of a single treatment, as macrophage infiltration was markedly elevated in tumors that were treated by multiple low dose nsPEFs [124]. Finally, nanosecond pulsed electric field (nsPEF), a technology targeting tumor cells with a non-thermal high-voltage electric field using ultra-short pulses, increased HCC cell phagocytosis by human macrophage cells (THP1) in vitro [125] (Table 3).

**Table 3.** Summary of studies evaluating the role of TAMs in HCC therapy.

| Study (Year) | Study Subjects | Outcome |
|---|---|---|
| Zhang et al. (2010) [109] | Animal | Depletion of macrophages by clodrolip or zoledronic acid, in combination with sorafenib, significantly inhibited HCC progression, angiogenesis and lung metastasis compared with the use of sorafenib alone. |
| Wu et al. (2019) [110] | Animal | Sorafenib, at a subpharmacologic level, augments the antitumor effects of mCAR-T cells, by promoting IL12 secretion in TAMs. |
| Sprinzl et al. (2013) [111] | Animal | Sorafenib triggers the proinflammatory activity of TAMs and subsequently induces antitumor NK cell responses in a cytokine- and NF-κB-dependent fashion. |
| Yao et al. (2017) [113] | Animal | The natural CCR2 antagonist 747 elevates the number of CD8+ T cells in HCC by blocking TAM-mediated immunosuppression and inhibiting HCC progression in a CD8+ T-cell-dependent manner. |
| Yang et al. (2012) [116] | Animal | E2 suppresses macrophage alternative activation and, as a result, HCC progression, by keeping ERβ away from interacting with ATP5J, thus inhibiting the JAK1-STAT6 signaling pathway. |
| Tsuchiyama et al. (2008) [117] | Animal | Recombinant adenovirus vector expressing MCP-1 enhances the antitumor effects of suicide gene therapy against HCC by M1 macrophage activation. |
| Guerra et al. (2017) [118] | Animal | Hydrogel-embedded M1 macrophages upregulate nitrite and TNF-α, activating caspase-3-induced apoptosis and HCC regression. |
| Xiao et al. (2015) [120] | Animal | Macrophage phagocytosis of HCC cells is increased after treatment with CD47 antibodies that block CD47 binding to SIRPα. |
| Tan et al. (2016) [121] | Animal | IRE1α inhibition by genipin on TAMs reduces XBP-1 splicing and NF-κB activation, suppressing HCC proliferation. |
| Wang et al. (2019) [122] | Animal | Co-delivery of glycyrrhizin and doxorubicin attenuates the activation of macrophages and their phagocytic activity, enhancing the therapeutic efficacy for HCC. |
| Sprinzl et al. (2015) [112] | Animal | Sorafenib lowers mCD163 and IGF-1 release by M2 macrophages, decelerating M2-macrophage-driven HepG2 proliferation. |
| Deng et al. (2016) [114] | Human/Animal | Sorafenib abolished polarized-macrophage-induced EMT and migration of HCC cells in vitro and also attenuated HGF secretion in polarized macrophages, decreasing plasma HGF in patients with HCC. |
| Wei et al. (2015) [115] | Animal | Sorafenib inhibited miR-101 expression, enhanced DUSP1 expression and lowered TGF-β and CD206 release in M2 cells, slowing macrophage-driven HCC. |
| Li et al. (2018) [123] | Animal | In mice with HCC, injection of LipC6 reduces the number of TAMs, their production of ROS and their ability to suppress the anti-tumor immune response. |
| Yin et al. (2014) [125] | Animal | nsPEFs enhance HCC cell phagocytosis by human macrophage cell (THP1) in vitro. |
| Chen et al.(2014) [124] | Animal | In vivo, low doses and multiple treatments of nsPEF significantly elevate macrophage infiltration in HCC tumors, contributing to tumor ablation. |

TAM: tumor-associated macrophage; HCC: hepatocellular carcinoma; TME: tumor microenvironment; IL: interleukin; mCAR: mouse chimeric antigen receptor; NK: natural killer; NF-kB: nuclear factor kappa-light-chain-enhancer of activated B cells; CCR2: C-C chemokine receptor type 2; E2: estradiol; ERβ: estrogen receptor beta; ATP5J: ATP synthase-coupling factor 6; JAK1: janus kinase 1; STAT6: signal transducer and activator of transcription 6; MCP: monocyte chemoattractant protein; TNF-α: tumor necrosis factor alpha; SIRPα: signal regulatory protein alpha; IRE1α: inositol-requiring endoribonuclease 1α; XBP-1: x-box-binding protein 1; IGF-1: insulin-like growth factor-1; EMT: epithelial–mesenchymal transition; HGF: hepatocyte growth factor; DUSP1: dual specificity phosphatase 1; TGF-β: transforming growth factor beta; LipC6: nanoliposome-loaded C6- ceramide; ROS: reactive oxygen species; nsPEF: nanosecond pulsed electric field.

Recent advances in tissue engineering research enabled the construction of three-dimensional (3D) in vitro tissue models, in order to recapitulate the TME without engaging in vivo animal models. In these 3D culture systems, different types of cancer cells and cells that form the TME, including CAFs and TAMs, are combined providing valuable tools for the discovery of new drugs [126,127]. This type of experimental approach has already been used in a model of lung carcinoma to test whether treatment can alter TAM density and spatial distribution and it revealed that the treatment had an important effect on the latter [128]. Regarding HCC, hydrogels loaded with M1-polarized macrophages had a tumor suppressing potential both in vitro and in vivo [118].

## 7. Critical Analysis of Data and Future Perspectives

The landscape and dynamics of macrophages have been studied alongside other cell populations in human HCC multiple tissue compartments using single cell-RNA sequencing analysis [72]. It was identified that the enrichment of TAM gene signatures was significantly associated with a survival disadvantage in HCC, rendering this type of tumor-infiltrating TAMs potential cellular candidates for therapeutic targeting in the TME. Importantly, two genes, SLC40A1, encoding ferroportin, and GPNMB, encoding type I transmembrane glycoprotein, were highly expressed as potential markers in these TAM-like cells. Furthermore, in a second human study of an early HCC relapse ecosystem, a different innate-like $CD8^+$ T cell population was described by single-cell profiling in recurrent tumors. These T cells were overexpressing the CD161 surface marker and displaying an innate-like low cytotoxic state, with low clonal expansion, unlike the classical $CD8^+$ T cell exhausted state observed in primary HCC. The selective relative enrichment of these cells in the TME was associated with a worse prognosis in patients [129]. These unique aspects of altered immune response associated with HCC relapse relative to the primary tumor underline the HCC immune micro-ecosystem complexity of heterogeneous spatiotemporal interactions between and within cell types, which may guide the development of rational precision oncology immune therapies, benefiting a wide range of patients.

In accordance with the above, the HCC microenvironment in human patients and mice is characterized by functionally distinct macrophage populations. There are four possible main interventions for TAM-based antitumor therapy: inhibition of macrophage recruitment, induction of TAM death or apoptosis, enhancement of M1 antitumoral activity of TAMs and, last but not least, inhibition of the functional axes of M2 tumor-promoting activity of TAMs. In addition, functional subsets of TAMs were analyzed in human HCC samples and, in a combined fibrosis–HCC mouse model, demonstrated that human CCR2+ TAMs accumulated at the highly vascularized HCC border and expressed the inflammatory marker S100A9, whereas a second subset of CD163+ immune-suppressive TAMs accumulated in the HCC epicenter. Inhibition of CCR2+ TAM infiltration using a CCL2 antagonist in the fibrosis–HCC model significantly reduced pathogenic angiogenesis alongside tumor growth [90]. Moreover, the dual CCR2/CCR5 inhibitor cenicriviroc is currently under phase 3 clinical trial evaluation in patients with NASH and advanced fibrosis, representing a high-risk group for liver cancer [130]. However, the exact role of CCR2 and CCR5 in macrophage function in the liver is rather obscure. Recently, it was demonstrated that both CCR2 and CCR5 deficiency/inhibition led to reduced fibrosis, and sole CCR5 deficiency increased steatosis and the incidence of HCC in the model of NEMO LPC-KO mice. While CCR2 controlled the recruitment of monocytes to injured livers, CCR5+ macrophages limited liver injury in NEMOLPC-KO mice (CCR5-dependent differential function), thereby reducing steatosis and hepatocarcinogenesis [131]. In the hypoxic environment of HCC, HIF-1a enhanced the expression of triggering receptor expressed on myeloid cells-1 (TREM-1) in TAMs, leading to immunosuppression through the impairment of the cytotoxic functions of CD8+ T cells. Mechanistically, TREM-1+ TAMs increased the expression of CCL20 via the extracellular signal-regulated kinase/NF-κβ pathway in response to hypoxia and tumor metabolites leading to CCR6+ Foxp3+ Treg accumulation. Inhibition of the TREM-1 pathway could hinder tumor progression, reduce CCR6+ Foxp3+ Treg recruitment and improve the therapeutic efficacy of PD-L1 blockade [132].

It has been also indicated that the function of CCR2+ myeloid cells depends on the developmental stage of liver tumors. Precancerous senescent hepatocytes produce CCL2, which attracts macrophages, eliminating precancerous lesions (antitumoral effect), while established HCCs can also attract monocytic macrophages, which can, in turn, block the antitumor activity of NK cells (tumor-promoting effect) [79]. It has also been shown that selective blocking of CCR5 induces antitumoral macrophage polarization, and anti-CCR5 therapy was reported to be efficient in treating metastases [133]. Therefore, a critical review of the aforementioned data indicates that not all patients with HCC might eventually benefit from a selective CCR2- or CCR5-directed axis-inhibiting tumor therapy, while combined CCR2 and CCR5 inhibition is only beneficial for certain subgroups of patients with HCC. Due to the complicated nature of myeloid inflammation, multiple target inhibition might be necessary in order to overcome myeloid-mediated immune suppression. In this context, it was found that GM-CSF- and TNFa-producing CD206+ macrophages accumulated in human fibrotic liver. GM-CSF potentiated monocytes to CD206+ macrophage conversion, while anti-GM-CSF therapy suppressed liver fibrosis and CD206+ macrophage accumulation [134]. Furthermore, it was identified that tumor-derived GM-CSF was the primary regulator of myeloid cell ARG1 expression and local immune suppression. STAT3, p38 mitogen-activated protein kinases and acid signaling through cAMP were required to activate myeloid cell ARG1 expression in a STAT6-independent manner. A blockade of the tumor-derived GM-CSF enhanced the efficacy of tumor-specific adaptive T-cell therapy and immune checkpoint blockade [135]. Taken together, it seems that either monocyte or tumor cell–derived GM-CSF significantly contributed to the development of the immunosuppressive TME by regulating myeloid cell ARG1 expression and could serve as a target in order to enhance the efficacy of cancer immunotherapy.

There are currently several clinical trials investigating the use of monoclonal antibodies to inhibit GM-CSF or GM-CSFR in patients with various diseases. In a completed phase IIb study, a 24-week treatment with mavrilimumab, a human monoclonal antibody targeting the GM-CSFR a-chain, significantly reduced rheumatoid arthritis disease activity compared to placebo [136]. Moreover, GSK3196165, a human monoclonal antibody inhibiting GM-CSF, has also shown evidence of rapid favorable clinical responses in a phase Ib/IIa trial of patients with moderate RA [137]. In addition, it was demonstrated that GM-CSF neutralization with lenzilumab results in the reduction of neuro-inflammation and cytokine release syndrome in a primary acute lymphoblastic leukemia patient-derived xenograft model following chimeric antigen receptor T-cell therapy [138]. Furthermore, a phase III trial is underway, investigating the potential use of lenzilumab to improve the likelihood of ventilator-free survival beyond standard supportive care, in hospitalized patients with severe SARS-CoV-2 [139]. Along the same line, recent evidence from studies of human and transplant mouse melanomas implicate CSF1 induction as a CD8+ T-cell–dependent adaptive resistance mechanism and demonstrate that simultaneous CSF1R targeting might be beneficial in melanomas refractory to immune checkpoint blockade and potentially, in other T-cell–based therapies [140]. Future findings from those ongoing trials and studies could provide insight into the potential use of GM-CSF-targeted therapies for the treatment of patients with HCC. Finally, subsets of tumor-associated innate immune cells, macrophages and neutrophils in particular, suppress the cytotoxic activity of innate and adaptive immune cells and interact with tumor cells to promote tumor growth and metastasis, suggesting that selectively targeting these sub-populations of TAMs and TANs holds therapeutic promise in treating metastatic disease [141].

## 8. Conclusions

Monocytes are highly adaptive cells that are influenced by the cytokine-chemokine milieu and are subsequently transformed by signals encountered upon entry into a tissue niche. Due to myeloid cell complexity and diversity, we are still far from understanding the complete set of internal and external signals that are sufficient to establish any particular monocyte-macrophage phenotype in the TME. Given the pro-metastatic role of monocytes in HCC, these shifts seem to have functional outcomes influencing disease state, rather than being simply epiphenomenal markers of the tumor and systemic environment.

Broadening our knowledge on how different signaling pathways regarding the recruitment and differentiation of monocytes interact with lineage-determining transcription factors, and how these factors interact with the overlaid differentiation factors within blood and tissue in cancer, might potentially enable us to intervene and define cell fate or phenotype for therapeutic purposes. Thus, insight into intra- and inter-cellular crosstalk may better showcase the role of monocytes and macrophages in tumor immunity. Some of these cells hinder tumor growth and are essential in effective tumor therapies, particularly immunotherapy. Such a delicate balance argues against systemic elimination of cells using a generic cell-type marker. It should also be emphasized that myeloid cells crosstalk with each other, and in many instances when one cell-type is removed (TAMs), there may be a subsequent increase of another (TANs). These complex cellular interactions within the TME, as well as those between the acquired and innate immune systems, although incompletely understood, should be at the forefront of future investigation, as immunotherapy is undoubtedly promising. A better understanding of the mechanisms and axes controlling tumor context-specific monocytes and tissue-resident macrophage phenotypes, is essential for the rational development of methods that can favorably alter their functions in HCC.

**Author Contributions:** Conceptualization, G.G. and I.M.; investigation, K.A. and I.M.; resources, K.A. and T.K.; data curation, K.A. and I.M.; writing—original draft preparation, K.A., T.K. and I.M.; writing—review and editing, K.A., I.M., T.K. and G.G.; visualization, G.G.; supervision, I.M. and G.G.; project administration, I.M. and G.G. All authors have read and agreed to the published version of the manuscript.

**Funding:** This research received no external funding.

**Conflicts of Interest:** The authors declare no conflict of interest.

## References

1. Bray, F.; Ferlay, J.; Soerjomataram, I.; Siegel, R.L.; Torre, L.A.; Jemal, A. Global cancer statistics 2018: GLOBOCAN estimates of incidence and mortality worldwide for 36 cancers in 185 countries. *CA Cancer J. Clin.* **2018**, *68*, 394–424. [CrossRef] [PubMed]
2. Global Burden of Disease Liver Cancer Collaboration; Akinyemiju, T.; Abera, S.F.; Ahmed, M.B.; Alam, N.; Alemayohu, M.A.; Allen, C.; Alraddadi, R.; Alvisguzman, N.; Amoako, Y.; et al. The Burden of Primary Liver Cancer and Underlying Etiologies From 1990 to 2015 at the Global, Regional, and National Level: Results From the Global Burden of Disease Study 2015. *JAMA Oncol.* **2017**, *3*, 1683–1691. [PubMed]
3. Fattovich, G.; Stroffolini, T.; Zagni, I.; Donato, F. Hepatocellular carcinoma in cirrhosis: Incidence and risk factors. *Gastroenterology* **2004**, *127*, S35–S50. [CrossRef]
4. Kew, M. Hepatocellular carcinoma: Epidemiology and risk factors. *J. Hepatocell. Carcinoma* **2014**, *1*, 115–125. [CrossRef]
5. Bruix, J.; Takayama, T.; Mazzaferro, V.; Chau, G.-Y.; Yang, J.; Kudo, M.; Cai, J.; Poon, R.T.; Han, K.-H.; Tak, W.Y.; et al. Adjuvant sorafenib for hepatocellular carcinoma after resection or ablation (STORM): A phase 3, randomised, double-blind, placebo-controlled trial. *Lancet Oncol.* **2015**, *16*, 1344–1354. [CrossRef]
6. Abou-Alfa, G.K.; Meyer, T.; Cheng, A.-L.; El-Khoueiry, A.B.; Rimassa, L.; Ryoo, B.-Y.; Cicin, I.; Merle, P.; Chen, Y.; Park, J.-W.; et al. Cabozantinib in Patients with Advanced and Progressing Hepatocellular Carcinoma. *N. Engl. J. Med.* **2018**, *379*, 54–63. [CrossRef]
7. Bruix, J.; Qin, S.; Merle, P.; Granito, A.; Huang, Y.-H.; Bodoky, G.; Pracht, M.; Yokosuka, O.; Rosmorduc, O.; Breder, V.; et al. Regorafenib for patients with hepatocellular carcinoma who progressed on sorafenib treatment (RESORCE): A randomised, double-blind, placebo-controlled, phase 3 trial. *Lancet* **2017**, *389*, 56–66. [CrossRef]
8. Kudo, M.; Finn, R.S.; Qin, S.; Han, K.-H.; Ikeda, K.; Piscaglia, F.; Baron, A.; Park, J.-W.; Han, G.; Jassem, J.; et al. Lenvatinib versus sorafenib in first-line treatment of patients with unresectable hepatocellular carcinoma: A randomised phase 3 non-inferiority trial. *Lancet* **2018**, *391*, 1163–1173. [CrossRef]
9. Llovet, J.M.; Ricci, S.; Mazzaferro, V.; Hilgard, P.; Gane, E.; Blanc, J.-F.; De Oliveira, A.C.; Santoro, A.; Raoul, J.-L.; Forner, A.; et al. Sorafenib in Advanced Hepatocellular Carcinoma. *N. Engl. J. Med.* **2008**, *359*, 378–390. [CrossRef] [PubMed]
10. Parikh, N.D.; Singal, A.G.; Hutton, D.W. Cost effectiveness of regorafenib as second-line therapy for patients with advanced hepatocellular carcinoma. *Cancer* **2017**, *123*, 3725–3731. [CrossRef]
11. Finn, R.S.; Qin, S.; Ikeda, M.; Galle, P.R.; Ducreux, M.; Kim, T.-Y.; Kudo, M.; Breder, V.; Merle, P.; Kaseb, A.O.; et al. Atezolizumab plus Bevacizumab in Unresectable Hepatocellular Carcinoma. *N. Engl. J. Med.* **2020**, *382*, 1894–1905. [CrossRef]
12. Pinter, M.; Scheiner, B.; Peck-Radosavljevic, M. Immunotherapy for advanced hepatocellular carcinoma: A focus on special subgroups. *Gut* **2021**, *70*, 204–214. [CrossRef] [PubMed]
13. Pfister, D.; Núñez, N.G.; Pinyol, R.; Govaere, O.; Pinter, M.; Szydlowska, M.; Gupta, R.; Qiu, M.; Deczkowska, A.; Weiner, A.; et al. NASH limits anti-tumour surveillance in immunotherapy-treated HCC. *Nature* **2021**, *592*, 450–456. [CrossRef] [PubMed]
14. Albillos, A.; Lario, M.; Álvarez-Mon, M. Cirrhosis-associated immune dysfunction: Distinctive features and clinical relevance. *J. Hepatol.* **2014**, *61*, 1385–1396. [CrossRef] [PubMed]
15. Aravalli, R.N. Role of innate immunity in the development of hepatocellular carcinoma. *World J. Gastroenterol.* **2013**, *19*, 7500–7514. [CrossRef]

16. Binnewies, M.; Roberts, E.W.; Kersten, K.; Chan, V.; Fearon, D.F.; Merad, M.; Coussens, L.M.; Gabrilovich, D.I.; Ostrand-Rosenberg, S.; Hedrick, C.C.; et al. Understanding the tumor immune microenvironment (TIME) for effective therapy. *Nat. Med.* **2018**, *24*, 541–550. [CrossRef]
17. Li, X.-F.; Chen, D.-P.; Ouyang, F.-Z.; Chen, M.-M.; Wu, Y.; Kuang, D.-M.; Zheng, L. Increased autophagy sustains the survival and pro-tumourigenic effects of neutrophils in human hepatocellular carcinoma. *J. Hepatol.* **2015**, *62*, 131–139. [CrossRef]
18. Rizvi, S.; Wang, J.; El-Khoueiry, A.B. Liver Cancer Immunity. *Hepatology* **2021**, *73* (Suppl. S1), 86–103. [CrossRef]
19. Veglia, F.; Perego, M.; Gabrilovich, D. Myeloid-derived suppressor cells coming of age. *Nat. Immunol.* **2018**, *19*, 108–119. [CrossRef]
20. Guilliams, M.; Dutertre, C.-A.; Scott, C.L.; McGovern, N.; Sichien, D.; Chakarov, S.; Van Gassen, S.; Chen, J.; Poidinger, M.; De Prijck, S.; et al. Unsupervised High-Dimensional Analysis Aligns Dendritic Cells across Tissues and Species. *Immunity* **2016**, *45*, 669–684. [CrossRef]
21. Lopez, B.G.; Tsai, M.S.; Baratta, J.L.; Longmuir, K.J.; Robertson, R.T. Characterization of Kupffer cells in livers of developing mice. *Comp. Hepatol.* **2011**, *10*, 2. [CrossRef] [PubMed]
22. David, B.A.; Rezende, R.M.; Antunes, M.; Santos, M.M.; Lopes, M.A.F.; Diniz, A.B.; Pereira, R.V.S.; Marchesi, S.C.; Alvarenga, D.M.; Nakagaki, B.N.; et al. Combination of Mass Cytometry and Imaging Analysis Reveals Origin, Location, and Functional Repopulation of Liver Myeloid Cells in Mice. *Gastroenterology* **2016**, *151*, 1176–1191. [CrossRef] [PubMed]
23. Surewaard, B.G.; Kubes, P. Measurement of bacterial capture and phagosome maturation of Kupffer cells by intravital microscopy. *Methods* **2017**, *128*, 12–19. [CrossRef] [PubMed]
24. Mossanen, J.C.; Krenkel, O.; Ergen, C.; Govaere, O.; Liepelt, A.; Puengel, T.; Heymann, F.; Kalthoff, S.; Lefebvre, E.; Eulberg, D.; et al. Chemokine (C-C motif) receptor 2-positive monocytes aggravate the early phase of acetaminophen-induced acute liver injury. *Hepatology* **2016**, *64*, 1667–1682. [CrossRef]
25. Scott, C.; Zheng, F.; De Baetselier, P.; Martens, L.; Saeys, Y.; De Prijck, S.; Lippens, S.; Abels, C.; Schoonooghe, S.; Raes, G.; et al. Bone marrow-derived monocytes give rise to self-renewing and fully differentiated Kupffer cells. *Nat. Commun.* **2016**, *7*, 10321. [CrossRef]
26. Ritz, T.; Krenkel, O.; Tacke, F. Dynamic plasticity of macrophage functions in diseased liver. *Cell. Immunol.* **2018**, *330*, 175–182. [CrossRef]
27. Wan, S.; Kuo, N.; Kryczek, I.; Zou, W.; Welling, T.H. Myeloid cells in hepatocellular carcinoma. *Hepatology* **2015**, *62*, 1304–1312. [CrossRef]
28. Kang, T.-W.; Yevsa, T.; Woller, N.; Hoenicke, L.; Wuestefeld, T.; Dauch, D.; Hohmeyer, A.; Gereke, M.; Rudalska, R.; Potapova, A.; et al. Senescence surveillance of pre-malignant hepatocytes limits liver cancer development. *Nature* **2011**, *479*, 547–551. [CrossRef]
29. Mossanen, J.C.; Tacke, F. Role of lymphocytes in liver cancer. *OncoImmunology* **2013**, *2*, e26468. [CrossRef]
30. Man, K.; Ng, K.T.; Xu, A.; Cheng, Q.; Lo, C.M.; Xiao, J.W.; Sun, B.S.; Lim, Z.X.; Cheung, J.S.; Wu, E.X.; et al. Suppression of Liver Tumor Growth and Metastasis by Adiponectin in Nude Mice through Inhibition of Tumor Angiogenesis and Downregulation of Rho Kinase/IFN-Inducible Protein 10/Matrix Metalloproteinase 9 Signaling. *Clin. Cancer Res.* **2010**, *16*, 967–977. [CrossRef]
31. Coussens, L.M.; Zitvogel, L.; Palucka, A.K. Neutralizing Tumor-Promoting Chronic Inflammation: A Magic Bullet? *Science* **2013**, *339*, 286–291. [CrossRef]
32. Gabrilovich, D.I.; Ostrand-Rosenberg, S.; Bronte, V. Coordinated regulation of myeloid cells by tumours. *Nat. Rev. Immunol.* **2012**, *12*, 253–268. [CrossRef] [PubMed]
33. Marigo, I.; Bosio, E.; Solito, S.; Mesa, C.; Fernandez, A.; Dolcetti, L.; Ugel, S.; Sonda, N.; Bicciato, S.; Falisi, E.; et al. Tumor-Induced Tolerance and Immune Suppression Depend on the C/EBPβ Transcription Factor. *Immunity* **2010**, *32*, 790–802. [CrossRef]
34. McAllister, S.S.; Weinberg, R.A. The tumour-induced systemic environment as a critical regulator of cancer progression and metastasis. *Nat. Cell Biol.* **2014**, *16*, 717–727. [CrossRef] [PubMed]
35. Wu, W.-C.; Sun, H.-W.; Chen, H.-T.; Liang, J.; Yu, X.-J.; Wu, C.; Wang, Z.; Zheng, L. Circulating hematopoietic stem and progenitor cells are myeloid-biased in cancer patients. *Proc. Natl. Acad. Sci. USA* **2014**, *111*, 4221–4226. [CrossRef] [PubMed]
36. Geissmann, F.; Manz, M.G.; Jung, S.; Sieweke, M.H.; Merad, M.; Ley, K. Development of Monocytes, Macrophages, and Dendritic Cells. *Science* **2010**, *327*, 656–661. [CrossRef] [PubMed]
37. Yáñez, A.; Coetzee, S.; Olsson, A.; Muench, D.; Berman, B.P.; Hazelett, D.J.; Salomonis, N.; Grimes, H.L.; Goodridge, H.S. Granulocyte-Monocyte Progenitors and Monocyte-Dendritic Cell Progenitors Independently Produce Functionally Distinct Monocytes. *Immunity* **2017**, *47*, 890–902.e4. [CrossRef]
38. Cortez-Retamozo, V.; Etzrodt, M.; Newton, A.; Rauch, P.; Chudnovskiy, A.; Berger, C.; Ryan, R.; Iwamoto, Y.; Marinelli, B.; Gorbatov, R.; et al. Origins of tumor-associated macrophages and neutrophils. *Proc. Natl. Acad. Sci. USA* **2012**, *109*, 2491–2496. [CrossRef]
39. Yousif, A.S.; Ronsard, L.; Shah, P.; Omatsu, T.; Sangesland, M.; Moreno, T.B.; Lam, E.C.; Vrbanac, V.D.; Balazs, A.B.; Reinecker, H.-C.; et al. The persistence of interleukin-6 is regulated by a blood buffer system derived from dendritic cells. *Immunity* **2021**, *54*, 235–246.e5. [CrossRef]
40. Jordan, K.R.; Kapoor, P.; Spongberg, E.; Tobin, R.P.; Gao, D.; Borges, V.F.; McCarter, M.D. Immunosuppressive myeloid-derived suppressor cells are increased in splenocytes from cancer patients. *Cancer Immunol. Immunother.* **2017**, *66*, 503–513. [CrossRef]
41. Wu, C.; Ning, H.; Liu, M.; Lin, J.; Luo, S.; Zhu, W.; Xu, J.; Wu, W.-C.; Liang, J.; Shao, C.-K.; et al. Spleen mediates a distinct hematopoietic progenitor response supporting tumor-promoting myelopoiesis. *J. Clin. Investig.* **2018**, *128*, 3425–3438. [CrossRef]

42. Li, H.; Zhou, Y.; Wang, H.; Zhang, M.; Qiu, P.; Zhang, R.; Zhao, Q.; Liu, J. Crosstalk Between Liver Macrophages and Surrounding Cells in Nonalcoholic Steatohepatitis. *Front. Immunol.* **2020**, *11*, 1169. [CrossRef] [PubMed]
43. Schwabe, R.F.; Tabas, I.; Pajvani, U.B. Mechanisms of Fibrosis Development in Nonalcoholic Steatohepatitis. *Gastroenterology* **2020**, *158*, 1913–1928. [CrossRef]
44. Guillot, A.; Tacke, F. Liver Macrophages: Old Dogmas and New Insights. *Hepatol. Commun.* **2019**, *3*, 730–743. [CrossRef]
45. Yang, J.D.; Nakamura, I.; Roberts, L.R. The tumor microenvironment in hepatocellular carcinoma: Current status and therapeutic targets. *Semin. Cancer Biol.* **2011**, *21*, 35–43. [CrossRef] [PubMed]
46. Mantovani, A.; Germano, G.; Marchesi, F.; Locatelli, M.; Biswas, S.K. Cancer-promoting tumor-associated macrophages: New vistas and open questions. *Eur. J. Immunol.* **2011**, *41*, 2522–2525. [CrossRef] [PubMed]
47. Solinas, G.; Germano, G.; Mantovani, A.; Allavena, P. Tumor-associated macrophages (TAM) as major players of the cancer-related inflammation. *J. Leukoc. Biol.* **2009**, *86*, 1065–1073. [CrossRef] [PubMed]
48. Mantovani, A.; Garlanda, C.; Allavena, P. Molecular pathways and targets in cancer-related inflammation. *Ann. Med.* **2010**, *42*, 161–170. [CrossRef] [PubMed]
49. Robinson, A.; Han, C.Z.; Glass, C.K.; Pollard, J.W. Monocyte Regulation in Homeostasis and Malignancy. *Trends Immunol.* **2021**, *42*, 104–119. [CrossRef] [PubMed]
50. Nielsen, S.R.; Schmid, M.C. Macrophages as Key Drivers of Cancer Progression and Metastasis. *Mediat. Inflamm.* **2017**, *2017*, 9624760. [CrossRef] [PubMed]
51. Sica, A.; Mantovani, A. Macrophage plasticity and polarization: In vivo veritas. *J. Clin. Investig.* **2012**, *122*, 787–795. [CrossRef]
52. Locati, M.; Curtale, G.; Mantovani, A. Diversity, Mechanisms, and Significance of Macrophage Plasticity. *Annu. Rev. Pathol. Mech. Dis.* **2020**, *15*, 123–147. [CrossRef]
53. Gordon, S.; Martinez, F.O. Alternative Activation of Macrophages: Mechanism and Functions. *Immunity* **2010**, *32*, 593–604. [CrossRef]
54. Hao, N.-B.; Lü, M.-H.; Fan, Y.-H.; Cao, Y.-L.; Zhang, Z.-R.; Yang, S.-M. Macrophages in Tumor Microenvironments and the Progression of Tumors. *Clin. Dev. Immunol.* **2012**, *2012*, 948098. [CrossRef] [PubMed]
55. Huang, Y.; Ge, W.; Zhou, J.; Gao, B.; Qian, X.; Wang, W. The Role of Tumor Associated Macrophages in Hepatocellular Carcinoma. *J. Cancer* **2021**, *12*, 1284–1294. [CrossRef] [PubMed]
56. Jayasingam, S.D.; Citartan, M.; Thang, T.H.; Mat Zin, A.A.; Ang, K.C.; Ch'ng, E.S. Evaluating the Polarization of Tumor-Associated Macrophages Into M1 and M2 Phenotypes in Human Cancer Tissue: Technicalities and Challenges in Routine Clinical Practice. *Front. Oncol.* **2020**, *9*, 1512. [CrossRef] [PubMed]
57. Comi, M.; Avancini, D.; De Sio, F.S.; Villa, M.; Uyeda, M.J.; Floris, M.; Tomasoni, D.; Bulfone, A.; Roncarolo, M.G.; Gregori, S. Coexpression of CD163 and CD141 identifies human circulating IL-10-producing dendritic cells (DC-10). *Cell. Mol. Immunol.* **2020**, *17*, 95–107. [CrossRef]
58. Bohn, T.; Rapp, S.; Luther, N.; Klein, M.; Bruehl, T.-J.; Kojima, N.; Lopez, P.A.; Hahlbrock, J.; Muth, S.; Endo, S.; et al. Tumor immunoevasion via acidosis-dependent induction of regulatory tumor-associated macrophages. *Nat. Immunol.* **2018**, *19*, 1319–1329. [CrossRef]
59. Yang, Y.; Ye, Y.-C.; Chen, Y.; Zhao, J.-L.; Gao, C.-C.; Han, H.; Liu, W.-C.; Qin, H.-Y. Crosstalk between hepatic tumor cells and macrophages via Wnt/β-catenin signaling promotes M2-like macrophage polarization and reinforces tumor malignant behaviors. *Cell Death Dis.* **2018**, *9*, 793. [CrossRef] [PubMed]
60. Chen, M.-M.; Xiao, X.; Lao, X.-M.; Wei, Y.; Liu, R.-X.; Zeng, Q.-H.; Wang, J.-C.; Ouyang, F.-Z.; Chen, D.-P.; Chan, K.-W.; et al. Polarization of Tissue-Resident TFH-Like Cells in Human Hepatoma Bridges Innate Monocyte Inflammation and M2b Macrophage Polarization. *Cancer Discov.* **2016**, *6*, 1182–1195. [CrossRef]
61. Zhou, J.; Ding, T.; Pan, W.; Zhu, L.-Y.; Li, L.; Zheng, L. Increased intratumoral regulatory T cells are related to intratumoral macrophages and poor prognosis in hepatocellular carcinoma patients. *Int. J. Cancer* **2009**, *125*, 1640–1648. [CrossRef]
62. Wang, H.; Zhang, H.; Wang, Y.; Brown, Z.J.; Xia, Y.; Huang, Z.; Shen, C.; Hu, Z.; Beane, J.; Ansa-Addo, E.A.; et al. Regulatory T-cell and neutrophil extracellular trap interaction contributes to carcinogenesis in non-alcoholic steatohepatitis. *J. Hepatol.* **2021**, *75*, 1271–1283. [CrossRef] [PubMed]
63. Zhang, Y.-L.; Li, Q.; Yang, X.-M.; Fang, F.; Li, J.; Wang, Y.-H.; Yang, Q.; Zhu, L.; Nie, H.-Z.; Zhang, X.; et al. SPON2 Promotes M1-like Macrophage Recruitment and Inhibits Hepatocellular Carcinoma Metastasis by Distinct Integrin–Rho GTPase–Hippo Pathways. *Cancer Res.* **2018**, *78*, 2305–2317. [CrossRef] [PubMed]
64. Liu, N.; Wang, X.; Steer, C.J.; Song, G. MicroRNA-206 promotes the recruitment of CD8+ T cells by driving M1 polarisation of Kupffer cells. *Gut* **2021**. [CrossRef] [PubMed]
65. Liu, R.-X.; Wei, Y.; Zeng, Q.-H.; Chan, K.-W.; Xiao, X.; Zhao, X.; Chen, M.-M.; Ouyang, F.-Z.; Chen, D.-P.; Zheng, L.; et al. Chemokine (C-X-C motif) receptor 3-positive B cells link interleukin-17 inflammation to protumorigenic macrophage polarization in human hepatocellular carcinoma. *Hepatology* **2015**, *62*, 1779–1790. [CrossRef]
66. Zhao, Q.; Kuang, D.-M.; Wu, Y.; Xiao, X.; Li, X.-F.; Li, T.-J.; Zheng, L. Activated CD69+ T Cells Foster Immune Privilege by Regulating IDO Expression in Tumor-Associated Macrophages. *J. Immunol.* **2012**, *188*, 1117–1124. [CrossRef] [PubMed]
67. Miura, K.; Ishioka, M.; Minami, S.; Horie, Y.; Ohshima, S.; Goto, T.; Ohnishi, H. Toll-like Receptor 4 on Macrophage Promotes the Development of Steatohepatitis-related Hepatocellular Carcinoma in Mice. *J. Biol. Chem.* **2016**, *291*, 11504–11517. [CrossRef]

68. Wang, Q.; Cheng, F.; Ma, T.-T.; Xiong, H.-Y.; Li, Z.-W.; Xie, C.-L.; Liu, C.-Y.; Tu, Z.-G. Interleukin-12 inhibits the hepatocellular carcinoma growth by inducing macrophage polarization to the M1-like phenotype through downregulation of Stat-3. *Mol. Cell. Biochem.* **2016**, *415*, 157–168. [CrossRef]
69. Wang, T.-T.; Yuan, J.-H.; Ma, J.-Z.; Yang, W.-J.; Liu, X.-N.; Yin, Y.-P.; Liu, Y.; Pan, W.; Sun, S.-H. CTGF secreted by mesenchymal-like hepatocellular carcinoma cells plays a role in the polarization of macrophages in hepatocellular carcinoma progression. *Biomed. Pharmacother.* **2017**, *95*, 111–119. [CrossRef] [PubMed]
70. Curiel, T.J.; Coukos, G.; Zou, L.; Alvarez, X.; Cheng, P.; Mottram, P.; Evdemon-Hogan, M.; Conejo-Garcia, J.R.; Zhang, L.; Burow, M.; et al. Specific recruitment of regulatory T cells in ovarian carcinoma fosters immune privilege and predicts reduced survival. *Nat. Med.* **2004**, *10*, 942–949. [CrossRef] [PubMed]
71. Hefetz-Sela, S.; Stein, I.; Klieger, Y.; Porat, R.; Sade-Feldman, M.; Zreik, F.; Nagler, A.; Pappo, O.; Quagliata, L.; Dazert, E.; et al. Acquisition of an immunosuppressive protumorigenic macrophage phenotype depending on c-Jun phosphorylation. *Proc. Natl. Acad. Sci. USA* **2014**, *111*, 17582–17587. [CrossRef]
72. Zhang, Q.; He, Y.; Luo, N.; Patel, S.J.; Han, Y.; Gao, R.; Modak, M.; Carotta, S.; Haslinger, C.; Kind, D.; et al. Landscape and Dynamics of Single Immune Cells in Hepatocellular Carcinoma. *Cell* **2019**, *179*, 829–845.e20. [CrossRef] [PubMed]
73. Song, G.; Shi, Y.; Zhang, M.; Goswami, S.; Afridi, S.; Meng, L.; Ma, J.; Chen, Y.; Lin, Y.; Zhang, J.; et al. Global immune characterization of HBV/HCV-related hepatocellular carcinoma identifies macrophage and T-cell subsets associated with disease progression. *Cell Discov.* **2020**, *6*, 90. [CrossRef]
74. Dong, L.-Q.; Peng, L.-H.; Ma, L.-J.; Liu, D.-B.; Zhang, S.; Luo, S.-Z.; Rao, J.-H.; Zhu, H.-W.; Yang, S.-X.; Xi, S.-J.; et al. Heterogeneous immunogenomic features and distinct escape mechanisms in multifocal hepatocellular carcinoma. *J. Hepatol.* **2020**, *72*, 896–908. [CrossRef] [PubMed]
75. Zhang, Q.; Wang, H.; Mao, C.; Sun, M.; Dominah, G.; Chen, L.; Zhuang, Z. Fatty acid oxidation contributes to IL-1$\beta$ secretion in M2 macrophages and promotes macrophage-mediated tumor cell migration. *Mol. Immunol.* **2018**, *94*, 27–35. [CrossRef]
76. Degroote, H.; Lefere, S.; Vandierendonck, A.; Vanderborght, B.; Meese, T.; Van Nieuwerburgh, F.; Verhelst, X.; Geerts, A.; Van Vlierberghe, H.; Devisscher, L. Characterization of the inflammatory microenvironment and hepatic macrophage subsets in experimental hepatocellular carcinoma models. *Oncotarget* **2021**, *12*, 562–577. [CrossRef] [PubMed]
77. Schneider, C.; Teufel, A.; Yevsa, T.; Staib, F.; Hohmeyer, A.; Walenda, G.; Zimmermann, H.W.; Vucur, M.; Huss, S.; Gassler, N.; et al. Adaptive immunity suppresses formation and progression of diethylnitrosamine-induced liver cancer. *Gut* **2012**, *61*, 1733–1743. [CrossRef]
78. Mano, Y.; Aishima, S.; Fujita, N.; Tanaka, Y.; Kubo, Y.; Motomura, T.; Taketomi, A.; Shirabe, K.; Maehara, Y.; Oda, Y. Tumor-Associated Macrophage Promotes Tumor Progression via STAT3 Signaling in Hepatocellular Carcinoma. *Pathobiology* **2013**, *80*, 146–154. [CrossRef]
79. Eggert, T.; Wolter, K.; Ji, J.; Ma, C.; Yevsa, T.; Klotz, S.; Medina-Echeverz, J.; Longerich, T.; Forgues, M.; Reisinger, F.; et al. Distinct Functions of Senescence-Associated Immune Responses in Liver Tumor Surveillance and Tumor Progression. *Cancer Cell* **2016**, *30*, 533–547. [CrossRef]
80. Wolf, M.J.; Adili, A.; Piotrowitz, K.; Abdullah, Z.; Boege, Y.; Stemmer, K.; Ringelhan, M.; Simonavicius, N.; Egger, M.; Wohlleber, D.; et al. Metabolic Activation of Intrahepatic CD8+ T Cells and NKT Cells Causes Nonalcoholic Steatohepatitis and Liver Cancer via Cross-Talk with Hepatocytes. *Cancer Cell* **2014**, *26*, 549–564. [CrossRef]
81. Guo, B.; Li, L.; Guo, J.; Liu, A.; Wu, J.; Wang, H.; Shi, J.; Pang, D.; Cao, Q. M2 tumor-associated macrophages produce interleukin-17 to suppress oxaliplatin-induced apoptosis in hepatocellular carcinoma. *Oncotarget* **2017**, *8*, 44465–44476. [CrossRef]
82. Wei, Y.; Shi, D.; Liang, Z.; Liu, Y.; Li, Y.; Xing, Y.; Liu, W.; Ai, Z.; Zhuang, J.; Chen, X.; et al. IL-17A secreted from lymphatic endothelial cells promotes tumorigenesis by upregulation of PD-L1 in hepatoma stem cells. *J. Hepatol.* **2019**, *71*, 1206–1215. [CrossRef]
83. Wu, J.; Gao, W.; Zuo, X.; Zhang, Y.; Chen, Z.; Ding, W.; Li, X.; Lin, F.; Shen, H.; Tang, J.; et al. M2 Macrophage–Derived Exosomes Facilitate HCC Metastasis by Transferring $\alpha_M\beta_2$ Integrin to Tumor Cells. *Hepatology* **2020**, *73*, 1365–1380. [CrossRef]
84. Li, X.; Yao, W.; Yuan, Y.; Chen, P.; Li, B.; Li, J.; Chu, R.; Song, H.; Xie, D.; Jiang, X.; et al. Targeting of tumour-infiltrating macrophages via CCL2/CCR2 signalling as a therapeutic strategy against hepatocellular carcinoma. *Gut* **2015**, *66*, 157–167. [CrossRef]
85. Zhang, J.; Zhang, Q.; Lou, Y.; Fu, Q.; Chen, Q.; Wei, T.; Yang, J.; Tang, J.; Wang, J.; Chen, Y.; et al. Hypoxia-inducible factor-1$\alpha$/interleukin-1$\beta$ signaling enhances hepatoma epithelial-mesenchymal transition through macrophages in a hypoxic-inflammatory microenvironment. *Hepatology* **2018**, *67*, 1872–1889. [CrossRef]
86. Zhao, Q.; Xiao, X.; Wu, Y.; Wei, Y.; Zhu, L.-Y.; Zhou, J.; Kuang, D.-M. Interleukin-17-educated monocytes suppress cytotoxic T-cell function through B7-H1 in hepatocellular carcinoma patients. *Eur. J. Immunol.* **2011**, *41*, 2314–2322. [CrossRef] [PubMed]
87. Wu, Y.; Kuang, D.-M.; Pan, W.-D.; Wan, Y.-L.; Lao, X.-M.; Wang, D.; Li, X.-F.; Zheng, L. Monocyte/macrophage-elicited natural killer cell dysfunction in hepatocellular carcinoma is mediated by CD48/2B4 interactions. *Hepatology* **2013**, *57*, 1107–1116. [CrossRef]
88. Zang, M.; Li, Y.; He, H.; Ding, H.; Chen, K.; Du, J.; Chen, T.; Wu, Z.; Liu, H.; Wang, D.; et al. IL-23 production of liver inflammatory macrophages to damaged hepatocytes promotes hepatocellular carcinoma development after chronic hepatitis B virus infection. *Biochim. Biophys. Acta (BBA)-Mol. Basis Dis.* **2018**, *1864*, 3759–3770. [CrossRef] [PubMed]

89. Zhang, Y.; Li, J.-Q.; Jiang, Z.-Z.; Li, L.; Wu, Y.; Zheng, L. CD169 identifies an anti-tumour macrophage subpopulation in human hepatocellular carcinoma. *J. Pathol.* **2016**, *239*, 231–241. [CrossRef]
90. Bartneck, M.; Schrammen, P.L.; Möckel, D.; Govaere, O.; Liepelt, A.; Krenkel, O.; Ergen, C.; McCain, M.V.; Eulberg, D.; Luedde, T.; et al. The CCR2+ Macrophage Subset Promotes Pathogenic Angiogenesis for Tumor Vascularization in Fibrotic Livers. *Cell. Mol. Gastroenterol. Hepatol.* **2019**, *7*, 371–390. [CrossRef] [PubMed]
91. Zhang, H.; Wang, X.; Shen, Z.; Xu, J.; Qin, J.; Sun, Y. Infiltration of diametrically polarized macrophages predicts overall survival of patients with gastric cancer after surgical resection. *Gastric Cancer* **2015**, *18*, 740–750. [CrossRef]
92. Zhang, M.; He, Y.; Sun, X.; Li, Q.; Wang, W.; Zhao, A.; Di, W. A high M1/M2 ratio of tumor-associated macrophages is associated with extended survival in ovarian cancer patients. *J. Ovarian Res.* **2014**, *7*, 19. [CrossRef]
93. Dong, P.; Ma, L.; Liu, L.; Zhao, G.; Zhang, S.; Dong, L.; Xue, R.; Chen, S. CD86+/CD206+, Diametrically Polarized Tumor-Associated Macrophages, Predict Hepatocellular Carcinoma Patient Prognosis. *Int. J. Mol. Sci.* **2016**, *17*, 320. [CrossRef]
94. Sia, D.; Jiao, Y.; Martinez-Quetglas, I.; Kuchuk, O.; Villacorta-Martin, C.; de Moura, M.C.; Putra, J.; Campreciós, G.; Bassaganyas, L.; Akers, N.; et al. Identification of an Immune-specific Class of Hepatocellular Carcinoma, Based on Molecular Features. *Gastroenterology* **2017**, *153*, 812–826. [CrossRef]
95. Liu, J.-Y.; Peng, C.-W.; Yang, G.-F.; Hu, W.-Q.; Yang, X.-J.; Huang, C.-Q.; Xiong, B.; Li, Y. Distribution pattern of tumor associated macrophages predicts the prognosis of gastric cancer. *Oncotarget* **2017**, *8*, 92757–92769. [CrossRef] [PubMed]
96. Yang, C.; Wei, C.; Wang, S.; Shi, D.; Zhang, C.; Lin, X.; Dou, R.; Xiong, B. Elevated CD163+/CD68+ Ratio at Tumor Invasive Front is Closely Associated with Aggressive Phenotype and Poor Prognosis in Colorectal Cancer. *Int. J. Biol. Sci.* **2019**, *15*, 984–998. [CrossRef] [PubMed]
97. Ke, M.; Zhang, Z.; Cong, L.; Zhao, S.; Li, Y.; Wang, X.; Lv, Y.; Zhu, Y.; Dong, J. MicroRNA-148b-colony-stimulating factor-1 signaling-induced tumor-associated macrophage infiltration promotes hepatocellular carcinoma metastasis. *Biomed. Pharmacother.* **2019**, *120*, 109523. [CrossRef] [PubMed]
98. Chen, D.-P.; Ning, W.-R.; Jiang, Z.-Z.; Peng, Z.-P.; Zhu, L.-Y.; Zhuang, S.-M.; Kuang, D.-M.; Zheng, L.; Wu, Y. Glycolytic activation of peritumoral monocytes fosters immune privilege via the PFKFB3-PD-L1 axis in human hepatocellular carcinoma. *J. Hepatol.* **2019**, *71*, 333–343. [CrossRef]
99. Li, Z.; Li, H.; Zhao, Z.-B.; Zhu, W.; Feng, P.-P.; Zhu, X.-W.; Gong, J.-P. SIRT4 silencing in tumor-associated macrophages promotes HCC development via PPARδ signalling-mediated alternative activation of macrophages. *J. Exp. Clin. Cancer Res.* **2019**, *38*, 469. [CrossRef]
100. Zhang, Q.-B.; Jia, Q.-A.; Wang, H.; Hu, C.-X.; Sun, D.; Jiang, R.-D.; Zhang, Z.-L. High-mobility group protein box1 expression correlates with peritumoral macrophage infiltration and unfavorable prognosis in patients with hepatocellular carcinoma and cirrhosis. *BMC Cancer* **2016**, *16*, 880. [CrossRef]
101. Zhao, Y.-M.; Wang, L.; Dai, Z.; Wang, D.-D.; Hei, Z.-Y.; Zhang, N.; Fu, X.-T.; Wang, X.-L.; Zhang, S.-C.; Qin, L.-X.; et al. Validity of plasma macrophage migration inhibitory factor for diagnosis and prognosis of hepatocellular carcinoma. *Int. J. Cancer* **2011**, *129*, 2463–2472. [CrossRef]
102. Kono, H.; Fujii, H.; Furuya, S.; Hara, M.; Hirayama, K.; Akazawa, Y.; Nakata, Y.; Tsuchiya, M.; Hosomura, N.; Sun, C. Macrophage colony-stimulating factor expressed in non-cancer tissues provides predictive powers for recurrence in hepatocellular carcinoma. *World J. Gastroenterol.* **2016**, *22*, 8779–8789. [CrossRef]
103. Ohno, A.; Yorita, K.; Haruyama, Y.; Kondo, K.; Kato, A.; Ohtomo, T.; Kawaguchi, M.; Marutuska, K.; Chijiiwa, K.; Kataoka, H. Aberrant expression of monocarboxylate transporter 4 in tumour cells predicts an unfavourable outcome in patients with hepatocellular carcinoma. *Liver Int.* **2014**, *34*, 942–952. [CrossRef] [PubMed]
104. Zhu, F.; Li, X.; Jiang, Y.; Zhu, H.; Zhang, H.; Zhang, C.; Zhao, Y.; Luo, F. GdCl3 suppresses the malignant potential of hepatocellular carcinoma by inhibiting the expression of CD206 in tumor-associated macrophages. *Oncol. Rep.* **2015**, *34*, 2643–2655. [CrossRef] [PubMed]
105. Zhu, X.-D.; Zhang, J.-B.; Zhuang, P.-Y.; Zhu, H.-G.; Zhang, W.; Xiong, Y.-Q.; Wu, W.-Z.; Wang, L.; Tang, Z.-Y.; Sun, H.-C. High Expression of Macrophage Colony-Stimulating Factor in Peritumoral Liver Tissue Is Associated With Poor Survival After Curative Resection of Hepatocellular Carcinoma. *J. Clin. Oncol.* **2008**, *26*, 2707–2716. [CrossRef] [PubMed]
106. Zhu, W.; Guo, L.; Zhang, B.; Lou, L.; Lin, Z.; Zhu, X.; Ren, N.; Dong, Q.; Ye, Q.; Qin, L. Combination of Osteopontin with Peritumoral Infiltrating Macrophages is Associated with Poor Prognosis of Early-Stage Hepatocellular Carcinoma after Curative Resection. *Ann. Surg. Oncol.* **2013**, *21*, 1304–1313. [CrossRef]
107. Zhou, T.-Y.; Zhou, Y.-L.; Qian, M.-J.; Fang, Y.-Z.; Ye, S.; Xin, W.-X.; Yang, X.-C.; Wu, H.-H. Interleukin-6 induced by YAP in hepatocellular carcinoma cells recruits tumor-associated macrophages. *J. Pharmacol. Sci.* **2018**, *138*, 89–95. [CrossRef]
108. Wang, X.; Wang, H.; Li, G.; Song, Y.; Wang, S.; Zhu, F.; Guo, C.; Zhang, L.; Shi, Y. Activated macrophages down-regulate expression of E-cadherin in hepatocellular carcinoma cells via NF-κB/Slug pathway. *Tumor Biol.* **2014**, *35*, 8893–8901. [CrossRef]
109. Zhang, W.; Zhu, X.-D.; Sun, H.-C.; Xiong, Y.-Q.; Zhuang, P.-Y.; Xu, H.-X.; Kong, L.-Q.; Wang, L.; Wu, W.-Z.; Tang, Z.-Y. Depletion of Tumor-Associated Macrophages Enhances the Effect of Sorafenib in Metastatic Liver Cancer Models by Antimetastatic and Antiangiogenic Effects. *Clin. Cancer Res.* **2010**, *16*, 3420–3430. [CrossRef]
110. Wu, X.; Luo, H.; Shi, B.; Di, S.; Sun, R.; Su, J.; Liu, Y.; Li, H.; Jiang, H.; Li, Z. Combined Antitumor Effects of Sorafenib and GPC3-CAR T Cells in Mouse Models of Hepatocellular Carcinoma. *Mol. Ther.* **2019**, *27*, 1483–1494. [CrossRef]

111. Sprinzl, M.F.; Reisinger, F.; Puschnik, A.; Ringelhan, M.; Ackermann, K.; Hartmann, D.; Schiemann, M.; Weinmann, A.; Galle, P.R.; Schuchmann, M.; et al. Sorafenib perpetuates cellular anticancer effector functions by modulating the crosstalk between macrophages and natural killer cells. *Hepatology* **2013**, *57*, 2358–2368. [CrossRef]
112. Sprinzl, M.F.; Puschnik, A.; Schlitter, A.M.; Schad, A.; Ackermann, K.; Esposito, I.; Lang, H.; Galle, P.R.; Weinmann, A.; Heikenwälder, M.; et al. Sorafenib inhibits macrophage-induced growth of hepatoma cells by interference with insulin-like growth factor-1 secretion. *J. Hepatol.* **2015**, *62*, 863–870. [CrossRef]
113. Yao, W.; Ba, Q.; Li, X.; Li, H.; Zhang, S.; Yuan, Y.; Wang, F.; Duan, X.; Li, J.; Zhang, W.; et al. A Natural CCR2 Antagonist Relieves Tumor-associated Macrophage-mediated Immunosuppression to Produce a Therapeutic Effect for Liver Cancer. *EBioMedicine* **2017**, *22*, 58–67. [CrossRef] [PubMed]
114. Deng, Y.-R.; Liu, W.-B.; Lian, Z.-X.; Li, X.; Hou, X. Sorafenib inhibits macrophage-mediated epithelial-mesenchymal transition in hepatocellular carcinoma. *Oncotarget* **2016**, *7*, 38292–38305. [CrossRef] [PubMed]
115. Wei, X.; Tang, C.; Lu, X.; Liu, R.; Zhou, M.; He, D.; Zheng, D.; Sun, C.; Wu, Z. MiR-101 targets DUSP1 to regulate the TGF-β secretion in sorafenib inhibits macrophage-induced growth of hepatocarcinoma. *Oncotarget* **2015**, *6*, 18389–18405. [CrossRef]
116. Yang, W.; Lu, Y.; Xu, Y.; Xu, L.; Zheng, W.; Wu, Y.; Li, L.; Shen, P. Estrogen Represses Hepatocellular Carcinoma (HCC) Growth via Inhibiting Alternative Activation of Tumor-associated Macrophages (TAMs)*. *J. Biol. Chem.* **2012**, *287*, 40140–40149. [CrossRef] [PubMed]
117. Tsuchiyama, T.; Nakamoto, Y.; Sakai, Y.; Mukaida, N.; Kaneko, S. Optimal amount of monocyte chemoattractant protein-1 enhances antitumor effects of suicide gene therapy against hepatocellular carcinoma by M1 macrophage activation. *Cancer Sci.* **2008**, *99*, 2075–2082. [CrossRef]
118. Guerra, A.D.; Yeung, O.W.; Qi, X.; Kao, W.J.; Man, K. The Anti-Tumor Effects of M1 Macrophage-Loaded Poly (ethylene glycol) and Gelatin-Based Hydrogels on Hepatocellular Carcinoma. *Theranostics* **2017**, *7*, 3732–3744. [CrossRef] [PubMed]
119. Huang, X.-H.; Jian, W.-H.; Wu, Z.-F.; Zhao, J.; Wang, H.; Li, W.; Xia, J.-T. Small interfering RNA (siRNA)-mediated knockdown of macrophage migration inhibitory factor (MIF) suppressed cyclin D1 expression and hepatocellular carcinoma cell proliferation. *Oncotarget* **2014**, *5*, 5570–5580. [CrossRef]
120. Xiao, Z.; Chung, H.; Banan, B.; Manning, P.T.; Ott, K.C.; Lin, S.; Capoccia, B.J.; Subramanian, V.; Hiebsch, R.R.; Upadhya, G.A.; et al. Antibody mediated therapy targeting CD47 inhibits tumor progression of hepatocellular carcinoma. *Cancer Lett.* **2015**, *360*, 302–309. [CrossRef]
121. Tan, H.-Y.; Wang, N.; Tsao, S.-W.; Che, C.-M.; Yuen, M.-F.; Feng, Y. IRE1α inhibition by natural compound genipin on tumour associated macrophages reduces growth of hepatocellular carcinoma. *Oncotarget* **2016**, *7*, 43792–43804. [CrossRef] [PubMed]
122. Wang, Q.-S.; Gao, L.-N.; Zhu, X.-N.; Zhang, Y.; Zhang, C.-N.; Xu, D.; Cui, Y.-L. Co-delivery of glycyrrhizin and doxorubicin by alginate nanogel particles attenuates the activation of macrophage and enhances the therapeutic efficacy for hepatocellular carcinoma. *Theranostics* **2019**, *9*, 6239–6255. [CrossRef] [PubMed]
123. Li, G.; Liu, D.; Kimchi, E.T.; Kaifi, J.T.; Qi, X.; Manjunath, Y.; Liu, X.; Deering, T.; Avella, D.M.; Fox, T.; et al. Nanoliposome C6-Ceramide Increases the Anti-tumor Immune Response and Slows Growth of Liver Tumors in Mice. *Gastroenterology* **2018**, *154*, 1024–1036.e9. [CrossRef]
124. Chen, X.; Yin, S.; Hu, C.; Chen, X.; Jiang, K.; Ye, S.; Feng, X.; Fan, S.; Xie, H.; Zhou, L.; et al. Comparative Study of Nanosecond Electric Fields In Vitro and In Vivo on Hepatocellular Carcinoma Indicate Macrophage Infiltration Contribute to Tumor Ablation In Vivo. *PLoS ONE* **2014**, *9*, e86421. [CrossRef] [PubMed]
125. Yin, S.; Chen, X.; Hu, C.; Zhang, X.; Hu, Z.; Yu, J.; Feng, X.; Jiang, K.; Ye, S.; Shen, K.; et al. Nanosecond pulsed electric field (nsPEF) treatment for hepatocellular carcinoma: A novel locoregional ablation decreasing lung metastasis. *Cancer Lett.* **2014**, *346*, 285–291. [CrossRef]
126. Nii, T.; Kuwahara, T.; Makino, K.; Tabata, P.Y. A Co-Culture System of Three-Dimensional Tumor-Associated Macrophages and Three-Dimensional Cancer-Associated Fibroblasts Combined with Biomolecule Release for Cancer Cell Migration. *Tissue Eng. Part A* **2020**, *26*, 1272–1282. [CrossRef] [PubMed]
127. Nii, T.; Makino, K.; Tabata, Y. Three-Dimensional Culture System of Cancer Cells Combined with Biomaterials for Drug Screening. *Cancers* **2020**, *12*, 2754. [CrossRef]
128. Cuccarese, M.F.; Dubach, J.M.; Pfirschke, C.; Engblom, C.; Garris, C.; Miller, M.; Pittet, M.J.; Weissleder, R. Heterogeneity of macrophage infiltration and therapeutic response in lung carcinoma revealed by 3D organ imaging. *Nat. Commun.* **2017**, *8*, 14293. [CrossRef]
129. Sun, Y.; Wu, L.; Zhong, Y.; Zhou, K.; Hou, Y.; Wang, Z.; Zhang, Z.; Xie, J.; Wang, C.; Chen, D.; et al. Single-cell landscape of the ecosystem in early-relapse hepatocellular carcinoma. *Cell* **2021**, *184*, 404–421.e16. [CrossRef]
130. Tacke, F. Cenicriviroc for the treatment of non-alcoholic steatohepatitis and liver fibrosis. *Expert Opin. Investig. Drugs* **2018**, *27*, 301–311. [CrossRef]
131. Bartneck, M.; Koppe, C.; Fech, V.; Warzecha, K.T.; Kohlhepp, M.; Huss, S.; Weiskirchen, R.; Trautwein, C.; Luedde, T.; Tacke, F. Roles of CCR2 and CCR5 for Hepatic Macrophage Polarization in Mice With Liver Parenchymal Cell-Specific NEMO Deletion. *Cell. Mol. Gastroenterol. Hepatol.* **2021**, *11*, 327–347. [CrossRef] [PubMed]
132. Wu, Q.; Zhou, W.; Yin, S.; Zhou, Y.; Chen, T.; Qian, J.; Su, R.; Hong, L.; Lu, H.; Zhang, F.; et al. Blocking Triggering Receptor Expressed on Myeloid Cells-1-Positive Tumor-Associated Macrophages Induced by Hypoxia Reverses Immunosuppression and Anti-Programmed Cell Death Ligand 1 Resistance in Liver Cancer. *Hepatology* **2019**, *70*, 198–214. [CrossRef] [PubMed]

133. Halama, N.; Zoernig, I.; Berthel, A.; Kahlert, C.; Klupp, F.; Suarez-Carmona, M.; Suetterlin, T.; Brand, K.; Krauss, J.; Lasitschka, F.; et al. Tumoral Immune Cell Exploitation in Colorectal Cancer Metastases Can Be Targeted Effectively by Anti-CCR5 Therapy in Cancer Patients. *Cancer Cell* **2016**, *29*, 587–601. [CrossRef] [PubMed]
134. Tan-Garcia, A.; Lai, F.; Yeong, J.P.S.; Irac, S.E.; Ng, P.Y.; Msallam, R.; Lim, J.C.T.; Wai, L.-E.; Tham, C.Y.; Choo, S.P.; et al. Liver fibrosis and CD206+ macrophage accumulation are suppressed by anti-GM-CSF therapy. *JHEP Rep.* **2020**, *2*, 100062. [CrossRef]
135. Su, X.; Xu, Y.; Fox, G.C.; Xiang, J.; Kwakwa, K.A.; Davis, J.L.; Belle, J.I.; Lee, W.-C.; Wong, W.H.; Fontana, F.; et al. Breast cancer–derived GM-CSF regulates arginase 1 in myeloid cells to promote an immunosuppressive microenvironment. *J. Clin. Investig.* **2021**, *131*. [CrossRef]
136. Burmester, G.R.; McInnes, I.B.; Kremer, J.; Miranda, P.; Korkosz, M.; Vencovsky, J.; Rubbert-Roth, A.; Mysler, E.; Sleeman, M.A.; Godwood, A.; et al. A randomised phase IIb study of mavrilimumab, a novel GM–CSF receptor alpha monoclonal antibody, in the treatment of rheumatoid arthritis. *Ann. Rheum. Dis.* **2017**, *76*, 1020–1030. [CrossRef]
137. Behrens, F.; Tak, P.P.; Østergaard, M.; Stoilov, R.; Wiland, P.; Huizinga, T.W.; Berenfus, V.; Vladeva, S.; Rech, J.; Rubbert-Roth, A.; et al. MOR103, a human monoclonal antibody to granulocyte–macrophage colony-stimulating factor, in the treatment of patients with moderate rheumatoid arthritis: Results of a phase Ib/IIa randomised, double-blind, placebo-controlled, dose-escalation trial. *Ann. Rheum. Dis.* **2014**, *74*, 1058–1064. [CrossRef]
138. Sterner, R.M.; Sakemura, R.; Cox, M.J.; Yang, N.; Khadka, R.H.; Forsman, C.L.; Hansen, M.J.; Jin, F.; Ayasoufi, K.; Hefazi, M.; et al. GM-CSF inhibition reduces cytokine release syndrome and neuroinflammation but enhances CAR-T cell function in xenografts. *Blood* **2019**, *133*, 697–709. [CrossRef]
139. Temesgen, Z.; Burger, C.D.; Baker, J.; Polk, C.; Libertin, C.; Kelley, C.; Marconi, V.C.; Orenstein, R.; Durrant, C.; Chappell, D.; et al. Lenzilumab Efficacy and Safety in Newly Hospitalized COVID-19 Subjects: Results from the Live-Air Phase 3 Randomized Double-Blind Placebo-Controlled Trial. *medRxiv* **2021**. [CrossRef]
140. Neubert, N.J.; Schmittnaegel, M.; Bordry, N.; Nassiri, S.; Wald, N.; Martignier, C.; Tillé, L.; Homicsko, K.; Damsky, W.; Hajjami, H.M.-E.; et al. T cell–induced CSF1 promotes melanoma resistance to PD1 blockade. *Sci. Transl. Med.* **2018**, *10*. [CrossRef]
141. Güç, E.; Pollard, J.W. Redefining macrophage and neutrophil biology in the metastatic cascade. *Immunity* **2021**, *54*, 885–902. [CrossRef] [PubMed]

cancers MDPI

*Review*

# The Role of the NLRP3 Inflammasome in HCC Carcinogenesis and Treatment: Harnessing Innate Immunity

**Stavros P. Papadakos [1,†], Nikolaos Dedes [1,†], Elias Kouroumalis [2] and Stamatios Theocharis [1,*]**

1 First Department of Pathology, Medical School, National and Kapodistrian University of Athens, 11527 Athens, Greece; stavrospapadakos@gmail.com (S.P.P.); ndedes@med.uoa.gr (N.D.)
2 Department of Gastroenterology, Medical School, University of Crete, 71500 Heraklion, Greece; kouroum@med.uoc.gr
* Correspondence: stamtheo@med.uoa.gr
† These authors contributed equally to this work.

**Simple Summary:** The Hepatocellular Carcinoma (HCC) remains a major concern for the public health. The pandemic of metabolic syndrome in the Western societies and the considerable amounts of viral hepatitis in the underdeveloped countries keep fueling s the development of HCC. The hepatotropic viruses evade the NLRP3 inflammasome in order to sustain the chronicity of infection leading to cirrhosis while in the established tumors the activation of NLRP3 promotes several pro-tumorigenic effects. That leads to substantial economic burden for the societies and alternative therapeutic targets should be investigated. Reviewing past and more recent literature it can be deduced that the NLRP3 inflammasome could be an ideal therapeutic effects and it should be studied in more depth.

**Abstract:** The HCC constitutes one of the most frequent cancers, with a non-decreasing trend in disease mortality despite advances in systemic therapy and surgery. This trend is fueled by the rise of an obesity wave which is prominent the Western populations and has reshaped the etiologic landscape of HCC. Interest in the nucleotide-binding domain leucine-rich repeat containing (NLR) family member NLRP3 has recently been revived since it would appear that, by generating inflammasomes, it participates in several physiologic processes and its dysfunction leads to disease. The NLRP3 inflammasome has been studied in depth, and its influence in HCC pathogenesis has been extensively documented during the past quinquennial. Since inflammation comprises a major regulator of carcinogenesis, it is of paramount importance an attempt to evaluate the contribution of the NLRP3 inflammasome to the generation and management of HCC. The aim of this review was to examine the literature in order to determine the impact of the NLRP3 inflammasome on, and present a hypothesis about its input in, HCC.

**Keywords:** HCC; NLRP3; innate immunity; inflammasome

**Citation:** Papadakos, S.P.; Dedes, N.; Kouroumalis, E.; Theocharis, S. The Role of the NLRP3 Inflammasome in HCC Carcinogenesis and Treatment: Harnessing Innate Immunity. *Cancers* **2022**, *14*, 3150. https://doi.org/10.3390/cancers14133150

Academic Editor: Jens Marquardt

Received: 5 May 2022
Accepted: 20 June 2022
Published: 27 June 2022



## 1. Introduction

### 1.1. Epidemiology of Hepatocellular Carcinoma

Hepatocellular carcinoma (HCC) comprises the most common primary liver cancer. Globally, it ranks fifth in terms of incidence and second in terms of cancer-related mortality, indicating that it will be at the forefront of public health awareness in the coming decades [1]. HCC arises in the vast majority of cases as the end-result of chronic infection with viral hepatitis B and C (HBV and HCV), alcohol abuse and in the context of non-alcoholic fatty liver disease (NAFLD)/non-alcoholic steatohepatitis (NASH) [2]. The latter is the hepatic manifestation of obesity and metabolic syndrome. Progress towards the augmentation of NAFLD-driven HCC diagnoses has been achieved [3] in recent decades. In fact, NAFLD-induced HCC is expected to become the leading indication of liver transplantation in the years to come [4]. The development of cirrhosis substantially enhances the risk of HCC development [5]. In terms of molecular classification, two main categories have been recognized: a proliferative group and a non-proliferative one. The first constitutes a more aggressive disease spectrum, with poor cellular differentiation, elevated alpha-fetoprotein (AFP) levels, the presence of *TP53* and chromosomal instability. It is associated with hepatitis B (HBV) infection and a poor clinical outcome. The non-proliferative class, on the other hand, is characterized by a better clinical course as a result of its genetic proximity with normal liver parenchyma. Its primary genetic characteristics are mutations on the *CTNNB1* gene, which encodes the catenin beta-1 protein, an immune-deserted phenotype and a low tumor grade. Clinically, there is an association between HCV infection and chronic alcohol abuse [6–9]. Inflammation has been recognized as one of the hallmarks of carcinogenesis [10] and multiple immunotherapeutic approaches have been designed to target cancer-specific inflammation. The combination of atezolizumab and bevacizumab has gained ground as a first line therapy over sorafenib, while several immune checkpoint inhibitors (ICIs) are being evaluated in clinical trials [11]. Additionally, PD-L1 and tumor mutational burden (TMB) constitute well-characterized biomarkers that could potentially target those patients that would benefit most from immunotherapy treatments [12,13]. These were ratified by epidemiologic studies which demonstrated a reduced risk of HCC in patients treated with low-dose aspirin [14]. The effects of inflammasomes in various cancers has gained importance in recent years [15,16], with interest in this subject expected to peak soon. Multiple single-nucleotide polymorphisms (SNPs) in inflammation-related genes such as the interleukin (IL)1B gene have been documented to promote the HCC [17]. This could further indicate that inflammasomes might influence carcinogenesis. In fact, references in the literature to the role of the NLRP3 inflammasome in HCC have begun to emerge [18].

### 1.2. Mechanisms of NLRP3 Inflammasome Activation

Inflammasomes are multimeric protein complexes which are a fundamental constituent of the innate immune system. Its responsiveness to stimuli is defined by their sensor molecule, which constitutes a member of the pattern recognition receptors (PRR) system. To date, only five members of the PRR system have been demonstrated to participate in the formation of inflammasomes: the nucleotide-binding oligomerization domain (NOD), NLRP1, NLRP3, NLRC4 and absent-in-melanoma 2 (AIM2) and pyrin [19]. Inflammasomes participate in various physiological processes including the orchestration of the immune response on mucosal surfaces, the shaping of resistance against microorganisms and the mediation of insulin signaling [20]. Their deregulation is linked to a variety of inflammatory [21,22] degenerative and metabolic diseases [23] as well as tumorigenesis [24].

The NLRP3 inflammasome has been identified as a mediator in a wide variety of diseases and as a potential novel therapeutic target. Since its role in the progression of atherosclerosis, heart failure, glomerulonephritis and several infectious diseases [25–29] has been acknowledged, it would be useful to highlight the most fundamental stages of NLRP3 activation. It resides in the cytoplasm and acts as an immune sensor to a plethora of stimuli, producing IL-1b and IL-18 and initiating the immune response. It is arranged in a dodecameric structure by three proteins: an NLRP3 scaffold, an apoptosis-associated speck-like protein (ASC), which functions as adaptor, and caspase-1. An exhaustive presentation of the NLRP3 inflammasome has been conducted elsewhere [30,31] and it goes beyond the scope of this manuscript [32]. The fundamental proposition regarding NLRP3 inflammasome activation constitutes the two-part hypothesis. Concisely, "Signal 1" emerges when pathogen-associated molecular patterns (PAMPs) from microbes and endogenous damage-associated molecular patterns (DAMPs) prime the cell through toll-like receptors (TLRs), the tumor necrosis factor receptor (TNFR) and the IL-1b receptor (IL-1R), activate the nuclear factor kappa B (NF-kB) signaling pathway and upregulate the expression of the NLRP3 oligomers pro-IL-1b and pro- IL-18. "Signal 2" derives mainly from a K+ efflux, which is the end-result of a wide variety of stimuli such as the activation of the P2 × 7 receptor, which is activated in the presence of elevated adenosine triphosphate (ATP) concentrations, the dissolution of lysosomes and the presence of pore-forming toxins. The fact that mitochondrial dysfunction is a source of reactive oxygen species (ROS) and oxidized mitochondrial DNA (mtDNA) also potentiates the assembly of inflammasome monomers into the activated form of the NLRP3 inflammasome. Consequently, the activation of the NLRP3 complex leads to the cleavage of pro-ILs and GasderminD to form pores in the cell membrane, resulting in pyroptosis [32,33].

As is apparent from the above, the stimulants of the NLRP3 inflammasome are extensive, and it is essential to investigate the importance of the NLRP3 inflammasome in the progression of liver disease.

## 2. The NLRP3 Inflammasome in Liver Disease

Our knowledge of the function of the liver has evolved considerably in recent decades. Classically, it was only perceived as the primary metabolism organ, with it regulating the lipid and cholesterol metabolism, the production of albumin and clotting factors, detoxifying the end-products of the metabolism and exerting the glycogen storage. Presently, it is widely recognized as an immune response-orchestrating organ given its role in the generation of acute phase reactants, proteins of complementary systems and anti-microbials. In parallel, the liver maintains an antigen-presenting capacity and can efficiently remove endotoxin [34]. Taking this a step further, due to its unique anatomy, the liver is exposed to a series of portal blood stream-derived signals from the diet (e.g., a high concentration of fats and carbohydrates) and the commensal flora (e.g., PAMPs). The nature of the liver's immune response is determined by the fact that a delicate balance has to be maintained between the tolerance towards the above mentioned ligands of the PRR system and the elimination of invading pathogens [35]. A synergy of inflammation-inducing and -resolving mechanisms is necessary in order to carry out physiologic functions such as liver regeneration, the resolution of fibrosis, the response to infection and PAMP tolerance, and this synergy is accomplished with the constitutive expression of both pro-inflammatory [IL-2, IL-7, IL-12, IL-15] and interferon-(IFN)γ and anti-inflammatory [IL-10, IL-13 and the transforming growth factor beta(TGFβ)] cytokines [34,36]. The effects of the NLRP3 inflammasome on certain liver diseases are presented below.

### *2.1. The NLRP3 Inflammasome in NAFLD–NASH*

The preeminent dysfunction in NASH derives from the accumulation of lipids (such as fatty acids and cholesterol) and the consequent activation of NLRP3 in hepatocytes, immune cells and hepatic stellate cells (HSCs) [37]. Another contributing mechanism comprises the upregulation of NF-κB by a lipopolysaccharide (LPS)-mediated stimulation of the TLR4 [37]. Since, as mentioned above, the NASH epidemic in the Western world is expected to promote NASH to the primary cause of HCC and that a great amount of research has been conducted to elucidate its role in pathogenesis and to target it therapeutically, it would be meaningful to present the relevant data in Table 1.

**Table 1.** Summarizes several cellular processes that can be therapeutically targeted.

| Drug/Therapeutic Target | Study/Year/Reference | Study Subjects | Pathway | Outcomes |
|---|---|---|---|---|
| Obeticholic acid | Huang S. (2021) [38] | BMDM cells, hepatocytes/DIO + CCl4 mice | Inhibition of NLRP3 inflammasome activation in macrophages | Reduction in steatosis, fibrosis and immune infiltration |
| | | | Inhibition of lipid-induced NLRP3 inflammasome activation in hepatocytes | |

**Table 1.** *Cont.*

| Drug/Therapeutic Target | Study/Year/Reference | Study Subjects | Pathway | Outcomes |
|---|---|---|---|---|
| Antcin A | Ruan S. (2021) [39] | KC cells/NAFLD mice | Inhibition of NLRP3 inflammasome activation in vitro/in vitro | Inhibition of immune infiltration |
| Auranofin | Hwangbo H. (2020) [40] | High-fat diet (HFD) NAFLD model | Inhibition of NLRP3 inflammasome, NOX4 and PPARγ activation | Inhibition of immune infiltration |
| Cardiolipin inhibitors (shRNA-CLS1) | Liu J. (2019) [41] | KC cells/methionine choline-deficient (MCD) diet mice | Inhibition of NLRP3 inflammasome activation in vitro/in vitro | Improvement in liver biochemistry |
| Cathepsin B inhibition | Tang Y. (2018) [42] | KC cells/MCD diet NASH mice model | Inhibition of NLRP3 inflammasome activation | Inhibition of immune infiltration and steatosis |
| Polyunsaturated fatty acid (PUFA) | Sui Y. (2016) [43] | HFD NASH mice model | Inhibition of NLRP3 inflammasome activation in vitro and in vivo | |
| Melatonin | Yu Y. (2021) [44] | db/m mice, db/db mice | Improvement in mitochondrial membrane potential (MMP)<br>Inhibition in NLRP3 inflammasome activation | Reduction in steatosis, fibrosis and immune infiltration |

Such concrete data about the influence of the NLRP3 inflammasome on the molecular pathogenesis of NASH contributed to the concept of the NLRP3 blockade being a therapeutic target. Mridha A. attempted to block the NLRP3 protein complex in two murine models of steatohepatitis. They demonstrated that cholesterol crystals can activate the NLRP3 inflammasome in TLR-4/Myd88-primed macrophages in vitro and in vivo. Treatment with MCC950, a direct NLRP3 inhibitor in liver tissue, reduced the immune infiltration and mitigated liver injury and progression to fibrosis. Importantly, it was documented that the administration of MCC950 could not only suppress the development of fibrosis but reverse that which was already generated [45]. The latter could prove to change the course of the disease more widely, with a striking impact on the overall survival (OS) of patients and their quality of life and a substantial lightening of the burden of hospitalization and insurance expenses.

### 2.2. The Role of NLRP3 in Viral Hepatitis

More recent evidence proposes that the NLRP3 inflammasome pathway is implicated in the natural course of both HCV and HBV infections.

The non-CD81 uptake of HCV by Kuppfer cells comprises a stimulus for cellular PRRs. The HCV RNA induces the production of IL-1b through the TLR7 MyD88-dependent signaling pathway. In parallel, an K+ efflux triggers the activation of NLRP3 inflammasome for the cleavage and secretion of IL-1b, which further orchestrates the immune response [46]. Negash A. et al. demonstrated an alternative mechanism of NLRP3 activation in HCV. According to their research, the HCV core protein in hepatic macrophages triggers a calcium efflux interceded by phospholipase C [47]. IL-18, the other concomitant factor from NLRP3 activation, appears to stimulate the anti-viral effects of NK cells [48]. On the other hand, HBV, during its natural course, advances strategies to escape from the immune response. Related to this, HBeAg downregulates the NF-kB pathway and the generation of ROS, inhibiting the NLRP3 inflammasome. The above contributes to the HBV viral persistence [49].

A concise report about the role NLRP3 in liver disease (viral hepatitis, alcohol-related liver disease (ARLD), NASH and NAFLD) is presented in Figure 1.

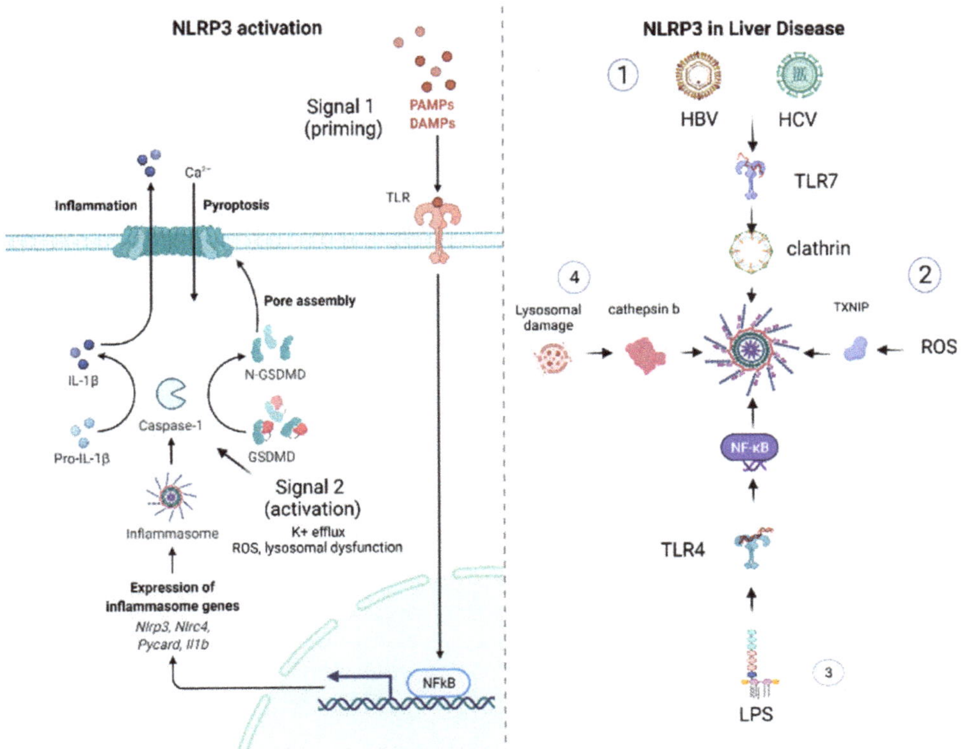


**Figure 1.** On the *left* is presented the activation of the NLRP3 inflammasome. PAMPs and DAMPs from the neighboring parenchyma stimulate the PRR system to upregulate the expression of NF-kB (Signal 1—priming), which triggers the expression of pro-ILs and the components of the inflammasome machinery. Additional signals from K+ efflux, ROS generation or lysosomal dysfunction activate the NLRP3 inflammasome. The activated caspase-1 potentiates the generation of active IL-1, IL-18 and gasdermin. The latter provokes pore formation in the cellular membrane, causing cell death and the release of inflammation mediators. On the *right* are reported the principal pathologic mechanisms in common liver diseases leading to NLRP3 activation: (1) HBV and HCV infection, (2) ARLD, (3) NASH and liver injury in sepsis and (4) the accumulation of lipid droplets.

## 3. The Role of NLRP3 in the Shaping of the HCC Tumor Microenvironment (TME)

As mentioned above, HCC develops in the vast majority of cases under the influence of chronic liver inflammation. Its microenvironment is comprised of a heterogeneous spectrum of cellular populations such as cancer-associated fibroblasts (CAFs) with the resultant extracellular matrix (ECM), HSCs, endothelial cells and various immunosuppressive aggregates. The latter are composed by an amalgam of myeloid-derived suppressor cells (MDSCs), tumor-associated macrophages (TAMs), tumor-associated neutrophils (TANs) and T regulatory cells (Tregs) [50]. In relapsed tumors, the microenvironment is characterized by the abundance of innate cells (e.g., dendritic cells and CD8+ lymphocytes with low toxicity) and diminished amounts of Tregs in comparison with primary tumors [51]. With the exception of MDSCs and Tregs, which consistently display a pro-tumorigenic phenotype exerting immune tolerance towards tumor cells, a sophisticated crosstalk within the tumor microenvironment (TME) polarizes the cellular populations in a context-specific manner, mediating tumor growth, invasion and metastasis as well as regulating angiogenesis and enforcing drug resistance [52]. It is of paramount importance to highlight the "double-edged sword" impact of inflammatory lytic programmed cell death during the initiation and progression of hepatocellular carcinogenesis. During the early stages of carcinogenesis, the immune response can repress tumorigenesis, potentiating tumor surveillance, e.g., IL-18 secretion activates natural killer (NK) cells to exert their cytotoxic potential against tumor cells; however, unprovoked continuous inflammasome activation and the presence of IL-1b in the TME attracts immunosuppressive cells such as MDSCs, which promote tumor invasion and metastasis [53]. These impacts are highlighted in a comprehensive study by Wei Q. et al., which investigated the expression of NLRP3 inflammasome in various stages of HCC [18].

Despite a recent surge in the literature of single cell analyses that demonstrate the existence of TAMs which synchronously express signatures from both the M1 and M2 macrophage clusters [54], TAMs and TANs are classically divided into two subtypes: on the one hand, M1 and N1 secrete IFN-$\gamma$ and IL-12, potentiating the effector compartment of the cellular immunity, evading tumor growth; on the other hand, the M2 and N2 phenotypes, which are associated with the secretion of IL-4, IL-10 and TGF-$\beta$, induce an Th2 immunologic response [50,55,56]. The exact mechanisms by which TAMs, TANs and their contexts regulate HCC progression have been extensively reviewed elsewhere [50,56,57]. A continuum of transitions in the synthesis and composition of HCC TME can be observed based on integral causes such as the stochastic accumulation of mutations or extrinsic stimuli (e.g., chemotherapy). For example, the enhancement of Wnt/$\beta$-catenin signaling in cancer cells induces a c-myc-potentiated differentiation into M2 macrophages [58], while the administration of sorafenib augments the infiltration of TANs, which is accountable for the further recruitment of TAMs and Tregs [59]. The significance of metabolic reprogramming in the establishment of TME cannot be overstated enough. Aerobic glycolysis (or the Warburg effect), which is indispensable in order to provide building blocks to support the enhanced anabolic demands of proliferating tumor cells, generates substantial amounts of lactate. The presence of an acidic, nutrient-poor and hypoxic microenvironment results in the exhaustion of effector T-cells, polarizes the macrophages towards an M2 phenotype and impairs tumor-antigen presentation by dendritic cells [60]. Additionally, neutrophils are enforced to secrete neutrophil extracellular traps (NETs), promoting metastasis [61]. By virtue of intrinsic tumor hypoxia and drug-imposed hypoxia, and the fact that angiogenesis blocking constitutes a first- and second-line therapeutic choice in HCC, a shift towards fatty acid instead of glucose metabolism has been evidenced [62,63]. The importance of lipid metabolism was summarized exhaustively by Hu B. et al. [64]. It is of paramount importance to comprehensively investigate the role of exosomes as a means of HCC progression [52] and as a therapeutic moratorium [65], while the impact of ECM composition on the regulation of carcinogenesis is increasingly being inferred [66].

The importance of NLRP3 as a regulatory molecule in the pre-malignant stages of liver disease has been highlighted. It is well established that the NLRP3 inflammasome is released by pyroptotic hepatocytes and incorporated by adjoining cells, promoting inflammation and ECM deposition [67]. A growing body of recent literature has investigated the influence of the NLRP3 inflammasome on the configuration of the HCC microenvironment. Ding Y. et al. recently demonstrated that NLRP3 might orchestrate the infiltration of cellular populations in innate and adaptive immunity. In more detail, NLRP3 expression was associated with the degree of B cell, CD4+ T cell, CD8+ T cell, neutrophils, and dendritic cell (DC) invasion. This was reflected by the correlation of the immune, stromal and estimate score with NLRP3 expression. The above might be mediated by the mutation burden of the mismatch repair system (MMR) and co-expression with various methyl-transferases and immune checkpoint inhibitors such as lymphocyte-activation gene 3 (LAG3), inducible T cell co-stimulator (ICOS), cytotoxic T lymphocyte antigen 4 (CTLA4), T cell immunoglobulin, mucin domain-containing protein 3 (TIM3), programmed cell death protein 1 (PD1), programmed death-ligand 1 (PD-L1), PD-L2, T cell immunoglobulin and the ITIM domain (TIGIT) [68]. The latter appears to intercede the cytotoxic capacity of NK cells. NK cells carry several receptors on their surface, e.g., NKG2D. The abundance of their ligands in the TME is regulated by enzymes such as matrix metalloproteinases

(MMPs), which regulate the composition of the ECM, among others. Lee H. et al. reported that the ablation of NLRP3 in HCC SK-Hep1 Luc cells resulted in elevated NK cytotoxicity in an IFN-g-independent mechanism. The involved mechanism constitutes the downregulation of MICA/B, a ligand of the activating NKG2D receptor, by the downregulation of MMP2, MMP9 and MMP14 in mice engrafted with NLRP3 KO(−/−) SK-Hep1 Luc cells [69]. Fatty acid oxidation (FAO) constitutes another process by which NLRP3 influences the HCC TME. The fibronectin type III structural domain-containing protein 5 (FNDC5) is frequently upregulated in HCC tissues, with a negative impact on the potential of tumor malignancy [70]. Liu H. et al. documented both in vitro and in vivo that FNDC5 induces the FAO-dependent M2 phenotype by blocking M1 polarization, downregulating the NF-κB/NLRP3 pathway [71]. Related to this, Zhang Q. et al. studied the contribution of FAO to the pro-inflammatory properties of an M2 subset in HCC, as they had previously documented the importance of the hypoxia-inducing factor (HIF)-1α/IL-1β signaling pathway [72]. They concluded that FAO-generated ROS drive, in an NLRP3-dependent manner, the expression of IL1-β, providing another potentially therapeutically targetable mechanism. In parallel, receptor-interacting protein kinase 3 (RIPK3), another regulator of fatty acid metabolism in TAMs, utilized the NLRP3 inflammasome machinery in order to become activated through caspase-1-mediated cleavage [73]. Finally, Tu C. et al. demonstrated that lactate and TGF-b exert their immunomodulatory effects in the TME partially by utilizing the NLRP3 inflammasome. Specifically, lactate can trigger inflammasome initiation on tumor-associated macrophages by building up ROS in parallel with tumor cell-derived TGF-b secretion. The latter exerts its function through the SMAD-autophagy-ROS signaling cascade [74]. It worth mentioning that evidence has begun to emerge about the regulatory role of the NLRP3 inflammasome in the tumor-stromal interconnection. Zan Y et al. demonstrated that the blockage of NEK7 in HCC cells could reduce the activation of HSCs [75]. As demonstrated above, the NLRP3 inflammasome, acting as a scavenging and effector system in the HCC TME, significantly shapes the anti-tumor immune response. Meanwhile, its effects in the tumor stroma requires further investigation.

## 4. The Role of the NLRP3 Inflammasome in the Therapeutic Management of HCC

The introduction of a wide variety of therapeutic approaches to the management of HCC has made the reformation of medical practice an experienced reality. Without exhaustive details, according to the latest European Society of Medical Oncology (ESMO) practice guidelines, patients who present with a single mass irrespectively of its size or with up to three nodules that are less than 3 cm each and have preserved liver function and Eastern Cooperative Oncology Group (ECOG) performance status (PS) are candidates for: hepatectomy, liver transplantation, thermal ablation or transarterial chemoembolization (TACE). Stereotactic body radiotherapy (SBRT), high dose rate (HDR)-brachytherapy and selective internal radiotherapy (SIRT) constitute alternative treatment approaches when management constrains arise. TACE comprises the standard of care in the treatment of patients presenting multinodular disease, ECOG PS 0 and efficient PS, with liver transplantation, resection, SIRT and systemic medical therapy being potent alternatives. Imaging-based signs of systemic disease, such as the invasion of portal veins or extrahepatic spread, render patients with preserved liver function and PS 0–2 as candidates for immunotherapeutic or anti-angiogenetic therapies with certain monoclonal antibodies and tyrosine kinase inhibitors (TKIs). Atezolizumab plus bevacizumab, and alternatively sorafenib or Lenvatinib, are considered the first line of treatment. Sorafenib, lenvatinib, cabozantinib, regorafenib and ramucirumab are authorized alternatives after the combination of atezolizumab and bevacizumab, while cabozantinib, regorafenib and ramucirumab constitute second line choices after sorafenib. For patients with end-stage liver disease, only supportive care measures are applied [76,77]. Concerning surgical techniques, an immense developmental evolution has radicalized surgical approaches. For minimal, unilobar or bilobar disease, "parenchyma-sparing liver surgery" constitutes an one-step resection. When the future liver remnant (FLR) is expected to be insufficient to sustain normal hepatic function after surgery, portal vein embolization (PVE) with the subsequent hypertrophy can be proposed as a treatment choice [78]. Two-stage hepatectomies (TSH)—i.e., conventional TSH with PVE only [79,80] and liver partition and portal vein ligation for staged hepatectomy (ALPPS) [81]—when performed in centers with sufficient experience, can offer significant survival advantages in patients with unresectable primary liver tumors [82,83].

### 4.1. The NLRP3 as Therapeutic Target

Given the above and taking into consideration a recent surge in clinical studies in the literature regarding the effects of the NLRP3 inflammasome system in HCC, it becomes evident that the NLRP3 inflammasome system could acquire a role in the therapeutic algorithm either as a therapeutic target or as a biomarker. In the first case, it could exert its usefulness as an adjuvant treatment in patients with low tumor burden or as part of systemic treatment in patients with greater tumor volume. The broad spectrum of NLRP3 activating stimuli, which was presented earlier, results in a diverse spectrum of signaling pathways and mechanisms that can influence HCC progression through NLRP3 activations such as ROS scavenging, immune cell reprogramming and metabolic reprogramming of the infiltrated immune cells. The vast majority the NLRP3 activations lead to tumor growth regression, with a few exceptions. The above is presented analytically in the Table 2.

**Table 2.** A brief summary of the preclinical data with respect to NLRP3 inflammasome.

| Drug/Therapeutic Target | Study/Year/Reference | Study Subjects | Pathway/Mechanism | Outcomes |
|---|---|---|---|---|
| Alpinumisoflavone (AIF) | Zhang Y. (2020) [84] | SMMC 7721, Huh7 cells | NLRP3-mediated pyroptosis | Reduction of tumor growth and metastatic potential |
| NEK7 inhibition | Yan Z. (2022) [75] | MHCC97L, HepG2 cell/mice | NLRP3-mediated pyroptosis | Reduction of tumor growth and metastatic potential<br>Promotion of cancer cell-stromal communication |
| Biejiajian pills (BJJ) | Feng M. (2020) [85] | Diethyl nitrosamine-mediated hepatocarcinogenesis in SD rats | Dose-dependent reduction in NLRP3 activation | Reduction of tumor growth |
| Luteoloside | Fan S. (2014) [86] | Hep3B, SNU-449, Huh-7, MHCC- LM3 and MHCC97-H cell lines/BALB/c-nu/nu male mice | Downregulation of NLRP3 activation | Reduction of tumor growth and metastatic potential in vitro and in vivo |
| Metformin | Shen Z. (2021) [87] | BALB/c nude male mice | FOXO3-dependent induction of the NLRP3 inflammasome and autophagy | Reduction of tumor growth |
| Geranylgeranoic acid (GGA) | Yabuta S. (2020) [88] | HuH-7 cells | TLR4-induced ROS generation activating both non-canonical and canonical phases of pyroptosis | Reduction of tumor growth |
| NLRP3 siRNA or CPT1A blockage or N-acetyl cysteine (NAC) or etomoxir | Zhang Q. (2018) [89] | HepG2, Hep3B cells | Reduction in NLRP3 activation by FAO-mediated ROS | Reduction of HCC metastatic potential |

**Table 2.** *Cont.*

| Drug/Therapeutic Target | Study/Year/Reference | Study Subjects | Pathway/Mechanism | Outcomes |
|---|---|---|---|---|
| 17β-estradiol (E2) | Wei Q. (2015) [90] | BEL7402, SMMC7721 and HepG2 cells | ERβ/MAPK/ERK-mediated activation of NLRP3 inflammasome | Reduction of tumor growth |
| 17β-estradiol (E2) | Wei Q. (2019) [91] | HepG2 cells | Autophagy reduction through E2/ERβ/AMPK/mTOR-induced NLRP3 activation | Reduction of tumor growth |
| IRAK1 blockage | Chen W. (2020) [92] | Huh7, Hep3B cells | Downregulation of NLRP3 activation through ERK/JNK pathway | Reduction of tumor growth |
| PPARγ inhibitors or FNDC5 blockage | Liu H. (2021) [71] | HepG2, SMCC7721 cells overexpressing FNDC5 | Activation of the NF-κB/NLRP3 pathway | M1 TAM polarization |
| NLRP3 blockage | Lee H. (2021) [69] | HCC SK-Hep1 Luc, NK-92 cells | Upregulation of MICA/B on the HCC cells induced by NK activation through NKG2D receptor | Reduction in tumor growth and metastasis |
| RIPK3 mimic or FAO blockage | Wu L. (2020) [73] | Human HCC tissues | Activation of the ROS–Caspase1–PPAR pathway reversed M2 programming | Reduction in tumor growth |

### 4.2. The NLRP3 as Biomarker

Biomarkers are invaluable tools in our efforts to implement a more personalized approach in medical practice. Their application extends from HCC diagnosis, staging and tumor grading to guiding therapeutic management. Several studies [9] considering the NLRP3 as a biomarker in several malignancies have begun to accumulate [29]. The reduced expression of NLRP3 in colorectal cancer (CRC) tissue holds a positive predictive value [93]. Likewise, the upregulation of NLRP3 expression in breast cancer heralds a poorer five-year survival rate [94]. The capability of the NLRP3 inflammasome to promote an immune response provides the potential to predict the response to immunotherapy. In melanoma, the mutational burden of the *NLRP3* gene could provide useful data about the response to immunotherapy [95]. Unfortunately, despite the necessity of detecting and implementing into clinical practice new biomarkers for HCC [77], the evidence regarding the role of NLRP3 as a biomarker in HCC is limited. Wei Q. et al. investigated the expression of NLRP3 inflammasome components in accordance with HCC progression. They reported the inhibition of the expression of the inflammasome in HCC tissues, which was inversely correlated with the pathological grading and the HCC clinical stage [18]. Wang J. et al. developed a pyroptosis-associated computational algorithm to assess patient prognosis. The *NLRP3* gene had the highest mutation incidence. Copy number amplification was detected while its expression was significantly differentiated among HCC and the neighboring healthy tissue. Concerning the predictive value of NLRP3, the highest expression of NLRP3 was associated with poor OS [96].

## 5. Conclusions—Future Perspectives

Inflammation comprises a cardinal processes in liver homeostasis and disease [34]. Its significance has been studied conscientiously in the context of carcinogenesis [10], and immunotherapy with atezolizumab and bevacizumab constitutes the first line of therapy in patients with unresectable disease [97].

The NLRP3 inflammasome appears to orchestrate the immune responses in liver diseases. It can be concluded that the activation of the NLRP3 inflammasome drives the progression from NAFLD to NASH, while hepatotropic viruses (e.g., HBV, HCV) downregulate it in order to evade immune invasion. Several candidate therapeutic targets that mediate the progression of NASH have been recognized, but clinical trials about their effects in humans are lacking. The generation of drugs that target the NLRP3 inflammasome would be game-changing in the management of NAFLD-driven HCC. On the other hand, the NLRP3 inflammasome appears to be an ideal therapeutic target in HCC. The NLRP3 inflammasome and its constituents are downregulated in HCC. This inhibition correlates with a greater disease stage and poorer differentiation [18]. Multiple mechanisms of HCC cytotoxicity by the therapeutic instrumentation of NLRP3 have been documented, such as direct hepatotoxicity, M1 polarization of TAM, the enhancement of NK cell activity and the regulation of the tumor–stromal communication. The pre-clinical data are encouraging, and their efficacy should be validated in clinical trials in the near future. The above therapeutic mechanisms are summarized in Figure 2.

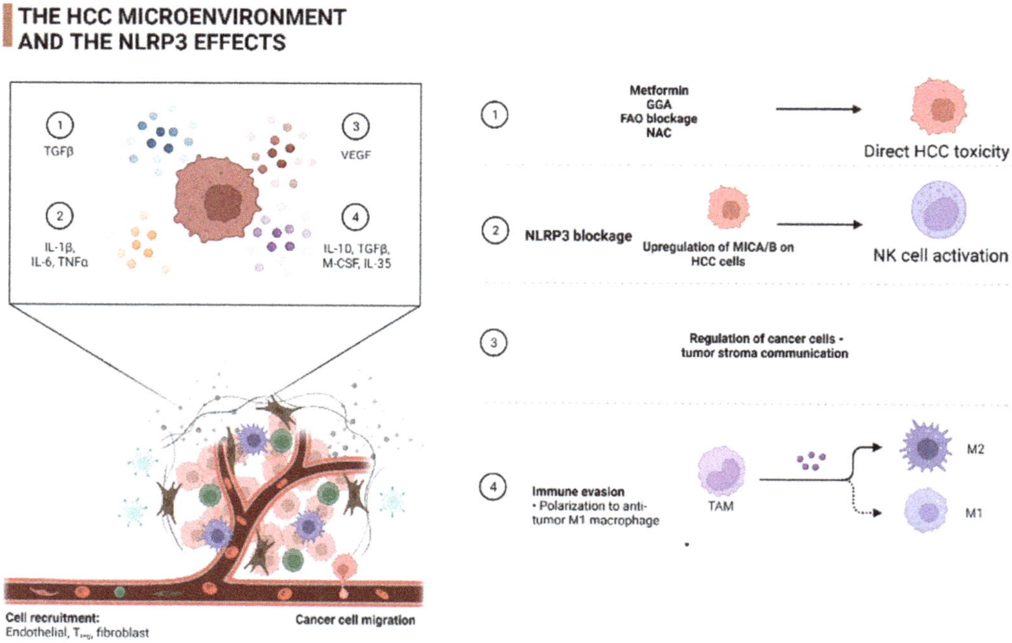


**Figure 2.** A brief outline of the therapeutic effects of the NLRP3 system in HCC.

Useful conclusions regarding role of the NLRP3 inflammasome in HCC carcinogenesis could be arrived at by analyzing the effects of other inflammasomes in different tissues. For example, the NLRP6 inflammasome constitutes a fundamental sensor of the intestinal mucosa and its downregulation is linked with alterations in gut microbiota [98] and a predisposition to inflammatory bowel disease and gastrointestinal cancer [24]. The latter is exerted by its assistance in the regulation of mucus secretion in goblet cells, where caspase-1 is required to activate the microtubule-associated protein 1A/1B-light chain 3 (LC3) and trigger mucus secretion [99]. Chen G. et al. demonstrated that *Nlrp6-knockdown* mice produce excessive amounts of inflammation and fail to heal damaged mucosa. The latter, in conjunction with the upregulation of TNF-a and IL-6, which stimulate the NF-kB and STAT3 pathways, promotes carcinogenesis [24]. A "steatohepatic subtype" has been recognized in HCC with alteration in the IL6/JAK-STAT signaling axis, and mutations in this pathway are frequent in HCC [6]. It would not be illogical to hypothesize that the downregulation of the NLRP3 inflammasome in HCC could potentiate carcinogenesis by the subsequent upregulation of tumor-promoting cytokines.

The role of the NLRP3 inflammasome is poorly investigated, and our scientific efforts should be pointed in this direction. A growing body of work in the literature substantiates that, in the years to come, the manipulation of the NLRP3 sensor pathway could prevent the progression to carcinogenesis in several patient groups.

**Author Contributions:** Conceptualization, S.P.P.; writing—original draft preparation, S.P.P. and N.D.; writing—review and editing, S.P.P. and N.D.; visualization, S.P.P.; supervision, E.K. and S.T. All authors have read and agreed to the published version of the manuscript.

**Funding:** This research received no external funding.

**Conflicts of Interest:** The authors declare no conflict of interest.

## References

1. Galle, P.R.; Forner, A.; Llovet, J.M.; Mazzaferro, V.; Piscaglia, F.; Raoul, J.L.; Schirmacher, P.; Vilgrain, V. EASL Clinical Practice Guidelines: Management of hepatocellular carcinoma. *J. Hepatol.* **2018**, *69*, 182–236. [CrossRef] [PubMed]
2. Papatheodoridi, A.; Papatheodoridis, G. Hepatocellular carcinoma: The virus or the liver? *Liver Int.* **2022**, 1–9. [CrossRef]
3. Markakis, G.E.; Koulouris, A.; Tampaki, M.; Cholongitas, E.; Deutsch, M.; Papatheodoridis, G.V.; Koskinas, J. The changing epidemiology of hepatocellular carcinoma in Greece. *Ann. Gastroenterol.* **2022**, *35*, 88–94. [CrossRef] [PubMed]
4. Younossi, Z.; Stepanova, M.; Ong, J.P.; Jacobson, I.M.; Bugianesi, E.; Duseja, A.; Eguchi, Y.; Wong, V.W.; Negro, F.; Yilmaz, Y.; et al. Nonalcoholic Steatohepatitis Is the Fastest Growing Cause of Hepatocellular Carcinoma in Liver Transplant Candidates. *Clin. Gastroenterol. Hepatol.* **2019**, *17*, 748–755.e3. [CrossRef] [PubMed]
5. Tarao, K.; Nozaki, A.; Ikeda, T.; Sato, A.; Komatsu, H.; Komatsu, T.; Taguri, M.; Tanaka, K. Real impact of liver cirrhosis on the development of hepatocellular carcinoma in various liver diseases—Meta-analytic assessment. *Cancer Med.* **2019**, *8*, 1054–1065. [CrossRef]
6. Rebouissou, S.; Nault, J.C. Advances in molecular classification and precision oncology in hepatocellular carcinoma. *J. Hepatol.* **2020**, *72*, 215–229. [CrossRef]
7. Villanueva, A. Hepatocellular Carcinoma. *N. Engl. J. Med.* **2019**, *380*, 1450–1462. [CrossRef]
8. Calderaro, J.; Ziol, M.; Paradis, V.; Zucman-Rossi, J. Molecular and histological correlations in liver cancer. *J. Hepatol.* **2019**, *71*, 616–630. [CrossRef]
9. Calderaro, J.; Couchy, G.; Imbeaud, S.; Amaddeo, G.; Letouzé, E.; Blanc, J.F.; Laurent, C.; Hajji, Y.; Azoulay, D.; Bioulac-Sage, P.; et al. Histological subtypes of hepatocellular carcinoma are related to gene mutations and molecular tumour classification. *J. Hepatol.* **2017**, *67*, 727–738. [CrossRef]
10. Hanahan, D.; Weinberg, R.A. Hallmarks of cancer: The next generation. *Cell* **2011**, *144*, 646–674. [CrossRef]
11. Rizzo, A.; Ricci, A.D.; Gadaleta-Caldarola, G.; Brandi, G. First-line immune checkpoint inhibitor-based combinations in unresectable hepatocellular carcinoma: Current management and future challenges. *Expert Rev. Gastroenterol. Hepatol.* **2021**, *15*, 1245–1251. [CrossRef] [PubMed]
12. Rizzo, A.; Ricci, A.D. PD-L1, TMB, and other potential predictors of response to immunotherapy for hepatocellular carcinoma: How can they assist drug clinical trials? *Expert Opin. Investig. Drugs* **2022**, *31*, 415–423. [CrossRef] [PubMed]
13. Choucair, K.; Morand, S.; Stanbery, L.; Edelman, G.; Dworkin, L.; Nemunaitis, J. TMB: A promising immune-response biomarker, and potential spearhead in advancing targeted therapy trials. *Cancer Gene Ther.* **2020**, *27*, 841–853. [CrossRef] [PubMed]
14. Simon, T.G.; Duberg, A.-S.; Aleman, S.; Chung, R.T.; Chan, A.T.; Ludvigsson, J.F. Association of Aspirin with Hepatocellular Carcinoma and Liver-Related Mortality. *N. Engl. J. Med.* **2020**, *382*, 1018–1028. [CrossRef] [PubMed]
15. Liu, C.; Huang, X.; Su, H. The role of the inflammasome and its related pathways in ovarian cancer. *Clin. Transl. Oncol.* **2022**. [CrossRef]
16. West, A.J.; Deswaerte, V.; West, A.C.; Gearing, L.J.; Tan, P.; Jenkins, B.J. Inflammasome-Associated Gastric Tumorigenesis Is Independent of the NLRP3 Pattern Recognition Receptor. *Front. Oncol.* **2022**, *12*, 830350. [CrossRef]
17. Caruso, S.; O'Brien, D.R.; Cleary, S.P.; Roberts, L.R.; Zucman-Rossi, J. Genetics of Hepatocellular Carcinoma: Approaches to Explore Molecular Diversity. *Hepatology* **2021**, *73*, 14–26. [CrossRef]
18. Wei, Q.; Mu, K.; Li, T.; Zhang, Y.; Yang, Z.; Jia, X.; Zhao, W.; Huai, W.; Guo, P.; Han, L. Deregulation of the NLRP3 inflammasome in hepatic parenchymal cells during liver cancer progression. *Lab. Investig.* **2014**, *94*, 52–62. [CrossRef]
19. Kelley, N.; Jeltema, D.; Duan, Y.; He, Y. The NLRP3 Inflammasome: An Overview of Mechanisms of Activation and Regulation. *Int. J. Mol. Sci.* **2019**, *20*, 3328. [CrossRef]
20. Strowig, T.; Henao-Mejia, J.; Elinav, E.; Flavell, R. Inflammasomes in health and disease. *Nature* **2012**, *481*, 278–286. [CrossRef]

21. Zhen, Y.; Zhang, H. NLRP3 inflammasome and inflammatory bowel disease. *Front. Immunol.* **2019**, *10*, 276. [CrossRef]
22. Ranson, N.; Kunde, D.; Eri, R. Regulation and Sensing of Inflammasomes and Their Impact on Intestinal Health. *Int. J. Mol. Sci.* **2017**, *18*, 2379. [CrossRef] [PubMed]
23. Pirzada, R.H.; Javaid, N.; Choi, S. The Roles of the NLRP3 Inflammasome in Neurodegenerative and Metabolic Diseases and in Relevant Advanced Therapeutic Interventions. *Genes* **2020**, *11*, 131. [CrossRef] [PubMed]
24. Chen, G.Y.; Liu, M.; Wang, F.; Bertin, J.; Núñez, G. A Functional Role for Nlrp6 in Intestinal Inflammation and Tumorigenesis. *J. Immunol.* **2011**, *186*, 7187–7194. [CrossRef]
25. Liao, Y.; Liu, K.; Zhu, L. Emerging Roles of Inflammasomes in Cardiovascular Diseases. *Front. Immunol.* **2022**, *13*, 834289. [CrossRef] [PubMed]
26. Anton-Pampols, P.; Diaz-Requena, C.; Martinez-Valenzuela, L.; Gomez-Preciado, F.; Fulladosa, X.; Vidal-Alabro, A.; Torras, J.; Lloberas, N.; Draibe, J. The Role of Inflammasomes in Glomerulonephritis. *Int. J. Mol. Sci.* **2022**, *23*, 4208. [CrossRef] [PubMed]
27. Saljic, A.; Heijman, J.; Dobrev, D. Emerging Antiarrhythmic Drugs for Atrial Fibrillation. *Int. J. Mol. Sci.* **2022**, *23*, 4096. [CrossRef]
28. Carvalho, A.M.; Bacellar, O.; Carvalho, E.M. Protection and Pathology in Leishmania braziliensis Infection. *Pathogens* **2022**, *11*, 466. [CrossRef]
29. Li, Z.; Chen, X.; Tao, J.; Shi, A.; Zhang, J.; Yu, P. Exosomes Regulate NLRP3 Inflammasome in Diseases. *Front. Cell Dev. Biol.* **2022**, *9*, 802509. [CrossRef] [PubMed]
30. Ohto, U.; Kamitsukasa, Y.; Ishida, H.; Zhang, Z.; Murakami, K.; Hirama, C. Structural basis for the oligomerization-mediated regulation of NLRP3 in fl ammasome activation. *Proc. Natl. Acad. Sci. USA* **2022**, *119*, e2121353119. [CrossRef]
31. Liu, X.; Zhang, Z.; Ruan, J.; Pan, Y.; Magupalli, V.G.; Wu, H.; Lieberman, J. Inflammasome-activated gasdermin D causes pyroptosis by forming membrane pores. *Nature* **2016**, *535*, 153–158. [CrossRef] [PubMed]
32. Mangan, M.S.J.; Olhava, E.J.; Roush, W.R.; Seidel, H.M.; Glick, G.D.; Latz, E. Targeting the NLRP3 inflammasome in inflammatory diseases: Current perspectives. *Nat. Rev. Drug Discov.* **2018**, *17*, 588–606. [CrossRef] [PubMed]
33. He, Y.; Hara, H.; Núñez, G. Mechanism and Regulation of NLRP3 Inflammasome Activation. *Trends Biochem. Sci.* **2016**, *41*, 1012–1021. [CrossRef]
34. Robinson, M.W.; Harmon, C.; O'Farrelly, C. Liver immunology and its role in inflammation and homeostasis. *Cell. Mol. Immunol.* **2016**, *13*, 267–276. [CrossRef]
35. Liaskou, E.; Wilson, D.V.; Oo, Y.H. Innate immune cells in liver inflammation. *Mediat. Inflamm.* **2012**, *2012*, 949157. [CrossRef]
36. Kelly, A.M.; Golden-Mason, L.; Traynor, O.; Geoghegan, J.; McEntee, G.; Hegarty, J.E.; O'Farrelly, C. Changes in hepatic immunoregulatory cytokines in patients with metastatic colorectal carcinoma: Implications for hepatic anti-tumour immunity. *Cytokine* **2006**, *35*, 171–179. [CrossRef]
37. Al Mamun, A.; Akter, A.; Hossain, S.; Sarker, T.; Safa, S.A.; Mustafa, Q.G.; Muhammad, S.A.; Munir, F. Role of NLRP3 inflammasome in liver disease. *J. Dig. Dis.* **2020**, *21*, 430–436. [CrossRef] [PubMed]
38. Huang, S.; Wu, Y.; Zhao, Z.; Wu, B.; Sun, K.; Wang, H.; Qin, L.; Bai, F.; Leng, Y.; Tang, W. A new mechanism of obeticholic acid on NASH treatment by inhibiting NLRP3 inflammasome activation in macrophage. *Metabolism* **2021**, *120*, 154797. [CrossRef]
39. Ruan, S.; Han, C.; Sheng, Y.; Wang, J.; Zhou, X.; Guan, Q.; Li, W.; Zhang, C.; Yang, Y. Antcin A alleviates pyroptosis and inflammatory response in Kupffercells of non-alcoholic fatty liver disease by targeting NLRP3. *Int. Immunopharmacol.* **2021**, *100*, 108126. [CrossRef]
40. Hwangbo, H.; Kim, M.Y.; Ji, S.Y.; Kim, S.Y.; Lee, H.; Kim, G.Y.; Park, C.; Keum, Y.S.; Hong, S.H.; Cheong, J.; et al. Auranofin attenuates non-alcoholic fatty liver disease by suppressing lipid accumulation and nlrp3 inflammasome-mediated hepatic inflammation in vivo and in vitro. *Antioxidants* **2020**, *9*, 1040. [CrossRef]
41. Liu, J.; Wang, T.; He, K.; Xu, M.; Gong, J.P. Cardiolipin inhibitor ameliorates the non-alcoholic steatohepatitis through suppressing IMLRP3 inflammasome activation. *Eur. Rev. Med. Pharmacol. Sci.* **2019**, *23*, 8158–8167. [CrossRef] [PubMed]
42. Tang, Y.; Cao, G.; Min, X.; Wang, T.; Sun, S.; Du, X.; Zhang, W. Cathepsin B inhibition ameliorates the non-alcoholic steatohepatitis through suppressing caspase-1 activation. *J. Physiol. Biochem.* **2018**, *74*, 503–510. [CrossRef] [PubMed]
43. Sui, Y.H.; Luo, W.J.; Xu, Q.Y.; Hua, J. Dietary saturated fatty acid and polyunsaturated fatty acid oppositely affect hepatic NOD-like receptor protein 3 inflammasome through regulating nuclear factor-kappa B activation. *World J. Gastroenterol.* **2016**, *22*, 2533–2544. [CrossRef]
44. Yu, Y.; Chen, D.; Zhao, Y.; Zhu, J.; Dong, X. Melatonin ameliorates hepatic steatosis by inhibiting NLRP3 inflammasome in db/db mice. *Int. J. Immunopathol. Pharmacol.* **2021**, *35*, 205873842110368. [CrossRef] [PubMed]
45. Mridha, A.R.; Wree, A.; Robertson, A.A.B.; Yeh, M.M.; Johnson, C.D.; Van Rooyen, D.M.; Haczeyni, F.; Teoh, N.C.H.; Savard, C.; Ioannou, G.N.; et al. NLRP3 inflammasome blockade reduces liver inflammation and fibrosis in experimental NASH in mice. *J. Hepatol.* **2017**, *66*, 1037–1046. [CrossRef]
46. Negash, A.A.; Ramos, H.J.; Crochet, N.; Lau, D.T.Y.; Doehle, B.; Papic, N.; Delker, D.A.; Jo, J.; Bertoletti, A.; Hagedorn, C.H.; et al. IL-1$\beta$ production through the NLRP3 inflammasome by hepatic macrophages links hepatitis C virus infection with liver inflammation and disease. *PLoS Pathog.* **2013**, *9*, e1003330. [CrossRef] [PubMed]
47. Negash, A.A.; Olson, R.M.; Griffin, S.; Gale, M. Modulation of calcium signaling pathway by hepatitis C virus core protein stimulates NLRP3 inflammasome activation. *PLoS Pathog.* **2019**, *15*, e1007593. [CrossRef]
48. Serti, E.; Werner, J.M.; Chattergoon, M.; Cox, A.L.; Lohmann, V.; Rehermann, B. Monocytes Activate Natural Killer Cells via Inflammasome-Induced Interleukin 18 in Response to Hepatitis C Virus Replication. *Gastroenterology* **2014**, *147*, 209–220.e3. [CrossRef]

49. Yu, X.; Lan, P.; Hou, X.; Han, Q.; Lu, N.; Li, T.; Jiao, C.; Zhang, J.; Zhang, C.; Tian, Z. HBV inhibits LPS-induced NLRP3 inflammasome activation and IL-1β production via suppressing the NF-κB pathway and ROS production. *J. Hepatol.* **2017**, *66*, 693–702. [CrossRef]
50. Arvanitakis, K.; Koletsa, T.; Mitroulis, I.; Germanidis, G. Tumor-Associated Macrophages in Hepatocellular Carcinoma Pathogenesis, Prognosis and Therapy. *Cancers* **2022**, *14*, 226. [CrossRef]
51. Sun, Y.; Wu, L.; Zhong, Y.; Zhou, K.; Hou, Y.; Wang, Z.; Zhang, Z.; Xie, J.; Wang, C.; Chen, D.; et al. Single-cell landscape of the ecosystem in early-relapse hepatocellular carcinoma. *Cell* **2021**, *184*, 404–421.e16. [CrossRef] [PubMed]
52. Wu, Q.; Zhou, L.; Lv, D.; Zhu, X.; Tang, H. Exosome-mediated communication in the tumor microenvironment contributes to hepatocellular carcinoma development and progression. *J. Hematol. Oncol.* **2019**, *12*, 53. [CrossRef] [PubMed]
53. Thi, H.T.H.; Hong, S. Inflammasome as a Therapeutic Target for Cancer Prevention and Treatment. *J. Cancer Prev.* **2017**, *22*, 62–73. [CrossRef] [PubMed]
54. Zhang, Q.; He, Y.; Luo, N.; Patel, S.J.; Han, Y.; Gao, R.; Modak, M.; Carotta, S.; Haslinger, C.; Kind, D.; et al. Landscape and Dynamics of Single Immune Cells in Hepatocellular Carcinoma. *Cell* **2019**, *179*, 829–845.e20. [CrossRef]
55. Hao, N.B.; Lü, M.H.; Fan, Y.H.; Cao, Y.L.; Zhang, Z.R.; Yang, S.M. Macrophages in tumor microenvironments and the progression of tumors. *Clin. Dev. Immunol.* **2012**, *2012*, 948098. [CrossRef]
56. Arvanitakis, K.; Mitroulis, I.; Germanidis, G. Tumor-associated neutrophils in hepatocellular carcinoma pathogenesis, prognosis, and therapy. *Cancers* **2021**, *13*, 2899. [CrossRef]
57. Peña-Romero, A.C.; Orenes-Piñero, E. Dual Effect of Immune Cells within Tumour Microenvironment: Pro- and Anti-Tumour Effects and Their Triggers. *Cancers* **2022**, *14*, 1681. [CrossRef]
58. Yang, Y.; Ye, Y.C.; Chen, Y.; Zhao, J.L.; Gao, C.C.; Han, H.; Liu, W.C.; Qin, H.Y. Crosstalk between hepatic tumor cells and macrophages via Wnt/β-catenin signaling promotes M2-like macrophage polarization and reinforces tumor malignant behaviors. *Cell Death Dis.* **2018**, *9*, 793. [CrossRef]
59. Zhou, S.L.; Zhou, Z.J.; Hu, Z.Q.; Huang, X.W.; Wang, Z.; Chen, E.B.; Fan, J.; Cao, Y.; Dai, Z.; Zhou, J. Tumor-Associated Neutrophils Recruit Macrophages and T-Regulatory Cells to Promote Progression of Hepatocellular Carcinoma and Resistance to Sorafenib. *Gastroenterology* **2016**, *150*, 1646–1658.e17. [CrossRef]
60. Du, D.; Liu, C.; Qin, M.; Zhang, X.; Xi, T.; Yuan, S.; Hao, H.; Xiong, J. Metabolic dysregulation and emerging therapeutical targets for hepatocellular carcinoma. *Acta Pharm. Sin. B* **2022**, *12*, 558–580. [CrossRef]
61. Jiang, Z.-Z.; Peng, Z.-P.; Liu, X.-C.; Guo, H.-F.; Zhou, M.-M.; Jiang, D.; Ning, W.-R.; Huang, Y.-F.; Zheng, L.; Wu, Y. Neutrophil extracellular traps induce tumor metastasis through dual effects on cancer and endothelial cells. *Oncoimmunology* **2022**, *11*, 2052418. [CrossRef] [PubMed]
62. Iwamoto, H.; Abe, M.; Yang, Y.; Cui, D.; Seki, T.; Nakamura, M.; Hosaka, K.; Lim, S.; Wu, J.; He, X.; et al. Cancer Lipid Metabolism Confers Antiangiogenic Drug Resistance. *Cell Metab.* **2018**, *28*, 104–117.e5. [CrossRef] [PubMed]
63. Berndt, N.; Eckstein, J.; Heucke, N.; Gajowski, R.; Stockmann, M.; Meierhofer, D.; Holzhütter, H.-G. Characterization of Lipid and Lipid Droplet Metabolism in Human HCC. *Cells* **2019**, *8*, 512. [CrossRef] [PubMed]
64. Hu, B.; Lin, J.Z.; Yang, X.B.; Sang, X.T. Aberrant lipid metabolism in hepatocellular carcinoma cells as well as immune microenvironment: A review. *Cell Prolif.* **2020**, *53*, e12772. [CrossRef] [PubMed]
65. Moris, D.; Beal, E.W.; Chakedis, J.; Burkhart, R.A.; Schmidt, C.; Dillhoff, M.; Zhang, X.; Theocharis, S.; Pawlik, T.M. Role of exosomes in treatment of hepatocellular carcinoma. *Surg. Oncol.* **2017**, *26*, 219–228. [CrossRef]
66. Gordon-Weeks, A.; Yuzhalin, A.E. Cancer Extracellular Matrix Proteins Regulate Tumour Immunity. *Cancers* **2020**, *12*, 3331. [CrossRef]
67. Gaul, S.; Leszczynska, A.; Alegre, F.; Kaufmann, B.; Johnson, C.D.; Adams, L.A.; Wree, A.; Damm, G.; Seehofer, D.; Calvente, C.J.; et al. Hepatocyte pyroptosis and release of inflammasome particles induce stellate cell activation and liver fibrosis. *J. Hepatol.* **2021**, *74*, 156–167. [CrossRef]
68. Ding, Y.; Yan, Y.; Dong, Y.; Xu, J.; Su, W.; Shi, W.; Zou, Q.; Yang, X. NLRP3 promotes immune escape by regulating immune checkpoints: A pan-cancer analysis. *Int. Immunopharmacol.* **2022**, *104*, 108512. [CrossRef]
69. Lee, H.H.; Kim, D.; Jung, J.; Kang, H.; Cho, H. NLRP3 Deficiency in Hepatocellular Carcinoma Enhances Surveillance of NK-92 through a Modulation of MICA/B. *Int. J. Mol. Sci.* **2021**, *22*, 9285. [CrossRef]
70. Shi, G.; Tang, N.; Qiu, J.; Zhang, D.; Huang, F.; Cheng, Y.; Ding, K.; Li, W.; Zhang, P.; Tan, X. Irisin stimulates cell proliferation and invasion by targeting the PI3K/AKT pathway in human hepatocellular carcinoma. *Biochem. Biophys. Res. Commun.* **2017**, *493*, 585–591. [CrossRef]
71. Liu, H.; Wang, M.; Jin, Z.; Sun, D.; Zhu, T.; Liu, X.; Tan, X.; Shi, G. FNDC5 induces M2 macrophage polarization and promotes hepatocellular carcinoma cell growth by affecting the PPARγ/NF-κB/NLRP3 pathway. *Biochem. Biophys. Res. Commun.* **2021**, *582*, 77–85. [CrossRef] [PubMed]
72. Zhang, J.; Zhang, Q.; Lou, Y.; Fu, Q.; Chen, Q.; Wei, T.; Yang, J.; Tang, J.; Wang, J.; Chen, Y.; et al. Hypoxia-inducible factor-1α/interleukin-1β signaling enhances hepatoma epithelial-mesenchymal transition through macrophages in a hypoxic-inflammatory microenvironment. *Hepatology* **2018**, *67*, 1872–1889. [CrossRef] [PubMed]
73. Wu, L.; Zhang, X.; Zheng, L.; Zhao, H.; Yan, G.; Zhang, Q.; Zhou, Y.; Lei, J.; Zhang, J.; Wang, J.; et al. RIPK3 orchestrates fatty acid metabolism in tumor-associated macrophages and hepatocarcinogenesis. *Cancer Immunol. Res.* **2020**, *8*, 710–721. [CrossRef] [PubMed]

74. Tu, C.E.; Hu, Y.; Zhou, P.; Guo, X.; Gu, C.; Zhang, Y.; Li, A.; Liu, S. Lactate and TGF-β antagonistically regulate inflammasome activation in the tumor microenvironment. *J. Cell. Physiol.* **2021**, *236*, 4528–4537. [CrossRef] [PubMed]
75. Yan, Z.; Da, Q.; Li, Z.; Lin, Q.; Yi, J.; Su, Y.; Yu, G.; Ren, Q.; Liu, X.; Lin, Z.; et al. Inhibition of NEK7 Suppressed Hepatocellular Carcinoma Progression by Mediating Cancer Cell Pyroptosis. *Front. Oncol.* **2022**, *12*, 812655. [CrossRef]
76. Vogel, A.; Martinelli, E.; Cervantes, A.; Chau, I.; Daniele, B.; Llovet, J.M.; Meyer, T.; Nault, J.C.; Neumann, U.; Ricke, J.; et al. Updated treatment recommendations for hepatocellular carcinoma (HCC) from the ESMO Clinical Practice Guidelines. *Ann. Oncol.* **2021**, *32*, 801–805. [CrossRef]
77. Llovet, J.M.; Pinyol, R.; Kelley, R.K.; El-Khoueiry, A.; Reeves, H.L.; Wang, X.W.; Gores, G.J.; Villanueva, A. Molecular pathogenesis and systemic therapies for hepatocellular carcinoma. *Nat. Cancer* **2022**, *3*, 386–401. [CrossRef]
78. Petrowsky, H.; Fritsch, R.; Guckenberger, M.; De Oliveira, M.L.; Dutkowski, P.; Clavien, P.A. Modern therapeutic approaches for the treatment of malignant liver tumours. *Nat. Rev. Gastroenterol. Hepatol.* **2020**, *17*, 755–772. [CrossRef]
79. Jaeck, D.; Oussoultzoglou, E.; Rosso, E.; Greget, M.; Weber, J.C.; Bachellier, P. A two-stage hepatectomy procedure combined with portal vein embolization to achieve curative resection for initially unresectable multiple and bilobar colorectal liver metastases. *Ann. Surg.* **2004**, *240*, 1037–1051. [CrossRef]
80. Clavien, P.-A.; Petrowsky, H.; DeOliveira, M.L.; Graf, R. Strategies for safer liver surgery and partial liver transplantation. *N. Engl. J. Med.* **2007**, *356*, 1545–1559. [CrossRef]
81. Schnitzbauer, A.A.; Lang, S.A.; Goessmann, H.; Nadalin, S.; Baumgart, J.; Farkas, S.A.; Fichtner-Feigl, S.; Lorf, T.; Goralcyk, A.; Hörbelt, R.; et al. Right portal vein ligation combined with in situ splitting induces rapid left lateral liver lobe hypertrophy enabling 2-staged extended right hepatic resection in small-for-size settings. *Ann. Surg.* **2012**, *255*, 405–414. [CrossRef] [PubMed]
82. Baili, E.; Tsilimigras, D.I.; Filippou, D.; Ioannidis, A.; Bakopoulos, A.; Machairas, N.; Papalampros, A.; Petrou, A.; Schizas, D.; Moris, D. Associating liver partition and portal vein ligation for staged hepatectomy in patients with primary liver malignancies: A systematic review of the literature. *J. Balk. Union Oncol. JBUON* **2019**, *24*, 1371–1381.
83. Wakabayashi, G.; Cherqui, D.; Geller, D.A.; Buell, J.F.; Kaneko, H.; Han, H.S.; Asbun, H.; O'Rourke, N.; Tanabe, M.; Koffron, A.J.; et al. Recommendations for laparoscopic liver resection: A report from the second international consensus conference held in Morioka. *Ann. Surg.* **2015**, *261*, 619–629. [CrossRef] [PubMed]
84. Zhang, Y.; Yang, H.; Sun, M.; He, T.; Liu, Y.; Yang, X.; Shi, X.; Liu, X. Alpinumisoflavone suppresses hepatocellular carcinoma cell growth and metastasis via NLRP3 inflammasome-mediated pyroptosis. *Pharmacol. Rep.* **2020**, *72*, 1370–1382. [CrossRef]
85. Feng, M.; He, S.; Huang, S.; Lin, J.; Yang, H.; Wang, J.; Pang, J. Inhibitory effect of Biejiajian pills against diethylnitrosamine-induced hepatocarcinogenesis in rats. *J. South. Med. Univ.* **2020**, *40*, 1148–1154. [CrossRef]
86. Fan, S.; Wang, Y.; Lu, J.; Zheng, Y.; Wu, D.; Li, M.; Hu, B.; Zhang, Z.; Cheng, W.; Shan, Q. Luteoloside Suppresses Proliferation and Metastasis of Hepatocellular Carcinoma Cells by Inhibition of NLRP3 Inflammasome. *PLoS ONE* **2014**, *9*, e89961. [CrossRef]
87. Shen, Z.; Zhou, H.; Li, A.; Wu, T.; Ji, X.; Guo, L.; Zhu, X.; Zhang, D.; He, X. Metformin inhibits hepatocellular carcinoma development by inducing apoptosis and pyroptosis through regulating FOXO3. *Aging* **2021**, *13*, 22120–22133. [CrossRef]
88. Yabuta, S.; Shidoji, Y. TLR4-mediated pyroptosis in human hepatoma-derived HuH-7 cells induced by a branched-chain polyunsaturated fatty acid, geranylgeranoic acid. *Biosci. Rep.* **2020**, *40*, BSR20194118. [CrossRef]
89. Zhang, Q.; Wang, H.; Mao, C.; Sun, M.; Dominah, G.; Chen, L.; Zhuang, Z. Fatty acid oxidation contributes to IL-1β secretion in M2 macrophages and promotes macrophage-mediated tumor cell migration. *Mol. Immunol.* **2018**, *94*, 27–35. [CrossRef]
90. Wei, Q.; Guo, P.; Mu, K.; Zhang, Y.; Zhao, W.; Huai, W.; Qiu, Y.; Li, T.; Ma, X.; Liu, Y.; et al. Estrogen suppresses hepatocellular carcinoma cells through ERβ-mediated upregulation of the NLRP3 inflammasome. *Lab. Investig.* **2015**, *95*, 804–816. [CrossRef]
91. Wei, Q.; Zhu, R.; Zhu, J.; Zhao, R.; Li, M. E2-induced activation of the NLRP3 inflammasome triggers pyroptosis and inhibits autophagy in HCC cells. *Oncol. Res.* **2019**, *27*, 827–834. [CrossRef] [PubMed]
92. Chen, W.; Wei, T.; Chen, Y.; Yang, L.; Wu, X. Downregulation of irak1 prevents the malignant behavior of hepatocellular carcinoma cells by blocking activation of the mapks/nlrp3/il-1β pathway. *OncoTargets Ther.* **2020**, *13*, 12787–12796. [CrossRef] [PubMed]
93. Shi, F.; Wei, B.; Lan, T.; Xiao, Y.; Quan, X.; Chen, J.; Zhao, C.; Gao, J. Low NLRP3 expression predicts a better prognosis of colorectal cancer. *Biosci. Rep.* **2021**, *41*, BSR20210280. [CrossRef] [PubMed]
94. Saponaro, C.; Scarpi, E.; Sonnessa, M.; Cioffi, A.; Buccino, F.; Giotta, F.; Pastena, M.I.; Zito, F.A.; Mangia, A. Prognostic Value of NLRP3 Inflammasome and TLR4 Expression in Breast Cancer Patients. *Front. Oncol.* **2021**, *11*, 705331. [CrossRef]
95. Wang, Q.; Lyu, J.; Zhang, W.; Shi, F.; Ren, Y.; Mao, Q.; Liu, Y.; Li, Y.; Wang, S. Immunological and clinical immunotherapy implications of NLRP3 mutations in melanoma. *Aging* **2021**, *13*, 24271–24289. [CrossRef]
96. Wang, J.; Huang, Z.; Lu, H.; Zhang, R.; Feng, Q.; He, A. A Pyroptosis-Related Gene Signature to Predict Patients' Prognosis and Immune Landscape in Liver Hepatocellular Carcinoma. *Comput. Math. Methods Med.* **2022**, *2022*, 1258480. [CrossRef]
97. Finn, R.S.; Qin, S.; Ikeda, M.; Galle, P.R.; Ducreux, M.; Kim, T.-Y.; Kudo, M.; Breder, V.; Merle, P.; Kaseb, A.O.; et al. Atezolizumab plus Bevacizumab in Unresectable Hepatocellular Carcinoma. *N. Engl. J. Med.* **2020**, *382*, 1894–1905. [CrossRef]
98. Elinav, E.; Strowig, T.; Kau, A.L.; Henao-Mejia, J.; Thaiss, C.A.; Booth, C.J.; Peaper, D.R.; Bertin, J.; Eisenbarth, S.C.; Gordon, J.I.; et al. NLRP6 inflammasome regulates colonic microbial ecology and risk for colitis. *Cell* **2011**, *145*, 745–757. [CrossRef]
99. Wlodarska, M.; Thaiss, C.A.; Nowarski, R.; Henao-Mejia, J.; Zhang, J.P.; Brown, E.M.; Frankel, G.; Levy, M.; Katz, M.N.; Philbrick, W.M.; et al. NLRP6 inflammasome orchestrates the colonic host-microbial interface by regulating goblet cell mucus secretion. *Cell* **2014**, *156*, 1045–1059. [CrossRef]

cancers

Review

# Management of Hepatocellular Carcinoma in Decompensated Cirrhotic Patients: A Comprehensive Overview

Maria Tampaki [1], George V. Papatheodoridis [1] and Evangelos Cholongitas [2,*]

1 Academic Department of Gastroenterology, General Hospital of Athens "Laiko", Medical School, National and Kapodistrian University of Athens, 11527 Athens, Greece
2 First Department of Internal Medicine, General Hospital of Athens "Laiko", Medical School, National and Kapodistrian University of Athens, 11527 Athens, Greece
* Correspondence: cholongitas@yahoo.gr; Tel.: +30-6936-378903; Fax: +30-2132061795

**Simple Summary:** Decompensated patients with hepatocellular carcinoma (HCC) are a wide patient category with limited therapeutic options, and are often excluded from existing trials. Liver transplantation is the best treatment option for such patients but is affected by strict selection criteria and liver donor shortages. The data regarding locoregional treatments in patients with impaired liver function are scarce but indicate a possible survival benefit provided they are well tolerated. Perhaps the systemic treatments, and particularly immunotherapy, are the safest option for such patients based on the available real-life data. Regardless of the type of treatment, close adverse event monitoring is mandatory due to the high risk of hepatic disease deterioration. The aim of this review is to analyze the existing data regarding the administration of treatment in decompensated patients with HCC, evaluate the effect of therapy on overall survival, highlight the potential risks in terms of tolerability and elucidate the optimal therapeutic management.

**Abstract:** Primary liver cancer is the sixth most common cancer and the fourth leading cause of cancer-related death. Hepatocellular carcinoma (HCC) accounts for 75% of primary liver cancer cases, mostly on the basis of cirrhosis. However, the data and therapeutic options for the treatment of HCC in patients with decompensated cirrhosis are rather limited. This patient category is often considered to be in a terminal stage without the possibility of a specific treatment except liver transplantation, which is restricted by several criteria and liver donor shortages. Systemic treatments may provide a solution for patients with Child Pugh class B or C since they are less invasive. Although most of the existing trials have excluded patients with decompensated cirrhosis, there are increasing data from real-life settings that show acceptable tolerability and satisfying efficacy in terms of response. The data on the administration of locoregional treatments in such patients are also limited, but the overall survival seems to be potentially prolonged when patients are carefully selected, and close adverse event monitoring is applied. The aim of this review is to analyze the existing data regarding the administration of treatments in decompensated patients with HCC, evaluate the effect of therapy on overall survival and highlight the potential risks in terms of tolerability.

**Keywords:** HCC; decompensated cirrhosis; treatment

**Citation:** Tampaki, M.; Papatheodoridis, G.V.; Cholongitas, E. Management of Hepatocellular Carcinoma in Decompensated Cirrhotic Patients: A Comprehensive Overview. *Cancers* **2023**, *15*, 1310. https://doi.org/10.3390/cancers15041310

Academic Editor: Georgios Germanidis

Received: 6 January 2023
Revised: 14 February 2023
Accepted: 16 February 2023
Published: 18 February 2023



## 1. Introduction

Primary liver cancer is the sixth most common cancer and the fourth leading cause of cancer-related mortality worldwide [1–3]. Hepatocellular carcinoma (HCC) is the predominant type of primary liver cancer, accounting for 70–85% of all cases [4] and associated with severe morbidity and mortality. Although there are increasing therapeutic options for HCC in patients with compensated liver disease, the management of HCC in patients with decompensated cirrhosis remains challenging and relatively unclear. The International Scientific Guidelines for the management of HCC have analyzed extensively the available therapeutic techniques for patients with Child Pugh (CP) class A or B according to HCC stage, while patients with CP class C are categorized as end-stage liver disease and are eligible only for palliative care or liver transplantation (LT) if they fulfill specific clinical and tumor criteria [5–7].

Patients with CP class B are a wide heterogenous patient group with borderline liver function and high-risk for post-treatment hepatic deterioration [8,9]. In particular, they are rarely able to undergo hepatic resection and cannot always receive locoregional treatments due to the risk of liver-related complications [6,10]. Additionally, despite the recent impressive progress in the availability of systemic treatments with different mechanisms of action against HCC, almost all registrational studies have excluded patients with decompensated cirrhosis [11–13]. In general, survival rates are considered to be lower in this patient group even after administration of systemic therapies, while the rates of adverse events (AE) are expected to be higher. Regarding HCC patients with CP class C cirrhosis, the treatment landscape is even more unclear since the overall survival (OS) is quite limited for these patients, mostly due to liver-related complications rather than HCC progression [14]. Nevertheless, the administration of HCC treatment to carefully selected patients with CP class C might prolong OS even for this fragile patient group [15].

The aim of this review is to analyze the existing data regarding the administration of treatment for HCC in decompensated cirrhotic patients, evaluate the effect of therapy on overall survival and highlight the potential risks in terms of tolerability.

## 2. Literature Search

A comprehensive literature search was conducted for relevant literature using the "PubMed" database, in which only studies written in the English language published until December 2022 were included. The following search terms were used: "Hepatocellular carcinoma" or "HCC" AND "decompensated cirrhosis" or "Child Pugh B" or "Child Pugh C" AND "liver transplantation" or "locoregional treatments" or "transarterial chemoembolization" or "TACE" or "radiofrequency ablation" or "RFA" or "liver resection" or "systemic treatments" or "immunotherapy" or every drug included in the two last categories. In addition, the references of the research articles were scrutinized for relevant studies.

## 3. Systemic Treatments

The administration of systemic treatments in cases of decompensated cirrhosis remains quite challenging since the safety of such agents has not been confirmed and the consequences from potentially more frequent hepatic toxicity and liver-associated AEs can be dramatic in such patients [16,17]. Consequently, the therapeutic decision should weigh the survival benefit provided by the available therapeutic agents and the risks of further hepatic impairment when the balance is already very fragile. Most clinical trials for systemic treatments do not include CP class B and C patients in order to avoid compromising the clinical outcomes of the studies, thus resulting in lack of scientific evidence to guide clinical management of HCC in this population [18,19].

### 3.1. *Multikinase Inhibitors (MKIs) (Table 1)*

#### 3.1.1. Sorafenib

Since sorafenib was the only available systemic treatment for HCC patients for almost a decade, the available data on its use in decompensated cirrhosis is possibly richer compared with other systemic treatments (Table 1) [20]. It is the only MKI with an approved indication for administration in CP B class patients [21]. The SHARP trial, the registrational study for sorafenib, did include 20 patients with advanced HCC of CP class B, which was associated with worse overall survival (OS) in the multivariate analysis [22]. The GIDEON study showed that median OS in 3213 real-life patients who received sorafenib for advanced HCC was 13.6 vs. 5.2 vs. 2.6 months in CP class A, B and C patients, respectively [23]. Interestingly, there was a higher incidence of serious adverse events (SAEs) and treatment discontinuation due to AEs in CP class B compared with CP class A patients. Liver- and

HCC-related death rates were similar between CP class A and B groups, but significantly higher in the CP class C group. Woo et al. reported that sorafenib treatment was significantly longer in CP class A patients compared with CP class B (233 ± 240 days vs. 100 ± 136 days, respectively; $p$ = 0.006) [24]. Liver failure was the most common cause of treatment discontinuation in CP class B patients, as opposed to HCC progression in CP class A patients. In a recent prospective study by Leal et al., the OS was significantly higher in CP class A patients (12 vs. 6 months), but still the reported OS in CP class B patients was considered satisfying enough for this patient group [25]. Finally, a meta-analysis of 30 studies examining the administration of sorafenib as first-line treatment in 8678 patients with advanced HCC further confirmed that OS was significantly worse in patients with CP class B than in class A patients (4.6 vs. 8.8 months, respectively, $p$ = 0.001) [26]. However, in the same meta-analysis clinical response, safety and tolerability did not differ significantly between CP class A and B patients. In fact, treatment-related death rates were similar across CP class groups. Based on this, CP class B patients with advanced HCC may respond to sorafenib and tolerate treatment but without prolongation of survival, as they often experience progression of liver disease that leads to treatment discontinuation and decreased life expectancy.

#### 3.1.2. Lenvatinib

Lenvatinib is an MKI approved as a first-line treatment based on the REFLECT study, which compared its efficacy to sorafenib in advanced HCC patients [27]. In a post hoc analysis of the REFLECT study that included CP class A patients and also patients who progressed to CP B during treatment, among the patients who received lenvatinib, the median OS was 6.8 months (95% CI 2.6–10.3) for CP class B patients and 13.3 months (95% CI 11.6–16.1) for CP class A patients [28]. For the sorafenib group, OS was 4.5 months (95% CI 2.9–6.1) for CP class B patients and 12 months (95% CI 10.2–14.0) for CP class A patients. Similarly, the median progression-free survival (PFS) in the lenvatinib group was 3.7 and 6.5 months in CP class B and A patients, respectively. In the same study, the AE rates (number of AE episodes per patient year) for grade 3 treatment-related AE (TRAE) episodes in the lenvatinib arm were 3.65 and 1.41 for CP class B and A patients, respectively, and more patients with CP class B discontinued therapy because of TRAEs (18.3% vs. 7.5% for CP class A). The AEs that most frequently led to lenvatinib dose reduction or discontinuation in the CP class B subgroup were hepatic encephalopathy (15%), decreased appetite (13%), and increased bilirubin (12%). In a retrospective multi-center study by Ogushi et al., the multivariate analysis showed that OS following lenvatinib treatment was significantly associated with CP class (A vs. B, $p$ = 0.007) and Barcelona clinic liver cancer (BCLC) stage (BCLC B vs. C, $p$ = 0.002) [29]. In particular, OS following 12 months of lenvatinib treatment was 66% and 30% in CP class A and B patients, respectively ($p$ = 0.002), while the objective response (OR) rate was markedly higher in CP A5 patients (44%) compared with CP A6 (25.5%), CP B7 (22.2%), and CP B8 patients (5.3%) ($p$ = 0.002). Finally, lenvatinib-associated AEs were also found to be higher in the CP class B group and included decreased appetite ($p$ = 0.034), diarrhea ($p$ = 0.040), vomiting ($p$ = 0.009) and increased serum bilirubin levels ($p$ = 0.016). In a recent retrospective study that compared the effectiveness of lenvatinib vs. sorafenib treatment in 94 patients with decompensated cirrhosis (CP class B and C), there was no significant difference in the OS between the treatment groups (4.2 in the lenvatinib group vs. 4.1 months in the sorafenib group) [30]. Furthermore, there was no significant difference concerning the AE rates between the regimens. In the real-life setting, the reports regarding the efficacy and safety of lenvatinib in decompensated patients seem to generally agree. In a US study that included 164 patients (49.4% with CP class B cirrhosis), clinical response rates were similar between CP class A and B patients ($p$ = 0.11), although dose reductions were higher in patients with CP class B [31]. In terms of OS, a real-life study from Japan including 276 CP class A and 67 CP class B patients showed that the survival rates were 21 and 9 months, respectively. In CP class B patients, drug discontinuation was observed in 47/67 (70%) because of disease progression ($n$ = 9), TRAEs ($n$ = 36), and worsening of other comorbidities ($n$ = 2) [32]. Finally, in a small study by Cosma et al. that included 12 CP A and 14 CP B patients with HCC, the calculated one-year survival rates were 59% and 27%, respectively, and liver disease deteriorated in two CP A and one CP B patient. The most frequent adverse event was fatigue, irrespective of CP status, and in general, the AE rates did not differ significantly between CP groups [33]. In conclusion, lenvatinib, similar to sorafenib, may be efficacious in CP class B patients in terms of HCC response, but higher rates of treatment dose reduction or discontinuation, TRAEs and liver-related deaths limit its effect on OS. Additional data from prospective studies may be needed to clarify if lenvatinib provides a clear benefit in terms of OS in CP class B patients with advanced HCC.

**Table 1.** Studies examining the efficacy and safety of Multikinase Inhibitors (MKIs) in patients with decompensated cirrhosis and hepatocellular carcinoma.

| | Systemic Treatment | Type of Study | Total Patients, *n* | CP B Patients, *n* | CP C Patients, *n* | Mean OS, Months | | | ORR, % | | | DCR, % | | | AEs, % | | |
|---|---|---|---|---|---|---|---|---|---|---|---|---|---|---|---|---|---|
| | | | | | | CP A | CP B | CP C | CP A | CP B | CP C | CP A | CP B | CP C | CP A | CP B | CP C |
| Lencioni 2014 [20] | Sorafenib | Prospective | 3213 | 361 | 35 | 13.6 | 5.2 | 2.6 | n/a | n/a | n/a | n/a | n/a | n/a | 82 | 89 | 86 |
| Leal 2018 [22] | Sorafenib | Prospective | 130 | 65 | 0 | 12 | 6 | n/a | n/a | n/a | n/a | n/a | n/a | n/a | 93.8 | 76.9 | n/a |
| Huynh 2022 [25] | Lenvatinib vs. Sorafenib | Post hoc analysis | 478<br>476 | 60<br>47 | 0<br>0 | 13.3<br>12 | 6.8<br>4.5 | | 42.9<br>12.9 | 28.3<br>8.5 | - | n/a | n/a | - | 10.4 *<br>11.6 * | 18.4 *<br>19.9 * | - |
| Ogushi 2020 [26] | Lenvatinib | Retrospective | 181 | 55 | 0 | 1-year 66% | 1-year 30% | n/a | 36.5 | 16.3 | - | n/a | n/a | - | 98.4 | 94.5 | - |
| Tsuchiya 2021 [29] | Lenvatinib | Retrospective | 343 | 67 | 0 | 21 | 9 | - | n/a | n/a | - | n/a | n/a | - | n/a | n/a | - |
| Cosma 2021 [32] | Lenvatinib | Retrospective | 28 | 14 | 2 | 1-year 59.3% | 1-year 26.9% | n/a | n/a | n/a | n/a | n/a | n/a | n/a | n/a | n/a | n/a |
| El-Khoueiry 2022 [31] | Cabozantinib vs. placebo | Retrospective | 51<br>22 | 51<br>22 | 0 | - | 8.5<br>3.8 | - | - | 0<br>0 | - | - | 57<br>23 | - | - | 100<br>100 | - |
| Bang 2022 [33] | Cabozantinib | Retrospective | 110 | 22 | 0 | 9 | 3.8 | - | 4.5 | 0 | - | 71.5 | 45.5 | - | 76.1 | 72.7 | - |
| Finkelmeier 2021 [34] | Cabozantinib | Retrospective | 88 | 22 | 0 | 9.7 | 3.4 | - | n/a | n/a | - | n/a | n/a | - | 43.3 | 72.7 | - |

CP, Child Pugh; OS, overall survival; ORR, objective response rate; DCR, disease control rate; AE, adverse event; n/a, not available. * Adjusted by patient-years.

#### 3.1.3. Cabozantinib

Data on cabozantinib in decompensated patients with advanced HCC, another MKI agent that can be administered as a second-line HCC treatment, are lately increasing (Table 1) [34]. In a retrospective analysis from the CELESTIAL study focusing on the patients who progressed to CP class B, there was no difference in terms of treatment safety and tolerability in this group compared with the overall population [35]. Moreover, the OS for CP class B patients in the cabozantinb group was 8.5 vs. 3.8 months in the placebo group (HR 0.32, 95% CI 0.18–0.58), and the median PFS was 3.7 vs. 1.9 months (HR 0.44, 95% CI 0.25–0.76), respectively. In a post hoc analysis based again on the CELESTIAL study, the investigators examined the association of albumin-bilirubin (ALBI) grade with OS [36]. The results of this analysis showed that cabozantinib efficacy in terms of OS and PFS was similar between the ALBI grade 1 (score $\leq -2.60$) and ALBI grade 2 (score $> -2.60$ to $\leq -1.39$) subgroups. Grade 3 AEs associated with hepatic decompensation were found to be higher in the ALBI grade 2 subgroup. However, the survival outcomes were significantly worse for patients with advanced HCC and CP class B cirrhosis (PFS: 4.3 vs. 2.2 months, $p < 0.001$; OS: 9.0 vs. 3.8 months, $p < 0.001$) according to a retrospective Korean study including 110 patients [37]. In the same study, AE rates were generally similar between CP class A and B patients. On the other hand, a recent international real-life study that included 60, 22 and 1 CP class A, B and C patients, respectively, showed that the AE rates were 73% in CP B compared with 43% in CP A patients ($p = 0.017$), and the OS was 7 months in CP B7 vs. 9.7 months in the CP class A group [38]. In contrast, the median OS in CP B8, B9 and C was limited to 3.4 months. The increased AE rate in the CP class B group was attributed to the possible decreased drug metabolism due to impaired liver function, leadingto higher drug levels and thus higher adverse reactions. Based on the above, the effect of cabozantinib on survival of patients with advanced HCC seems to depend significantly on the severity of hepatic impairment, with more favorable outcomes in lower score-CP class B patients. AE and toxicity rates could be limited by administrating lower drug doses provided that efficacy is not compromised.

### 3.2. *Immunotherapy (Table 2)*

In recent years, immune checkpoint inhibitors (ICIs) have drastically changed the treatment landscape in HCC [39]. CheckMate-040 was the first ICI study that included CP class B patients with advanced HCC [40]. In this phase 1/2 trial, 49 CP class B patients received nivolumab intravenously, resulting in a reported objective response rate (ORR) of 12% and a disease control rate (DCR) of 55%. Interestingly, the safety profile of nivolumab in this population was comparable to the CP class A patients, as TRAEs were reported in 51% of patients and led to discontinuation in 4% (83% and 6% in the CP class A patients, respectively) [41]. These results were further confirmed by a recent real-life study that recruited 431 CP class B patients from the Veteran Affairs medical centers in the USA to compare efficacy and safety of sorafenib vs. nivolumab [42]. The median OS was 5 months for the 79 patients that received nivolumab vs. 4 months for the sorafenib group, and treatment was discontinued due to toxicity in 12% of patients receiving nivolumab compared with 36% receiving sorafenib ($p = 0.001$). A Korean study by Choi et al. showed that ORR with nivolumab treatment was significantly lower in CP class B compared to CP A patients (2.8% vs. 15.9%; $p = 0.010$) and that OS was specifically worse in patients with CP B8 or B9 than in those with CP B7 (7.4 vs. 15.3 weeks; $p < 0.020$) [43]. Additionally, AE rates were similar between the CP class A and B groups, and immune-mediated AEs were more frequent in CP class A patients (2 patients with hepatitis and 3 with pneumonitis vs. 0 patients in the CP class B group). Lower response rates in the CP class B group in this study were explained by the reduced capacity of immune response because of advanced cirrhosis. The same explanation was provided for lower immune-mediated adverse events in the CP class B population. Additionally, a real-life study examining the efficacy and safety of nivolumab or pembrolizumab treatment in 32 CP class A and 28 CP class B patients with advanced HCC showed that the ORR and DCR for CP class A vs. B was 9% vs. 14%

($p = 0.438$) and 56% vs. 46% ($p = 0.947$), respectively [44]. Median OS values of 16.7 (95% CI, 8.2–25.2) months for CP class A and 8.6 (95% CI, 4.8–12.4) months for CP class B ($p = 0.065$) were reported, whereas there was no significant difference in terms of AE rates (CP class A vs. B, 31% vs. 43%; $p = 0.352$) or high-grade AEs (CP A vs. B, 16% vs. 18%; $p = 1.000$).

Concerning first-line treatments, atezolizumab-bevacizumab is a combination with high efficacy in non-resectable HCC based on the IMbrave 150 study results [5,45]. A recent multi-center retrospective study by D'Alessio et al. examined the safety and efficacy of the combination treatment in 154 CP class A and 48 CP class B patients [46]. According to this study, the treatment-related AE rates were similar between the two groups, while ORR and DCR were 25% and 73%, respectively, without significant difference across CP classes (Table 1). However, the median OS was 16.8 months (95% CI, 14.1–23.9) for CP class A patients, vs. 6.7 months (95% CI, 4.3–15.6) for patients with CP class B ($p = 0.0003$). This difference was attributed to deaths due to the underlying liver impairment. Interestingly, TRAEs and bleeding events were also comparable between the CP groups. Interestingly, Chen et al. examined the OS of patients after failure of first line treatment with atezolizumb/bevacizumab. All the patients had CP A cirrhosis at treatment initiation, but 15% and 17% of them progressed to CP B and C, respectively, after failure. The median OS after treatment discontinuation was 9.6 vs. 3.8 vs. 1.2 months, for CP A, B, and C patients, respectively. After therapy failure, the tumor burden was increased, and 32% of the patients did not maintain CP A liver reserve. Not receiving second-line treatment was associated with liver deterioration and poorer OS [47]. Moreover, Persano et al. compared efficacy of atezolizumab/bevacizumab vs. lenvatinib in HCC patients with HCC and showed that ORR was similar for CP B patients (22% vs. 36%, respectively, $p = 0.43$). In the same study, the ORR rates for CP A patients were 28% for the atezolizumab/bevacizumab group and 39% for the lenvatinib group [48]. Another real-life study by de Castro et al. reported that the median OS was 12 months (95% CI: 8.2–15.8), 6.8 months (95% CI: 3.1–10.5; $p = 0.04$) and 1 month (95% CI: 0.0–3.9; $p < 0.001$) for CP class A, B and C patients, respectively [49]. Apart from the CP score, the ALBI score was also significantly associated with OS. Moreover, patients with ALBI grade $\geq 2$ ($p = 0.002$) and decreased performance status ($p < 0.001$) at baseline were at highest risk for developing ascites and hepatic encephalopathy. Intriguingly, mono-immunotherapy was not associated with the above adverse events. Based on the previous findings, immunotherapy may be a relatively safe option for patients with impaired liver function, although close monitoring is mandatory, especially when it is combined with anti-VEGF agents. The combination of two ICI agents, durvalumab plus tremelimumab, has also been tested in patients with unresectable HCC and has yielded positive results in terms of OS, but the majority of the study population were CP A patients [50]. The fact that ICIs metabolism does not depend on liver function possibly provides an advantage for CP class B patients. Furthermore, ICIs especially when combined with anti-VEGF agents, seem to offer a survival benefit that may not be as high as that offered in CP class A, but still, they may prolong life expectancy in carefully selected CP class B patients (Table 2). In other words, the unmet need that exists for CP class B patients with advanced HCC could be covered by this type of treatment in the future.

**Table 2.** Studies examining the efficacy and safety of Immune Checkpoint Inhibitors (ICIs) in patients with decompensated cirrhosis and hepatocellular carcinoma.

| | Systemic Treatment | Type of Study | Total Patients, *n* | CP B Patients, *n* | CP C Patients, *n* | Mean OS, Months | | | ORR, % | | | DCR, % | | | AEs, % | | |
|---|---|---|---|---|---|---|---|---|---|---|---|---|---|---|---|---|---|
| | | | | | | CP A | CP B | CP C | CP A | CP B | CP C | CP A | CP B | CP C | CP A | CP B | CP C |
| Kudo 2021 [36] | Nivolumab | I/II phase | 49 | 49 | 0 | - | 7.6 | - | - | 12 | - | - | 55 | - | - | 51 | - |
| Chapin 2022 [38] | Sorafenib vs. Nivolumab | Retrospective | 439<br>79 | 439<br>79 | 0 | - | 4<br>5 | - | - | n/a | - | - | n/a | - | - | n/a | - |
| Choi 2020 [39] | Nivolumab | Retrospective | 203 | 71 | 0 | 10 | 3.6 | - | 15.9 | 2.8 | - | 42.4 | 22.5 | - | n/a | n/a | - |
| Scheiner 2019 [40] | Nivolumab vs. Pem-brolizumab | Retrospective | 65 | 28 | 5 | 16.7 | 8.6 | n/a | 9 | 14 | n/a | 56 | 46 | n/a | 31 | 43 | n/a |
| D'Alessio 2022 [42] | Atezolizumab/ bevacizumab | Retrospective | 216 | 48 | 0 | 16.8 | 6.7 | - | 26 | 21 | - | 74 | 68 | - | 48 | 46 | - |
| de Castro 2022 [43] | Atezolizumab/ bevacizumab | Retrospective | 147 | 35 | 6 | 12 | 6.8 | 1 | n/a | n/a | n/a | n/a | n/a | n/a | n/a | n/a | n/a |
| Persano 2022 [47] | Atezolizumab/ bevacizumab vs. lenvatinib | Retrospective | 823<br>1312 | n/a | n/a | n/a | n/a | n/a | 27.7<br>38.8 | 22.2<br>36.3 | - | n/a | n/a | n/a | n/a | n/a | n/a |

CP, Child Pugh; OS, overall survival; ORR, objective response rate; DCR, disease control rate; AE, adverse event; n/a, not available.

## 4. Locoregional Treatments (Table 3)

There is no doubt that LT is the optimal treatment option for decompensated patients with HCC tumors within the Milan criteria [5,6]. However, the low availability of liver transplants in many cases requires alternative therapeutic strategies for patients with advanced liver disease. Based on the current International Guidelines for the management of HCC, patients with CP class B can benefit from local ablation treatments (LAT), transarterial chemoembolization (TACE) or transarterial radioembolization (TARE) based on their BCLC status, provided that their liver function is adequately preserved and the risk for hepatic disease deterioration is not high [5,6]. Especially for patients with CP B8 or B9, when surgical treatments are not applicable due to the risk of hepatic deterioration, the less invasive locoregional techniques may provide survival benefit [51]. However, the availability of studies for the administration of locoregional treatments in patients with CP class C cirrhosis and their effect on OS is restricted to some small retrospective trials (Table 3).

### *4.1. Local Ablation Treatments (LAT)*

For patients with CP class A or B and small single tumors < 3 cm who are not eligible for surgery, radiofrequency ablation (RFA) is the indicated treatment option [6]. However, according to a meta-analysis by Casadei-Gardini et al., CP class B is predictive of poor OS and RFS ($p < 0.0001$) compared with CP class A in HCC patients treated with RFA alone [52]. Regarding the effect of RFA on decompensated patients, a retrospective study in 19 patients with mean CP score 10.7 (range 10–12) showed a median survival time of $12.0 \pm 1.7$ months and reported two deaths of hepatic failure, one at two months and one at four months after treatment [53]. Another study that examined the effects of non-surgical treatments in HCC patients with CP class C reported zero survival benefit of LAT, possibly because hepatic deterioration progressed faster than HCC [54]. On the other hand, Kudo et al. showed that the administration of non-transplant techniques in HCC patients with CP class C who exceeded the Milan criteria or could not be transplanted due to liver donor shortage was a significant prognostic factor of better survival compared to non-treated patients [55].

### *4.2. TACE*

TACE is currently the indicated treatment for patients with intermediate stage HCC (BCLC B) and CP class A or B [5,6]. Since it is a therapeutic technique based on the transarterial administration of chemotherapeutic agents and the occlusion of the tumor-feeding vessel, it has been associated with deterioration of the hepatic function and decompensation when applied to patients with advanced liver disease [56,57]. Therefore, several prognostic models have been developed in order to select carefully the patients that are suitable to receive TACE without risking their hepatic reserve [58,59]. These models are based on biochemical parameters, ALBI score, CP score and tumor characteristics, among others. In a large study by Takayasu et al. that included 4966 HCC patients who underwent TACE, the survival rate was significantly associated with CP score, with the lowest rates being reported in CP class B patients with three lesions $\geq$ 5.1 cm [60]. Specifically, the 3-year survival rate was 53% in CP class B patients with a single tumor (vs. 73% in CP class A, $p = 0.0001$) and 22% in patients with $\geq$4 tumors (vs. 46% in CP class A, $p = 0.0001$). Treatment-related death occurred in 19 (0.38%) out of 4966 patients (10/3229 CP class A, 8/1296 CP class B and 1/167 CP class C). In another study including 100 CP class A and 90 CP class B/C cirrhotic patients, the administration of TACE was equally efficacious in both groups in terms of tumor necrosis, but the OS was significantly higher in the CP class A group (21.9 vs. 13.7 months, $p = 0.03$) [61]. There was no significant difference in terms of 30-day or 90-day post-treatment mortality between the two groups. The survival curves for the two CP groups separated at 3 months and reached maximal separation at 12 months. According to the Cox proportional hazards model, post-treatment mortality for the CP class B group was significantly associated only with total tumor diameter (hazard ratio 1.26, 95% CI 1.10–1.44, $p < 0.001$). According to Piscaglia et al., TACE treatment offered a survival rate of 22 and 8 months in patients with CP B7 and B8, respectively [51]. The study concluded that TACE should be performed in CP class B patients with compensated disease (B7), but it could be detrimental for liver function in patients with more advanced hepatic impairment. According to the above, it can be assumed that even within the group of CP class B patients, the OS rates may differ significantly based on the exact CP score, so TACE is probably a good therapeutic option for those who have borderline liver function and BCLC B HCC.

The available studies examining the administration of TACE specifically to CP class C patients are very limited due to the mentioned risks. In a retrospective study comparing best supportive care (BSC) to locoregional treatments in CP class C patients with multinodular HCC, the OS rates were found to be significantly higher in patients treated with TACE compared with non-treated patients (14 vs. 2 months, respectively, $p < 0.0001$) [62]. In the same study, the propensity score matching for patients with tumors within the Milan criteria showed no significant differences in the clinical characteristics between the two treatment groups, so it was concluded that the patient selection bias was relatively low. Similarly, in the above-mentioned study of Kudo et al., superselective TACE in decompensated patients with multinodular HCC provided better survival compared with palliative care [55], which is the currently recommended treatment for such patients. However, the most benefited patients in this study were those with the lowest CP C scores, while most patients with 14- or 15-point scores remained untreated. In another Japanese study by Nouso et al., the OS rates after superselective TACE were superior to BSC in matched CP C patients ($p < 0.009$) [54]. Regarding the selection of decompensated patients eligible for TACE in this study, the therapeutic technique was specifically performed in patients without severe portal vein thrombus, with median bilirubin of 1.9 mg/dL and median prothrombin time of 64%. Consequently, TACE could prolong OS in carefully selected patients with CP C scores in specialized centers performing superselective embolization. However, it is crucial to understand the limitations of this therapeutic technique for patients with advanced liver disease and HCC and seek for other therapeutic options, such as systemic treatments, when required. The timely switch to a more tolerable therapeutic option for patients with higher CP scores such as immunotherapy, even in earlier HCC stages, could benefit decompensated patients who are not eligible for more invasive treatments.

### 4.3. *TARE*

TARE, also called selective internal radiation therapy (SIRT), is another locoregional technique based on the transarterial administration of microspheres loaded with radioactive compounds such as yttrium-90 or lipiodol labeled with iodine$^{131}$ or rhenium$^{188}$ [63]. The indication of this treatment for HCC patients is not clearly defined but it is mostly administered in patients with BCLC stage B or C [64]. One main advantage of TARE in terms of safety is the absence of macro-embolic effects, so it does not affect hepatic blood flow, which is beneficial for patients with advanced liver disease, especially in cases of macrovascular invasion [65]. However, several AEs are reported following TARE, such as liver failure or radio-induced liver disease (4%), biliary complications (<10%) and post-radioembolization syndrome (20–55%).

**Table 3.** Studies examining the efficacy and safety of locoregional treatments in patients with decompensated cirrhosis and hepatocellular carcinoma.

| | Treatment | Type of Study | Total Patients, n | CP B Patients, n | CP C Patients, n | Mean OS, Months | | | ORR, % | | | DCR, % | | | AE % | | |
|---|---|---|---|---|---|---|---|---|---|---|---|---|---|---|---|---|---|
| | | | | | | CP A | CP B | CP C | CP A | CP B | CP C | CP A | CP B | CP C | CP A | CP B | CP C |
| Kim et al. [53] | RFA | Retrospective | 19 | 0 | 19 | - | - | 12 | - | - | 88.5% | - | - | n/a | - | - | n/a |
| Nouso et al. [54] | LAT | Retrospective | 23 | 0 | 23 | - | - | 1-year 69.1% | - | - | n/a | - | - | n/a | - | - | n/a |
| Kudo et al. [55] | RFA | Retrospective | 60 | 0 | 60 | - | - | 1-year 67% | - | - | n/a | - | - | n/a | - | - | n/a |
| Takayasu et al. [60] | TACE | Prospective | 4966 | 1296 | 167 | 1-year 61% | 1-year 43% | 1-year 23% | n/a | n/a | n/a | n/a | n/a | n/a | n/a | n/a | n/a |
| Dorn et al. [61] | TACE | Retrospective | 190 | 90 | | 21.9 | 13.7 | | 79% | 84% | | n/a | n/a | | n/a | n/a | n/a |
| Piscaglia et al. [51] | TACE | Retrospective | 86 | 86 | 0 | - | 21 | - | - | n/a | - | - | n/a | - | - | n/a | - |
| Nouso et al. [54] | TACE | Retrospective | 27 | 0 | 27 | - | - | 1-year 62.5% | - | - | n/a | - | - | n/a | - | - | n/a |
| Kudo et al. [55] | TACE | Retrospective | 79 | 0 | 79 | - | - | 1-year 69% | - | - | n/a | - | - | n/a | - | - | n/a |
| Zu et al. [66] | TARE | Retrospective | 106 | 27 | 0 | 20.2 | 5.5–6 | - | - | - | n/a | - | - | n/a | - | - | n/a |
| Abouchaleh et al. [67] | TARE | Retrospective | 185 | 60 | 32 | 13.3 | 6.9 | 3.9 | n/a | n/a | n/a | n/a | n/a | n/a | 8% | 8% | 32% |
| Memon et al. [68] | TARE | Retrospective | 63 | 35 | 0 | 13.8 | 6.5 | - | 37% | 32% | - | 37% | 57% | - | n/a | n/a | - |

RFA, radiofrequency ablation; TACE, transarterial chemoembolization; TARE, transarterial radioembolization; CP, Child Pugh; OS, overall survival; ORR, objective response rate; DCR, disease control rate; AE, adverse event; n/a, not available.

In a study by Zu et al. [66] that examined the effect of TARE across CP groups and included a total of 106 patients with BCLC stage C, it was shown that the median OS was 20.2 months for patients with CP A class, 6 months for CP B7 and 5.5 for CP B8/9 groups. Despite the assumed lower risk of liver ischemia with TARE, this did not translate into clinical survival benefits for CP B7 in the study. On the contrary, since the majority of patients had impaired liver function due to portal vein invasion, the authors concluded that the severity of the hepatic disease had a greater impact on OS than the tumor burden. In another study that included 185 HCC patients with portal vein thrombosis, the median OS was 13.3 months for CP A class patients, 6.9 months for CP B7 and 3.3 for CP $\geq$ 8 groups [67]. Interestingly, when the latter group of patients was stratified by location of portal vein thrombosis, the median OS was 8.4 months for segmental, 4.4 months for lobar, and 3.4 months for main portal vein thrombosis ($p = 0.015$). In the multivariate analysis, ECOG status and extent of portal vein thrombosis were significant predictors of survival, but CP status was not. Additionally, for patients with CP B7 and CP B8 class, tumor size (>5 cm) was associated with worse survival. Finally, in a study by Memon et al., survival following TARE was 13.8 months for the group with CP A and 6.5 months for CP B7 class patients [68]. At the time of HCC progression, 50% of patients who had CP B7 scores at baseline had progressed to CP score $\geq$ 8, while a similar percentage (55%) of CP A score patients had progressed to CP B. In conclusion, TARE could be a safer choice for HCC patients with borderline hepatic function (CP B7 scores), especially with macrovascular involvement, although its benefit in patients with more advanced cirrhosis is still debatable. Furthermore, the extent of portal vein thrombosis seems to affect significantly patients' survival [67].

## 5. Surgical Treatments and Liver Transplantation

HCC surgical resection in eligible patients with very early or early tumor stage may lead to a 5-year survival rate of 50–68% in specialized centers [69–71]. However, impaired liver function and severe portal hypertension are the most common contraindications for hepatic surgery in cirrhotic patients, since they are associated with significant postoperative morbidity and liver decompensation [72,73]. The selection of patients that can tolerate surgery with minimal risk for hepatic deterioration is based on CP/MELD scores and the evaluation of the functional capacity of the future liver remnant [72,74]. Liver function tests, such as the LiMax test, which is based on the hepatocyte-specific metabolism of the $^{13}$C-labelled substrate by the cytochrome P450 1A2 enzyme, have been developed as surrogate parameters of liver function capacity to guide treatment decisions [75]. In recent years, patients with portal hypertension and low risk for hepatic decompensation have been carefully selected based on an algorithm that predicts the exact risk for liver deterioration [76]. Laparoscopic techniques have further extended the criteria for choosing patients who are able to withstand hepatic resection without severe morbidity [77,78]. In fact, a recent study by Azoulay et al. showed that cirrhotic patients with hepatic venous pressure gradient $\geq$ 10 mmHg can undergo laparoscopic liver resection with low rates of mortality and hepatic decompensation [75]. However, almost all patients of the study were compensated with CP class A and a median MELD score of 8. In other words, the criteria for the eligibility of HCC patients for hepatic surgery may have been widened, but still the possibility to include patients with CP class B/C is out of the question.

Liver surgery may be the optimal choice for compensated patients with HCC [79], but LT remains the best option for those with CP class B or C provided that they fulfill the strict transplantation criteria [80,81]. As a therapeutic technique, LT not only removes the tumor, but also offers a cure for the underlying hepatic disease. The main limitation of this curative treatment is the shortage of liver donors and consequently the frequent dropout of HCC patients from the transplantation lists due to disease progression [82]. Rigorous patient selection based on specific clinical and tumor criteria (MELD score and Milan/San Francisco criteria) and waiting time in transplant list < 6 months are pivotal to achieve minimal HCC recurrence rates and maximize OS [79]. Locoregional therapeutic techniques have been administered as a bridge to transplantation in order to downstage the tumor or delay HCC progression [83,84]. However, patients with decompensated cirrhosis may not be eligible for these treatments due to the risk of hepatic disease deterioration, and thus, they should be prioritized based on their MELD score [80]. Recently, there have been increasing reports on the use of systemic treatments such as sorafenib or cabozantinib as bridging therapies for HCC patients before LT [85,86]. ICIs have also been used as neoadjuvant treatment before LT, but perhaps their administration should be paused at least 3 months before transplantation in order to avoid the risk of graft rejection after LT [80]. If transplanted within the Milan criteria, patients with HCC have excellent OS rates with an estimated 5-year survival close to 60–75% [87–89]. In conclusion, LT represents the best treatment choice for HCC patients with CP class B or C scores and tumor characteristics within the Milan criteria (or the extended UCSF criteria), but the low availability of liver grafts and the high dropout rates or deaths within the transplant lists indicate the necessity of alternative/bridging treatments to optimize OS rates for such patients.

## 6. Palliative Care

Unfortunately, 15–20% of HCC patients present with end stage HCC and an estimated median survival close to 3–4 months [90]. This patient category consists not only of patients with a high tumor burden and metastatic disease but also of patients with CP class C and affected physical performance status. The available therapeutic techniques cannot offer survival benefit for this patient group, so the healthcare services that can be provided focus on management of the complications of cirrhosis and pain, nutrition and psychological support [91].

## 7. Discussion

Patients with CP class B or C cirrhosis are a heterogenous group with various clinical characteristics, different degrees of performance status and hepatic function [92]. Therefore, the decision for HCC treatment should be based on individual patient status according to liver function scores such as CP, MELD and ALBI scores, and of course, the tumor stage of each patient [50]. Based on our research, the existing clinical data on the safety and efficacy of HCC treatments for this patient group are scarce, but they generally show that OS is lower compared with CP class A patients, regardless of treatment, while TRAEs are more frequent and can lead to hepatic deterioration (Table 4). Immunotherapy seems to be the safest option among systemic treatments since it has not been associated with liver-related AEs in the existing studies [42,43,45]. Recent data examining the sequential use of ICIs (first-line and second-line) in compensated patients with HCC have shown promising results in terms of efficacy without high-grade TRAEs [93]. Since immunotherapy is the safest option among systemic treatments for decompensated patients, ICI rechallenge in this patient group should be studied as it may offer additional therapeutic choices.

**Table 4.** Advantages and disadvantages per treatment category in patients with decompensated cirrhosis and hepatocellular carcinoma.

| Type of Treatment | Advantages | Disadvantages |
|---|---|---|
| Systemic therapy | • Acceptable tolerance and safety<br>• Bridging therapy to liver transplantation *<br>• Alternative therapeutic option when locoregional treatments are not tolerated | • Lower OS rates in CP B compared with CP A patients<br>• More frequent AEs in CP B compared with CP A<br>• Scarce data in CP C patients |
| Radiofrequency ablation | • Minimally invasive<br>• Low AE rates<br>• Bridging therapy to liver transplantation | • No clear benefit for OS in CP C patients |

**Table 4.** *Cont.*

| Type of Treatment | Advantages | Disadvantages |
| --- | --- | --- |
| Transarterial Chemoembolization | • Bridging therapy to liver transplantation<br>• May prolong OS compared with palliative care in decompensated patients | • Risk of hepatic deterioration<br>• Specialized center required<br>• CP C patients are rarely eligible |
| Transarterial radioembolization | • Safer in cases of portal vein thrombosis | • Risk of hepatic deterioration<br>• CP C patients not eligible |
| Liver resection | • High OS rates<br>• Laparoscopic techniques are more tolerated<br>• Can be applied in selected patients with portal hypertension | • Restricted to CP A patients |
| Liver transplantation | • Best available therapeutic option<br>• Optimal OS | • Strict criteria<br>• Liver donor shortage<br>• Frequent drop-out from the transplantation list due to HCC progression |

OS, overall survival; CP, Child Pugh; AE, adverse event; HCC, hepatocellular carcinoma. * Caution that immunotherapy may be associated with higher rejection rates after liver transplantation.

For HCC patients with small HCC (BCLC 0/A) and CP class B or C cirrhosis, the therapeutic management clearly leans towards LT, as it provides the highest OS rates [87,88]. With regards to other surgical techniques, a few patients with CP class B cirrhosis and well-compensated disease may benefit from laparoscopic surgery if chosen properly [70,76,77]. Since RFA is a minimally invasive technique with low risk for liver related AEs, it can also prolong OS for the patients with early HCC who cannot undergo surgery and can also be a bridging therapy until a liver transplant is available [52]. Moreover, it can be an alternative option for patients with small tumors that are not eligible for LT [54]. However, its effect on the OS of patients with CP class C is under debate, and consequently, patients of this category should be selected very carefully in specialized centers.

Intermediate HCC (BCLC B) is the classic indication for TACE, especially in patients with adequate hepatic function [6]. As mentioned above, a few studies have shown that TACE may prolong OS, even in decompensated patients, when compared with palliative care [50,54]. Perhaps carefully selected patients with lower scores within CP class B groups and clinical characteristics that permit the administration of TACE may be benefited in specialized centers.

TARE is also an available option perhaps suitable for patients who cannot undergo TACE due to portal vein invasion [67] or as an intermediate treatment between locoregional therapies and systemic agents [68]. Its non-occlusive effect may be an asset for patients with borderline liver function [65], while its role as a bridge therapy to LT is also under discussion.

For those who do not respond or cannot undergo TACE, systemic treatments are the next available therapeutic option. According to small, mainly retrospective studies, systemic therapies may provide a survival benefit for patients with intermediate or advanced HCC and decompensated cirrhosis with relatively adequate tolerability [24,30,34]. Of course, larger prospective studies are needed to further support these indications.

In general, OS rates seem to be significantly lower in decompensated patients with HCC compared with Child Pugh A class patients, mainly due to hepatic disease progression and liver-related mortality. Therefore, the benefit of the administrated HCC treatment is not always clear since the prolongation of survival is relatively low. Perhaps ORR and PFS rates are a more accurate reflection of the efficacy of the provided therapies and their effect on HCC patients as they focus on the immediate impact of anti-cancer treatment on the disease course.

## 8. Conclusions

Although the International Guidelines and the registrational studies of HCC treatments focus on patients with compensated cirrhosis, decompensated patients are a large population with heterogenous clinical characteristics and limited therapeutic options. HCC treatment should be individualized for this fragile patient group in order to provide maximum survival with minimal risk for liver toxicity. Good performance status should be recognized as a possible indication for treatment administration despite liver impairment. The use of specific scores for the evaluation of hepatic function and the assessment of TRAE risk may also support clinical decision making. Additionally, the identification of the correct parameters of treatment response such as ORR or PFS instead of OS may improve the therapeutic management. Nevertheless, the inclusion of patients with CP class B or C in larger studies is needed to clarify to what extent these patients can benefit from the available anti-cancer treatments. It is possible that the treatment landscape will evolve drastically in the coming years since the number of the latest clinical trials that include this patient category is significantly increasing.

**Author Contributions:** Conceptualization, E.C.; methodology, M.T. and E.C.; validation, E.C. and G.V.P.; formal analysis, M.T.; investigation, M.T.; writing—original draft preparation, M.T.; writing—review and editing, E.C. and G.V.P.; visualization, M.T.; supervision, E.C. and G.V.P.; project administration, E.C. All authors have read and agreed to the published version of the manuscript.

**Funding:** This research received no external funding.

**Conflicts of Interest:** The authors declare no conflict of interest.

## References


1. Yang, J.D.; Hainaut, P.; Gores, G.J.; Amadou, A.; Plymoth, A.; Roberts, L.R. A global view of hepatocellular carcinoma: Trends, risk, prevention and management. *Nat. Rev. Gastroenterol. Hepatol.* **2019**, *16*, 589–604. [CrossRef]
2. Global Burden of Disease Cancer Collaboration; Fitzmaurice, C.; Abate, D.; Abbasi, N.; Abbastabar, H.; Abd-Allah, F.; Abdel-Rahman, O.; Abdelalim, A.; Abdoli, A.; Abdollahpour, I.; et al. Global, Regional, and National Cancer Incidence, Mortality, Years of Life Lost, Years Lived with Disability, and Disability-Adjusted Life-Years for 29 Cancer Groups, 1990 to 2017: A Systematic Analysis for the Global Burden of Disease Study. *JAMA Oncol.* **2019**, *5*, 1749. [PubMed]
3. Sung, H.; Ferlay, J.; Siegel, R.L.; Laversanne, M.; Soerjomataram, I.; Jemal, A.; Bray, F. Global Cancer Statistics 2020: GLOBOCAN Estimates of Incidence and Mortality Worldwide for 36 Cancers in 185 Countries. *CA A Cancer J Clin.* **2021**, *71*, 209–249. [CrossRef] [PubMed]
4. Torre, L.A.; Bray, F.; Siegel, R.L.; Ferlay, J.; Lortet-Tieulent, J.; Jemal, A. Global cancer statistics, 2012: Global Cancer Statistics, 2012. *CA A Cancer J. Clin.* **2015**, *65*, 87–108. [CrossRef]
5. Vogel, A.; Martinelli, E.; Cervantes, A.; Chau, I.; Daniele, B.; Llovet, J.; Meyer, T.; Nault, J.-C.; Neumann, U.; Ricke, J.; et al. Updated treatment recommendations for hepatocellular carcinoma (HCC) from the ESMO Clinical Practice Guidelines. *Ann. Oncol.* **2021**, *32*, 801–805. [CrossRef] [PubMed]
6. European Association for the Study of the Liver. EASL Clinical Practice Guidelines: Management of hepatocellular carcinoma. *J. Hepatol.* **2018**, *69*, 182–236. [CrossRef]
7. Heimbach, J.K.; Kulik, L.M.; Finn, R.S.; Sirlin, C.B.; Abecassis, M.M.; Roberts, L.R.; Zhu, A.X.; Murad, M.H.; Marrero, J.A. AASLD guidelines for the treatment of hepatocellular carcinoma: Heimbach. *Hepatology* **2018**, *67*, 358–380. [CrossRef]
8. Chen, P.C.; Chen, B.H.; Huang, C.H.; Jeng, W.J.; Hsieh, Y.C.; Teng, W.; Chen, Y.C.; Ho, Y.P.; Sheen, I.S.; Lin, C.Y. Integrated model for end-stage liver disease maybe superior to some other model for end-stage liver disease-based systems in addition to Child-Turcotte-Pugh and albumin-bilirubin scores in patients with hepatitis B virus-related liver cirrhosis and spontaneous bacterial peritonitis. *Eur. J. Gastroenterol. Hepatol.* **2019**, *31*, 1256–1263.
9. Wan, S.Z.; Nie, Y.; Zhang, Y.; Liu, C.; Zhu, X. Assessing the Prognostic Performance of the Child-Pugh, Model for End-Stage Liver Disease, and Albumin-Bilirubin Scores in Patients with Decompensated Cirrhosis: A Large Asian Cohort from Gastroenterology Department. *Dis. Markers* **2020**, *2020*, 5193028. [CrossRef]
10. Cho, H.C.; Jung, H.Y.; Sinn, D.H.; Choi, M.S.; Koh, K.C.; Paik, S.W.; Yoo, B.C.; Kim, S.W.; Lee, J.H. Mortality after surgery in patients with liver cirrhosis: Comparison of Child–Turcotte–Pugh, MELD and MELDNa score. *Eur. J. Gastroenterol. Hepatol.* **2011**, *23*, 51–59. [CrossRef]
11. Wang, J.; Pillai, A. Systemic Therapy for Hepatocellular Carcinoma. *Clin. Liver Dis.* **2021**, *17*, 337–340. [CrossRef] [PubMed]
12. Bruix, J.; Chan, S.L.; Galle, P.R.; Rimassa, L.; Sangro, B. Systemic treatment of hepatocellular carcinoma: An EASL position paper. *J. Hepatol.* **2021**, *75*, 960–974. [CrossRef] [PubMed]

13. Su, G.L.; Altayar, O.; O'Shea, R.; Shah, R.; Estfan, B.; Wenzell, C.; Sultan, S.; Falck-Ytter, Y. AGA Clinical Practice Guideline on Systemic Therapy for Hepatocellular Carcinoma. *Gastroenterology* **2022**, *162*, 920–934. [CrossRef] [PubMed]
14. Harrison, P.M. Management of patients with decompensated cirrhosis. *Clin. Med.* **2015**, *15*, 201–203. [CrossRef]
15. Nouso, K.; Kokudo, N.; Tanaka, M.; Kuromatsu, R.; Nishikawa, H.; Toyoda, H.; Oishi, N.; Kuwaki, K.; Kusanaga, M.; Sakaguchi, T.; et al. Treatment of Hepatocellular Carcinoma with Child-Pugh C Cirrhosis. *Oncology* **2014**, *87*, 99–103. [CrossRef]
16. Zhang, H.; Zhang, W.; Jiang, L.; Chen, Y. Recent advances in systemic therapy for hepatocellular carcinoma. *Biomark Res.* **2022**, *10*, 3. [CrossRef]
17. Pelizzaro, F.; Ramadori, G.; Farinati, F. Systemic Therapies for Hepatocellular Carcinoma: An Evolving Landscape. HR.2021. Available online: https://hrjournal.net/article/view/4021 (accessed on 6 November 2022).
18. Shimose, S.; Hiraoka, A.; Tanaka, M.; Iwamoto, H.; Tanaka, T.; Noguchi, K.; Aino, H.; Yamaguchi, T.; Itano, S.; Suga, H.; et al. Deterioration of liver function and aging disturb sequential systemic therapy for unresectable hepatocellular carcinoma. *Sci. Rep.* **2022**, *12*, 17018. [CrossRef]
19. Graziadei, I. Systemic therapy in advanced-stage hepatocellular carcinoma. *Memo-Mag. Eur. Med. Oncol.* **2020**, *13*, 212–217. [CrossRef]
20. Marisi, G.; Cucchetti, A.; Ulivi, P.; Canale, M.; Cabibbo, G.; Solaini, L.; Foschi, F.G.; De Matteis, S.; Ercolani, G.; Valgiusti, M.; et al. Ten years of sorafenib in hepatocellular carcinoma: Are there any predictive and/or prognostic markers? *WJG* **2018**, *24*, 4152–4163. [CrossRef]
21. "Nexavar EPAR". European Medicines Agency (EMA). Archived from the original on 14 October 2021. Retrieved 18 September 2022. Available online: www.ema.europa.eu/en/medicines/human/EPAR/nexavar (accessed on 18 June 2014).
22. Llovet, J.M.; Ricci, S.; Mazzaferro, V.; Hilgard, P.; Gane, E.; Blanc, J.F.; de Oliveira, A.C.; Santoro, A.; Raoul, J.L.; Forner, A.; et al. Sorafenib in Advanced Hepatocellular Carcinoma. *N. Engl. J. Med.* **2008**, *359*, 378–390. [CrossRef]
23. Lencioni, R.; Kudo, M.; Ye, S.-L.; Bronowicki, J.-P.; Chen, X.-P.; Dagher, L.; Furuse, J.; Geschwind, J.F.; de Guevara, L.L.; Papandreou, C.; et al. GIDEON (Global Investigation of therapeutic DE cisions in hepatocellular carcinoma and Of its treatment with sorafeNib): Second interim analysis. *Int. J. Clin. Pract.* **2014**, *68*, 609–617. [CrossRef]
24. Woo, H.Y.; Heo, J.; Yoon, K.T.; Kim, G.H.; Kang, D.H.; Song, G.A.; Cho, M. Clinical course of sorafenib treatment in patients with hepatocellular carcinoma. *Scand. J. Gastroenterol.* **2012**, *47*, 809–819. [CrossRef] [PubMed]
25. Leal, C.R.G.; Magalhães, C.; Barbosa, D.; Aquino, D.; Carvalho, B.; Balbi, E.; Pacheco, L.; Perez, R.; de Tarso Pinto, P.; Setubal, S. Survival and tolerance to sorafenib in Child-Pugh B patients with hepatocellular carcinoma: A prospective study. *Investig. New Drugs* **2018**, *36*, 911–918. [CrossRef] [PubMed]
26. McNamara, M.G.; Slagter, A.E.; Nuttall, C.; Frizziero, M.; Pihlak, R.; Lamarca, A.; Tariq, N.; Valle, J.W.; Hubner, R.A.; Knox, J.J.; et al. Sorafenib as first-line therapy in patients with advanced Child-Pugh B hepatocellular carcinoma—A meta-analysis. *Eur. J. Cancer* **2018**, *105*, 1–9. [CrossRef]
27. Kudo, M.; Finn, R.S.; Qin, S.; Han, K.H.; Ikeda, K.; Piscaglia, F.; Baron, A.; Park, J.W.; Han, G.; Jassem, J.; et al. Lenvatinib versus sorafenib in first-line treatment of patients with unresectable hepatocellular carcinoma: A randomised phase 3 non-inferiority trial. *Lancet* **2018**, *391*, 1163–1173. [CrossRef]
28. Huynh, J.; Cho, M.T.; Kim, E.J.H.; Ren, M.; Ramji, Z.; Vogel, A. Lenvatinib in patients with unresectable hepatocellular carcinoma who progressed to Child-Pugh B liver function. *Ther. Adv. Med. Oncol.* **2022**, *14*, 175883592211166. [CrossRef] [PubMed]
29. Ogushi, K.; Chuma, M.; Uojima, H.; Hidaka, H.; Numata, K.; Kobayashi, S.; Hirose, S.; Hattori, N.; Fujikawa, T.; Nakazawa, T.; et al. Safety and Efficacy of Lenvatinib Treatment in Child–Pugh A and B Patients with Unresectable Hepatocellular Carcinoma in Clinical Practice: A Multicenter Analysis. *CEG* **2020**, *13*, 385–396. [CrossRef]
30. Park, M.K.; Lee, Y.B.; Moon, H.; Choi, N.R.; Kim, M.A.; Jang, H.; Nam, J.Y.; Cho, E.J.; Lee, J.H.; Yu, S.J.; et al. Effectiveness of Lenvatinib Versus Sorafenib for Unresectable Hepatocellular Carcinoma in Patients with Hepatic Decompensation. *Dig. Dis. Sci.* **2022**. Available online: https://link.springer.com/10.1007/s10620-021-07365-9 (accessed on 20 September 2022). [CrossRef]
31. Singal, A.G.; Nagar, S.P.; Hitchens, A.; Davis, K.L.; Iyer, S. REAL-WORLD effectiveness of lenvatinib monotherapy in previously treated unresectable hepatocellular carcinoma in US clinical practice. *Cancer Rep.* **2021**, *17*, 2759–2768. [CrossRef]
32. Tsuchiya, K.; Kurosaki, M.; Sakamoto, A.; Marusawa, H.; Kojima, Y.; Hasebe, C.; Arai, H.; Joko, K.; Kondo, M.; Tsuji, K.; et al. The Real-World Data in Japanese Patients with Unresectable Hepatocellular Carcinoma Treated with Lenvatinib from a Nationwide Multicenter Study. *Cancers* **2021**, *13*, 2608. [CrossRef]
33. Cosma, L.S.; Weigand, K.; Müller-Schilling, M.; Kandulski, A. Lenvatinib as First-line Treatment of Hepatocellular Carcinoma in Patients with Impaired Liver Function in Advanced Liver Cirrhosis: Real World Data and Experience of a Tertiary Hepatobiliary Center. *J. Gastrointest. Liver Dis.* **2021**, *30*, 247–253. Available online: https://www.jgld.ro/jgld/index.php/jgld/article/view/3345 (accessed on 21 December 2022). [CrossRef] [PubMed]
34. Abou-Alfa, G.K.; Meyer, T.; Cheng, A.L.; El-Khoueiry, A.B.; Rimassa, L.; Ryoo, B.Y.; Cicin, I.; Merle, P.; Chen, Y.; Park, J.W.; et al. Cabozantinib in Patients with Advanced and Progressing Hepatocellular Carcinoma. *N. Engl. J. Med.* **2018**, *379*, 54–63. [CrossRef] [PubMed]
35. El-Khoueiry, A.B.; Meyer, T.; Cheng, A.L.; Rimassa, L.; Sen, S.; Milwee, S.; Kelley, R.K.; Abou-Alfa, G.K. Safety and efficacy of cabozantinib for patients with advanced hepatocellular carcinoma who advanced to Child–Pugh B liver function at study week 8: A retrospective analysis of the CELESTIAL randomised controlled trial. *BMC Cancer* **2022**, *22*, 377. [CrossRef]

36. Kelley, R.K.; Miksad, R.; Cicin, I.; Chen, Y.; Klümpen, H.J.; Kim, S.; Lin, Z.Z.; Youkstetter, J.; Hazra, S.; Sen, S.; et al. Efficacy and safety of cabozantinib for patients with advanced hepatocellular carcinoma based on albumin-bilirubin grade. *Br. J. Cancer* **2022**, *26*, 569–575. [CrossRef]
37. Bang, Y.H.; Lee, C.K.; Yoo, C.; Chon, H.J.; Hong, M.; Kang, B.; Kim, H.D.; Park, S.R.; Choi, W.M.; Choi, J.; et al. Real-world efficacy and safety of cabozantinib in Korean patients with advanced hepatocellular carcinoma: A multicenter retrospective analysis. *Ther. Adv. Med. Oncol.* **2022**, *14*, 175883592210979. [CrossRef]
38. Finkelmeier, F.; Scheiner, B.; Leyh, C.; Best, J.; Fründt, T.W.; Czauderna, C.; Beutel, A.; Bettinger, D.; Weiß, J.; Meischl, T.; et al. Cabozantinib in Advanced Hepatocellular Carcinoma: Efficacy and Safety Data from an International Multicenter Real-Life Cohort. *Liver Cancer* **2021**, *10*, 360–369. [CrossRef] [PubMed]
39. Llovet, J.M.; Castet, F.; Heikenwalder, M.; Maini, M.K.; Mazzaferro, V.; Pinato, D.J.; Pikarsky, E.; Zhu, A.X.; Finn, R.S. Immunotherapies for hepatocellular carcinoma. *Nat. Rev. Clin. Oncol.* **2022**, *19*, 151–172. [CrossRef] [PubMed]
40. Kudo, M.; Matilla, A.; Santoro, A.; Melero, I.; Gracián, A.C.; Acosta-Rivera, M.; Choo, S.P.; El-Khoueiry, A.B.; Kuromatsu, R.; El-Rayes, B.; et al. CheckMate 040 cohort 5: A phase I/II study of nivolumab in patients with advanced hepatocellular carcinoma and Child-Pugh B cirrhosis. *J. Hepatol.* **2021**, *75*, 600–609. [CrossRef]
41. El-Khoueiry, A.B.; Sangro, B.; Yau, T.; Crocenzi, T.S.; Kudo, M.; Hsu, C.; Kim, T.Y.; Choo, S.P.; Trojan, J.; Welling, T.H.R.; et al. Nivolumab in patients with advanced hepatocellular carcinoma (CheckMate 040): An open-label, non-comparative, phase 1/2 dose escalation and expansion trial. *Lancet* **2017**, *389*, 2492–2502. [CrossRef]
42. Chapin, W.J.; Hwang, W.; Karasic, T.B.; McCarthy, A.M.; Kaplan, D.E. Comparison of nivolumab and sorafenib for first systemic therapy in patients with hepatocellular carcinoma and Child-Pugh B cirrhosis. *Cancer Med.* **2022**, *1*, 189–199. [CrossRef]
43. Choi, W.M.; Lee, D.; Shim, J.H.; Kim, K.M.; Lim, Y.S.; Lee, H.C.; Yoo, C.; Park, S.R.; Ryu, M.H.; Ryoo, B.Y.; et al. Effectiveness and Safety of Nivolumab in Child–Pugh B Patients with Hepatocellular Carcinoma: A Real-World Cohort Study. *Cancers* **2020**, *12*, 1968. [CrossRef]
44. Scheiner, B.; Kirstein, M.M.; Hucke, F.; Finkelmeier, F.; Schulze, K.; von Felden, J.; Koch, S.; Schwabl, P.; Hinrichs, J.B.; Waneck, F.; et al. Programmed cell death protein-1 (PD-1)-targeted immunotherapy in advanced hepatocellular carcinoma: Efficacy and safety data from an international multicentre real-world cohort. *Aliment. Pharmacol. Ther.* **2019**, *49*, 1323–1333. [CrossRef]
45. Finn, R.S.; Qin, S.; Ikeda, M.; Galle, P.R.; Ducreux, M.; Kim, T.Y.; Kudo, M.; Breder, V.; Merle, P.; Kaseb, A.O.; et al. Atezolizumab plus Bevacizumab in Unresectable Hepatocellular Carcinoma. *N. Engl. J. Med.* **2020**, *382*, 1894–1905. [CrossRef]
46. D'Alessio, A.; Fulgenzi, C.A.M.; Nishida, N.; Schönlein, M.; von Felden, J.; Schulze, K.; Wege, H.; Gaillard, V.E.; Saeed, A.; Wietharn, B.; et al. Preliminary evidence of safety and tolerability of atezolizumab plus bevacizumab in patients with hepatocellular carcinoma and Child-Pugh A and B cirrhosis: A real-world study. *Hepatology* **2022**, *4*, 1000–1012. [CrossRef] [PubMed]
47. Chen, C.T.; Feng, Y.H.; Yen, C.J.; Chen, S.C.; Lin, Y.T.; Lu, L.C.; Hsu, C.H.; Cheng, A.L.; Shao, Y.Y. Prognosis and treatment pattern of advanced hepatocellular carcinoma after failure of first-line atezolizumab and bevacizumab treatment. *Hepatol. Int.* **2022**, *16*, 1199–1207. [CrossRef] [PubMed]
48. Persano, M.; Rimini, M.; Tada, T.; Suda, G.; Shimose, S.; Kudo, M.; Cheon, J.; Finkelmeier, F.; Lim, H.Y.; Rimassa, L.; et al. Clinical outcomes with atezolizumab plus bevacizumab or lenvatinib in patients with hepatocellular carcinoma: A multicenter real-world study. *J. Cancer Res. Clin. Oncol. J.* **2022**. *ahead of print*. [CrossRef]
49. de Castro, T.; Jochheim, L.S.; Bathon, M.; Welland, S.; Scheiner, B.; Shmanko, K.; Roessler, D.; Ben Khaled, N.; Jeschke, M.; Ludwig, J.M.; et al. Atezolizumab and bevacizumab in patients with advanced hepatocellular carcinoma with impaired liver function and prior systemic therapy: A real-world experience. *Ther. Adv. Med. Oncol.* **2022**, *14*, 175883592210802. [CrossRef]
50. Abou-Alfa, G.K.; Lau, G.; Kudo, M.; Chan, S.L.; Kelley, R.K.; Furuse, J.; Sukeepaisarnjaroen, W.; Kang, Y.K.; Dao, T.V.; De Toni, E.N.; et al. Tremelimumab plus Durvalumab in Unresectable Hepatocellular Carcinoma. *NEJM* **2022**, *1*, EVIDoa2100070. [CrossRef]
51. Piscaglia, F.; Terzi, E.; Cucchetti, A.; Trimarchi, C.; Granito, A.; Leoni, S.; Marinelli, S.; Pini, P.; Bolondi, L. Treatment of hepatocellular carcinoma in Child-Pugh B patients. *Dig. Liver Dis.* **2013**, *45*, 852–858. [CrossRef] [PubMed]
52. CasadeiGardini, A.; Marisi, G.; Canale, M.; Foschi, F.G.; Donati, G.; Ercolani, G.; Valgiusti, M.; Passardi, A.; Frassineti, G.L.; Scarpi, E. Radiofrequency Ablation of hepatocellular carcinoma: A meta-analysis of overall survival and recurrence-free survival. *OTT* **2018**, *11*, 6555–6567. [CrossRef]
53. Kim, Y.K.; Kim, C.S.; Chung, G.H.; Han, Y.M.; Lee, S.Y.; Jin, G.Y.; Lee, J.M. Radiofrequency Ablation of Hepatocellular Carcinoma in Patients with Decompensated Cirrhosis: Evaluation of Therapeutic Efficacy and Safety. *Am. J. Roentgenol.* **2006**, *186*, 261–268. [CrossRef]
54. Nouso, K.; Ito, Y.M.; Kuwaki, K.; Kobayashi, Y.; Nakamura, S.; Ohashi, Y.; Yamamoto, K. Prognostic factors and treatment effects for hepatocellular carcinoma in Child C cirrhosis. *Br. J. Cancer* **2008**, *98*, 1161–1165. [CrossRef] [PubMed]
55. Kudo, M.; Osaki, Y.; Matsunaga, T.; Kasugai, H.; Oka, H.; Seki, T. Hepatocellular Carcinoma in Child-Pugh C Cirrhosis: Prognostic Factors and Survival Benefit of Nontransplant Treatments. *Dig. Dis.* **2013**, *31*, 490–498. [CrossRef]
56. Garwood, E.R.; Fidelman, N.; Hoch, S.E.; Kerlan, R.K.; Yao, F.Y. Morbidity and mortality following transarterial liver chemoembolization in patients with hepatocellular carcinoma and synthetic hepatic dysfunction: High-Risk Transarterial Chemoembolization Outcomes. *Liver Transpl.* **2013**, *19*, 164–173. [CrossRef] [PubMed]

57. Miksad, R.A.; Ogasawara, S.; Xia, F.; Fellous, M.; Piscaglia, F. Liver function changes after transarterial chemoembolization in US hepatocellular carcinoma patients: The LiverT study. *BMC Cancer* **2019**, *19*, 795. [CrossRef] [PubMed]
58. Wang, Q.; Xia, D.; Bai, W.; Wang, E.; Sun, J.; Huang, M.; Mu, W.; Yin, G.; Li, H.; Zhao, H.; et al. Development of a prognostic score for recommended TACE candidates with hepatocellular carcinoma: A multicentre observational study. *J. Hepatol.* **2019**, *70*, 893–903. [CrossRef] [PubMed]
59. Jia, K.F.; Wang, H.; Yu, C.L.; Yin, W.L.; Zhang, X.D.; Wang, F.; Sun, C.; Shen, W. ASARA, a prediction model based on Child-Pugh class in hepatocellular carcinoma patients undergoing transarterial chemoembolization. *Hepatobiliary Pancreat. Dis. Int.* **2022**, *22*, S1499387222000157. [CrossRef] [PubMed]
60. Takayasu, K.; Arii, S.; Kudo, M.; Ichida, T.; Matsui, O.; Izumi, N.; Izumi, N.; Matsuyama, Y.; Sakamoto, M.; Nakashima, O.; et al. Superselectivetransarterial chemoembolization for hepatocellular carcinoma. Validation of treatment algorithm proposed by Japanese guidelines. *J. Hepatol.* **2012**, *56*, 886–892. [CrossRef] [PubMed]
61. Dorn, D.P.; Bryant, M.K.; Zarzour, J.; Smith, J.K.; Redden, D.T.; Saddekni, S.; Abdel Aal, A.K.; Gray, S.; White, J.; Eckhoff, D.E.; et al. Chemoembolization outcomes for hepatocellular carcinoma in cirrhotic patients with compromised liver function. *HPB* **2014**, *16*, 648–655. [CrossRef]
62. Kitai, S.; Kudo, M.; Nishida, N.; Izumi, N.; Sakamoto, M.; Matsuyama, Y.; Ichida, T.; Nakashima, O.; Matsui, O.; Ku, Y.; et al. Survival Benefit of Locoregional Treatment for Hepatocellular Carcinoma with Advanced Liver Cirrhosis. *Liver Cancer* **2016**, *5*, 175–189. [CrossRef]
63. Sacco, R.; Mismas, V.; Marceglia, S.; Romano, A.; Giacomelli, L.; Bertini, M.; Federici, G.; Metrangolo, S.; Parisi, G.; Tumino, E.; et al. Transarterial radioembolization for hepatocellular carcinoma: An update and perspectives. *World J. Gastroenterol.* **2015**, *21*, 6518–6525. [CrossRef]
64. Rahman, S.I.; Nunez-Herrero, L.; Berkes, J.L. Position 2: Transarterial Radioembolization Should Be the Primary Locoregional Therapy for Unresectable Hepatocellular Carcinoma. *Clin. Liver Dis. (Hoboken)* **2020**, *15*, 74–76. [CrossRef]
65. Cho, Y.Y.; Lee, M.; Kim, H.C.; Chung, J.W.; Kim, Y.H.; Gwak, G.Y.; Bae, S.H.; do Kim, Y.; Heo, J.; Kim, Y.J. Radioembolization Is a Safe and Effective Treatment for Hepatocellular Carcinoma with Portal Vein Thrombosis: A Propensity Score Analysis. *PLoS ONE* **2016**, *11*, e0154986. [CrossRef] [PubMed]
66. Zu, Q.; Schenning, R.C.; Jahangiri, Y.; Tomozawa, Y.; Kolbeck, K.J.; Kaufman, J.A.; Al-Hakim, R.; Naugler, W.E.; Nabavizadeh, N.; Kardosh, A. Yttrium-90 Radioembolization for BCLC Stage C Hepatocellular Carcinoma Comparing Child-Pugh A Versus B7 Patients: Are the Outcomes Equivalent? *Cardiovasc. Interv. Radiol.* **2020**, *43*, 721–731. [CrossRef]
67. Abouchaleh, N.; Gabr, A.; Ali, R.; Al Asadi, A.; Mora, R.A.; Kallini, J.R.; Mouli, S.; Riaz, A.; Lewandowski, R.J.; Salem, R. $^{90}$Y Radioembolization for Locally Advanced Hepatocellular Carcinoma with Portal Vein Thrombosis: Long-Term Outcomes in a 185-Patient Cohort. *J. Nucl. Med.* **2018**, *59*, 1042–1048. [CrossRef]
68. Memon, K.; Kulik, L.; Lewandowski, R.J.; Mulcahy, M.F.; Benson, A.B.; Ganger, D.; Riaz, A.; Gupta, R.; Vouche, M.; Gates, V.L.; et al. Radioembolization for hepatocellular carcinoma with portal vein thrombosis: Impact of liver function on systemic treatment options at disease progression. *J. Hepatol.* **2013**, *58*, 73–80. [CrossRef] [PubMed]
69. Llovet, J.M.; Fuster, J.; Bruix, J.; Barcelona Clinic Liver Cancer (BCLC) Group. Intention-to-treat analysis of surgical treatment for early hepatocellular carcinoma: Resection versus transplantation. *Hepatology* **1999**, *30*, 1434–1440. [CrossRef] [PubMed]
70. Utsunomiya, T.; Shimada, M.; Kudo, M.; Ichida, T.; Matsui, O.; Izumi, N.; Matsuyama, Y.; Sakamoto, M.; Nakashima, O.; Ku, Y.; et al. A Comparison of the Surgical Outcomes Among Patients With HBV-positive, HCV-positive, and Non-B Non-C Hepatocellular Carcinoma: A Nationwide Study of 11,950 Patients. *Ann. Surg.* **2015**, *261*, 513–520. [CrossRef]
71. Krenzien, F.; Schmelzle, M.; Struecker, B.; Raschzok, N.; Benzing, C.; Jara, M.; Bahra, M.; Öllinger, R.; Sauer, I.M.; Pascher, A.; et al. Liver Transplantation and Liver Resection for Cirrhotic Patients with Hepatocellular Carcinoma: Comparison of Long-Term Survivals. *J. Gastrointest. Surg.* **2018**, *22*, 840–848. [CrossRef]
72. Sugawara, Y.; Hibi, T. Surgical treatment of hepatocellular carcinoma. *BST* **2021**, *15*, 138–141. [CrossRef]
73. Bruix, J.; Castells, A.; Bosch, J.; Feu, F.; Fuster, J.; Garcia-Pagan, J.; Visa, J.; Bru, C.; Rodés, J. Surgical resection of hepatocellular carcinoma in cirrhotic patients: Prognostic value of preoperative portal pressure. *Gastroenterology* **1996**, *111*, 1018–1022. [CrossRef] [PubMed]
74. Azoulay, D.; Ramos, E.; Casellas-Robert, M.; Salloum, C.; Lladó, L.; Nadler, R.; Caula-Freixa, C.; Mils, K.; Lopez-Ben, S.; Figueras, J.; et al. Liver resection for hepatocellular carcinoma in patients with clinically significant portal hypertension. *JHEP Rep.* **2021**, *3*, 100190. [CrossRef]
75. Leyh, C.; Heucke, N.; Schotten, C.; Büchter, M.; Bechmann, L.P.; Wichert, M.; Dechêne, A.; Herrmann, K.; Heider, D.; Sydor, S.; et al. LiMAx Prior to Radioembolization for Hepatocellular Carcinoma as an Additional Tool for Patient Selection in Patients with Liver Cirrhosis. *Cancers* **2022**, *14*, 4584. [CrossRef] [PubMed]
76. Citterio, D.; Facciorusso, A.; Sposito, C.; Rota, R.; Bhoori, S.; Mazzaferro, V. Hierarchic Interaction of Factors Associated With Liver Decompensation After Resection for Hepatocellular Carcinoma. *JAMA Surg.* **2016**, *15*, 846. [CrossRef]
77. Han, H.S.; Shehta, A.; Ahn, S.; Yoon, Y.S.; Cho, J.Y.; Choi, Y. Laparoscopic versus open liver resection for hepatocellular carcinoma: Case-matched study with propensity score matching. *J. Hepatol.* **2015**, *63*, 643–650. [CrossRef] [PubMed]
78. Ciria, R.; Ocaña, S.; Gomez-Luque, I.; Cipriani, F.; Halls, M.; Fretland, Å.A.; Okuda, Y.; Aroori, S.; Briceño, J.; Aldrighetti, L.; et al. A systematic review and meta-analysis comparing the short- and long-term outcomes for laparoscopic and open liver resections for liver metastases from colorectal cancer. *Surg. Endosc.* **2020**, *34*, 349–360. [CrossRef] [PubMed]

79. Parikh, N.D.; Yopp, A.; Singal, A.G. Controversies in criteria for liver transplantation in hepatocellular carcinoma. *Curr. Opin. Gastroenterol.* **2016**, *32*, 182–188. [CrossRef]
80. Llovet, J.M.; Burroughs, A.; Bruix, J. Hepatocellular carcinoma. *Lancet* **2003**, *36*, 1907–1917. [CrossRef]
81. Durand, F.; Antoine, C.; Soubrane, O. Liver Transplantation in France. *Liver Transpl.* **2019**, *25*, 763–770. [CrossRef]
82. Yao, F.Y.; Bass, N.M.; Nikolai, B.; Davern, T.J.; Kerlan, R.; Wu, V.; Ascher, N.L.; Roberts, J.P. Liver transplantation for hepatocellular carcinoma: Analysis of survival according to the intention-to-treat principle and dropout from the waiting list. *Liver Transplant.* **2002**, *8*, 873–883. [CrossRef]
83. Crocetti, L.; Bozzi, E.; Scalise, P.; Bargellini, I.; Lorenzoni, G.; Ghinolfi, D.; Campani, D.; Balzano, E.; De Simone, P.; Cioni, R. Locoregional Treatments for Bridging and Downstaging HCC to Liver Transplantation. *Cancers* **2021**, *13*, 5558. [CrossRef]
84. Hibi, T.; Sugawara, Y. Locoregional therapy as a bridge to liver transplantation for hepatocellular carcinoma within Milan criteria: From a transplant oncology viewpoint. *Hepatobiliary Surg. Nutr.* **2018**, *7*, 134–135. [CrossRef]
85. Bhardwaj, H.; Fritze, D.; Mais, D.; Kadaba, V.; Arora, S.P. Neoadjuvant Therapy With Cabozantinib as a Bridge to Liver Transplantation in Patients With Hepatocellular Carcinoma (HCC): A Case Report. *Front Transplant.* **2022**, *1*, 863086. [CrossRef]
86. Coletta, M.; Nicolini, D.; Cacciaguerra, A.B.; Mazzocato, S.; Rossi, R.; Vivarelli, M. Bridging patients with hepatocellular cancer waiting for liver transplant: All the patients are the same? *Transl. Gastroenterol. Hepatol.* **2017**, *2*, 78. [CrossRef] [PubMed]
87. Schnickel, G.T.; Fabbri, K.; Hosseini, M.; Misel, M.; Berumen, J.; Parekh, J.; Mekeel, K.; Dehghan, Y.; Kono, Y.; Ajmera, V. Liver transplantation for hepatocellular carcinoma following checkpoint inhibitor therapy with nivolumab. *Am. J Transplant.* **2022**, *22*, 1699–1704. [CrossRef] [PubMed]
88. Yao, F. Liver transplantation for hepatocellular carcinoma: Expansion of the tumor size limits does not adversely impact survival. *Hepatology* **2001**, *33*, 1394–1403. [CrossRef] [PubMed]
89. Santopaolo, F.; Lenci, I.; Milana, M.; Manzia, T.M.; Baiocchi, L. Liver transplantation for hepatocellular carcinoma: Where do we stand? *WJG* **2019**, *25*, 2591–2602. [CrossRef] [PubMed]
90. Kumar, M.; Panda, D. Role of Supportive Care for Terminal Stage Hepatocellular Carcinoma. *J. Clin. Exp. Hepatol.* **2014**, *4*, S130–S139. [CrossRef]
91. Woodrell, C.D.; Hansen, L.; Schiano, T.D.; Goldstein, N.E. Palliative Care for People With Hepatocellular Carcinoma, and Specific Benefits for Older Adults. *Clin. Ther.* **2018**, *40*, 512–525. [CrossRef]
92. Peng, Y.; Qi, X.; Guo, X. Child–Pugh Versus MELD Score for the Assessment of Prognosis in Liver Cirrhosis: A Systematic Review and Meta-Analysis of Observational Studies. *Medicine* **2016**, *95*, e2877. [CrossRef]
93. Scheiner, B.; Roessler, D.; Phen, S.; Lim, M.; Pomej, K.; Pressiani, T.; Cammarota, A.; Fründt, T.W.; von Felden, J.; Schulze, K.; et al. Efficacy and safety of immune checkpoint inhibitor rechallenge in individuals with hepatocellular carcinoma. *JHEP Rep.* **2022**, *5*, 100620. [CrossRef] [PubMed]

cancers

Review

# Local and Regional Therapies for Hepatocellular Carcinoma and Future Combinations

Adam Hatzidakis [1,*], Lukas Müller [2], Miltiadis Krokidis [3] and Roman Kloeckner [2]

1 Department of Radiology, AHEPA University Hospital of Thessaloniki, Faculty of Health Sciences, School of Medicine, Aristotle University of Thessaloniki, 54124 Thessaloniki, Greece
2 Department of Diagnostic and Interventional Radiology, University Medical Center of the Johannes Gutenberg University Mainz, 55131 Mainz, Germany; lukas.mueller@unimedizin-mainz.de (L.M.); roman.kloeckner@unimedizin-mainz.de (R.K.)
3 1st Department of Radiology, Areteion Hospital, School of Medicine, National and Kapodistrian University of Athens, 11528 Athens, Greece; mkrokidis@hotmail.com
* Correspondence: adamhatz@hotmail.com

**Simple Summary:** Percutaneous interventional radiological techniques offer many alternatives for treatment of Hepatocellular Carcinoma (HCC) using local anesthesia and sedation. These methods aim to destroy the malignant tumors locally without affecting the non-malignant liver. In this way, complications are kept low and patient recovery is quick. Indications depend on tumor size, type and stage, as well as patient's condition, liver function and co-morbidities. In recent years, a lot of research has been made in combining such approaches with immune therapy, but there is still much work to be done. This manuscript tries to analyze where we stand today and explain, using a comprehensive algorithm, the treatment options for each different clinical condition.

**Abstract:** Background: Hepatocellular carcinoma (HCC) can be treated by local and regional methods of percutaneous interventional radiological techniques. Indications depend on tumor size, type and stage, as well as patient's condition, liver function and co-morbidities. According to international classification systems such as Barcelona Clinic Liver Cancer (BCLC) classification, very early, early or intermediate staged tumors can be treated either with ablative methods or with transarterial chemoembolization (TACE), depending on tumor characteristics. The combination of both allows for individualized forms of treatment with the ultimate goal of improving response and survival. In recent years, a lot of research has been carried out in combining locoregional approaches with immune therapy. Although recent developments in systemic treatment, especially immunotherapy, seem quite promising and have expanded possible combined treatment options, there is still not enough evidence in their favor. The aim of this review is to provide a comprehensive up-to-date overview of all these techniques, explaining indications, contraindications, technical problems, outcomes, results and complications. Moreover, combinations of percutaneous treatment with each other or with immunotherapy and future options will be discussed. Use of all those methods as down-staging or bridging solutions until surgery or transplantation are taken into consideration will also be reviewed. Conclusion: Local and regional therapies remain a mainstay of curative and palliative treatment of patients with HCC. Currently, evidence on potential combination of the local and regional treatment options with each other as well as with other treatment modalities is growing and has the potential to further individualize HCC therapy. To identify the most suitable treatment option out of these new various options, a repeated interdisciplinary discussion of each case by the tumor board is of utmost importance.

**Keywords:** hepatocellular carcinoma; percutaneous treatment; locoregional treatment; chemoembolization; tumor ablation; radioembolization; immunotherapy

**Citation:** Hatzidakis, A.; Müller, L.; Krokidis, M.; Kloeckner, R. Local and Regional Therapies for Hepatocellular Carcinoma and Future Combinations. *Cancers* **2022**, *14*, 2469. https://doi.org/10.3390/cancers14102469

Academic Editor: Georgios Germanidis

Received: 2 April 2022
Accepted: 11 May 2022
Published: 17 May 2022

## 1. Transarterial Chemoembolization (TACE)

### 1.1. Background

According to the current EASL guidelines, TACE is the recommended first-line treatment for patients within the intermediate stage, which is defined as multinodular tumor burden, preserved liver function and good performance status [1]. The recent 2022 update of the BCLC criteria limits the recommendation for TACE to patients with well-defined tumor nodules, preserved portal flow and selective vascular access (namely the second subgroup of BCLC stage B), that are not eligible for liver transplantation with extended Milan criteria (first subgroup of BCLC stage B) [2]. Following these recommendations, patients with diffuse, infiltrative and extensive bilobar liver involvement do not benefit from TACE (third subgroup of BCLC stage B patients), and should be better candidates for systemic treatment in the first line. However, no clear cut-off criteria when to prefer systemic treatment over TACE can be provided up to date [2]. The suggestions for TACE as recommended therapy in intermediate stage mainly rely on two randomized trials, which showed a survival benefit of TACE compared to best supportive care [3,4]. However, patient selection in both trials followed very strict inclusion criteria. In clinical reality, the intermediate stage comprises a heterogeneous subgroup of patients with distinct differences in tumor burden and remaining liver function [5]. Furthermore, the concept of stage-migration is commonly applied, meaning that patients in earlier or more advanced stages are referred to TACE according to individual treatment concepts.

### 1.2. Biological Rational for TACE

Chemoembolization relies on important tumor characteristics of HCC: the tumor tissue has, in most cases, a strong arterial tumor supply, which is based on an intense neo-angiogenesis during tumor development and progression. Embolization takes advantage of HCC's strong arterial supply, aiming for complete anoxia inside the malignancy, therefore inducing an ischemic reaction and leading to tumor necrosis. Theoretically, the surrounding liver parenchyma is spared out from these necrotic effects as it receives its blood supply mainly from the portal venous system. However, the most peripheral part of the tumor nodules does receive blood supply from the branches of the portal vein. Thus, there is always a rest supply of the peripheral tumor that could not be stopped with TACE. Therefore, TACE is not considered as curative treatment option in patients with HCC. With the additional local application of chemotherapeutic agents (most commonly the anthracyclines doxorubicin and epirubicin or the platinum-derivates cisplatin or miriplatin), TACE has theoretically the benefit of a synergistic effect with a high local chemotherapy concentration leading to a higher rate of tumor cell necrosis.

### 1.3. Technical Considerations

Up to date, two main TACE techniques are the standard of care: The conventional TACE (cTACE) was already used before 1990 as injection of an emulsion consisting of chemotherapy and lipiodol followed by an embolizing agent through diagnostic catheters and was further developed after 2000 as a superselective embolization technique with the use of flexible microcatheters [3]. In 2006, drug-eluting beads TACE (DEB-TACE) was introduced, which is based on a slower release of the chemotherapeutic agents to improve its therapeutic effect while reducing side effects [6]. Since then, several trials compared the outcome of both TACE types [7–10]. However, no significant difference has been observed regarding tumor response or survival. Both types only differed regarding post-procedural pain and chemotherapy-associated systemic side effects, which were less often observed in patients undergoing DEB-TACE [7,10]. Regardless of the applied chemotherapeutic agent, TACE should be performed in a superselective manner, as this leads to an increased rate of tumor necrosis and minimizes the damage in the surrounding liver tissue [11,12].

### 1.4. Patient Selection

In more than 80% percent of the cases, HCC develops in a cirrhotic liver [1]. Thus, the patients suffer from a combination of two chronic illnesses: progressing liver damage resulting in an impaired synthesis function and the HCC tumor burden itself. Therefore, the rationale for initiation and continuation should be based on both the tumor and the remaining liver function.

### 1.5. Patient Selection: Tumor Burden

Regarding the tumor burden, the image-derived features tumor size and tumor number correlate with the prognosis [1,4,13–16]. Furthermore, a large tumor size is related to a higher risk for complications [12,16–19]. Although the current guidelines do not recommend a clear cut-off regarding the tumor size, in clinical reality, a commonly used cut-off is a single nodule size of 5 cm as a response to TACE is significantly lower if this threshold is exceeded [12]. Furthermore, a total tumor burden of more than half of the total liver volume is a contraindication for TACE as radiologic response is unlikely and the risk of post-interventional hepatic decompensation is substantially increased [1]. Several systems for the stratification of patients according to the sum of tumor size and the number of nodules have been proposed [20–22]. However, their prognostic value in external validation was only moderate and a direct head-to-head comparison of the various scores and cut-offs is missing. Apart from tumor size, number and growth type, pre-procedural imaging offers an insight into various prognosis-related tumor features [23]. One of the most important features is the degree and anatomy of tumor vascularity, which highly influences the treatment success. Furthermore, the growth pattern of the tumor correlates with the biological aggressiveness and infiltrative and diffuse tumor growth is linked to an impaired OS [24,25]. Concretely, the current guidelines recommend a critical discussion on the indication for TACE in patients with severe hepatic decompensation (Child-Pugh B with signs of decompensation and Child-Pugh C) as those are at high risk for post-interventional liver failure. Furthermore, patients with bilirubin levels above 2 mg/dl have a high potential for post-interventional hepatic decompensation [1,2].

### 1.6. Patient Selection: Remaining Liver Function

Regarding the liver function, particularly the albumin-bilirubin (ALBI) score has gained importance as an easy-to-use estimate of the patients' remaining liver function in the recent decade [26,27]. In comparison to the Child-Pugh score, the ALBI score does not contain subjectively estimated parameters such as the degree of ascites or the degree of encephalopathy and is only composed of objective laboratory parameters. Due to the high predictive performance for patients undergoing TACE, the ALBI score, as well as other liver function-related parameters such as INR, platelet count and cholinesterase play an important role in treatment decision making [28–30]. Despite laboratory parameters, other factors related to an impaired liver function such as, e.g., ascites or a reduction in the overall patient status should be considered in decision-making. The presence of other surrogates for an elevated portal pressure, however, seem to play a minor role in the initial patient selection [31].

### 1.7. Unmet Problems

Although extensively evaluated, several problems prior to and during TACE treatment remain unclear. One common problem is the planning of re-treatment in patients with remaining vital tumor after the initial TACE. Up to date, it remains unclear whether an on-demand treatment repetition or a fixed schedule is superior. Scheduled treatment in regular intervals may lead to improved patient compliance and monitoring of the case, but an aggressive schedule may also increase the risk of liver failure [32]. Thus, liver function and performance status have to be monitored closely in order to avoid "overtreatment" [33]. From our point of view, regardless of the re-treatment type, continuous re-evaluation of each patient by an interdisciplinary tumor board is of the utmost importance [23]. Each indication should be jointly approved by this board.

Another common problem during follow-up is the optimal time-point to switch to other treatment modalities in case of tumor progress and/or ongoing decrease in the remaining liver function. In clinical routine, the optimal time-point for a therapy switch may be hard to determine, particularly as defined cut-off criteria are missing [23].

Additional TACE sessions over a certain point may not lead to a survival benefit for the patient, while delaying the switch to systemic therapy or even completely impeding this switch due to a deterioration of the remaining liver function caused by repetitive TACE could lead to an impaired survival outcome [34–38]. One prospective trial currently investigating the role of systemic treatment following TACE is the OPTIMIS trial (NCT01933945). In cases of tumor progression, decrease in remaining liver function, early recurrence, incomplete necrosis and occurrence of extrahepatic spread and/or vascular invasion during TACE, but also in the case of unbearable side effects (e.g., severe forms of post-embolization syndrome or hepatorenal syndrome), TACE treatment should be immediately stopped.

### 1.8. Long-Term Outcome and Risk Prediction

For patients with intermediate-stage HCC, a median OS of 2.5 years can be expected [1]. A median OS of up 4 years can be achieved in case of strict patient selection [13]. However, especially because of the above-mentioned heterogeneity of patients undergoing TACE in conjunction with the concept of stage migration, the individual prognosis prediction remains difficult. Although several risk scores for patient selection prior to or during TACE have shown promising results initially, all showed only moderate performance in external validation [19,21,23,28,39–54]. The underperformance of these scoring systems might not only be explained through the heterogeneity of the patient cohort, but also by the complex interplay of co-existing liver cirrhosis and HCC as two synchronous diseases. All these issues are reasons why none of the scoring systems play a significant role in clinical reality—although the need for precise risk estimation is tremendous.

Thus, current attempts try to include novel risk factors to improve the predictive performance of risk scoring systems. In particular, the knowledge on the interplay of tumor development and progression as well as immune response and inflammatory reaction has been growing continuously [55–59]. Especially the role of neutrophils for HCC development and progression on the tumor microenvironment as well as in systemic reaction is under investigation. Moreover, the relations of neutrophils to lymphocytes and platelets to lymphocytes, namely NLR and PLR, have been identified as important prognostic marker in HCC and particularly in patients undergoing TACE [58–64]. Based on the first experiences regarding the potential of immune response and inflammation, several other indices based on various laboratory markers have been investigated in patients with HCC undergoing TACE (Table 1).

**Table 1.** Overview on currently applied immune- and inflammation based prognostic indices in patients with HCC undergoing TACE. Modified according to [57].

| Index | Concept and Characteristics | Included Parameters | Pros | Cons | Current Research Status |
|---|---|---|---|---|---|
| NLR | -captures shifts in the relationships between blood cells, due to immune response effects | -neutrophil count<br>-lymphocyte count | -simple calculation<br>-well investigated | -nutritional status not included<br>-divergent results in studies that compared NLR to other immune-based indices | -designed for the stratification of critically ill patients, and validated in patients with colorectal cancer, in an oncologic context [63]<br>-extensively validated for various cancer entities, including patients with HCC undergoing TACE |

**Table 1.** *Cont.*

| Index | Concept and Characteristics | Included Parameters | Pros | Cons | Current Research Status |
|---|---|---|---|---|---|
| PLR | -captures shifts in the relationships between blood cells, due to immune response effects | -platelet count -lymphocyte count | -simple calculation -well investigated | -nutritional status not included -divergent results in studies that compared PLR to other immune-based indices | -designed for the stratification of patients with pancreatic cancer [64] -extensively validated for patients with HCC undergoing TACE |
| CALLY | -combines inflammation, immune response, and nutritional status markers (aspects of the PNI) -for liver disease, albumin functions as an indicator of liver function | -CRP -albumin -lymphocyte count | -novel combination of inflammation, immune response, nutritional status, and liver function markers provides a more holistic assessment | -CALLY was not superior to previously established scoring systems | -designed for a cohort of patients with HCC undergoing resections [65] -only validated in one study for patients with HCC undergoing TACE |
| PNI | -combines immune response and nutritional status markers | -albumin -lymphocyte count | -combination of immune response and nutritional status markers | -few studies available on patients with HCC undergoing TACE -divergent results regarding the predictive ability of PNI -the mathematical calculation may require improvement | -designed for patients with gastric cancer [66] -extensively validated for various cancer entities -few studies available for patients with HCC undergoing TACE -divergent results on its predictive ability -PNI combined with ALBI was identified as a novel, feasible stratification system for patients with HCC undergoing TACE [55] |
| CONUT | -combines immune response and nutritional status markers | -albumin -lymphocyte count -cholesterin | -combination of immune response and nutritional status markers | -few studies available on patients with HCC undergoing TACE -not superior to PNI [56] | -Only few validation results in patients with HCC undergoing TACE |
| SII | -combines inflammation and immune response markers | -lymphocyte count -neutrophil count -platelet count | -extensively validated for patients with HCC | -nutritional status not included - literature is scarce for patients undergoing TACE | -designed for the stratification of patients with HCC undergoing resections [67] -extensively validated for various cancer entities -few studies on the role of the SII in patients undergoing TACE |
| ILIS | -combines inflammation, liver function, and tumor markers -specifically developed for patients with HCC | -albumin -bilirubin -alkaline phosphatase -neutrophil count | -index is specific for HCC -includes tumor and liver function markers | -complex calculation -scarce literature for patients with HCC, particularly for patients undergoing TACE | -specifically designed for patients with HCC [68] -only one external validation study available |

Furthermore, current achievements in the field of AI-based risk prediction automatically allow the inclusion of a great number of risk factors simultaneously. First studies have shown the feasibility of this approach for patients with HCC undergoing TACE, which outperformed conventional risk scoring distinctly [69].

### 1.9. Combination of TACE and Thermal Tumor Ablation

One treatment combination, which has been extensively investigated in recent years, is the combination of TACE and thermal tumor ablation. Such a combination might lead to a reduction of the TACE-induced neo-angiogenesis and therefore reduces the risk of tumor recurrence and metastatic growth [70]. Furthermore, the combination of radiofrequency ablation and TACE increases the coagulated zone, which led to a significantly reduced rate of local tumor progression [71,72]. For patients within the early tumor stage, several meta-analyses showed a survival benefit and a better regression-free survival when combining TACE and ablation in patients unsuitable for resection [73–75]. Specifically, patients with a large tumor size could benefit [76–78]. For patients with intermediate-stage HCC, ablation might be a suitable addition in selected patients with a favorable tumor location [70]. Liu et al. suggest that for HCCs > 5 cm combination of both methods could lead to better outcome results [75]. They propose that for such large tumors, first-line TACE followed by ablation 1 month later is better than TACE alone. However, clear evidence on the combination of TACE and ablation is, particularly for Western patients, missing [1].

### 1.10. Combination of TACE and TARE

Studies investigating the sequential combination of transarterial radioembolization (TARE) with TACE are rare. Comparative analysis and RCTs on the sequential or parallel use of both techniques are missing. Preliminary results, however, indicate an acceptable safety profile and good treatment effect for specific patient conditions [79]. Through a different biologic rationale in comparison to TACE, TARE might be particularly an alternative to systemic treatment in patients with extensive liver progress during TACE.

### 1.11. Combination of TACE and SBRT

Another option for patients with unresectable HCC is the combination of TACE and stereotactic body radiotherapy (SBRT). SBRT is particularly effective in tumor areas with high oxygenation namely the tumor periphery, where TACE itself is less effective [80]. On the other hand, cytotoxic agents used for TACE could lead to a higher radiosensitivity [81]. A recent meta-analysis yielded a prolonged survival and an improved response for the combination of SBRT and TACE in comparison to SBRT alone for unresectable HCC. However, no survival benefit was observed for patients with portal vein tumor thrombosis (PVTT). Contrary, results of a phase 2 trial (NCT01901692), which compared combined SBRT and TACE against sorafenib for patients with macroscopic vascular invasion, showed a significant survival benefit for patients treated with SBRT and TACE [82]. Currently, several phase 3 RCTs on the combination of TACE and external radiotherapy for patients with unresectable HCC are recruiting (NCT03116984, NCT02794337, NCT03939845). Up to date, no recommendation can be made towards a generalized use for the combination of TACE with internal (TARE) or external (SBRT) radiotherapy for patients with unresectable HCC.

### 1.12. Combination of TACE with Systemic Treatment Agents and Future Directions

Multiple trials have investigated the combination of TACE and sorafenib and other tyrosine-kinase inhibitors. Recently, the TACTICS trial was the first to show an improved progression-free survival (PFS), while other trials have consistently failed to show a significant survival benefit [83–89]. The results of the TACTICS trial, which included intermediate-stage HCC patients, however, have to be interpreted with the background of its specific study characteristics [84]. In comparison to previous studies, the patients included in the TACTICS trial had mostly a better remaining liver function and in most cases no pre-existing liver cirrhosis [90]. Furthermore, the response was assessed using the RECICL criteria and not the commonly used mRECIST or RECIST1.1 criteria. These RECICL do not define new intrahepatic lesions as progressive disease and therefore not as an endpoint for

PFS, which is rather special. Furthermore, a recent post-hoc analysis yielded no differences in the overall survival [83]. Thus, through the unconventional methodology in combination with the lacking survival benefit, no treatment recommendations can be made based on the results of the TACTICS trial for intermediate-stage HCC.

Currently, there are a few phase 2/3 trials investigating the combination of TACE and immunotherapeutic agents [91] (Table 2).

**Table 2.** Overview on ongoing or planned randomized clinical phase 2/3 trials currently investing the combination of TACE and immunotherapeutic agents.

| Trial Name | Identifier | Phase | BCLC Stage | Treatment Arms | Primary Endpoint(s) |
|---|---|---|---|---|---|
| LEAP-012 | NCT04246177 | Phase 3 | B | • Lenvatinib + pembrolizumab + TACE<br>• TACE alone | • PFS per RECIST1.1<br>• OS |
| EMERALD-1 | NCT03778957 | Phase 3 | B | • Durvalumab + TACE<br>• Durvalumab + bevacizumab + TACE<br>• TACE alone | • PFS per RECIST1.1 |
| CheckMate 74W | NCT04340193 | Phase 3 | B | • Nivolumab + ipilimumab + TACE<br>• Nivolumab + TACE<br>• TACE alone | • Time to unTACEble progression<br>• OS |
| TACE-3 | NCT04268888 | Phase 2/3 | B | • Nivolumab + TACE/TAE<br>• TACE/TAE alone | • Time to unTACEble progression<br>• OS |
| TALENTACE | N/A | Phase 3 | B | • Atezolizumab + bevacizumab + TACE<br>• TACE alone | • TACE-PFS<br>• OS |

Theoretically, locoregional treatment might be the perfect partner for immunotherapy. Tumor necrosis induced through TACE might lead to a release of tumor-associated antigens, which could activate the tumor microenvironment and stimulate the specific immune response [11,92]. This could enhance the effect of immunotherapy agents programming the immune system against cancer cells again [93]. Furthermore, TACE-induced hypoxia increases the production of vascular endothelial growth factor (VEGF), which catalyzes recurrent tumor growth due to an increase in re-vascularization [94–96]. VEGF inhibitors could play a counterpart and inhibit the re-vascularization [93]. However, up to date, optimal patient selection remains difficult as evidence for biomarkers in patients with HCC and immunotherapy is low and has only been evaluated in small, retrospective studies. Particularly evidence on biomarkers in combined locoregional treatment and immunotherapy is scarce [97,98]. Besides the RCTs currently focusing on the combination of TACE and immunotherapy, the ABC-HCC trial (NCT04803994) is comparing TACE versus AtezoBev head-to-head and therefore investigating this promising combination which has become standard of care for patients within the advanced stage after the positive IMBRAVE150 results [2,99]. The RENOTACE (NCT04777851) trial is a second RCT on immunotherapy (regorafenib + nivolumab) versus TACE in the intermediate stage. However, this trial has not started recruitment yet.

### 1.13. Current Recommendations

In summary, the recommendations for patients with intermediate-stage HCC rely on two phase 3 RCTs [3,4]. Based on these results and those of several meta-analyses, the current guidelines strongly recommend TACE in patients within the intermediate stage [1]. Two RCTs and one meta-analysis compared cTACE and DEB-TACE but did not find significant outcome differences. Thus, there is strong evidence that neither technique has to be favored [1]. All of the other recommendations in patients with HCC undergoing TACE, particularly on possible prognostic factors, are mainly based on retrospective studies.

## 2. Transarterial Radioembolization (TARE)

### *2.1. Background*

Transarterial radioembolization (TARE), also known as selective internal radiation therapy (SIRT), is based on an application of radioactive particles directly into the liver artery. The BCLC 2022 treatment scheme does not include TARE as standard-of-care for patients within the intermediate or the advanced stage [2]. However, TARE has been specifically named as a relevant alternative to tumor ablation and resection in patients with BCLC stage 0 and A (TARE could be considered in patients with single nodules <8 cm), based on the recent results of the LEGACY study, which indicates a prolonged duration of response and a clinically meaningful response rate for these patients [100]. The study included patients with single nodules less than 8 cm, Child-Pugh A and ECOG-PS 0/1. It is important to emphasize that the median tumor size of the patients included in that study was 2.6 cm (range 0.9–8.1). The current EASL guideline as well as the updated ESMO guideline both entitle TARE as an alternative treatment option in patients within early, intermediate and advanced stage [1,101]. However, no clear criteria for patient selection have been identified so far. Thus, the role of TARE in the treatment of unresectable HCC remains unclear and is mainly part of individual treatment concepts besides the standard recommendations.

### *2.2. Biological Rational for TARE*

Similar to TACE, due to the higher proportion of arterial supply of the tumor tissue in comparison to the surrounding liver parenchyma, a high local concentration in the tumor tissue is intended. In comparison to TACE, however, the embolizing component is only minimal and the main effect of TARE is based on the radiation effect [102]. This absence of a vessel occlusion is of special importance in patients with portal venous occlusion. Under these circumstances, TARE might spare the remaining liver function and lower the risk for a post-procedural liver failure [1].

Prior to the actual TARE procedure, all patients have to undergo a pre-TARE angiography with application of technetium 99-labeled ($^{99}$Tc) macroaggregated human albumin (MAA), which is followed by a SPECT/CT scan afterwards. This screening method is used to identify the patients with a relevant pulmonary shunt fraction not feasible for TARE. Furthermore, non-targeted vessels of the gastrointestinal tract can also be embolized during angiography to prevent gastrointestinal radiation damage [103]. Apart from TARE planning, the $^{9}$Tc-MAA-SPECT/CT yields important prognostic information for the post-interventional outcome: A recent subsequent analysis of the patients included in the SARAH trial showed a significant survival outcome benefit and a positive association with disease control in patients who had higher dose levels in the $^{9}$Tc-MAA-SPECT/CT [104]. However, estimation of the actual dose delivered to the tumor tissue during the TARE procedure remains difficult and is, in most cases, estimated using a combination of both, the body surface area of the patient as well as the hepatic tumor burden [105]. These dose calculations rely on a uniform blood supply of the tumor. However, most HCC lesions have an unequal, inhomogeneous tumor supply. Thus, a low dose could lead to undertreated tumor areas, while a high dose could cause damage of the surrounding liver parenchyma [103].

### *2.3. Patient Selection: TARE in Intermediate Stage*

For patients within the intermediate stage, no results of large-scale randomized trials evaluating TACE vs. TARE are available [1]. Retrospective comparisons report less toxicity, better local tumor control and a longer progression-free survival as well as a better quality of life [106–108]. However, a survival benefit has not been observed so far, neither in retrospective comparisons nor in prospective pilot studies [109,110]. In comparison to TACE, TARE has several significant drawbacks: TARE requires a preprocedural angiographic evaluation, is less cost-effective and has a high personnel expenditure [103]. Thus, TARE has not become a standard procedure for patients within the intermediate stage. However, specifically in patients with a large tumor, for whom TACE is not recommendable due to a high risk of postembolization syndrome, TARE is a highly valuable treatment option. Furthermore, as the procedure relies on a different biological effect, TARE is a treatment option for patients with hepatic tumor progress after TACE. However, the role of TARE has not been fully defined yet. Thus, TARE as an individual treatment procedure after TACE failure requires an extensive interdisciplinary discussion.

### 2.4. Patient Selection: TARE in Advanced Stages

For patients within the advanced stage and particularly for patients with portal vein infiltration but without distant metastases, TARE has been considered a potential treatment option. Although benefits in quality of life and a better toxicity profile for TARE were reported, no survival benefit was observed in two phase 3 trials comparing TARE and sorafenib (SARAH trial and SIRveNIB trial) [111,112]. Up to date and similar to the intermediate stage, TARE remains an individual treatment option requiring an intense interdisciplinary evaluation. Furthermore, with immunotherapy as a novel treatment option for patients within the advanced stage, the role of TARE requires a re-definition supported by results of well-conducted RCTs.

### 2.5. Patient Selection: TARE for Bridging

For patients within the early stages, TARE is an option for bridging-to-transplant in selected patients. Initial clinical results indicate a better local tumor control leading to a higher transplantation rate in comparison to TACE [108]. However, the overall evidence is low and RCTs comparing TARE, TACE and ablation head-to-head are missing.

Besides bridging-to-transplant, the potential of TARE for downsizing has been reported in several studies. Downsizing through TARE in patients that initially did not meet the Milan criteria may lead to a tumor shrinkage enabling liver transplantation [113]. Furthermore, selective radioembolization could also facilitate a subsequent liver resection in patients with initially unresectable HCC as it causes hypertrophy of the future liver remnant [113].

### 2.6. Patient Selection: Radiation Segmentectomy

Furthermore, TARE has the potential to function as a curative treatment when performed in a specific manner: radiation segmentectomy, which is defined as highly selective TARE in one or two liver segments with a very high radiation dose. Radiation segmentectomy might serve as an additional treatment option in patients with a challenging tumor location, who are not amenable to thermal ablation or curative resection [114–116]. Furthermore, patients with comorbidities and limited remaining liver function could also benefit [116]. Initial clinical results in patients with a single HCC of 5 cm or smaller have been promising with a reported median OS between 4.4 and 6.5 years [115–117].

In summary, the few available RCTs evaluating TARE in intermediate stage and advanced stage did not find any survival benefit compared to standard treatment. Evidence on prognostic factors is low as it is mostly based on results from retrospective reports and meta-analyses as well as RCTs are missing. However, the current EASL guideline as well as the updated ESMO guideline do both entitle TARE as an alternative treatment option in patients within the early, intermediate and advanced stages. Thus, no clear recommendations and only weak suggestions on criteria for selecting patients likely to benefit from TARE can be made.

### 2.7. Combination of TARE with Other Locoregional Treatment Modalities

Evidence for the combination of TARE with TACE, SBRT and ablation is low. As mentioned above, the combination of TARE and TACE is scarcely investigated. The combination of TARE and SBRT in patients with portal vein infiltration seems to be safe and led to an improved prognosis [118,119]. However, these results are currently of experimental character and have been reported only for a very limited number of patients. Future validation is mandatory prior to clear recommendation.

### 2.8. Combination of TARE with Systemic Therapy and Future Directions

Due to its feasibility and beneficial toxicity profile in patients with high local tumor burden, TARE is potentially a combination partner to systemic treatment options. However, evidence for combined therapy is low. In the past, retrospective reports indicated a survival benefit for patients with an advanced-stage HCC treated with a combination of sorafenib and TARE in comparison to sorafenib alone [120]. However, the results of the phase 2 SORAMIC trial comparing sorafenib in combination with TARE versus sorafenib alone did not yield a survival benefit for the cohort undergoing combination therapy [121]. Nevertheless, post hoc subgroup analysis yielded a survival benefit for specific subgroups (patients without liver cirrhosis, nonalcoholic cirrhosis and patients younger than 65 years). Although promising, those subgroups were too small; therefore, additional trials are needed for clear recommendations [11]. Currently, the large phase 3 STOP-HCC trial with a similar design and the same treatment arms is recruiting (NCT01556490) [122].

Furthermore, several phase I and II trials on the combination of various immunotherapy agents and TARE are currently running [93]. Similar to other locoregional treatments, the release of tumor-related antigens during treatment could lead to a stimulation of the immune response [11]. This, again, could enhance the effect of immunotherapy agents programming the immune system against cancer cells [93]. As mentioned above, the locoregional treatment-induced hypoxia increases the production of vascular endothelial growth factor (VEGF), which catalyzes recurrent tumor growth due to an increase in re-vascularization [94–96]. Thus, VEGF inhibitors could antagonize this effect by impeding tumor re-vascularization. Apart from that, VEGF protects endothelial cells from radiation damage leading to a decrease in radio-sensitivity [93]. Thus, the blockade of VEGF potentially improves the response to radiation [93,123]. Therefore, VEGF inhibitors as a third component may further increase the duration of treatment response, ultimately leading to a survival benefit [93]. Hence, results of the currently recruiting trials are urgently awaited.

## 3. Ablation

### 3.1. Background

Chemical ablation for HCC has been historically performed by ethanol; however, this method has now been surpassed by energy-based ablation due to the significant recurrence rate [124,125]. Ethanol ablation as a loco-regional treatment method for small HCCs is nowadays replaced by thermal ablation. Ablation is an established treatment for HCC and, as described in the recently published BCLC guidelines [2], is the indicated treatment for the very early-stage lesions (BCLC 0: HCCs $\leq$ 2 cm, without vascular invasion or extrahepatic spread in patients with preserved liver function and no cancer-related symptoms). So, unless there is an option for liver transplantation, percutaneous ablation should be considered as the first-line treatment for such patients [124–128]. Ablation can also be proposed as a valid treatment option for BCLC A lesions, given the minimal invasiveness and lower cost of the method in comparison to surgery unless the lesion is in a non-accessible location [2].

One novelty of the new guidelines is the endorsement of treatment stage migration that permits a more flexible approach according to local expertise. Therefore, the indication for ablation might be expanded to all BCLC A cases, even for lesions larger than 3 cm [2]. For the same reason, even patients with up to three lesions that are smaller than 3 cm and who have no transplant option, may undergo ablation in conjunction with TACE.

### 3.2. Rationale for Ablation and Modalities

Ablation offers local tumor destruction either due to chemical or due to energy deposition [126]. It is expected to induce necrosis of the lesion in the specific area where energy is applied and minimizing the damage to the surrounding parenchyma. A safety margin of at least 0.5 cm around the treated lesion is required in order to prevent local recurrence [11].

Energy-based ablation includes radiofrequency ablation (RFA), microwave ablation (MWA), cryoablation (CRA), high-intensity ultrasound ablation (HIFU), laser ablation and irreversible electroporation (IRE). Most of the energy modalities have been used extensively over the years. Among them, RFA has been the most used and studied, even with a "no-touch technique" [127–129]. This technique consists of inserting more than one RFA electrode into the periphery of the tumor and performing the ablation after sequential activation of the electrodes. The number of electrodes depends on the size and the geometry of the tumor but usually varies between 2 and 4 [129]. Energy is delivered from the periphery to the center; therefore, the technique ensures that a safety margin is present. However, RFA outcomes are subject to limitations that are mainly related to the location and size of the lesion. Similar modalities such as MWA appear as a very attractive alternative offering larger ablation zones in shorter ablation time and without any heat loss due to surrounding structures.

Cryoablation, on the other hand, is extensively used in the treatment of other lesions such as renal tumors, but has also been introduced in the treatment of HCC in the last few years [130,131]. There are some limitations in the use of cryoablation in the liver. However, this method is also gaining ground in clinical practice. It offers the advantage of concurrent placement of multiple probes and precise intra-procedural monitoring of the created ice-ball. Laser ablation, although rather effective, has not been that extensively used mainly because of the multiple needles required and the relatively smaller ablation zones [132]. On that note, HIFU has also been mainly used empirically and as a bridging treatment for transplant. Another modality that is becoming popular is IRE, a non-thermal energy ablation method that induces cellular apoptosis, based on the irreversible electroporation of cellular membranes. The main advantage of IRE is that it does not cause any damage to the epithelial lining and can be therefore used in more challenging anatomical locations, i.e., close to large vessels or to bile ducts or the gallbladder. However, the cost is still high, general anesthesia and muscular relaxation are required and multiple needles in a precise geometric conformation need to be placed [133].

Nevertheless, in the BCLC criteria, when the term "ablation" is mentioned, this is mainly referring to RFA or MWA. Other ablation techniques like cryoablation and IRE have not been incorporated yet in the published guidelines due to low evidence. Furthermore, in real-life clinical practice, ablation, with or without combined TACE, might be used also for BCLC C patients; however, this is also not included in the guidelines and can be recommended only after multi-disciplinary discussion and if no other treatment option is available.

### 3.3. Outcomes—Complications

The outcomes that measure the performance of ablation are mainly local tumor control, recurrence rate and overall survival. Recurrence rate is expected to be around 70–80% at five years; however, this was shown to be conditioned by systemic treatment, both with sorafenib and immunotherapy [134–136]. Potential complications include bleeding, tumor seeding, adjacent organ accidental damage or thermal injury of biliary ducts [11].

### 3.4. RFA vs. MWA

Technically, MWA is expected to offer several advantages over RFA, including faster heating over a larger volume and less susceptibility to heat sink and local perfusion [137]. However, clinical outcomes appear still to be very similar between the two modalities [138]. A comparison between the outcomes of the two percutaneous modalities was published in a meta-analysis on 774 patients [139]. Surprisingly, while complete response rate was marginally higher for MWA, the 3-year survival rate was slightly higher for RFA. MWA obtained better results in terms of local recurrence rate for larger lesions but also led to a higher rate of major complications. Overall, the authors concluded that the two percutaneous modalities offer similar efficiency. However, in the meta-analysis, randomized controlled trials and case-control studies were mixed, when they should have been interpreted separately. On this "eternal debate", another meta-analysis, published in 2019 on a more extensive population of 1816 patients that were analyzed in 4 randomized control trials and 10 cohort studies, concluded that for the percutaneous treatment the two modalities appear to offer similar therapeutic effect [138]. A third meta-analysis published in 2020 [140] offered the same conclusion for the 3-year follow-up. Perhaps the only paper favoring MWA is that of Bouda et al. [141], where in relation to RFA, a lower local progression rate was found (23% versus 36% for RFA).

### *3.5. Puncture Technique and Navigation Assistance*

Ablation can be performed under US- or CT-guidance. Combination of both modalities offer some advantages, as well. CT is not a real-time imaging modality; thus, puncturing a relatively small HCC can be more challenging than with the use of US-guidance, where precise lesion puncture can be made with the needle placed in the middle of the nodule. This is the reason why Hermida et al. propose that US should be the first-line guidance modality for 2–3 cm HCCs [142]. Therefore, the exact needle position can lead to lower local recurrence rate, which is an interesting outcome factor. With US guidance the ablative process can be better followed, while with CT the ablation result can be monitored immediately after the procedure.

Novel navigation systems offer valuable assistance for placing one or more needles in the correct position, especially if HCCs are in difficult positions, such as sub-diaphragmatic locations in the liver dome [143].

### *3.6. Ablation vs. Surgery*

The comparison between the two treatment approaches has been previously performed given the very similar indications for the BCLC 0 and BCLC A cases. Indications in everyday clinical practice would favor ablation in presence of portal hypertension with a gradient higher than 10 mmHg or other comorbidities that would contribute towards a more complex surgical approach [144]. However, in the absence of clinically significant portal hypertension, resection should be considered as stated by the recent BCLC criteria [2]. Transplant should also be addressed if microvascular invasion and/or satellite nodules are confirmed after resection [145,146]. Nevertheless, given that the survival rate for lesions smaller than 3 cm appears to be similar if not better for ablation, with lower overall cost and hospital stay, ablation is gaining more ground as first-line treatment for such lesions [124,126–128]. For larger lesions, RFA has not performed very well; however, MWA appears to offer better results and is the preferable option if the lesion reaches 4 cm [137].

The location of the lesion also plays a significant role in the clinical decision-making; for intra-parenchymal tumors, ablation may be preferable, whereas if the lesion is close to a thermo-sensitive structure, such as the gallbladder or bowel, then resection might be considered [147]. However, it needs to be taken into account that, with hydro-dissection and use of non-thermal ablation, these boundaries are now reduced. Moreover, in cases of larger HCCs or in location next to major vascular or biliary structures, careful intraoperative ablation is an alternative.

There have been a few randomized trials comparing RFA with resection showing mainly more adverse effects from surgery and a higher recurrence rate from ablation [148,149]. However, this higher recurrence rate of ablation is not conditioning the overall survival [150–153]. In any case, the complication rate was lower after RFA than laparoscopic resection for small single HCC nodules (5.1 vs. 10%) [152].

Cost-effectiveness is also another aspect where ablation prevails [126]. In essence, ablation appears to be superior to surgical resection for HCC up to 3 cm in patients with Child-Pugh class A or B cirrhosis.

### 3.7. RFA vs. Cryoablation

Even though cryoablation is a very promising modality, no clear superiority is shown in the treatment of HCC vs. the other thermal modalities [154]. It needs to be considered that cryoablation is not offering the hemostatic option that both RFA and MWA offer when the electrode is retracted; therefore, provision of the coaxial approach needs to be made, to be in position to embolize the access tract with a hemostatic sponge. Cryoablation also is expected to cause significantly less pain than both RFA and MWA and does not produce the "oven effect," where heat is trapped within the tumor, as seen in RFA [155].

A multicenter RCT compared the two modalities in 360 patients with Child-Pugh class A or B and up to two lesions of ≥4 cm [156]. Local tumor progression was lower for cryoablation at 3 years (7% vs. 11% for RFA); the deference was more pronounced for lesions >3 cm (7.7% vs. 18.2%). However, 5-year overall survival, tumor free survival and complication rates were similar between the two modalities. The evidence that cryoablation could offer satisfactory outcomes for lesions larger than 3 cm triggered significant interest on the modality and could potentially change the position of ablation at the BCLC Group HCC management guidelines. However, given the small number of patients, statistical power was subject to condition from the outcome of individual patients.

### 3.8. IRE

Irreversible electroporation offers the advantage of ablating lesions close to vital structures [157]. In a recently published single-center study of patients that could not be treated differently due to the anatomical location of the lesion, local recurrence-free survival at 12 months was 83.6% for a median tumor size of 2 cm [158]. In a comparative study of IRE [159] with RFA and MWA, in terms of complications no significant difference was noticed among thermal and non-thermal modalities even though more needles are required with IRE and no tract cauterization may be performed. Another interesting aspect as suggested by few recent publications is that IRE could be more suitable for patients with cirrhosis [160–163].

### 3.9. Ablation and Immunotherapy

Immunotherapy appears to be very promising in the treatment of HCC. Ablation is the process of releasing tumor-associated antigens that enhance the immune response against the tumor itself [164]. In a study of 32 patients that underwent treatment with tremelimumab followed by ablation, intratumoral accumulation of CD8+ T cells was detected [165]. There are multiple open questions about the timing of ablation and immunotherapy, e.g., if tumor ablation must be complete or partial and many more [165,166]. Therefore further research is required. Clinical trials investigating the role of ablation on HCC and that are currently recruiting are shown on Table 3.

### 3.10. Prediction Models and Artificial Intelligence

Attempts for predictive models after RFA have been developed in the effort to standardize the approach and improve outcomes. In a recently published study on 238 patients that underwent ablation for early-stage HCC, several factors, i.e., tumor size and a-fetoprotein levels, were related to ablation outcomes [143]. Outcome-predictive models with the use of artificial intelligence have also been developed. In a study of 83 HCC patients, who received ablation as first treatment, features that would predict the outcome were analyzed via five different feature-selection methods [167]. In another study of 252 patients who received RFA, artificial neural network models among 15 clinical variables were created to predict 1- and 2-year disease-free survival, achieving an acceptable prediction performance [168].

Table 3. Overview on currently recruiting clinical trials investing the use of ablation for the treatment of HCC.

| Trial Name | Identifier | Phase | BCLC Stage | Treatment Arms | Primary Endpoint(s) |
|---|---|---|---|---|---|
| IMMULAB | NCT03753659 | Phase 2 | A | • Pembrolizumab + RFA/MWA/brachytherapy or TACE | • ORR per RECIST1.1 |
| | NCT04663035 | Phase 2 | A | • Ablation + Tislelizumab vs.<br>• Ablation Alone | • RFS |
| AB-LATE02 | NCT04727307 | Phase 2 | A | • RFA + Atezolizumab + Bevacizumab vs.<br>• RFA Alone | • RFS |
| | NCT04652440 | Phase 2 | A/B | • RFA + Tislelizumab | • TRAEs<br>• SAEs<br>• Tolerability |
| | NCT02964260 | Phase 2 | B | • TAE + Ablation (simultaneously)<br>• TACE + Ablation (1 month interval) | • OS |
| | NCT04365751 | N/A | B | • MWA vs.<br>• Laparoscopic hepatectomy | • OS |
| | NCT03898921 | 3 | A/B | • SBRT vs.<br>• RFA | • OS |
| | NCT04220944 | 1 | B/C | • MWA + TACE + Sintilimab | • PFS |

ORR—objective response rate; TTR—time to recurrence; RFS—recurrence-free survival; TRAEs—treatment-related adverse events; SAEs—serious adverse events; OS—overall survival; SBRT—Stereotactic Body Radiotherapy.

## 4. Stereotactic Body Radiation Therapy (SBRT)

### *4.1. Background and Biological Rationale*

Through technical advances in the field of radiation therapy, SBRT has become an alternative regional treatment modality for patients with HCC in various stages. The rationale behind SBRT is that it induces DNA damage leading to an inhibition of the cancer cell replication [169]. However, radiation therapy in the treatment of liver cancer was limited only to patients with a very high tumor burden and individualized treatment concepts as the radiation led to great damage in the tumor-surrounding liver parenchyma. With improvements in image guidance and conformal radiation techniques, nowadays, radiation can be applied more precisely to the tumor tissue in high doses, while sparing the surrounding liver tissue [170]. Compared to other local and regional treatment options, SBRT could be beneficial in complex anatomical situations and a high local tumor burden.

A rare complication of SBRT is the radiation-induced liver disease (RILD). RILD is associated with cholestasis, hepatomegaly, increase liver enzymes, impairment of the remaining liver function and development of ascites [171,172]. RILD typically appears in the first 2 months after radiation therapy, but reports vary between 2 weeks and 7 months for the appearance of typical symptoms [173]. To lower the risk for a post-interventional RILD, liver function has to be evaluated carefully prior to SBRT.

### *4.2. Patient Selection: SBRT in Early Stages*

In this stage, SBRT is a potential treatment option for patients not suitable for liver resection or transplantation due comorbidities and not suitable for ablation due to lesions near to liver vessels, biliary structures and adjacent organs [174]. Several phase I and II trials have shown promising results for patients with few HCC lesions not suitable for ablation [174]. In these studies, SBRT led to high local tumor control rates. Additionally, larger phase I and II trials showed that SBRT could be applied for bridging to transplant [168]. However, the number of available results from phase III RCTs in patients with early-stage HCC is scarce. In one available non-inferiority study, Kim et al. observed no differences in the local PFS for patients with recurrent HCC compared to RFA [175]. In a second available phase III study, patients with HCC and PVTT undergoing surgery, who had neoadjuvant radiation therapy, had a significantly better postoperative overall survival than patients that did not underwent adjuvant therapy [176]. Although these results for selected patients were promising, more evidence is needed prior clear treatment recommendations can be made.

### 4.3. Patient Selection: SBRT in Intermediate Stage

As the intermediate stage is defined as a group of patients with a high local tumor burden, SBRT has a large potential for improving local tumor control rates compared to other treatment modalities. Several retrospective studies as well as a few phase I and II trials demonstrated high local tumor control rates [174]. Up to date, however, no phase III trial results comparing SBRT to TACE are available. A recent meta-analysis consisting of ten retrospective studies showed a higher complete response rate as well as longer overall survival for patients receiving combined TACE and SBRT compared to patients receiving only SBRT [177].

### 4.4. Patient Selection: SBRT in Advanced Stages

Several retrospective studies and a few phase I and II trials have shown promising results for SBRT in patients with advanced HCC and macroscopic vascular invasion or impaired liver function [174]. However, phase III trials comparing SBRT with standard systemic treatment are currently not available. Only for the combined treatment of TACE and radiation therapy, a recent RCT reported superior outcome results including a significantly prolonged overall survival for patients with HCC and macrovascular invasion compared to a treatment with sorafenib [82]. Nevertheless, the role of SBRT compared to systemic treatment and particularly to the novel standard treatment combining atezolizumab and bevacizumab remains unclear.

### 4.5. SBRT Combined with Other Treatment Modalities

As mentioned above, the combination of TACE and SBRT has a high potential for patients with unresectable HCC, despite the fact that evidence for the combination of SBRT and TARE is low. Several retrospective analyses and phase I trials have yielded promising results for the combined use of sorafenib and SBRT [81], while a phase III trial of the Radiation Therapy Oncology Group is currently recruiting (NCT01730937). In this context, preclinical investigations have demonstrated the potential of sorafenib to function as a radiosensitizer [81]. The same seems to be apparent for immunotherapeutic agents. Vice versa, local radiotherapy might interfere with the immune reaction of the tumor microenvironment leading to immunogenic cell death [81]. Thus, radiation and immunotherapy could function as a strong synergistic treatment. However, up to date evidence is low and only a few small retrospective studies have been published so far [178]. Several prospective phase I and II studies investigating the combination of SBRT and various immunotherapeutic agents in various patient subgroups are currently recruiting [81].

### 4.6. Current Recommendations

Current guidelines show a significant disparity in terms of their recommendations for SBRT. While the current EASL guideline does not recommend SBRT for patients with HCC due to the low amount of evidence, the currently updated BCLC classification does not even mention SBRT as treatment option. Contrary, the European Society for Medical Oncology (ESMO) as well as the American Association for the Study of Liver Diseases (AASLD) guidelines do recommend radiation therapy as a treatment option for selected patients with HCC [1,2,101,179]. However, these guidelines also argue for a need of more evidence before stronger recommendations can be made. In contrast to these rather soft recommendations, the American Society for Radiation Oncology (ASTRO) has recommended the use of SBRT strongly for patients within the early stages for whom surgery and ablation is no treatment option [180]. Furthermore, the ASTRO guideline recommends a sequenced use of SBRT for patients with multifocal, unresectable HCC, in patients with macrovascular tumor invasion and for relief of symptoms in the palliative setting of best supportive care. Thus, the range of recommendations is significantly different between the various guidelines. It remains to be seen how the recommendations of the different societies will adapt to the results of the increasingly available RCTs results with the upcoming updates of the guidelines.

## 5. The Potential of Combined Treatment—Current State and Future Directions

In this article, we aimed to summarize the current state of locoregional HCC treatment with a special focus on combined treatment options. From our point of view, an important aspect for the future of HCC treatment is a better understanding of how we can combine locoregional and systemic treatment. While expert knowledge and techniques for local and regional treatment options have continuously increased over the last decade, immunotherapy has had a rapid rise in the recent years and opened new doors for treatment of more advanced stages. In the context of combined treatment, immunotherapy has the potential to compensate various drawbacks of local and regional treatment modalities, which have been mentioned above. Furthermore, compared to previous studies combining locoregional treatment with tyrosine kinase inhibitors, the combination with immunotherapeutic agents offers a significantly improved side-effect profile. Thus, these novel options have the potential to facilitate HCC treatment towards highly individualized treatment approaches in the context of personalized medicine. To enable individualized treatment approaches, the identification of patients likely to benefit from such combined approaches remains of the utmost important. Although the knowledge on predictive biomarkers in patients with HCC and immunotherapy is continuously investigated, the current knowledge is still limited and, for combined locoregional and immunotherapeutic treatment, evidence on predictive biomarkers is missing completely. One fact attributing to the lack of biomarkers is the limited number of patients currently treated with combined locoregional and immunotherapeutic treatment. However, with the ongoing of the above-mentioned trials investigating various combinations, potential collectives for biomarker evaluation will be available.

## 6. Limitations

With this review, we aimed to give the reader a comprehensive overview on the aspects of utmost importance when evaluating local and regional treatment options in patients with HCC undergoing TACE. This review article particularly focused on the potential of combined treatment options. Of course, our review has the typical drawbacks of an expert review that need to be addressed. First, the structure of the article as well as the recommendations and literature selection were biased by the subjective estimations of the authors. Secondly, this review is not a systematic review. Thus, some relevant articles could have been missed. Thirdly, expert reviews always have a potential bias of authors interpreting the original data leading to different conclusions and recommendations. One important aspect that has to be mentioned is that patient selection in the cited retrospective or single-arm studies were based on very heterogeneous inclusion criteria, and the patient cohort varied strongly. Thus, a strong influence of the inclusion bias has to be estimated and, particularly, reported survival results should not be generalized.

## 7. Conclusions

In summary, local and regional therapies remain a mainstay of curative and palliative treatment of patients with HCC. Currently, evidence on potential combination of the local and regional treatment options with each other as well as with other treatment modalities is growing and has the potential to further individualize HCC therapy. To identify the most suitable treatment out of these new various options, a repeated interdisciplinary discussion of each case by the tumor board is of the utmost importance.

**Author Contributions:** Conceptualization, A.H. and R.K.; methodology, A.H. and R.K.; software, L.M.; validation, A.H. and R.K.; formal analysis, L.M and M.K.; investigation, L.M and M.K.; resources, M.K.; data curation, L.M and M.K.; writing—original draft preparation, L.M and M.K.; writing—review and editing, A.H. and R.K.; visualization, A.H.; supervision, A.H. and R.K.; project administration, A.H.; funding acquisition, M.K. All authors have read and agreed to the published version of the manuscript.

**Funding:** This research received no external funding.

**Conflicts of Interest:** The authors declare no conflict of interest.

## References

1. Galle, P.R.; Forner, A.; Llovet, J.M.; Mazzaferro, V.; Piscaglia, F.; Raoul, J.-L.; Schirmacher, P.; Vilgrain, V. EASL Clinical Practice Guidelines: Management of hepatocellular carcinoma. *J. Hepatol.* **2018**, *69*, 182–236. [CrossRef] [PubMed]
2. Reig, M.; Forner, A.; Rimola, J.; Ferrer-Fábrega, J.; Burrel, M.; Garcia-Criado, Á.; Kelley, R.K.; Galle, P.R.; Mazzaferro, V.; Salem, R.; et al. BCLC strategy for prognosis prediction and treatment recommendation: The 2022 update. *J. Hepatol.* **2022**, *76*, 681–693. [CrossRef]
3. Llovet, J.M.; Real, M.I.; Montaña, X.; Planas, R.; Coll, S.; Aponte, J.; Ayuso, C.; Sala, M.; Muchart, J.; Solà, R.; et al. Arterial embolisation or chemoembolisation versus symptomatic treatment in patients with unresectable hepatocellular carcinoma: A randomised controlled trial. *Lancet* **2002**, *359*, 1734–1739. [CrossRef]
4. Lo, C.M.; Ngan, H.; Tso, W.K.; Liu, C.L.; Lam, C.M.; Poon, R.T.P.; Fan, S.T.; Wong, J. Randomized controlled trial of transarterial Lipiodol chemoembolization for unresectable hepatocellular carcinoma. *Hepatology* **2002**, *35*, 1164–1171. [CrossRef] [PubMed]
5. Bolondi, L.; Burroughs, A.; Dufour, J.-F.; Galle, P.R.; Mazzaferro, V.; Piscaglia, F.; Raoul, J.L.; Sangro, B. Heterogeneity of patients with intermediate (BCLC B) Hepatocellular Carcinoma: Proposal for a subclassification to facilitate treatment decisions. In *Seminars in Liver Disease*; Thieme Medical Publishers: New York, NY, USA, 2012; Volume 32, pp. 348–359.
6. Melchiorre, F.; Patella, F.; Pescatori, L.; Pesapane, F.; Fumarola, E.; Biondetti, P.; Brambillasca, P.; Monaco, C.; Ierardi, A.M.; Franceschelli, G.; et al. DEB-TACE: A standard review. *Futur. Oncol.* **2018**, *14*, 2969–2984. [CrossRef] [PubMed]
7. Golfieri, R.; Giampalma, E.; Renzulli, M.; Cioni, R.; Bargellini, I.; Bartolozzi, C.; Breatta, A.D.; Gandini, G.; Nani, R.; Gasparini, D.; et al. Randomised controlled trial of doxorubicin-eluting beads vs. conventional chemoembolisation for hepatocellular carcinoma. *Br. J. Cancer* **2014**, *111*, 255–264. [CrossRef]
8. Sacco, R.; Bargellini, I.; Bertini, M.; Bozzi, E.; Romano, A.; Petruzzi, P.; Tumino, E.; Ginanni, B.; Federici, G.; Cioni, R.; et al. Conventional versus Doxorubicin-eluting Bead Transarterial Chemoembolization for Hepatocellular Carcinoma. *J. Vasc. Interv. Radiol.* **2011**, *22*, 1545–1552. [CrossRef]
9. Kloeckner, R.; Weinmann, A.; Prinz, F.; Pinto dos Santos, D.; Ruckes, C.; Dueber, C.; Pitton, M.B. Conventional transarterial chemoembolization versus drug-eluting bead transarterial chemoembolization for the treatment of hepatocellular carcinoma. *BMC Cancer* **2015**, *15*, 465. [CrossRef]
10. Lammer, J.; Malagari, K.; Vogl, T.; Pilleul, F.; Denys, A.; Watkinson, A.; Pitton, M.; Sergent, G.; Pfammatter, T.; Terraz, S.; et al. Prospective Randomized Study of Doxorubicin-Eluting-Bead Embolization in the Treatment of Hepatocellular Carcinoma: Results of the PRECISION V Study. *Cardiovasc. Interv. Radiol.* **2010**, *33*, 41–52. [CrossRef]
11. Kloeckner, R.; Galle, P.R.; Bruix, J. Local and regional therapies for hepatocellular carcinoma. *Hepatology* **2021**, *73*, 137–149. [CrossRef]
12. Golfieri, R.; Cappelli, A.; Cucchetti, A.; Piscaglia, F.; Carpenzano, M.; Peri, E.; Ravaioli, M.; D'Errico-Grigioni, A.; Pinna, A.D.; Bolondi, L. Efficacy of selective transarterial chemoembolization in inducing tumor necrosis in small (<5 cm) hepatocellular carcinomas. *Hepatology* **2011**, *53*, 1580–1589. [CrossRef]
13. Burrel, M.; Reig, M.; Forner, A.; Barrufet, M.; de Lope, C.R.; Tremosini, S.; Ayuso, C.; Llovet, J.M.; Real, M.I.; Bruix, J. Survival of patients with hepatocellular carcinoma treated by transarterial chemoembolisation (TACE) using Drug Eluting Beads. Implications for clinical practice and trial design. *J. Hepatol.* **2012**, *56*, 1330–1335. [CrossRef] [PubMed]
14. Malagari, K.; Pomoni, M.; Kelekis, A.; Pomoni, A.; Dourakis, S.; Spyridopoulos, T.; Moschouris, H.; Emmanouil, E.; Rizos, S.; Kelekis, D. Prospective Randomized Comparison of Chemoembolization with Doxorubicin-Eluting Beads and Bland Embolization with BeadBlock for Hepatocellular Carcinoma. *Cardiovasc. Interv. Radiol.* **2010**, *33*, 541–551. [CrossRef] [PubMed]
15. Herber, S.C.A.; Otto, G.; Schneider, J.; Schuchmann, M.; Düber, C.; Pitton, M.B.; Kummer, I.; Manzl, N. Transarterial Chemoembolization in Patients Not Eligible for Liver Transplantation: Single-Center Results. *Am. J. Roentgenol.* **2008**, *190*, 1035–1042. [CrossRef]
16. Kirchhoff, T.D.; Bleck, J.S.; Dettmer, A.; Chavan, A.; Rosenthal, H.; Merkesdal, S.; Frericks, B.; Zender, L.; Malek, N.P.; Greten, T.F.; et al. Transarterial chemoembolization using degradable starch microspheres and iodized oil in the treatment of advanced hepatocellular carcinoma: Evaluation of tumor response, toxicity, and survival. *Hepatobiliary Pancreat. Dis. Int.* **2007**, *6*, 259–266.
17. Jun, C.H.; Ki, H.S.; Lee, H.K.; Park, K.J.; Park, S.Y.; Cho, S.B.; Park, C.H.; Joo, Y.E.; Kim, H.S.; Choi, S.K.; et al. Clinical significance and risk factors of postembolization fever in patients with hepatocellular carcinoma. *World J. Gastroenterol.* **2013**, *19*, 284–289. [CrossRef] [PubMed]

18. Arslan, M.; Degirmencioglu, S. Risk Factors for Postembolization Syndrome After Transcatheter Arterial Chemoembolization. *Curr. Med. Imaging* **2019**, *15*, 380–385. [CrossRef] [PubMed]
19. Terzi, E.; Terenzi, L.; Venerandi, L.; Croci, L.; Renzulli, M.; Mosconi, C.; Allegretti, G.; Granito, A.; Golfieri, R.; Bolondi, L.; et al. The ART score is not effective to select patients for transarterial chemoembolization retreatment in an Italian series. *Dig. Dis.* **2014**, *32*, 711–716. [CrossRef] [PubMed]
20. Hung, Y.-W.; Lee, I.-C.; Chi, C.-T.; Lee, R.-C.; Liu, C.-A.; Chiu, N.-C.; Hwang, H.-E.; Chao, Y.; Hou, M.-C.; Huang, Y.-H. Redefining tumor burden in patients with intermediate-stage hepatocellular carcinoma: The seven-eleven criteria. *Liver Cancer* **2021**, *10*, 629–640. [CrossRef]
21. Wang, Q.; Xia, D.; Bai, W.; Wang, E.; Sun, J.; Huang, M.; Mu, W.; Yin, G.; Li, H.; Zhao, H.; et al. Development of a prognostic score for recommended TACE candidates with hepatocellular carcinoma: A multicentre observational study. *J. Hepatol.* **2019**, *70*, 893–903. [CrossRef]
22. Ho, S.-Y.; Liu, P.-H.; Hsu, C.-Y.; Ko, C.-C.; Huang, Y.-H.; Su, C.-W.; Lee, R.-C.; Tsai, P.-H.; Hou, M.-C.; Huo, T.-I. Tumor burden score as a new prognostic marker for patients with hepatocellular carcinoma undergoing transarterial chemoembolization. *J. Gastroenterol. Hepatol.* **2021**, *36*, 3196–3203. [CrossRef]
23. Müller, L.; Stoehr, F.; Mähringer-Kunz, A.; Hahn, F.; Weinmann, A.; Kloeckner, R. Current Strategies to Identify Patients That Will Benefit from TACE Treatment and Future Directions a Practical Step-by-Step Guide. *J. Hepatocell. Carcinoma* **2021**, *8*, 403. [CrossRef]
24. Benvegnu, L.; Noventa, F.; Bernardinello, E.; Pontisso, P.; Gatta, A.; Alberti, A. Evidence for an association between the aetiology of cirrhosis and pattern of hepatocellular carcinoma development. *Gut* **2001**, *48*, 110–115. [CrossRef] [PubMed]
25. Kneuertz, P.J.; Demirjian, A.; Firoozmand, A.; Corona-Villalobos, C.; Bhagat, N.; Herman, J.; Cameron, A.; Gurakar, A.; Cosgrove, D.; Choti, M.A.; et al. Diffuse infiltrative hepatocellular carcinoma: Assessment of presentation, treatment, and outcomes. *Ann. Surg. Oncol.* **2012**, *19*, 2897–2907. [CrossRef]
26. Johnson, P.J.; Berhane, S.; Kagebayashi, C.; Satomura, S.; Teng, M.; Reeves, H.L.; O'Beirne, J.; Fox, R.; Skowronska, A.; Palmer, D.; et al. Assessment of liver function in patients with hepatocellular carcinoma: A new evidence-based approach—The ALBI grade. *J. Clin. Oncol.* **2015**, *33*, 550. [CrossRef]
27. Pinato, D.J.; Sharma, R.; Allara, E.; Yen, C.; Arizumi, T.; Kubota, K.; Bettinger, D.; Jang, J.W.; Smirne, C.; Kim, Y.W.; et al. The ALBI grade provides objective hepatic reserve estimation across each BCLC stage of hepatocellular carcinoma. *J. Hepatol.* **2017**, *66*, 338–346. [CrossRef]
28. Cappelli, A.; Cucchetti, A.; Cabibbo, G.; Mosconi, C.; Maida, M.; Attardo, S.; Pettinari, I.; Pinna, A.D.; Golfieri, R. Refining prognosis after trans-arterial chemo-embolization for hepatocellular carcinoma. *Liver Int.* **2016**, *36*, 729–736. [CrossRef] [PubMed]
29. Sposito, C.; Brunero, F.; Spreafico, C.; Mazzaferro, V. External validation of an individual prognostic calculator after transarterial chemoembolization for hepatocellular carcinoma. *Liver Int.* **2016**, *8*, 1231. [CrossRef] [PubMed]
30. Garwood, E.R.; Fidelman, N.; Hoch, S.E.; Kerlan Jr, R.K.; Yao, F.Y. Morbidity and mortality following transarterial liver chemoembolization in patients with hepatocellular carcinoma and synthetic hepatic dysfunction. *Liver Transplant.* **2013**, *19*, 164–173. [CrossRef]
31. Müller, L.; Hahn, F.; Mähringer-Kunz, A.; Stoehr, F.; Gairing, S.J.; Foerster, F.; Weinmann, A.; Galle, P.R.; Mittler, J.; Pinto dos Santos, D.; et al. Prevalence and clinical significance of clinically evident portal hypertension in patients with hepatocellular carcinoma undergoing transarterial chemoembolization. *UEGJ* **2022**, *10*, 41–53. [CrossRef]
32. Georgiades, C.; Geschwind, J.-F.; Harrison, N.; Hines-Peralta, A.; Liapi, E.; Hong, K.; Wu, Z.; Kamel, I.; Frangakis, C. Lack of Response after Initial Chemoembolization for Hepatocellular Carcinoma: Does It Predict Failure of Subsequent Treatment? *Radiology* **2012**, *265*, 115–123. [CrossRef] [PubMed]
33. Miksad, R.A.; Ogasawara, S.; Xia, F.; Fellous, M.; Piscaglia, F. Liver function changes after transarterial chemoembolization in US hepatocellular carcinoma patients: The LiverT study. *BMC Cancer* **2019**, *19*, 795. [CrossRef] [PubMed]
34. Galle, P.R.; Tovoli, F.; Foerster, F.; Wörns, M.A.; Cucchetti, A.; Bolondi, L. The treatment of intermediate stage tumours beyond TACE: From surgery to systemic therapy. *J. Hepatol.* **2017**, *67*, 173–183. [CrossRef] [PubMed]
35. Kudo, M. A Paradigm Change in the Treatment Strategy for Hepatocellular Carcinoma. *Liver Cancer* **2020**, *9*, 367–377. [CrossRef]
36. Peck-Radosavljevic, M.; Kudo, M.; Raoul, J.-L.; Lee, H.C.; Decaens, T.; Heo, J.; Lin, S.-M.; Shan, H.; Yang, Y.; Bayh, I.; et al. Outcomes of patients (pts) with hepatocellular carcinoma (HCC) treated with transarterial chemoembolization (TACE): Global OPTIMIS final analysis. *J. Clin. Oncol.* **2018**, *36*, 4018. [CrossRef]
37. Ogasawara, S.; Chiba, T.; Ooka, Y.; Kanogawa, N.; Motoyama, T.; Suzuki, E.; Tawada, A.; Kanai, F.; Yoshikawa, M.; Yokosuka, O. Efficacy of Sorafenib in Intermediate-Stage Hepatocellular Carcinoma Patients Refractory to Transarterial Chemoembolization. *Oncology* **2014**, *87*, 330–341. [CrossRef]
38. Arizumi, T.; Ueshima, K.; Chishina, H.; Kono, M.; Takita, M.; Kitai, S.; Inoue, T.; Yada, N.; Hagiwara, S.; Minami, Y.; et al. Validation of the Criteria of Transcatheter Arterial Chemoembolization Failure or Refractoriness in Patients with Advanced Hepatocellular Carcinoma Proposed by the LCSGJ. *Oncology* **2014**, *87* (Suppl. S1), 32–36. [CrossRef]
39. Hucke, F.; Pinter, M.; Graziadei, I.; Bota, S.; Vogel, W.; Müller, C.; Heinzl, H.; Waneck, F.; Trauner, M.; Peck-Radosavljevic, M.; et al. How to STATE suitability and START transarterial chemoembolization in patients with intermediate stage hepatocellular carcinoma. *J. Hepatol.* **2014**, *61*, 1287–1296. [CrossRef]

40. Kadalayil, L.; Benini, R.; Pallan, L.; O'Beirne, J.; Marelli, L.; Yu, D.; Hackshaw, A.; Fox, R.; Johnson, P.; Burroughs, A.K.; et al. A simple prognostic scoring system for patients receiving transarterial embolisation for hepatocellular cancer. *Ann. Oncol.* **2013**, *24*, 2565–2570. [CrossRef]
41. Park, Y.; Kim, S.U.; Kim, B.K.; Park, J.Y.; Kim, D.Y.; Ahn, S.H.; Park, Y.E.; Park, J.H.; Lee, Y.I.; Yun, H.R.; et al. Addition of tumor multiplicity improves the prognostic performance of the hepatoma arterial-embolization prognostic score. *Liver Int.* **2016**, *36*, 100–107. [CrossRef]
42. Mähringer-Kunz, A.; Kloeckner, R.; Pitton, M.B.; Düber, C.; Schmidtmann, I.; Galle, P.R.; Koch, S.; Weinmann, A. Validation of the Risk Prediction Models STATE-Score and START-Strategy to Guide TACE Treatment in Patients with Hepatocellular Carcinoma. *Cardiovasc. Interv. Radiol.* **2017**, *40*, 1017–1025. [CrossRef] [PubMed]
43. Bourlière, M.; Pénaranda, G.; Adhoute, X.; Bronowicki, J.-P. The "six-and-twelve score" for TACE treatment: Does it really help us? *J. Hepatol.* **2019**, *71*, 1051–1052. [CrossRef] [PubMed]
44. Zamparelli, M.S.; Burrel, M.; Darnell, A.; Sapena, V.; Barrufet, M.; Bermudez, P.; Sotomayot, A.; Llarch, N.; Iserte, G.; Belmonte, E.; et al. SAT503—The "six-and-twelve" score in a prospective cohort of patients with hepatocellular carcinoma treated with trans-arterial chemoembolization following a fixed schedule. *J. Hepatol.* **2020**, *73*, S907. [CrossRef]
45. Adhoute, X.; Penaranda, G.; Naude, S.; Raoul, J.L.; Perrier, H.; Bayle, O.; Monnet, O.; Beaurain, P.; Bazin, C.; Pol, B.; et al. Retreatment with TACE: The ABCR SCORE, an aid to the decision-making process. *J. Hepatol.* **2015**, *62*, 855–862. [CrossRef] [PubMed]
46. Kim, B.K.; Shim, J.H.; Kim, S.U.; Park, J.Y.; Kim, D.Y.; Ahn, S.H.; Kim, K.M.; Lim, Y.; Han, K.; Lee, H.C. Risk prediction for patients with hepatocellular carcinoma undergoing chemoembolization: Development of a prediction model. *Liver Int.* **2016**, *36*, 92–99. [CrossRef] [PubMed]
47. Sieghart, W.; Hucke, F.; Pinter, M.; Graziadei, I.; Vogel, W.; Müller, C.; Heinzl, H.; Trauner, M.; Peck-Radosavljevic, M. The ART of decision making: Retreatment with transarterial chemoembolization in patients with hepatocellular carcinoma. *Hepatology* **2013**, *57*, 2261–2273. [CrossRef]
48. Fatourou, E.M.; Tsochatzis, E.A. ART and science in using transarterial chemoembolization for retreating patients with hepatocellular carcinoma. *Hepatobiliary Surg. Nutr.* **2014**, *3*, 415.
49. Kudo, M.; Arizumi, T.; Ueshima, K. Assessment for retreatment (ART) score for repeated transarterial chemoembolization in patients with hepatocellular carcinoma. *Hepatology* **2014**, *59*, 2424–2425. [CrossRef]
50. Arizumi, T.; Ueshima, K.; Iwanishi, M.; Minami, T.; Chishina, H.; Kono, M.; Takita, M.; Kitai, S.; Inoue, T.; Yada, N.; et al. Evaluation of ART scores for repeated transarterial chemoembolization in Japanese patients with hepatocellular carcinoma. *Oncology* **2015**, *89*, 4–10. [CrossRef]
51. Pinato, D.J.; Arizumi, T.; Jang, J.W.; Allara, E.; Suppiah, P.I.; Smirne, C.; Tait, P.; Pai, M.; Grossi, G.; Kim, Y.W.; et al. Combined sequential use of HAP and ART scores to predict survival outcome and treatment failure following chemoembolization in hepatocellular carcinoma: A multi-center comparative study. *Oncotarget* **2016**, *7*, 44705. [CrossRef]
52. Yin, W.; Ye, Q.; Wang, F.; Liang, J.; Xu, B.; Zhang, X.; Zhang, Q.; Liu, Y.; Li, G.; Han, T. ART score and hepatocellular carcinoma: An appraisal of its applicability. *Clin. Res. Hepatol. Gastroenterol.* **2016**, *40*, 705–714. [CrossRef]
53. Kloeckner, R.; Pitton, M.B.; Dueber, C.; Schmidtmann, I.; Galle, P.R.; Koch, S.; Wörns, M.A.; Weinmann, A. Validation of clinical scoring systems ART and ABCR after transarterial chemoembolization of hepatocellular carcinoma. *J. Vasc. Interv. Radiol.* **2017**, *28*, 94–102. [CrossRef] [PubMed]
54. Mähringer-Kunz, A.; Weinmann, A.; Schmidtmann, I.; Koch, S.; Schotten, S.; Pinto dos Santos, D.; Pitton, M.B.; Dueber, C.; Galle, P.R.; Kloeckner, R. Validation of the SNACOR clinical scoring system after transarterial chemoembolisation in patients with hepatocellular carcinoma. *BMC Cancer* **2018**, *18*, 489. [CrossRef] [PubMed]
55. Müller, L.; Hahn, F.; Mähringer-Kunz, A.; Stoehr, F.; Gairing, S.J.; Foerster, F.; Weinmann, A.; Galle, P.R.; Mittler, J.; Pinto dos Santos, D.; et al. Refining Prognosis in Chemoembolization for Hepatocellular Carcinoma: Immunonutrition and Liver Function. *Cancers* **2021**, *13*, 3961. [CrossRef]
56. Müller, L.; Hahn, F.; Mähringer-Kunz, A.; Stoehr, F.; Gairing, S.J.; Foerster, F.; Weinmann, A.; Galle, P.R.; Mittler, J.; Pinto dos Santos, D.; et al. Immunonutritive Scoring in Patients with Hepatocellular Carcinoma Undergoing Transarterial Chemoembolization: Prognostic Nutritional Index or Controlling Nutritional Status Score? *Front. Oncol.* **2021**, *11*, 2205. [CrossRef] [PubMed]
57. Müller, L.; Hahn, F.; Mähringer-Kunz, A.; Stoehr, F.; Gairing, S.J.; Michel, M.; Foerster, F.; Weinmann, A.; Galle, P.R.; Mittler, J.; et al. Immunonutritive Scoring for Patients with Hepatocellular Carcinoma Undergoing Transarterial Chemoembolization: Evaluation of the CALLY Index. *Cancers* **2021**, *13*, 5018. [CrossRef] [PubMed]
58. Li, S.; Feng, X.; Cao, G.; Wang, Q.; Wang, L. Prognostic significance of inflammatory indices in hepatocellular carcinoma treated with transarterial chemoembolization: A systematic review and meta-analysis. *PLoS ONE* **2020**, *15*, e0230879. [CrossRef] [PubMed]
59. He, C.-B.; Lin, X.-J. Inflammation scores predict the survival of patients with hepatocellular carcinoma who were treated with transarterial chemoembolization and recombinant human type-5 adenovirus H101. *PLoS ONE* **2017**, *12*, e0174769. [CrossRef]
60. Arvanitakis, K.; Mitroulis, I.; Germanidis, G. Tumor-associated neutrophils in hepatocellular carcinoma pathogenesis, prognosis, and therapy. *Cancers* **2021**, *13*, 2899. [CrossRef]

61. Chu, H.H.; Kim, J.H.; Shim, J.H.; Gwon, D.I.; Ko, H.-K.; Shin, J.H.; Ko, G.-Y.; Yoon, H.-K.; Kim, N. Neutrophil-to-Lymphocyte Ratio as a Biomarker Predicting Overall Survival after Chemoembolization for Intermediate-Stage Hepatocellular Carcinoma. *Cancers* **2021**, *13*, 2830. [CrossRef]
62. Geh, D.; Leslie, J.; Rumney, R.; Reeves, H.L.; Bird, T.G.; Mann, D.A. Neutrophils as potential therapeutic targets in hepatocellular carcinoma. *Nat. Rev. Gastroenterol. Hepatol.* **2022**, *19*, 257–273. [CrossRef] [PubMed]
63. Walsh, S.R.; Cook, E.J.; Goulder, F.; Justin, T.A.; Keeling, N.J. Neutrophil-lymphocyte ratio as a prognostic factor in colorectal cancer. *J. Surg. Oncol.* **2005**, *91*, 181–184. [CrossRef] [PubMed]
64. Smith, R.A.; Bosonnet, L.; Raraty, M.; Sutton, R.; Neoptolemos, J.P.; Campbell, F.; Ghaneh, P. Preoperative platelet-lymphocyte ratio is an independent significant prognostic marker in resected pancreatic ductal adenocarcinoma. *Am. J. Surg.* **2009**, *197*, 466–472. [CrossRef]
65. Iida, H.; Tani, M.; Komeda, K.; Nomi, T.; Matsushima, H.; Tanaka, S.; Ueno, M.; Nakai, T.; Maehira, H.; Mori, H.; et al. Superiority of CRP-Albumin-Lymphocyte index (CALLY index) as a non-invasive prognostic biomarker after hepatectomy for hepatocellular carcinoma. *HPB* **2021**, *24*, 101–115. [CrossRef] [PubMed]
66. Sun, K.; Chen, S.; Xu, J.; Li, G.; He, Y. The prognostic significance of the prognostic nutritional index in cancer: A systematic review and meta-analysis. *J. Cancer Res. Clin. Oncol.* **2014**, *140*, 1537–1549. [CrossRef]
67. Hu, B.; Yang, X.-R.; Xu, Y.; Sun, Y.-F.; Sun, C.; Guo, W.; Zhang, X.; Wang, W.-M.; Qiu, S.-J.; Zhou, J.; et al. Systemic immune-inflammation index predicts prognosis of patients after curative resection for hepatocellular carcinoma. *Clin. Cancer Res.* **2014**, *20*, 6212–6222. [CrossRef] [PubMed]
68. Chan, S.L.; Wong, L.-L.; Chan, K.-C.A.; Chow, C.; Tong, J.H.-M.; Yip, T.C.-F.; Wong, G.L.-H.; Chong, C.C.-N.; Liu, P.-H.; Chu, C.-M.; et al. Development of a novel inflammation-based index for hepatocellular carcinoma. *Liver Cancer* **2020**, *9*, 167–181. [CrossRef]
69. Mähringer-Kunz, A.; Wagner, F.; Hahn, F.; Weinmann, A.; Brodehl, S.; Schotten, S.; Hinrichs, J.B.; Düber, C.; Galle, P.R.; Pinto dos Santos, D.; et al. Predicting survival after transarterial chemoembolization for hepatocellular carcinoma using a neural network: A Pilot Study. *Liver Int.* **2020**, *40*, 694–703. [CrossRef]
70. Li, W.; Ni, C.-F. Current status of the combination therapy of transarterial chemoembolization and local ablation for hepatocellular carcinoma. *Abdom. Radiol.* **2019**, *44*, 2268–2275. [CrossRef]
71. Kitamoto, M.; Imagawa, M.; Yamada, H.; Watanabe, C.; Sumioka, M.; Satoh, O.; Shimamoto, M.; Kodama, M.; Kimura, S.; Kishimoto, K.; et al. Radiofrequency ablation in the treatment of small hepatocellular carcinomas: Comparison of the radiofrequency effect with and without chemoembolization. *Am. J. Roentgenol.* **2003**, *181*, 997–1003. [CrossRef]
72. Morimoto, M.; Numata, K.; Kondou, M.; Nozaki, A.; Morita, S.; Tanaka, K. Midterm outcomes in patients with intermediate-sized hepatocellular carcinoma: A randomized controlled trial for determining the efficacy of radiofrequency ablation combined with transcatheter arterial chemoembolization. *Cancer* **2010**, *116*, 5452. [CrossRef] [PubMed]
73. Bucher, H.C.; Guyatt, G.H.; Griffith, L.E.; Walter, S.D. The results of direct and indirect treatment comparisons in meta-analysis of randomized controlled trials. *J. Clin. Epidemiol.* **1997**, *50*, 683–691. [CrossRef]
74. Li, L.; Tian, J.; Liu, P.; Wang, X.; Zhu, Z. Transarterial chemoembolization combination therapy vs. monotherapy in unresectable hepatocellular carcinoma: A meta-analysis. *Tumori J.* **2016**, *102*, 301–310. [CrossRef] [PubMed]
75. Wang, X.; Hu, Y.; Ren, M.; Lu, X.; Lu, G.; He, S. Efficacy and safety of radiofrequency ablation combined with transcatheter arterial chemoembolization for hepatocellular carcinomas compared with radiofrequency ablation alone: A time-to-event meta-analysis. *Korean J. Radiol.* **2016**, *17*, 93–102. [CrossRef] [PubMed]
76. Li, Z.; Li, Q.; Wang, X.; Chen, W.; Jin, X.; Liu, X.; Ye, F.; Dai, Z.; Zheng, X.; Li, P.; et al. Hyperthermia ablation combined with transarterial chemoembolization versus monotherapy for hepatocellular carcinoma: A systematic review and meta-analysis. *Cancer Med.* **2021**, *10*, 8432–8450. [CrossRef]
77. Liu, C.; Li, T.; He, J.; Shao, H. TACE combined with microwave ablation therapy vs. TACE alone for treatment of early-and intermediate-stage hepatocellular carcinomas larger than 5 cm: A meta-analysis. *Diagnostic Interv. Radiol.* **2020**, *26*, 575. [CrossRef] [PubMed]
78. Wang, L.; Ke, Q.; Lin, N.; Huang, Q.; Zeng, Y.; Liu, J. The efficacy of transarterial chemoembolization combined with microwave ablation for unresectable hepatocellular carcinoma: A systematic review and meta-analysis. *Int. J. Hyperth.* **2019**, *36*, 1287–1295. [CrossRef]
79. Kwon, J.H.; Kim, G.M.; Han, K.; Won, J.Y.; Kim, M.D.; Lee, D.Y.; Lee, J.; Choi, W.; Kim, Y.S.; Han, K.-H. Safety and efficacy of transarterial radioembolization combined with chemoembolization for bilobar hepatocellular carcinoma: A single-center retrospective study. *Cardiovasc. Intervent. Radiol.* **2018**, *41*, 459–465. [CrossRef]
80. Jacob, R.; Turley, F.; Redden, D.T.; Saddekni, S.; Aal, A.K.A.; Keene, K.; Yang, E.; Zarzour, J.; Bolus, D.; Smith, J.K.; et al. Adjuvant stereotactic body radiotherapy following transarterial chemoembolization in patients with non-resectable hepatocellular carcinoma tumours of $\geq$3 cm. *HPB* **2015**, *17*, 140–149. [CrossRef]
81. Pérez-Romasanta, L.A.; Portillo, G.-D.; Rodríguez-Gutiérrez, A.; Matías-Pérez, Á. Stereotactic Radiotherapy for Hepatocellular Carcinoma, Radiosensitization Strategies and Radiation-Immunotherapy Combination. *Cancers* **2021**, *13*, 192. [CrossRef]
82. Yoon, S.M.; Ryoo, B.-Y.; Lee, S.J.; Kim, J.H.; Shin, J.H.; An, J.H.; Lee, H.C.; Lim, Y.-S. Efficacy and safety of transarterial chemoembolization plus external beam radiotherapy vs. sorafenib in hepatocellular carcinoma with macroscopic vascular invasion: A randomized clinical trial. *JAMA Oncol.* **2018**, *4*, 661–669. [CrossRef]

83. Kudo, M.; Ueshima, K.; Ikeda, M.; Torimura, T.; Aikata, H.; Izumi, N.; Yamasaki, T.; Hino, K.; Kuzuya, T.; Isoda, N.; et al. TACTICS: Final overall survival (OS) data from a randomized, open label, multicenter, phase II trial of transcatheter arterial chemoembolization (TACE) therapy in combination with sorafenib as compared with TACE alone in patients (pts) with hepatocellular. *J. Clin. Oncol.* **2021**, *39*, 270. [CrossRef]
84. Kudo, M.; Ueshima, K.; Ikeda, M.; Torimura, T.; Tanabe, N.; Aikata, H.; Izumi, N.; Yamasaki, T.; Nojiri, S.; Hino, K.; et al. Randomised, multicentre prospective trial of transarterial chemoembolisation (TACE) plus sorafenib as compared with TACE alone in patients with hepatocellular carcinoma: TACTICS trial. *Gut* **2020**, *69*, 1492–1501. [CrossRef] [PubMed]
85. Park, J.-W.; Kim, Y.J.; Bae, S.-H.; Paik, S.W.; Lee, Y.-J.; Kim, H.Y.; Lee, H.C.; Han, S.Y.; Cheong, J.Y.; Kwon, O.S.; et al. Sorafenib with or without concurrent transarterial chemoembolization in patients with advanced hepatocellular carcinoma: The phase III STAH trial. *J. Hepatol.* **2019**, *70*, 684–691. [CrossRef] [PubMed]
86. Meyer, T.; Fox, R.; Ma, Y.T.; Ross, P.J.; James, M.W.; Sturgess, R.; Stubbs, C.; Stocken, D.D.; Wall, L.; Watkinson, A.; et al. Sorafenib in combination with transarterial chemoembolisation in patients with unresectable hepatocellular carcinoma (TACE 2): A randomised placebo-controlled, double-blind, phase 3 trial. *Lancet Gastroenterol. Hepatol.* **2017**, *2*, 565–575. [CrossRef]
87. Lencioni, R.; Llovet, J.M.; Han, G.; Tak, W.Y.; Yang, J.; Guglielmi, A.; Paik, S.W.; Reig, M.; Chau, G.-Y.; Luca, A.; et al. Sorafenib or placebo plus TACE with doxorubicin-eluting beads for intermediate stage HCC: The SPACE trial. *J. Hepatol.* **2016**, *64*, 1090–1098. [CrossRef]
88. Kudo, M.; Han, G.; Finn, R.S.; Poon, R.T.P.; Blanc, J.; Yan, L.; Yang, J.; Lu, L.; Tak, W.; Yu, X.; et al. Brivanib as adjuvant therapy to transarterial chemoembolization in patients with hepatocellular carcinoma: A randomized phase III trial. *Hepatology* **2014**, *60*, 1697–1707. [CrossRef]
89. Kudo, M.; Cheng, A.-L.; Park, J.-W.; Park, J.H.; Liang, P.-C.; Hidaka, H.; Izumi, N.; Heo, J.; Lee, Y.J.; Sheen, I.-S.; et al. Orantinib versus placebo combined with transcatheter arterial chemoembolisation in patients with unresectable hepatocellular carcinoma (ORIENTAL): A randomised, double-blind, placebo-controlled, multicentre, phase 3 study. *Lancet Gastroenterol. Hepatol.* **2018**, *3*, 37–46. [CrossRef]
90. Radu, P.; Dufour, J.-F. Changing TACTICS in intermediate HCC: TACE plus sorafenib. *Gut* **2020**, *69*, 1374–1376. [CrossRef]
91. Brown, Z.J.; Hewitt, D.B.; Pawlik, T.M. Combination therapies plus transarterial chemoembolization in hepatocellular carcinoma: A snapshot of clinical trial progress. *Expert Opin. Investig. Drugs* **2021**, *25*, 379–391. [CrossRef]
92. Llovet, J.M.; De Baere, T.; Kulik, L.; Haber, P.K.; Greten, T.F.; Meyer, T.; Lencioni, R. Locoregional therapies in the era of molecular and immune treatments for hepatocellular carcinoma. *Nat. Rev. Gastroenterol. Hepatol.* **2021**, *18*, 293–313. [CrossRef]
93. Di Federico, A.; Rizzo, A.; Carloni, R.; De Giglio, A.; Bruno, R.; Ricci, D.; Brandi, G. Atezolizumab-bevacizumab plus Y-90 TARE for the treatment of hepatocellular carcinoma: Preclinical rationale and ongoing clinical trials. *Expert Opin. Investig. Drugs* **2022**, *31*, 361–369. [CrossRef] [PubMed]
94. Li, X.; Feng, G.S.; Zheng, C.S.; Zhuo, C.K.; Liu, X. Expression of plasma vascular endothelial growth factor in patients with hepatocellular carcinoma and effect of transcatheter arterial chemoembolization therapy on plasma vascular endothelial growth factor level. *World J. Gastroenterol.* **2004**, *10*, 2878. [CrossRef]
95. Carmeliet, P.; Jain, R.K. Angiogenesis in cancer and other diseases. *Nature* **2000**, *407*, 249–257. [CrossRef] [PubMed]
96. Wang, B.; Xu, H.; Gao, Z.Q.; Ning, H.F.; Sun, Y.Q.; Cao, G.W. Increased expression of vascular endothelial growth factor in hepatocellular carcinoma after transcatheter arterial chemoembolization. *Acta Radiol.* **2008**, *49*, 523–529. [CrossRef] [PubMed]
97. Rizzo, A.; Brandi, G. Biochemical predictors of response to immune checkpoint inhibitors in unresectable hepatocellular carcinoma. *Cancer Treat. Res. Commun.* **2021**, *27*, 100328. [CrossRef] [PubMed]
98. Rizzo, A.; Ricci, A.D. PD-L1, TMB, and other potential predictors of response to immunotherapy for hepatocellular carcinoma: How can they assist drug clinical trials? *Expert Opin. Investig. Drugs* **2022**, *31*, 415–423. [CrossRef] [PubMed]
99. Finn, R.S.; Qin, S.; Ikeda, M.; Galle, P.R.; Ducreux, M.; Kim, T.-Y.; Kudo, M.; Breder, V.; Merle, P.; Kaseb, A.O.; et al. Atezolizumab plus Bevacizumab in Unresectable Hepatocellular Carcinoma. *N. Engl. J. Med.* **2020**, *382*, 1894–1905. [CrossRef]
100. Salem, R.; Johnson, G.E.; Kim, E.; Riaz, A.; Bishay, V.; Boucher, E.; Fowers, K.; Lewandowski, R.; Padia, S.A. Yttrium-90 Radioembolization for the Treatment of Solitary, Unresectable Hepatocellular Carcinoma: The LEGACY Study. *Hepatology* **2021**, *74*, 2342–2352. [CrossRef]
101. Vogel, A.; Martinelli, E.; Cervantes, A.; Chau, I.; Daniele, B.; Llovet, J.M.; Meyer, T.; Nault, J.-C.; Neumann, U.; Ricke, J.; et al. Updated treatment recommendations for hepatocellular carcinoma (HCC) from the ESMO Clinical Practice Guidelines. *Ann. Oncol.* **2021**, *32*, 801–805. [CrossRef]
102. Salem, R.; Lewandowski, R.J. Chemoembolization and radioembolization for hepatocellular carcinoma. *Clin. Gastroenterol. Hepatol.* **2013**, *11*, 604–611. [CrossRef] [PubMed]
103. Kim, D.Y.; Han, K.-H. Transarterial chemoembolization versus transarterial radioembolization in hepatocellular carcinoma: Optimization of selecting treatment modality. *Hepatol. Int.* **2016**, *10*, 883–892. [CrossRef] [PubMed]
104. Hermann, A.-L.; Dieudonné, A.; Ronot, M.; Sanchez, M.; Pereira, H.; Chatellier, G.; Garin, E.; Castera, L.; Lebtahi, R.; Vilgrain, V. Relationship of tumor radiation–absorbed dose to survival and response in hepatocellular carcinoma treated with transarterial radioembolization with 90Y in the SARAH study. *Radiology* **2020**, *296*, 673–684. [CrossRef] [PubMed]
105. Kallini, J.R.; Gabr, A.; Salem, R.; Lewandowski, R.J. Transarterial radioembolization with yttrium-90 for the treatment of hepatocellular carcinoma. *Adv. Ther.* **2016**, *33*, 699–714. [CrossRef]

106. Salem, R.; Gilbertsen, M.; Butt, Z.; Memon, K.; Vouche, M.; Hickey, R.; Baker, T.; Abecassis, M.M.; Atassi, R.; Riaz, A.; et al. Increased quality of life among hepatocellular carcinoma patients treated with radioembolization, compared with chemoembolization. *Clin. Gastroenterol. Hepatol.* **2013**, *11*, 1358–1365. [CrossRef]
107. Salem, R.; Lewandowski, R.J.; Kulik, L.; Wang, E.; Riaz, A.; Ryu, R.K.; Sato, K.T.; Gupta, R.; Nikolaidis, P.; Miller, F.H.; et al. Radioembolization results in longer time-to-progression and reduced toxicity compared with chemoembolization in patients with hepatocellular carcinoma. *Gastroenterology* **2011**, *140*, 497–507. [CrossRef]
108. Salem, R.; Gordon, A.C.; Mouli, S.; Hickey, R.; Kallini, J.; Gabr, A.; Mulcahy, M.F.; Baker, T.; Abecassis, M.; Miller, F.H.; et al. Y90 radioembolization significantly prolongs time to progression compared with chemoembolization in patients with hepatocellular carcinoma. *Gastroenterology* **2016**, *151*, 1155–1163. [CrossRef]
109. Pitton, M.B.; Kloeckner, R.; Ruckes, C.; Wirth, G.M.; Eichhorn, W.; Wörns, M.A.; Weinmann, A.; Schreckenberger, M.; Galle, P.R.; Otto, G.; et al. Randomized Comparison of Selective Internal Radiotherapy (SIRT) Versus Drug-Eluting Bead Transarterial Chemoembolization (DEB-TACE) for the Treatment of Hepatocellular Carcinoma. *Cardiovasc. Interv. Radiol.* **2015**, *38*, 352–360. [CrossRef]
110. Kolligs, F.T.; Bilbao, J.I.; Jakobs, T.; Iñarrairaegui, M.; Nagel, J.M.; Rodriguez, M.; Haug, A.; D'Avola, D.; op den Winkel, M.; Martinez-Cuesta, A.; et al. Pilot randomized trial of selective internal radiation therapy vs. chemoembolization in unresectable hepatocellular carcinoma. *Liver Int.* **2015**, *35*, 1715–1721. [CrossRef]
111. Vilgrain, V.; Pereira, H.; Assenat, E.; Guiu, B.; Ilonca, A.D.; Pageaux, G.-P.; Sibert, A.; Bouattour, M.; Lebtahi, R.; Allaham, W.; et al. Efficacy and safety of selective internal radiotherapy with yttrium-90 resin microspheres compared with sorafenib in locally advanced and inoperable hepatocellular carcinoma (SARAH): An open-label randomised controlled phase 3 trial. *Lancet Oncol.* **2017**, *18*, 1624–1636. [CrossRef]
112. Chow, P.K.H.; Gandhi, M.; Tan, S.B.; Khin, M.W.; Khasbazar, A.; Ong, J.; Choo, S.P.; Cheow, P.C.; Chotipanich, C.; Lim, K.; et al. SIRveNIB: Selective internal radiation therapy versus sorafenib in Asia-Pacific patients with hepatocellular carcinoma. *J. Clin. Oncol.* **2018**, *36*, 1913–1921. [CrossRef]
113. Miller, F.H.; Lopes Vendrami, C.; Gabr, A.; Horowitz, J.M.; Kelahan, L.C.; Riaz, A.; Salem, R.; Lewandowski, R.J. Evolution of Radioembolization in Treatment of Hepatocellular Carcinoma: A Pictorial Review. *RadioGraphics* **2021**, *41*, 1802–1818. [CrossRef] [PubMed]
114. Riaz, A.; Gates, V.L.; Atassi, B.; Lewandowski, R.J.; Mulcahy, M.F.; Ryu, R.K.; Sato, K.T.; Baker, T.; Kulik, L.; Gupta, R.; et al. Radiation Segmentectomy: A Novel Approach to Increase Safety and Efficacy of Radioembolization. *Int. J. Radiat. Oncol.* **2011**, *79*, 163–171. [CrossRef] [PubMed]
115. Vouche, M.; Habib, A.; Ward, T.J.; Kim, E.; Kulik, L.; Ganger, D.; Mulcahy, M.; Baker, T.; Abecassis, M.; Sato, K.T.; et al. Unresectable solitary hepatocellular carcinoma not amenable to radiofrequency ablation: Multicenter radiology-pathology correlation and survival of radiation segmentectomy. *Hepatology* **2014**, *60*, 192–201. [CrossRef]
116. Prachanronarong, K.; Kim, E. Radiation Segmentectomy. In *Seminars in Interventional Radiology*; Thieme Medical Publishers Inc.: New York, NY, USA, 2021; Volume 38, pp. 425–431.
117. Lewandowski, R.J.; Gabr, A.; Abouchaleh, N.; Ali, R.; Al Asadi, A.; Mora, R.A.; Kulik, L.; Ganger, D.; Desai, K.; Thornburg, B.; et al. Radiation segmentectomy: Potential curative therapy for early hepatocellular carcinoma. *Radiology* **2018**, *287*, 1050–1058. [CrossRef] [PubMed]
118. Liu, J.; Ladbury, C.; Amini, A.; Glaser, S.; Kessler, J.; Lee, A.; Chen, Y.-J. Combination of yttrium-90 radioembolization with stereotactic body radiation therapy in the treatment of portal vein tumor thrombosis. Radiat. *Oncol. J.* **2021**, *39*, 113.
119. Hardy-Abeloos, C.; Lazarev, S.; Ru, M.; Kim, E.; Fischman, A.; Moshier, E.; Rosenzweig, K.; Buckstein, M. Safety and efficacy of liver stereotactic body radiation therapy for hepatocellular carcinoma after segmental transarterial radioembolization. *Int. J. Radiat. Oncol. Biol. Phys.* **2019**, *105*, 968–976. [CrossRef] [PubMed]
120. Mahvash, A.; Murthy, R.; Odisio, B.C.; Raghav, K.P.; Girard, L.; Cheung, S.; Nguyen, V.; Ensor, J.; Gadani, S.; Elsayes, K.M.; et al. Yttrium-90 resin microspheres as an adjunct to sorafenib in patients with unresectable hepatocellular carcinoma. *J. Hepatocell. Carcinoma* **2016**, *3*, 1–7.
121. Ricke, J.; Klümpen, H.J.; Amthauer, H.; Bargellini, I.; Bartenstein, P.; de Toni, E.N.; Gasbarrini, A.; Pech, M.; Peck-Radosavljevic, M.; Popovič, P.; et al. Impact of combined selective internal radiation therapy and sorafenib on survival in advanced hepatocellular carcinoma. *J. Hepatol.* **2019**, *71*, 1164–1174. [CrossRef]
122. Chauhan, N.; Bukovcan, J.; Boucher, E.; Cosgrove, D.; Edeline, J.; Hamilton, B.; Kulik, L.; Master, F.; Salem, R. Intra-arterial TheraSphere yttrium-90 glass microspheres in the treatment of patients with unresectable hepatocellular carcinoma: Protocol for the STOP-HCC phase 3 randomized controlled trial. *JMIR Res. Protoc.* **2018**, *7*, e11234. [CrossRef]
123. Geng, L.; Donnelly, E.; McMahon, G.; Lin, P.C.; Sierra-Rivera, E.; Oshinka, H.; Hallahan, D.E. Inhibition of vascular endothelial growth factor receptor signaling leads to reversal of tumor resistance to radiotherapy. *Cancer Res.* **2001**, *61*, 2413–2419.
124. Cho, Y.K.; Kim, J.K.; Kim, W.T.; Chung, J.W. Hepatic resection versus radio-frequency ablation for very early stage hepatocellular carcinoma: A Markov model analysis. *Hepatology* **2010**, *51*, 1284–1290. [CrossRef] [PubMed]
125. Germani, G.; Pleguezuelo, M.; Gurusamy, K.; Meyer, T.; Isgrò, G.; Burroughs, A.K. Clinical outcomes of radiofrequency ablation, percutaneous alcohol and acetic acid injection for hepatocelullar carcinoma: A meta-analysis. *J. Hepatol.* **2010**, *52*, 380–388. [CrossRef]

126. Cucchetti, A.; Piscaglia, F.; Cescon, M.; Colecchia, A.; Ercolani, G.; Bolondi, L.; Pinna, A.D. Cost-effectiveness of hepatic resection versus percutaneous radiofrequency ablation for early hepatocellular carcinoma. *J. Hepatol.* **2013**, *59*, 300–307. [CrossRef] [PubMed]
127. Izumi, N.; Hasegawa, K.; Nishioka, Y.; Takayama, T.; Yamanaka, N.; Kudo, M.; Shimada, M.; Inomata, M.; Kaneko, S.; Baba, H.; et al. A multicenter randomized controlled trial to evaluate the efficacy of surgery vs. radiofrequency ablation for small hepatocellular carcinoma (SURF trial). *J. Clin. Oncol.* **2019**, *37* (Suppl. S15), 4002. [CrossRef]
128. Doyle, A.; Gorgen, A.; Muaddi, H.; Aravinthan, A.D.; Issachar, A.; Mironov, O.; Zhang, W.; Kachura, J.; Beecroft, R.; Cleary, S.P.; et al. Outcomes of radiofrequency ablation as first-line therapy for hepatocellular carcinoma less than 3 cm in potentially transplantable patients. *J. Hepatol.* **2019**, *70*, 866–873. [CrossRef]
129. Hocquelet, A.; Aubé, C.; Rode, A.; Cartier, V.; Sutter, O.; Manichon, A.F.; Boursier, J.; N'kontchou, G.; Merle, P.; Blanc, J.F.; et al. Comparison of no-touch multi-bipolar vs. monopolar radiofrequency ablation for small HCC. *J. Hepatol.* **2017**, *66*, 67–74. [CrossRef] [PubMed]
130. Orlacchio, A.; Bazzocchi, G.; Pastorelli, D.; Bolacchi, F.; Angelico, M.; Almerighi, C.; Masala, S.; Simonetti, G. Percutaneous cryoablation of small hepatocellular carcinoma with US guidance and CT monitoring: Initial experience. *Cardiovasc. Interv. Radiol.* **2008**, *31*, 587–594. [CrossRef] [PubMed]
131. Shimizu, T.; Sakuhara, Y.; Abo, D.; Hasegawa, Y.; Kodama, Y.; Endo, H.; Shirato, H.; Miyasaka, K. Outcome of MR-guided percutaneous cryoablation for hepatocellular carcinoma. *J. Hepatobil. Pancreat. Surg.* **2009**, *16*, 816–823. [CrossRef]
132. Di Costanzo, G.G.; Tortora, R.; D'Adamo, G.; De Luca, M.; Lampasi, F.; Addario, L.; Galeota Lanza, A.; Picciotto, F.P.; Tartaglione, M.T.; Cordone, G.; et al. Radiofrequency ablation versus laser ablation for the treatment of small hepatocellular carcinoma in cirrhosis: A randomized trial. *J. Gastroenterol. Hepatol.* **2015**, *30*, 559–565. [CrossRef]
133. Sutter, O.; Calvo, J.; N'Kontchou, G.; Nault, J.C.; Ourabia, R.; Nahon, P.; Ganne-Carrié, N.; Bourcier, V.; Zentar, N.; Bouhafs, F.; et al. Safety and efficacy of irreversible electroporation for the treatment of hepatocellular carcinoma not amenable to thermal ablation techniques: A retrospective single-center case series. *Radiology* **2017**, *284*, 877–886. [CrossRef] [PubMed]
134. Bruix, J.; Takayama, T.; Mazzaferro, V.; Chau, G.Y.; Yang, J.; Kudo, M.; Cai, J.; Poon, R.T.; Han, K.H.; Tak, W.Y.; et al. STORM investigators. Adjuvant sorafenib for hepatocellular carcinoma after resection or ablation (STORM): A phase 3, randomised, double-blind, placebo-controlled trial. *Lancet Oncol.* **2015**, *16*, 1344–1354. [CrossRef]
135. Pinyol, R.; Montal, R.; Bassaganyas, L.; Sia, D.; Takayama, T.; Chau, G.Y.; Mazzaferro, V.; Roayaie, S.; Lee, H.C.; Kokudo, N.; et al. Molecular predictors of prevention of recurrence in HCC with sorafenib as adjuvant treatment and prognostic factors in the phase 3 STORM trial. *Gut* **2019**, *68*, 1065–1075. [CrossRef] [PubMed]
136. Lee, J.H.; Lee, J.H.; Lim, Y.S.; Yeon, J.E.; Song, T.J.; Yu, S.J.; Gwak, G.Y.; Kim, K.M.; Kim, Y.J.; Lee, J.W.; et al. Adjuvant immunotherapy with autologous cytokine-induced killer cells for hepatocellular carcinoma. *Gastroenterology* **2015**, *148*, 1383–1391. [CrossRef]
137. Facciorusso, A.; Serviddio, G.; Muscatiello, N. Local ablative treatments for hepatocellular carcinoma: An updated review. *WJGPT* **2016**, *7*, 477–489. [CrossRef]
138. Tan, W.; Deng, Q.; Lin, S.; Wang, Y.; Xu, G. Comparison of microwave ablation and radiofrequency ablation for hepatocellular carcinoma: A systematic review and meta-analysis. *Int. J. Hyperth.* **2019**, *36*, 264–272. [CrossRef]
139. Facciorusso, A.; Di Maso, M.; Muscatiello, N. Microwave ablation versus radiofrequency ablation for the treatment of hepatocellular carcinoma: A systematic review and meta-analysis. *Int. J. Hyperth.* **2016**, *32*, 339–344. [CrossRef]
140. Ricci, A.D.; Rizzo, A.; Bonucci, C.; Tavolari, S.; Palloni, A.; Frega, G.; Mollica, V.; Tober, N.; Mazzotta, E.; Felicani, C.; et al. The (Eternal) Debate on Microwave Ablation versus Radiofrequency Ablation in BCLC-A Hepatocellular Carcinoma. *In Vivo* **2020**, *34*, 3421–3429. [CrossRef]
141. Bouda, D.; Barrau, V.; Raynaud, L.; Dioguardi Burgio, M.; Paulatto, L.; Roche, V.; Sibert, A.; Moussa, N.; Vilgrain, V.; Ronot, M. Factors associated with tumor progression after percutaneous ablation of HCC: Comparison between monopolar Radiofrequency and Microwaves. Results of a propensity score matching analysis. *Cardiovasc. Interv. Radiol.* **2020**, *43*, 1608–1618. [CrossRef]
142. Hermida, M.; Cassinotto, C.; Piron, L.; Aho-Glélé, S.; Guillot, C.; Schembri, V.; Allimant, C.; Jaber, S.; Pageaux, G.P.; Assenat, E.; et al. Multimodal Percutaneous Thermal Ablation of Small Hepatocellular Carcinoma: Predictive Factors of Recurrence and Survival in Western Patients. *Cancers* **2020**, *12*, 313. [CrossRef]
143. Lachenmayer, A.; Tinguely, P.; Maurer, M.H.; Frehner, L.; Knöpfli, M.; Peterhans, M.; Weber, S.; Dufour, J.F.; Candinas, D.; Banz, V. Stereotactic image-guided microwave ablation of hepatocellular carcinoma using a computer-assisted navigation system. *Liver Int.* **2019**, *39*, 1975–1985. [CrossRef] [PubMed]
144. Boleslawski, E.; Petrovai, G.; Truant, S.; Dharancy, S.; Duhamel, A.; Salleron, J.; Deltenre, P.; Lebuffe, G.; Mathurin, P.; Pruvot, F.R. Hepatic venous pressure gradient in the assessment of portal hypertension before liver resection in patients with cirrhosis. *Br. J. Surg.* **2012**, *99*, 855–863. [CrossRef]
145. Fuks, D.; Dokmak, S.; Paradis, V.; Diouf, M.; Durand, F.; Belghiti, J. Benefit of initial resection of hepatocellular carcinoma followed by transplantation in case of recurrence: An intention-to-treat analysis. *Hepatology* **2012**, *55*, 132–140. [CrossRef] [PubMed]
146. Ferrer-Fàbrega, J.; Forner, A.; Liccioni, A.; Miquel, R.; Molina, V.; Navasa, M.; Fondevila, C.; García-Valdecasas, J.C.; Bruix, J.; Fuster, J. Prospective validation of ab initio liver transplantation in hepatocellular carcinoma upon detection of risk factors for recurrence after resection. *Hepatology* **2016**, *63*, 839–849. [CrossRef] [PubMed]

147. Nault, J.-C.; Sutter, O.; Nahon, P.; Ganne-Carrié, N.; Séror, O. Percutaneous treatment of hepatocellular carcinoma: State of the art and innovations. *J. Hepatol.* **2018**, *68*, 783–797. [CrossRef]
148. Chen, M.S.; Li, J.Q.; Zheng, Y.; Guo, R.P.; Liang, H.H.; Zhang, Y.Q.; Lin, X.J.; Lau, W.Y. A prospective randomized trial comparing percutaneous local ablative therapy and partial hepatectomy for small hepatocellular carcinoma. *Ann. Surg.* **2006**, *243*, 321–328. [CrossRef]
149. Feng, K.; Yan, J.; Li, X.; Guo, R.P.; Liang, H.H.; Zhang, Y.Q.; Lin, X.J.; Lau, W.Y. A randomized controlled trial of radiofrequency ablation and surgical resection in the treatment of small hepatocellular carcinoma. *J. Hepatol.* **2012**, *57*, 794–802. [CrossRef]
150. Ueno, S.; Sakoda, M.; Kubo, F.; Hiwatashi, K.; Tateno, T.; Baba, Y.; Hasegawa, S.; Tsubouchi, H. Kagoshima Liver Cancer Study Group. Surgical resection versus radiofrequency ablation for small hepatocellular carcinomas within the Milan criteria. *J. Hepatobiliary Pancreat. Surg.* **2009**, *16*, 359–366. [CrossRef]
151. Uhlig, J.; Sellers, C.M.; Stein, S.M.; Kim, H.S. Radiofrequency ablation versus surgical resection of hepatocellular carcinoma: Contemporary treatment trends and outcomes from the United States National Cancer Database. *Eur. Radiol.* **2019**, *29*, 2679–2689. [CrossRef]
152. Lee, D.H.; Kim, J.W.; Lee, J.M.; Kim, J.M.; Lee, M.W.; Rhim, H.; Hur, Y.H.; Suh, K.S. Laparoscopic liver resection versus percutaneous radiofrequency ablation for small single nodular hepatocellular carcinoma: Comparison of treatment outcomes. *Liver Cancer* **2021**, *10*, 25–37. [CrossRef]
153. Ogiso, S.; Seo, S.; Eso, Y.; Yoh, T.; Kawai, T.; Okumura, S.; Ishii, T.; Fukumitsu, K.; Taura, K.; Seno, H.; et al. Laparoscopic liver resection versus percutaneous radiofrequency ablation for small hepatocellular carcinoma. *HPB* **2021**, *23*, 533–537. [CrossRef]
154. Huang, Y.Z.; Zhou, S.C.; Zhou, H.; Tong, M. Radiofrequency ablation versus cryosurgery ablation for hepatocellular carcinoma: A meta-analysis. *Hepatogastroenterology* **2013**, *60*, 1131–1135. [PubMed]
155. Hinshaw, J.L.; Lee, F.T., Jr. Cryoablation for liver cancer. *Tech. Vasc. Interv. Radiol.* **2007**, *10*, 47–57. [CrossRef] [PubMed]
156. Wang, C.; Wang, H.; Yang, W.; Hu, K.; Xie, H.; Hu, K.Q.; Bai, W.; Dong, Z.; Lu, Y.; Zeng, Z.; et al. Multicenter randomized controlled trial of percutaneous cryoablation versus radiofrequency ablation in hepatocellular carcinoma. *Hepatology* **2015**, *61*, 1579–1590. [CrossRef]
157. Cannon, R.; Ellis, S.; Hayes, D.; Narayanan, G.; Martin, R.C.G. Safety and early efficacy of irreversible electroporation for hepatic tumors in proximity to vital structures. *J. Surg. Oncol.* **2013**, *107*, 544–549. [CrossRef] [PubMed]
158. Freeman, E.; Cheung, W.; Kavnoudias, H.; Majeed, A.; Kemp, W.; Roberts, S.K. Irreversible Electroporation for Hepatocellular Carcinoma: Longer-Term Outcomes At A Single Centre. *Cardiovasc. Interv. Radiol.* **2021**, *44*, 247–253. [CrossRef] [PubMed]
159. Verloh, N.; Jensch, I.; Lürken, L.; Haimerl, M.; Dollinger, M.; Renner, P.; Wiggermann, P.; Werner, J.M.; Zeman, F.; Stroszczynski, C.; et al. Similar complication rates for irreversible electroporation and thermal ablation in patients with hepatocellular tumors. *Radiol. Oncol.* **2019**, *53*, 116–122. [CrossRef] [PubMed]
160. Bhutiani, N.; Philips, P.; Scoggins, C.R.; McMasters, K.M.; Potts, M.H.; Martin, R.C.G. Evaluation of tolerability and efficacy of irreversible electroporation (IRE) in treatment of Child-Pugh B (7/8) hepatocellular carcinoma (HCC). *HPB* **2016**, *18*, 593–599. [CrossRef]
161. Niessen, C.; Igl, J.; Pregler, B.; Beyer, L.; Noeva, E.; Dollinger, M.; Schreyer, A.G.; Jung, E.M.; Stroszczynski, C.; Wiggermann, P. Factors associated with short-term local recurrence of liver cancer after percutaneous ablation using irreversible electroporation: A prospective single-center study. *J. Vasc. Interv. Radiol.* **2015**, *26*, 694–702. [CrossRef]
162. Padia, S.A.; Johnson, G.E.; Yeung, R.S.; Park, J.O.; Hippe, D.S.; Kogut, M.J. Irreversible electroporation in patients with hepatocellular carcinoma: Immediate versus delayed findings at MR imaging. *Radiology* **2016**, *278*, 285–294. [CrossRef]
163. Cheng, R.G.; Bhattacharya, R.; Yeh, M.M.; Padia, S.A. Irreversible electroporation can effectively ablate hepatocellular carcinoma to complete pathologic necrosis. *J. Vasc. Interv. Radiol.* **2015**, *26*, 1184–1188. [CrossRef] [PubMed]
164. Zerbini, A.; Pilli, M.; Laccabue, D.; Pelosi, G.; Molinari, A.; Negri, E.; Cerioni, S.; Fagnoni, F.; Soliani, P.; Ferrari, C.; et al. Radiofrequency thermal ablation for hepatocellular carcinoma stimulates autologous NK-cell response. *Gastroenterology* **2010**, *138*, 1931–1942. [CrossRef] [PubMed]
165. Duffy, A.G.; Ulahannan, S.V.; Makorova-Rusher, O.; Rahma, O.; Wedemeyer, H.; Pratt, D.; Davis, J.L.; Hughes, M.S.; Heller, T.; ElGindi, M.; et al. Tremelimumab in combination with ablation in patients with advanced hepatocellular carcinoma. *J. Hepatol.* **2017**, *66*, 545–551. [CrossRef] [PubMed]
166. Erinjeri, J.P.; Fine, G.C.; Adema, G.J.; Ahmed, M.; Chapiro, J.; den Brok, M.; Duran, R.; Hunt, S.J.; Johnson, D.T.; Ricke, J.; et al. Immunotherapy and the interventional oncologist: Challenges and opportunities—A society of interventional oncology white paper. *Radiology* **2019**, *292*, 25–34. [CrossRef] [PubMed]
167. Liang, J.D.; Ping, X.O.; Tseng, Y.J.; Huang, G.T.; Lai, F.; Yang, P.M. Recurrence predictive models for patients with hepatocellular carcinoma after radiofrequency ablation using support vector machines with feature selection methods. *Comput. Methods Programs Biomed.* **2014**, *117*, 425–434. [CrossRef]
168. Wu, C.F.; Wu, Y.J.; Liang, P.C.; Wu, C.H.; Peng, S.F.; Chiu, H.W. Disease-free survival assessment by artificial neural networks for hepatocellular carcinoma patients after radiofrequency ablation. *J. Formos Med. Assoc.* **2017**, *116*, 765–773. [CrossRef] [PubMed]
169. Yacoub, J.H.; Mauro, D.; Moon, A.; He, A.R.; Bashir, M.R.; Hsu, C.C.; Fishbein, T.M.; Burke, L. Therapies for hepatocellular carcinoma: Overview, clinical indications, and comparative outcome evaluation. Part two: Noncurative intention. *Abdom. Radiol.* **2021**, *46*, 3540–3548. [CrossRef]

170. Miften, M.; Vinogradskiy, Y.; Moiseenko, V.; Grimm, J.; Yorke, E.; Jackson, A.; Tomé, W.A.; Ten Haken, R.K.; Ohri, N.; Romero, A.M.; et al. Radiation dose-volume effects for liver SBRT. *Int. J. Radiat. Oncol. Biol. Phys.* **2021**, *110*, 196–205. [CrossRef]
171. Dreher, C.; Høyer, K.I.; Fode, M.M.; Habermehl, D.; Combs, S.E.; Høyer, M. Metabolic liver function after stereotactic body radiation therapy for hepatocellular carcinoma. *Acta Oncol.* **2016**, *55*, 886–891. [CrossRef]
172. Jung, J.; Yoon, S.M.; Kim, S.Y.; Cho, B.; Park, J.; Kim, S.S.; Song, S.Y.; Lee, S.; Do Ahn, S.; Choi, E.K.; et al. Radiation-induced liver disease after stereotactic body radiotherapy for small hepatocellular carcinoma: Clinical and dose-volumetric parameters. *Radiat. Oncol.* **2013**, *8*, 249. [CrossRef]
173. Kim, J.; Jung, Y. Radiation-induced liver disease: Current understanding and future perspectives. *Exp. Mol. Med.* **2017**, *49*, e359. [CrossRef] [PubMed]
174. Thomas, H.R.; Feng, M. Stereotactic Body Radiation Therapy (SBRT) in Hepatocellular Carcinoma. *Curr. Hepatol. Rep.* **2021**, *20*, 12–22. [CrossRef]
175. Kim, T.H.; Koh, Y.H.; Kim, B.H.; Kim, M.J.; Lee, J.H.; Park, B.; Park, J.-W. Proton beam radiotherapy vs. radiofrequency ablation for recurrent hepatocellular carcinoma: A randomized phase III trial. *J. Hepatol.* **2021**, *74*, 603–612. [CrossRef] [PubMed]
176. Wei, X.; Jiang, Y.; Zhang, X.; Feng, S.; Zhou, B.; Ye, X.; Xing, H.; Xu, Y.; Shi, J.; Guo, W.; et al. Neoadjuvant three-dimensional conformal radiotherapy for resectable hepatocellular carcinoma with portal vein tumor thrombus: A randomized, open-label, multicenter controlled study. *J. Clin. Oncol.* **2019**, *37*, 2141. [CrossRef]
177. Zhao, J.; Zeng, L.; Wu, Q.; Wang, L.; Lei, J.; Luo, H.; Yi, F.; Wei, Y.; Yu, J.; Zhang, W. Stereotactic body radiotherapy combined with transcatheter arterial chemoembolization versus stereotactic body radiotherapy alone as the first-line treatment for unresectable hepatocellular carcinoma: A meta-analysis and systematic review. *Chemotherapy* **2019**, *64*, 248–258. [CrossRef]
178. Lee, Y.H.; Tai, D.; Yip, C.; Choo, S.P.; Chew, V. Combinational immunotherapy for hepatocellular carcinoma: Radiotherapy, immune checkpoint blockade and beyond. *Front. Immunol.* **2020**, *11*, 2577. [CrossRef]
179. Heimbach, J.K.; Kulik, L.M.; Finn, R.S.; Sirlin, C.B.; Abecassis, M.M.; Roberts, L.R.; Zhu, A.X.; Murad, M.H.; Marrero, J.A. AASLD guidelines for the treatment of hepatocellular carcinoma. *Hepatology* **2018**, *67*, 358–380. [CrossRef]
180. Apisarnthanarax, S.; Barry, A.; Cao, M.; Czito, B.; DeMatteo, R.; Drinane, M.; Hallemeier, C.L.; Koay, E.J.; Lasley, F.; Meyer, J.; et al. External Beam Radiation Therapy for Primary Liver Cancers: An ASTRO Clinical Practice Guideline. *Pract. Radiat. Oncol.* **2022**, *12*, 28–51. [CrossRef]

cancers

MDPI

*Article*

# Causes of Death among Patients with Hepatocellular Carcinoma According to Chronic Liver Disease Etiology

**Yi-Hao Yen [1,*,†], Kwong-Ming Kee [1], Wei-Feng Li [2], Yueh-Wei Liu [2] Chih-Chi Wang [2,*,†], Tsung-Hui Hu [1], Ming-Chao Tsai [1], Yuan-Hung Kuo [1] and Chih-Yun Lin [3]**

1 Division of Hepatogastroenterology, Department of Internal Medicine, Kaohsiung Chang Gung Memorial Hospital, Kaohsiung 83301, Taiwan
2 Liver Transplantation Center, Department of Surgery, Kaohsiung Chang Gung Memorial Hospital, Kaohsiung 83301, Taiwan
3 Biostatistics Center, Kaohsiung Chang Gung Memorial Hospital, Kaohsiung 83301, Taiwan
* Correspondence: cassellyen@yahoo.com.tw (Y.-H.Y.); ufel4996@ms26.hinet.net (C.-C.W.); Tel.: +886-7-7317123 (Y.-H.Y. & C.-C.W.); Fax: +886-7-7318762 (Y.-H.Y. & C.-C.W.)
† These authors contributed equally to this work.

**Simple Summary:** Hepatocellular carcinoma (HCC) is a highly aggressive and lethal form of liver cancer, and most patients with HCC die due to HCC-related causes. Although most patients die of HCC-related causes, non–HCC-related death represents a competing event among patients who engage in alcohol use and receive curative treatment and among patients 75 years and older in the hepatitis B virus and all-negative groups who receive curative treatments. All negative was defined as negative for hepatitis C virus, hepatitis B virus, and alcohol-related causes. The results of the current study underscore the importance of assessing and managing underlying comorbidities, especially among certain subgroups of patients with HCC.

**Abstract:** This study was conducted to determine whether the causes of death among patients with hepatocellular carcinoma (HCC) differ according to chronic liver disease (CLD) etiology. Between 2011 and 2020, 3977 patients who were newly diagnosed with HCC at our institution were enrolled in this study. We determined whether the cause of death was HCC-related and non-HCC-related. For patients with multiple CLD etiologies, etiology was classified using the following hierarchy: hepatitis C virus (HCV) > hepatitis B virus (HBV) > alcohol-related causes > all negative. All negative was defined as negative for HCV, HBV, and alcohol-related causes. Among 3977 patients, 1415 patients were classified as HCV-related, 1691 patients were HBV-related, 145 patients were alcohol-related, and 725 patients were all negative. HCC-related mortality was the leading cause of death, irrespective of etiology. Among patients who underwent curative treatment, HCC-related mortality was the leading cause of death for patients in the HCV, HBV, and all-negative groups, but not for patients in the alcohol-related group. Among patients 75 years and older who underwent curative treatment, HCC-related mortality was the leading cause of death in the HCV but not HBV or all-negative groups. In conclusion, although most patients with HCC die due to HCC-related causes, non-HCC-related mortality represents a competing event in certain patient subgroups. The current study results underscore the importance of assessing and managing underlying comorbidities, particularly among patients with HCC at risk of non-HCC-related mortality.

**Keywords:** liver cancer; cause-specific mortality; chronic liver disease (CLD)

**Citation:** Yen, Y.-H.; Kee, K.-M.; Li, W.-F.; Liu, Y.-W.; Wang, C.-C.; Hu, T.-H.; Tsai, M.-C.; Kuo, Y.-H.; Lin, C.-Y. Causes of Death among Patients with Hepatocellular Carcinoma According to Chronic Liver Disease Etiology. *Cancers* **2023**, *15*, 1687. https://doi.org/10.3390/cancers15061687

Academic Editor: Georgios Germanidis

Received: 5 February 2023
Revised: 28 February 2023
Accepted: 7 March 2023
Published: 9 March 2023



## 1. Introduction

Hepatocellular carcinoma (HCC) is a common cause of cancer-related death worldwide [1]. Non-alcoholic fatty liver disease (NAFLD) is the fastest growing HCC etiology globally due to the obesity epidemic [2,3]. Although cardiovascular disease is the leading cause of death among patients with NAFLD without HCC [4], whether this remains true among patients with NAFLD-related HCC remains unclear. A retrospective study including a cohort of patients with NAFLD-related HCC who were diagnosed at Veterans Administration (VA) facilities found that 40% of all deaths occurring 3–5 years after treatment were due to non-HCC-related causes among patients with NAFLD-related HCC who received curative treatment. All-cause mortality was driven by non-HCC-related mortality in patients 75 years and older who underwent curative treatments, as the cumulative number of non-HCC-related deaths was higher than HCC-related deaths during the treatment follow-up period. This study enrolled patients treated by the VA, who were mostly men, limiting the generalizability of study findings to women with NAFLD-related HCC [5]. Furthermore, the proportion of non-HCC-related deaths that occur among patients with NAFLD-related HCC compared with other chronic liver disease (CLD) etiologies remains unclear. We aim to clarify this issue.

## 2. Materials and Methods

The Institutional Review Board of Kaohsiung Chang Gung Memorial Hospital, Taiwan, approved this study (reference number: 202201189B0) and waived the need for informed consent due to the retrospective and observational nature of the study design. Data were extracted from Kaohsiung Chang Gung Memorial Hospital's HCC registry database of prospectively collected and annually updated data.

From 2011 to 2020, 3977 patients who were newly diagnosed with HCC at the institution were enrolled in this study.

### *2.1. Variables of Interest*

Patient clinical data, including tumor number and size, imaging-diagnosed tumor–node–metastasis (TNM) stage (based on the 7th edition of the American Joint Committee on Cancer [AJCC]) [6], Barcelona Clinic Liver Cancer (BCLC) stage [7], serum alpha-fetoprotein (AFP) level, presence of liver cirrhosis, Child–Pugh class [8], international normalized ratio (INR), creatinine level, bilirubin level, presence of hepatitis B surface antigen (HBsAg), presence of anti-HCV antibody, alcohol intake (assessed by asking patients how often they have a drink containing alcohol), and HCC diagnostic method (i.e., clinical vs. pathological diagnosis), were prospectively collected from the HCC registry data. Patients with HCV infection were defined based on anti-HCV antibody positivity. Patients with HBV infection were defined based on HBsAg positivity. Patients were defined as consuming alcohol if they reported regularly partaking in alcoholic beverages. Patients were classified as all negative if they were negative for HCV, HBV, and alcohol intake. For patients with multiple etiologies, classification was performed using the following hierarchy: HCV > HBV > alcohol-related causes > all negative [9]. Demographic information included height, weight, age, and sex. Tumor number (solitary vs. multiple) was determined based on imaging results. Tumor size was determined according to pathological examinations in patients who underwent surgery or imaging findings in patients who underwent non-surgical treatments. The presence of cirrhosis was indicated by an Ishak score [10] of 5 or 6 in patients who underwent surgery or imaging results in patients who underwent non-surgical treatments. Cirrhosis was indicated if imaging results showed small liver size, nodular liver surface, and the presence of regeneration nodules [11]. BCLC stages were defined according to the original definitions, and BCLC stage A was defined using the Milan criteria [12]. The cause of death was derived from death certificates. Curative treatment was defined as liver transplantation, liver resection, or ablation. Non-curative treatments included transcatheter arterial embolization (TAE)/transcatheter arterial chemoembolization (TACE), target therapy (i.e., sorafenib or lenvatinib), systemic chemotherapy, radiation therapy, and best supportive care.

Raw data for the cohort involved in this study is available via the following digital object identifier: https://www.dropbox.com/scl/fi/9y5d6re7n4cjt6viqakx4/raw-data-for-submission-n-3977.xlsx?dl=0&rlkey=szktna5d1pmnho47afrfrmoti (accessed on 6 March 2023).

### 2.2. Statistical Analysis

Variables are presented as the number and percentage or the median and interquartile range. The Chi-square test was used to compare categorical variables. The Kruskal–Wallis test was used to compare continuous variables. We calculated cumulative and cause-specific mortality within 5 years after HCC diagnosis. The cumulative HCC-related mortality analysis considered non-HCC-related and unknown causes of death as competing risks. All statistical analyses were performed using SPSS version 25.0 and SigmaPlot 14.0. Two-tailed significance values were applied, with significance defined as $p < 0.05$.

## 3. Results

### 3.1. Characteristics of Patients According to CLD Etiology

The all-negative group was older ($p < 0.001$) than the other groups. The proportion of men was smaller in the HCV group ($p < 0.001$) than in the other groups. The proportion of HCC diagnosed pathologically was smaller in the alcohol-related group ($p < 0.001$) than in the other groups. The tumor size was larger in the all-negative group ($p < 0.001$) than in the other groups. The proportions of TNM stages I and II were larger in the HCV group ($p < 0.001$) than in the other groups. The proportions of BCLC stages 0 and A were larger in the HCV group ($p < 0.001$) than in the other groups. The body mass index (BMI) was higher in the all-negative group ($p = 0.048$) than in the other groups. The proportion of AFP $\geq$20 ng/dl was larger in the HBV group ($p = 0.001$) than in the other groups. The proportion without cirrhosis was larger in the all-negative group ($p < 0.001$) than in the other groups. The creatinine level was higher in the all-negative group ($p < 0.001$) than in the other groups. The bilirubin level was higher in the alcohol-related group ($p < 0.001$) than in the other groups. The INR level was higher in the alcohol-related group ($p < 0.001$) than in the other groups. The proportion of Child–Pugh class A was smaller in the alcohol-related group ($p < 0.001$) than in the other groups. The proportion of patients who underwent resection was larger in the HBV group than in the other groups. The proportion of patients who underwent ablation was larger in the HCV group than in the other groups. The proportion of patients who underwent TAE/TACE was larger in the alcohol-related group than in the other groups ($p < 0.001$). However, no significant difference in tumor number was observed between the groups (Table 1).

### 3.2. Cause-Specific Mortality According to CLD Etiology

Figure 1 shows the cumulative mortality risk according to CLD etiology for all patients. HCC-related mortality was the leading cause of death among all patients, irrespective of CLD etiology (Figure 1A–D). The proportion of HCC-related death was largest among patients with HBV-related HCC and smallest among patients in the alcohol-related and all-negative groups. Among patients with HBV-related HCC, HCC contributed to 73.5% (424 of 577) of all deaths within 3 years. Approximately 12.5% of deaths within 3 years were attributed to non-HCC-related causes, and 14.0% were attributed to unknown causes. We found similar mortality outcomes among patients with HBV-related HCC at 1, 3, and 5 years. In the all-negative group, HCC contributed to 62.2% (186 of 299) of deaths within 3 years, non-HCC-related mortality contributed to approximately 25.8% of deaths within 3 years, and 12.0% was attributed to unknown causes. We found similar mortality outcomes among all-negative patients at 1, 3, and 5 years (Table 2).

### 3.3. Cause-Specific Mortality among Patients Who Received Curative Treatments According to CLD Etiology

Figure 2 shows the cumulative mortality risk according to CLD etiology among patients who underwent curative treatments. HCC-related mortality was the leading cause of death in the HCV (Figure 2A), HBV (Figure 2B), and all negative (Figure 2D) groups. However, the cumulative non-HCC-related mortality rate was higher than the HCC-related mortality rate for deaths that occurred within 5 years of treatment in the alcohol-related group (Figure 2C). The non-HCC-related mortality rates were higher than the HCC-related mortality rates at 1, 3, and 5 years in the alcohol-related group. By contrast, in the remaining three groups, the HCC-related mortality rates were higher than the non-HCC-related mortality rates at 1, 3, and 5 years (Table 3).

**Table 1.** Characteristics of patients according to etiologies of chronic liver disease.

| | HCV, n = 1415 | HBV, n = 1691 | Alcohol, n = 145 | All Negative, n = 725 | p |
|---|---|---|---|---|---|
| Age (years) | 66 (60–73) | 60 (52–67) | 59 (51.5–65) | 68 (60–76) | <0.001 |
| Male | 833 (58.9%) | 1394 (82.4%) | 141 (97.2%) | 503 (69.4%) | <0.001 |
| Method of HCC diagnosis | | | | | <0.001 |
| Clinical | 559 (39.5%) | 631 (37.3%) | 68 (46.9%) | 231 (31.9%) | |
| Pathological | 856 (60.5%) | 1061 (62.7%) | 77 (53.1%) | 494 (68.1%) | |
| Tumor size (mm) | 30 (21–50) | 35 (23–75) | 32 (21.5–82.5) | 48 (28–95) | <0.001 |
| 7th edition AJCC stage | | | | | <0.001 |
| 1 | 742 (52.4%) | 826 (48.8%) | 60 (41.4%) | 328 (45.2%) | |
| 2 | 317 (22.4%) | 283 (16.7%) | 39 (26.9%) | 115 (15.9%) | |
| 3 | 250 (17.7%) | 391 (23.1%) | 26 (17.9%) | 183 (25.2%) | |
| 4 | 89 (6.3%) | 164 (9.7%) | 17 (11.7%) | 81 (11.2%) | |
| Unknown | 17 (1.2%) | 28 (1.7%) | 3 (2.1%) | 18 (2.5%) | |
| Tumor number by imaging studies | | | | | 0.436 |
| Single | 870 (61.5%) | 1032 (61.0%) | 79 (54.5%) | 443 (61.1%) | |
| Multiple | 545 (38.5%) | 660 (39.0%) | 66 (45.5%) | 282 (38.9%) | |
| BCLC stage | | | | | <0.001 |
| 0 | 214 (15.1%) | 220 (13.0%) | 18 (12.4%) | 47 (6.5%) | |
| A | 581 (41.1%) | 592 (35.0%) | 45 (31.0%) | 205 (28.3%) | |
| B | 240 (17.0%) | 344 (20.3%) | 34 (23.4%) | 190 (26.2%) | |
| C | 283 (20.0%) | 421 (24.9) | 36 (24.8) | 215 (29.7%) | |
| D | 69 (4.9%) | 82 (4.8%) | 9 (6.2%) | 45 (6.2%) | |
| Unknown | 28 (2.0%) | 33 (2.0%) | 3 (2.1%) | 23 (3.2%) | |
| BMI (kg/m$^2$) | 24.5 (22.3–27.3) | 24.5 (22.1–27.3) | 24.8 (22.0–27.4) | 25.0 (22.6–28.0) | 0.048 |
| AFP | | | | | 0.001 |
| ≥20 ng/ml | 741 (52.4%) | 902 (53.3%) | 64 (44.1%) | 329 (45.4%) | |
| <20 ng/ml | 674 (47.6%) | 790 (46.7%) | 81 (55.9%) | 396 (54.6%) | |
| Cirrhosis | | | | | <0.001 |
| Yes | 1062 (75.3%) | 1157 (68.4%) | 100 (69.4%) | 444 (61.8%) | |
| No | 348 (24.7%) | 534 (31.6%) | 44 (30.6%) | 275 (38.2%) | |
| Unknown | | | | | |
| Creatinine (mg/dL) | 1.0 (0.8–1.3) | 1.0 (0.8–1.2) | 1.1 (0.9–1.4) | 1.1 (0.8–1.5) | <0.001 |
| Total bilirubin (mg/dL) | 1.1 (0.8–1.6) | 1.0 (0.8–1.6) | 1.3 (0.8–2.3) | 1.0 (0.7–1.5) | <0.001 |
| INR | 1.0 (1.0–1.1) | 1.0 (1.0–1.1) | 1.0 (1.1–1.2) | 1.0 (1.0–1.1) | <0.001 |
| Child Pugh class | | | | | <0.001 |
| A | 1125 (79.5%) | 1398 (82.6%) | 100 (69.0%) | 591 (81.5%) | |
| B | 229 (16.2%) | 212 (12.5%) | 39 (26.9%) | 100 (13.8%) | |
| C | 38 (2.7%) | 62 (3.7%) | 6 (4.1%) | 15 (2.1%) | |
| Unknown | 23 (1.6%) | 20 (1.2%) | 0 | 19 (2.6%) | |
| Treatment | | | | | <0.001 |
| Transplant | 54 (3.8%) | 58 (3.4%) | 5 (3.4%) | 16 (2.2%) | |
| Resection | 398 (28.1%) | 640 (37.8%) | 38 (26.2%) | 243 (33.5%) | |
| Ablation | 414 (29.3%) | 316 (18.7%) | 33 (22.8%) | 133 (18.3%) | |
| Best supportive care | 73 (5.2%) | 87 (5.1%) | 12 (8.3%) | 43 (5.9%) | |
| Chemotherapy | 10 (0.7%) | 35 (2.1%) | 2 (1.4%) | 14 (1.9%) | |
| TAE/TACE | 328 (23.2%) | 318 (18.8%) | 39 (26.9%) | 168 (23.2%) | |
| Target therapy | 95 (6.7%) | 186 (11.0%) | 11 (7.6%) | 77 (10.6%) | |
| Radiation therapy | 43 (3.0%) | 52 (3.1%) | 5 (3.4%) | 31 (4.3%) | |

AFP, alpha-fetoprotein; BMI, body mass index; HCV, hepatitis C virus; HBV, hepatitis B virus; INR, international normalized ratio; AJCC, American Joint Committee on Cancer; BCLC, Barcelona clinic liver cancer; TACE, Transcatheter Arterial Chemoembolization; TAE, Transcatheter Arterial embolization.

**Table 2.** Cause-specific mortality among all patients.

| Group | Cause of Mortality | 1-Year | 3-Year | 5-Year |
|---|---|---|---|---|
| Total, $N = 3976$ | Any cause | 883 (22.2) | 1426 (35.9) | 1593 (40.1) |
| | HCC-related | 639 (72.4) | 987 (69.2) | 1098 (68.9) |
| | Non-HCC-related | 158 (17.9) | 283 (19.8) | 339 (21.3) |
| | Unknown | 86 (9.7) | 156 (10.9) | 156 (9.8) |
| HCV, $n = 1415$ | Any cause | 269 (19.0) | 493 (34.8) | 561 (39.6) |
| | HCC-related | 189 (70.3) | 339 (68.8) | 385 (68.6) |
| | Non-HCC-related | 62 (23.0) | 118 (23.9) | 140 (25.0) |
| | Unknown | 18 (6.7) | 36 (7.3) | 36 (6.4) |
| HBV, $n = 1691$ | Any cause | 384 (22.7) | 577 (34.1) | 644 (38.1) |
| | HCC-related | 295 (76.8) | 424 (73.5) | 472 (73.3) |
| | Non-HCC-related | 42 (10.9) | 72 (12.5) | 91 (14.1) |
| | Unknown | 47 (12.2) | 81 (14.0) | 81 (12.6) |
| Alcohol, $n = 145$ | Any cause | 34 (23.4) | 57 (39.3) | 66 (45.5) |
| | HCC-related | 23 (67.6) | 38 (66.7) | 41 (62.1) |
| | Non-HCC-related | 10 (29.4) | 16 (28.1) | 22 (33.3) |
| | Unknown | 1 (2.9) | 3 (5.3) | 3 (4.5) |
| All negative, $n = 725$ | Any cause | 196 (27.0) | 299 (41.2) | 322 (44.4) |
| | HCC-related | 132 (67.3) | 186 (62.2) | 200 (62.1) |
| | Non-HCC-related | 44 (22.4) | 77 (25.8) | 86 (26.7) |
| | Unknown | 20 (10.2) | 36 (12.0) | 36 (11.2) |

HCC, hepatocellular carcinoma; HCV, hepatitis C virus; HBV, hepatitis B virus.

**Table 3.** Cause-specific mortality among patients who underwent curative treatment.

| Group | Cause of Mortality | 1-Year | 3-Year | 5-Year |
|---|---|---|---|---|
| Total, $N = 2347$ | Any cause | 154 (6.6) | 402 (17.1) | 496 (21.1) |
| | HCC-related | 88 (57.1) | 242 (60.2) | 302 (60.9) |
| | Non-HCC-related | 51 (33.1) | 122 (30.3) | 156 (31.5) |
| | Unknown | 15 (9.7) | 38 (9.5) | 38 (7.7) |
| HCV, $n = 866$ | Any cause | 58 (6.7) | 164 (18.9) | 202 (23.3) |
| | HCC-related | 34 (58.6) | 101 (61.6) | 126 (62.4) |
| | Non-HCC-related | 20 (34.5) | 50 (30.5) | 63 (31.2) |
| | Unknown | 4 (6.9) | 13 (7.9) | 13 (6.4) |
| HBV, $n = 1013$ | Any cause | 61 (6.0) | 144 (14.2) | 184 (18.2) |
| | HCC-related | 39 (63.9) | 94 (65.3) | 122 (66.3) |
| | Non-HCC-related | 15 (24.6) | 35 (24.3) | 47 (25.5) |
| | Unknown | 7 (11.5) | 15 (10.4) | 15 (8.2) |
| Alcohol, $n = 76$ | Any cause | 5 (6.6) | 14 (18.4) | 19 (25.0) |
| | HCC-related | 0 (0.0) | 6 (42.9) | 7 (36.8) |
| | Non-HCC-related | 4 (80.0) | 7 (50.0) | 11 (57.9) |
| | Unknown | 1 (20.0) | 1 (7.1) | 1 (5.3) |
| All negative, $n = 392$ | Any cause | 30 (7.7) | 80 (20.4) | 91 (23.2) |
| | HCC-related | 15 (50.0) | 41 (51.3) | 47 (51.6) |
| | Non-HCC-related | 12 (40.0) | 30 (37.5) | 35 (38.5) |
| | Unknown | 3 (10.0) | 9 (11.3) | 9 (9.9) |

HCC, hepatocellular carcinoma; HCV, hepatitis C virus; HBV, hepatitis B virus.

### *3.4. Cause-Specific Mortality in Patients Who Received Curative Treatments According to CLD Etiology and Stratified by Age*

Figure 3 shows the cumulative mortality risk according to CLD etiology in patients younger than 75 years who underwent curative treatments. HCC-related mortality was the leading cause of death for HCV (Figure 3A), HBV (Figure 3B), and all-negative groups (Figure 3D). By contrast, HCC-related mortality and non-HCC-related mortality were similar in the alcohol-related group (Figure 3C).

Figure 4 shows the cumulative mortality risk according to CLD etiology in patients 75 years and older who underwent curative treatments. HCC-related mortality was the leading cause of death for the HCV group (Figure 4A). However, HCC-related mortality and non-HCC-related mortality rates were similar in the HBV (Figure 4B) and all-negative groups (Figure 4C). Only three patients were included in the alcohol-related group, preventing the analysis of this group.

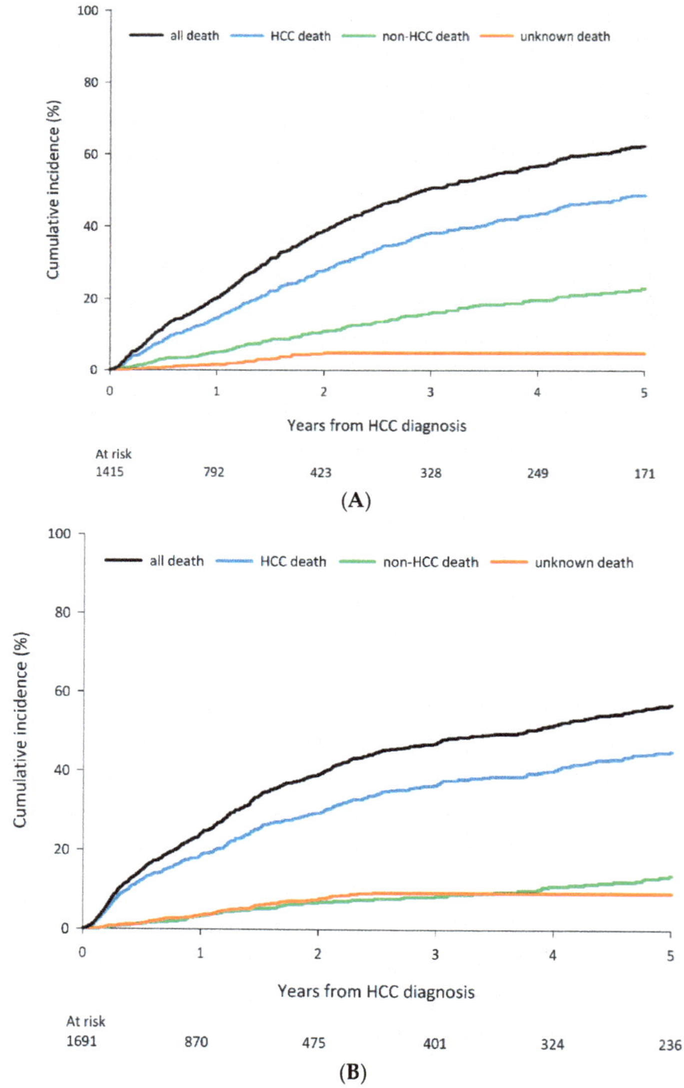


**Figure 1.** *Cont.*

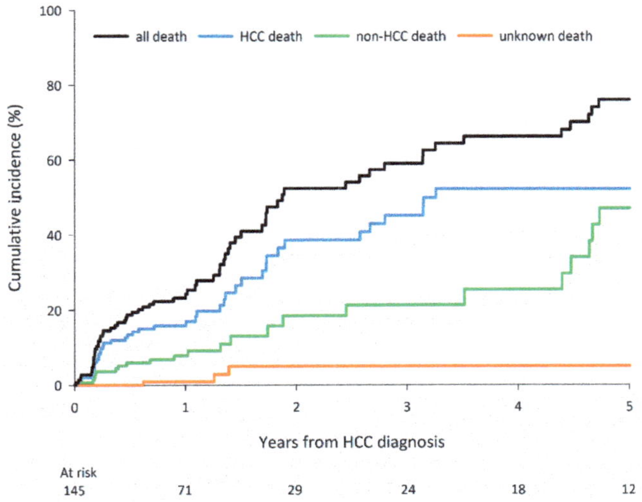


(C)

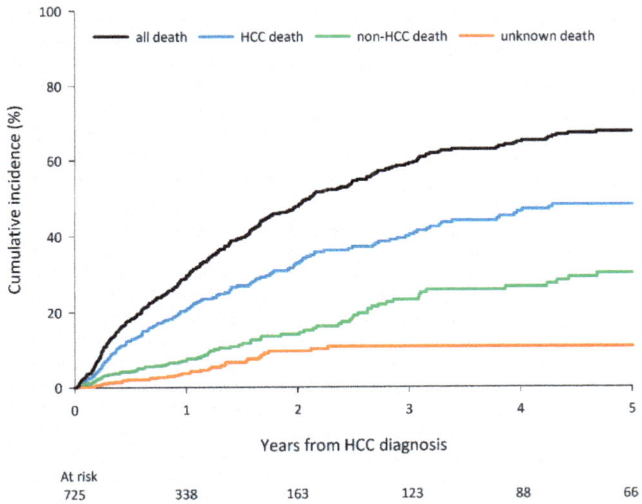


(D)

**Figure 1.** Cumulative mortality risk for hepatocellular carcinoma according to chronic liver disease etiology: (**A**) hepatitis C virus; (**B**) hepatitis B virus; (**C**) alcohol use; and (**D**) all negative (defined as no evidence of virus infection or alcohol use).

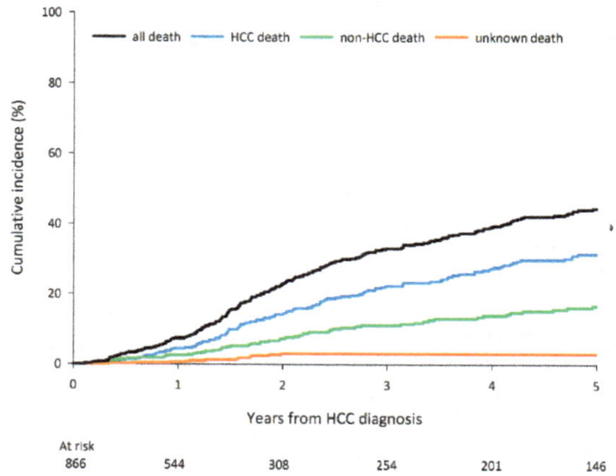


**(A)**

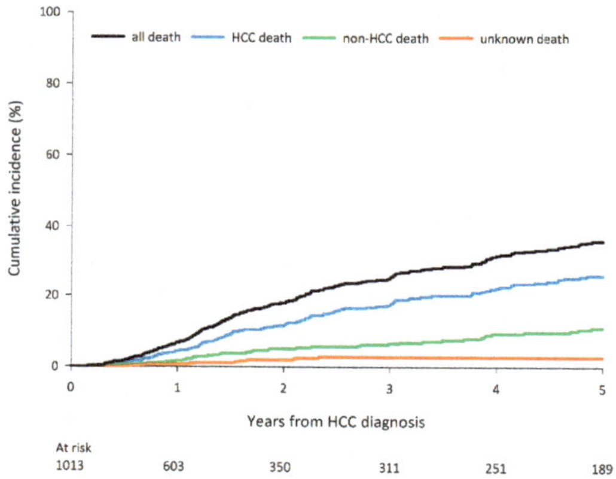


**(B)**

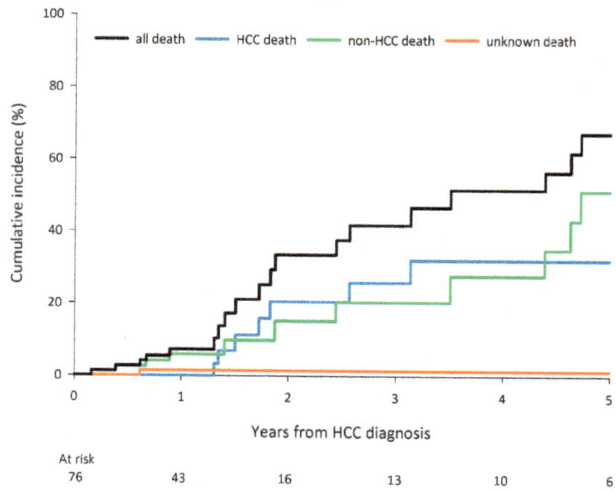


**(C)**

**Figure 2.** *Cont.*

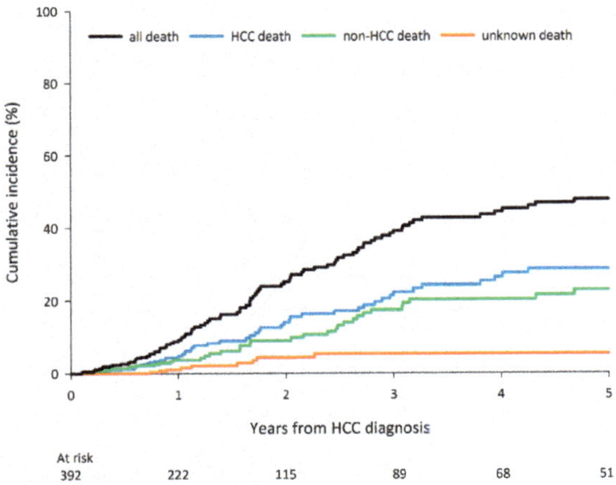


(**D**)

**Figure 2.** Cumulative mortality risk for hepatocellular carcinoma among patients who received curative treatments according to chronic liver disease etiology: (**A**) hepatitis C virus; (**B**) hepatitis B virus; (**C**) alcohol use; and (**D**) all negative (defined as no evidence of virus infection or alcohol use).

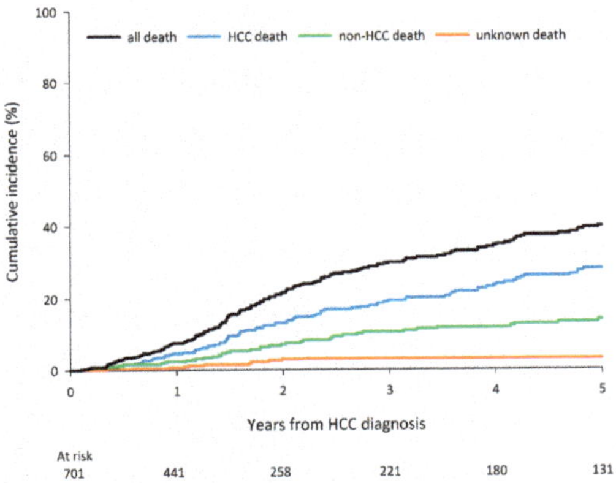


(**A**)

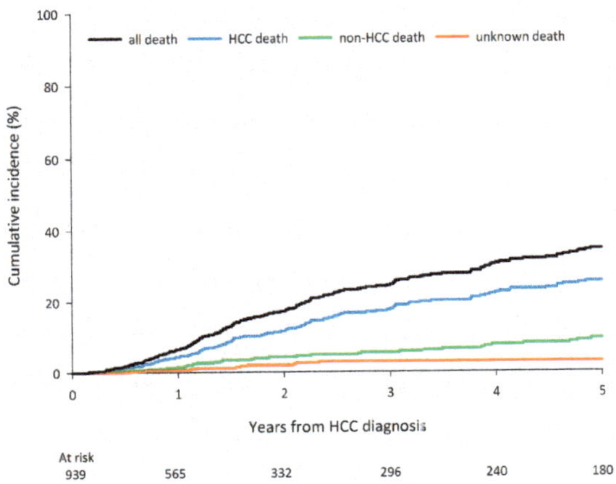


(**B**)

**Figure 3.** *Cont.*

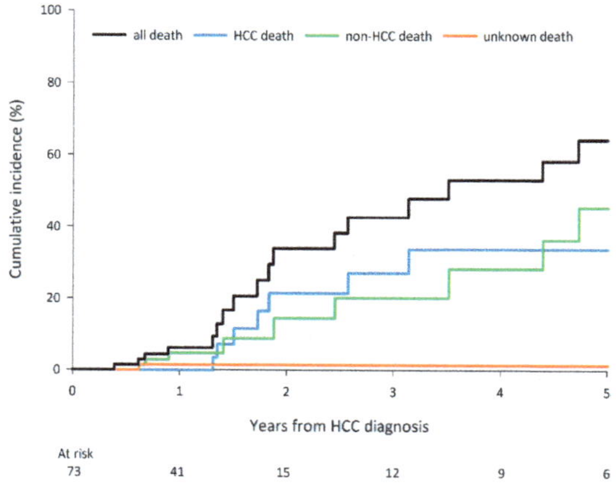


(C)

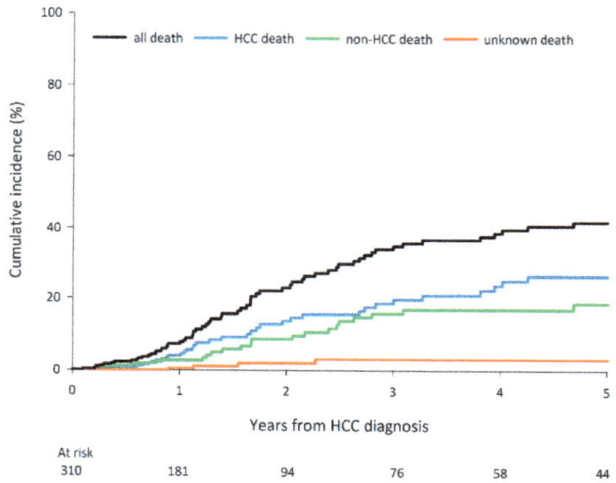


(D)

**Figure 3.** Cumulative mortality risk for hepatocellular carcinoma among patients younger than 75 years who received curative treatments according to chronic liver disease etiology: (**A**) hepatitis C virus; (**B**) hepatitis B virus; (**C**) alcohol use; and (**D**) all negative (defined as no evidence of virus infection or alcohol use).

### *3.5. Cause-Specific Mortality in Patients Who Received Non-Curative Treatments According to CLD Etiology*

HCC-related mortality was the leading cause of death among patients who underwent non-curative treatments, irrespective of CLD etiology (Figure 5A–D).

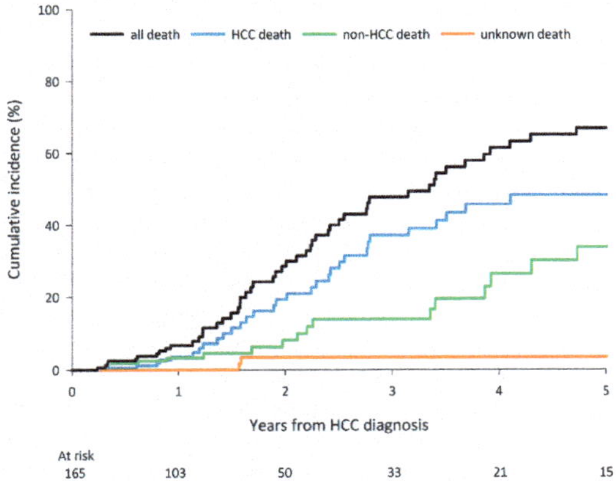


**(A)**

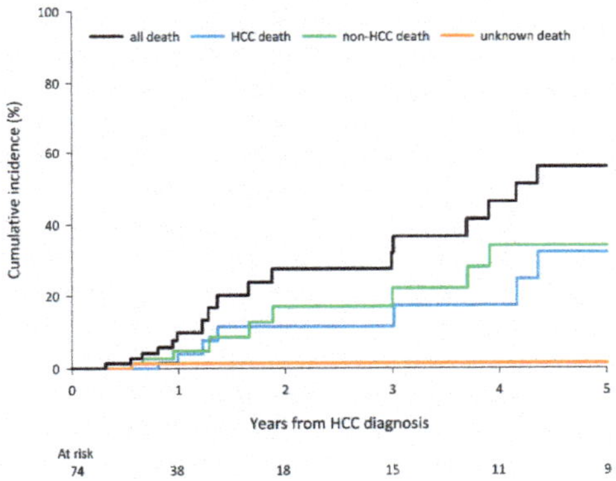


**(B)**

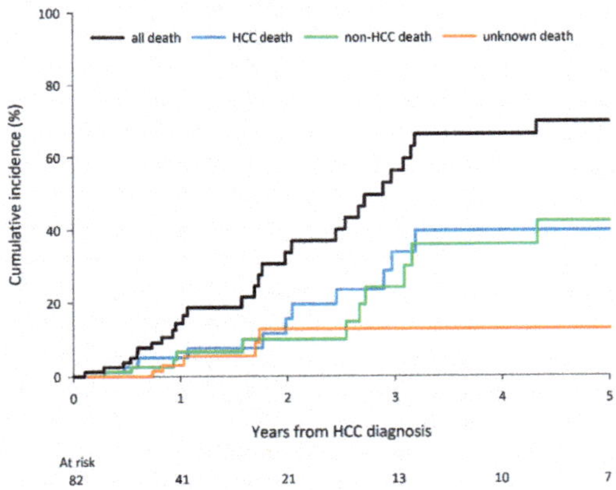


**(C)**

**Figure 4.** Cumulative mortality risk for hepatocellular carcinoma among patients 75 years and older who received curative treatments according to chronic liver disease etiology: (**A**) hepatitis C virus; (**B**) hepatitis B virus; and (**C**) all negative (defined as no evidence of virus infection or alcohol use).

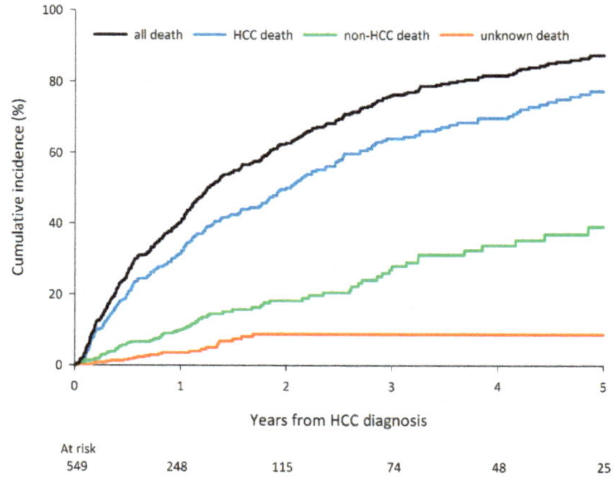


(A)

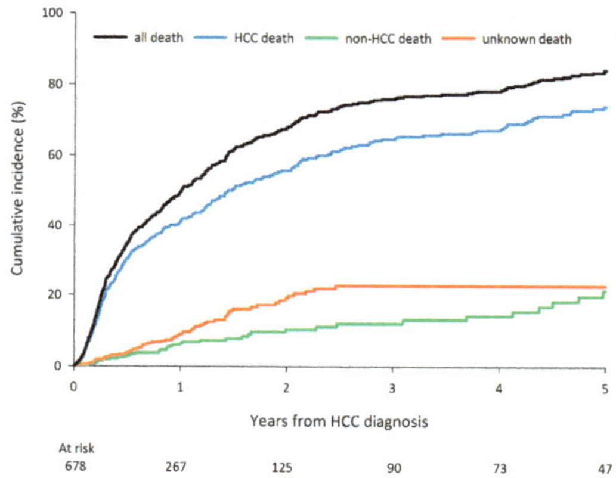


(B)

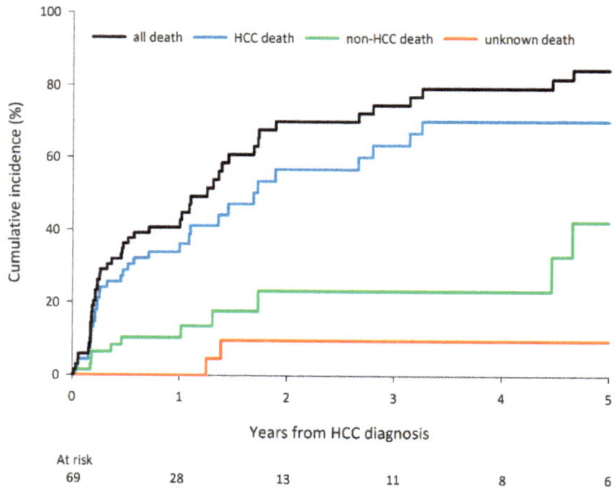


(C)

**Figure 5.** *Cont.*

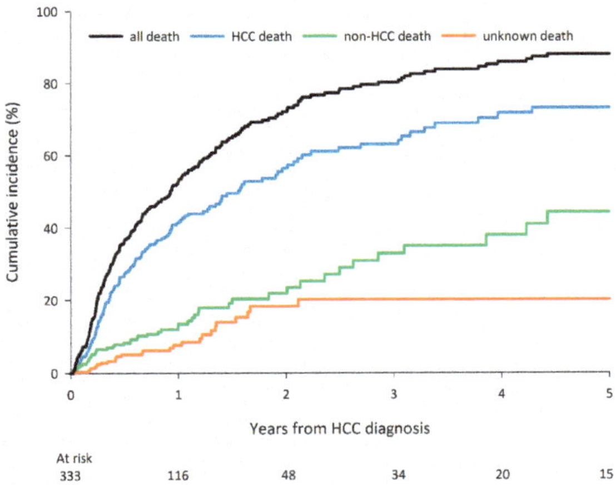


(D)

**Figure 5.** Cumulative mortality risk for hepatocellular carcinoma among patients who received non-curative treatments according to chronic liver disease etiology: (**A**) hepatitis C virus; (**B**) hepatitis B virus; (**C**) alcohol use; and (**D**) all negative (defined as no evidence of virus infection or alcohol use).

## 4. Discussion

In the current study, we found that although most patients with HCC die due to HCC-related causes, non-HCC-related mortality represents a competing event for certain subgroups of patients with HCC, including patients who partake in regular alcohol use who receive curative treatment and HBV and all-negative patients 75 years and older who receive curative treatment.

Tumor stage, the Eastern Cooperative Oncology Group performance status, treatment strategy, and liver disease severity are associated with mortality in patients with HCC according to the BCLC guidelines [7]. A recent study enrolled 10,826 patients with HCC from the Surveillance, Epidemiology, and End Results-Medicare database and found that the receipt of curative treatment was the strongest predictor of survival beyond 5 years among HCC patients [13]. In the current study, we analyzed the cause of death in patients stratified according to the receipt of curative treatment. As expected, HCC-related mortality was the leading cause of death for all patients and in patients who received non-curative treatments, irrespective of CLD etiology. However, among patients who received curative treatments, HCC-related mortality was the leading cause of death in the HCV, HBV, and all-negative groups but not in the alcohol-related group. Alcohol use impacts the outcomes of several diseases and injuries. The largest numbers of deaths attributable to heavy alcohol intake are associated with cardiovascular disease, followed by injuries, cirrhosis, and cancer [14]; the higher risks of these etiologies among alcohol users may explain the higher rate of non-HCC-related mortality observed for patients in the alcohol-related group who underwent curative treatment.

Aging is a prognostic factor for poor outcomes in most chronic diseases, including HCC [15]. The association between aging and poor prognosis in HCC patients could be due to older patients having higher risks of severe comorbidities. The World Health Organization has stated that people 65 years old and older can be defined as elderly in developed countries (https://www.who.int/healthinfo/survey/ageingdefnolder/en, accessed on 6 March 2023). Moreover, the definitions of "elderly" are changing, especially as life expectancies in many developed countries, including Taiwan, now exceed 80 years. Most people in their 60s and early 70s remain active, whereas people usually become frail after passing 75 years of age [16].

In this study, we defined elderly patients as those 75 years or older. Older patients are associated with higher risks of severe comorbidities, which could explain the similar rates of HCC-related and non-HCC-related death among patients 75 years and older who underwent curative treatments in the all-negative and HBV groups. However, the leading cause of death in the HCV group remained HCC. The proportion of cirrhotic patients was larger in the HCV group than in the other groups, and the risk of recurrent HCC has been associated with cirrhosis, suggesting that the risk of recurrent HCC may also be higher in the HCV group than in the other groups, as cirrhosis is a well-known risk factor for HCC recurrence after curative treatment [17].

The results of the current study underscore the importance of assessing and managing underlying comorbidities in patients with HCC, especially among patients who engage in alcohol use who receive curative treatments and all-negative and HBV-positive patients 75 years and older who receive curative treatments.

Compared to groups with other CLD etiologies, the all-negative group was the oldest and presented with the largest tumor size, the smallest proportion of early-stage HCC (i.e., BCLC stages 0 and A), the highest creatinine level, the highest BMI, and the largest proportion of non-cirrhotic livers, which is consistent with the characteristics of NAFLD-related HCC [9,18,19]. Patients with NAFLD-related HCC presented with larger tumors at later disease stages than patients with virus-related HCC [19], which may be due to a low HCC surveillance rate among NAFLD patients. A previous study reported that NAFLD is the leading cause of non-cirrhotic HCC [9]. NAFLD is a multisystem disease affecting extra-hepatic organs and increasing the risks of developing cardiovascular and chronic kidney disease [20], which could explain the characteristics of NAFLD-related HCC patients. The etiologies of non-viral HCC are most commonly alcohol use and NAFLD, although rarer etiologies have been identified, such as primary biliary cirrhosis, primary sclerosing cholangitis, and autoimmune hepatitis [9]. Therefore, we expected the majority of patients in the all-negative group would have NAFLD-related HCC. A recent study reported that among NAFLD-related HCC patients, most mortality (72.2% at 3 years) was attributable to HCC, although, among NAFLD-related HCC patients who underwent curative treatment, non-HCC-related mortality accounted for 40% of all deaths between 3 and 5 years after treatment [5]. In the current study, 62.2% of all death in the all-negative group were attributable to HCC at 3 years, and non-HCC-related mortality accounted for 37.5%–38.5% of all deaths between 3 and 5 years following curative treatment in the all-negative group, which is comparable to the previous study [5].

A strength of the present study was the use of a large cohort of patients with HCC associated with prospectively collected data and limited missing data. However, the study also has several limitations. First, the study lacked a complete list of comorbidities, which are the leading causes of non-HCC-related mortality. Multiple comorbidity indices (e.g., the cirrhosis-related comorbidity score, the National Cancer Institute Comorbidity Index, etc.) have been shown to have value for predicting mortality among HCC patients [5,19]. In the current study, we used older age (75 years or older) as a simple and objective surrogate for severe comorbidities. Furthermore, we did not have data on hepatic steatosis and metabolic risk factors (e.g., hypertension, dyslipidemia, central obesity, and hyperglycemia) [4]. Therefore, we could not define NAFLD in the present study. Third, this study was conducted as a retrospective and monocentric study. Fourth, the case number in the alcohol-related group was limited. Fifth, the etiology of liver disease is not properly defined. We defined the viral etiology just on the basis of anti-HCV antibodies or of HBsAg positivity, without giving information about antiviral therapies that might have caused suppression of HBV replication or HCV clearance. Finally, causes of death were derived from death certificates, which are not enough accurate for the study purposes.

## 5. Conclusions

Although most patients die of HCC-related causes, non-HCC-related death represents a competing event among patients who engage in alcohol use and receive curative treatment and among patients 75 years and older in the HBV and all-negative groups who receive curative treatments. The results of the current study underscore the importance of assessing and managing underlying comorbidities, especially among certain subgroups of patients with HCC.

**Author Contributions:** Conceptualization, Y.-H.Y. and C.-C.W.; Methodology, Y.-H.Y.; Software, C.-Y.L.; Validation, all authors; Formal Analysis, C.-Y.L.; Investigation, Y.-H.Y.; Resources, C.-C.W.; Data Curation, all authors; Writing—Original Draft Preparation, Y.-H.Y.; Writing—Review & Editing, C.-C.W.; Visualization, all authors; Supervision, all authors; Project Administration, Y.-H.Y.; Funding Acquisition, Y.-H.Y. All authors have read and agreed to the published version of the manuscript.

**Funding:** This study was supported by Grant CMRPG8L0181 from the Chang Gung Memorial Hospital-Kaohsiung Medical Center, Taiwan.

**Institutional Review Board Statement:** The Institutional Review Board of Kaohsiung Chang Gung Memorial Hospital approved this study (Reference number: 202201189B0, 8 August 2022) and waived the need for informed consent due to the retrospective and observational nature of the study design.

**Informed Consent Statement:** Patient consent was waived due to the retrospective and observational nature of the study design.

**Data Availability Statement:** The data presented in this study are available in this article.

**Acknowledgments:** The authors thank Cancer Center, Kaohsiung Chang Gung Memorial Hospital for the provision of HCC registry data. The authors thank Chih-Yun Lin and Nien-Tzu Hsu and the Biostatistics Center, Kaohsiung Chang Gung Memorial Hospital for statistics work. This study was supported by Grant CMRPG8L0181 from the Chang Gung Memorial Hospital-Kaohsiung Medical Center, Taiwan.

**Conflicts of Interest:** The authors have no conflict of interest to disclose for all authors.

## References

1. Bray, F.; Ferlay, J.; Soerjomataram, I.; Siegel, R.L.; Torre, L.A.; Jemal, A. Global cancer statistics 2018: GLOBOCAN estimates of incidence and mortality worldwide for 36 cancers in 185 countries. *CA Cancer J. Clin.* **2018**, *68*, 394–424. [CrossRef] [PubMed]
2. Estes, C.; Razavi, H.; Loomba, R.; Younossi, Z.; Sanyal, A.J. Modeling the epidemic of nonalcoholic fatty liver disease demonstrates an exponential increase in burden of disease. *Hepatology* **2018**, *67*, 123–133. [CrossRef] [PubMed]
3. Li, J.; Zou, B.; Yeo, Y.H.; Feng, Y.; Xie, X.; Lee, D.H.; Fujii, H.; Wu, Y.; Kam, L.Y.; Ji, F.; et al. Prevalence, incidence, and outcome of non-alcoholic fatty liver disease in Asia, 1999-2019: A systematic review and meta-analysis. *Lancet Gastroenterol. Hepatol.* **2019**, *4*, 389–398. [CrossRef] [PubMed]
4. Chalasani, N.; Younossi, Z.; Lavine, J.E.; Charlton, M.; Cusi, K.; Rinella, M.; Harrison, S.A.; Brunt, E.M.; Sanyal, A.J. The diagnosis and management of nonalcoholic fatty liver disease: Practice guidance from the American Association for the Study of Liver Diseases. *Hepatology* **2018**, *67*, 328–357. [CrossRef] [PubMed]
5. Chu, J.; Cholankeril, G.; Yu, X.; Rana, A.; Natarajan, Y.; El-Serag, H.B.; Kramer, J.; Kanwal, F. Clinical Course and Outcomes of Patients with Nonalcoholic Fatty Liver Disease-Related Hepatocellular Cancer (NAFLD-HCC). *Dig. Dis. Sci.* **2022**. [CrossRef] [PubMed]
6. American Joint Committee on Cancer. *American Joint Committee on Cancer Staging Manual*, 7th ed.; Edge, S.B., Byrd, D.R., Compton, C.C., Fritz, A.G., Greene, F.L., Trotti, A., III, Eds.; Springer: New York, NY, USA, 2010; p. 175.
7. Llovet, J.M.; Bru, C.; Bruix, J. Prognosis of hepatocellular carcinoma: The BCLC staging classification. *Semin. Liver Dis.* **1999**, *19*, 329–338. [CrossRef] [PubMed]
8. Pugh, R.N.; Murray-Lyon, I.M.; Dawson, J.L.; Pietroni, M.C.; Williams, R. Transection of the oesophagus for bleeding oesophageal varices. *Br. J. Surg.* **1973**, *60*, 646–649. [CrossRef] [PubMed]
9. Gawrieh, S.; Dakhoul, L.; Miller, E.; Scanga, A.; deLemos, A.; Kettler, C.; Burney, H.; Liu, H.; Abu-Sbeih, H.; Chalasani, N.; et al. Characteristics, aetiologies and trends of hepatocellular carcinoma in patients without cirrhosis: A United States multicentre study. *Aliment. Pharmacol. Ther.* **2019**, *50*, 809–821. [CrossRef] [PubMed]
10. Everhart, J.E.; Wright, E.C.; Goodman, Z.D.; Dienstag, J.L.; Hoefs, J.C.; Kleiner, D.E.; Ghany, M.G.; Mills, A.S.; Nash, S.R.; Govindarajan, S.; et al. Prognostic value of Ishak fibrosis stage: Findings from the hepatitis C antiviral long-term treatment against cirrhosis trial. *Hepatology* **2010**, *51*, 585–594. [CrossRef] [PubMed]
11. Shiha, G.; Ibrahim, A.; Helmy, A.; Sarin, S.K.; Omata, M.; Kumar, A.; Bernstien, D.; Maruyama, H.; Saraswat, V.; Chawla, Y.; et al. Asian-Pacific Association for the Study of the Liver (APASL) consensus guidelines on invasive and non-invasive assessment of hepatic fibrosis: A 2016 update. *Hepatol. Int.* **2017**, *11*, 1–30. [CrossRef] [PubMed]
12. Mazzaferro, V.; Regalia, E.; Doci, R.; Andreola, S.; Pulvirenti, A.; Bozzetti, F.; Montalto, F.; Ammatuna, M.; Morabito, A.; Gennari, L. Liver transplantation for the treatment of small hepatocellular carcinomas in patients with cirrhosis. *N. Engl. J. Med.* **1996**, *334*, 693–699. [CrossRef] [PubMed]

13. Zhang, X.; El-Serag, H.B.; Thrift, A.P. Predictors of five-year survival among patients with hepatocellular carcinoma in the United States: An analysis of SEER-Medicare. *Cancer Causes Control* **2021**, *32*, 317–325. [CrossRef] [PubMed]
14. European Association for the Study of the Liver. EASL Clinical Practice Guidelines: Management of alcohol-related liver disease. *J. Hepatol.* **2018**, *69*, 154–181. [CrossRef] [PubMed]
15. Kaibori, M.; Yoshii, K.; Yokota, I.; Hasegawa, K.; Nagashima, F.; Kubo, S.; Kon, M.; Izumi, N.; Kadoya, M.; Liver Cancer Study Group of Japan; et al. Impact of Advanced Age on Survival in Patients Undergoing Resection of Hepatocellular Carcinoma: Report of a Japanese Nationwide Survey. *Ann. Surg.* **2019**, *269*, 692–699. [CrossRef] [PubMed]
16. Torpy, J.M.; Lynm, C.; Glass, R.M. Frailty in older adults. *JAMA* **2006**, *296*, 2280. [CrossRef] [PubMed]
17. Liu, Y.W.; Yong, C.C.; Lin, C.C.; Wang, C.C.; Chen, C.L.; Cheng, Y.F.; Wang, J.H.; Yen, Y.H. Six months as a cutoff time point to define early recurrence after liver resection of hepatocellular carcinoma based on post-recurrence survival. *Updates Surg.* **2021**, *73*, 399–409. [CrossRef] [PubMed]
18. Hester, C.A.; Rich, N.E.; Singal, A.G.; Yopp, A.C. Comparative Analysis of Nonalcoholic Steatohepatitis- Versus Viral Hepatitis- and Alcohol-Related Liver Disease-Related Hepatocellular Carcinoma. *J. Natl. Compr. Cancer Netw.* **2019**, *17*, 322–329. [CrossRef] [PubMed]
19. Karim, M.A.; Singal, A.G.; Kum, H.C.; Lee, Y.T.; Park, S.; Rich, N.E.; Noureddin, M.; Yang, J.D. Clinical Characteristics and Outcomes of Nonalcoholic Fatty Liver Disease-Associated Hepatocellular Carcinoma in the United States. *Clin. Gastroenterol. Hepatol.* **2023**, *21*, 670–680.e18. [CrossRef] [PubMed]
20. Byrne, C.D.; Targher, G. NAFLD: A multisystem disease. *J. Hepatol.* **2015**, *62*, S47–S64. [CrossRef] [PubMed]

cancers

MDPI

*Review*

# New Challenges in the Management of Cholangiocarcinoma: The Role of Liver Transplantation, Locoregional Therapies, and Systemic Therapy

**Ezequiel Mauro [1,2], Joana Ferrer-Fàbrega [1,2,3,4], Tamara Sauri [1,4,5], Alexandre Soler [1,6], Amparo Cobo [1,7], Marta Burrel [1,8], Gemma Iserte [1,2,9] and Alejandro Forner [1,2,4,9,*]**

1 Barcelona Clinic Liver Cancer (BCLC) Group, IDIBAPS, 08036 Barcelona, Spain
2 Centro de Investigación Biomédica en Red de Enfermedades Hepáticas y Digestivas (CIBERehd), Av. Monforte de Lemos, 3-5. Pabellón 11, Planta 0, 28029 Madrid, Spain
3 Hepatobiliopancreatic Surgery and Liver and Pancreatic Transplantation Unit, Department of Surgery, ICMDM, Hospital Clinic Barcelona, 08036 Barcelona, Spain
4 Faculty of Medicine, University of Barcelona, C/ de Casanova, 143, 08036 Barcelona, Spain
5 Medical Oncology Department, ICMHO, Hospital Clinic Barcelona, 08036 Barcelona, Spain
6 Radiology Department, CDI, Hospital Clinic Barcelona, 08036 Barcelona, Spain
7 Nuclear Medicine Department, CDI, Hospital Clinic Barcelona, 08036 Barcelona, Spain
8 Department of Interventional Radiology, CDI, Hospital Clinic Barcelona, 08036 Barcelona, Spain
9 Liver Unit, Liver Oncology Unit, ICMDM, Hospital Clinic Barcelona, 08036 Barcelona, Spain
* Correspondence: aforner@clinic.cat; Tel.: +34-93-2279803

**Simple Summary:** Cholangiocarcinoma (CCA) is a highly lethal neoplasia, which incidence has steadily increased in the last years. Although surgical resection remains the cornerstone treatment for CCA, complete resection is only achieved in one third of patients, and the risk of recurrence exceeds 60%, which impacts the long-term outcome. In this context, the use of other therapeutic strategies such as liver transplantation in selected candidates, locoregional treatments or new chemotherapy schemes based on immunotherapy or targeted therapies may contribute in improving the overall survival of patients with CCA. The development of new treatment strategies forces us to redouble collaborative efforts to conduct prospective, high-quality studies that shed light on their use and applicability. The purpose of this review is discussing the actual controversies and future perspectives in the management of CCA.

**Abstract:** Cholangiocarcinoma (CCA) is a neoplasm with high mortality that represents 15% of all primary liver tumors. Its worldwide incidence is on the rise, and despite important advances in the knowledge of molecular mechanisms, diagnosis, and treatment, overall survival has not substantially improved in the last decade. Surgical resection remains the cornerstone therapy for CCA. Unfortunately, complete resection is only possible in less than 15–35% of cases, with a risk of recurrence greater than 60%. Liver transplantation (LT) has been postulated as an effective therapeutic strategy in those intrahepatic CCA (iCCA) smaller than 3 cm. However, the low rate of early diagnosis in non-resectable patients justifies the low applicability in clinical practice. The evidence regarding LT in locally advanced iCCA is scarce and based on small, retrospective, and, in most cases, single-center case series. In this setting, the response to neoadjuvant chemotherapy could be useful in identifying a subgroup of patients with biologically less aggressive tumors in whom LT may be successful. The results of LT in pCCA are promising, however, we need a very careful selection of patients and adequate experience in the transplant center. Locoregional therapies may be relevant in unresectable, liver-only CCA. In iCCA smaller than 2 cm, particularly those arising in patients with advanced chronic liver disease in whom resection or LT may not be feasible, thermal ablation may become a reliable alternative. The greatest advances in the management of CCA occur in systemic treatment. Immunotherapy associated with chemotherapy has emerged as the gold standard in the first-line treatment. Likewise, the most encouraging results have been obtained with targeted therapies, where the use of personalized treatments has shown high rates of objective and durable tumor response, with clear signs of survival benefit. In conclusion, the future of CCA treatment seems to be marked

**Citation:** Mauro, E.; Ferrer-Fàbrega, J.; Sauri, T.; Soler, A.; Cobo, A.; Burrel, M.; Iserte, G.; Forner, A. New Challenges in the Management of Cholangiocarcinoma: The Role of Liver Transplantation, Locoregional Therapies, and Systemic Therapy. *Cancers* **2023**, *15*, 1244. https://doi.org/10.3390/cancers15041244

Academic Editor: Georgios Germanidis

Received: 17 December 2022
Revised: 3 February 2023
Accepted: 11 February 2023
Published: 15 February 2023

by the development of new treatment strategies but high-quality, prospective studies that shed light on their use and applicability are mandatory.

**Keywords:** cholangiocarcinoma; liver transplant; locoregional therapies; systemic treatment

## 1. Introduction

Cholangiocarcinoma (CCA) is a highly lethal neoplasia comprising approximately 15% of all primary liver tumors. Its incidence is increasing worldwide, and despite significant advancements in the knowledge of molecular mechanisms, diagnosis, management, and survival have not substantially improved in the past decade [1,2]. These cancers are heterogeneous and are best classified according to the primary anatomic origin as intrahepatic CCA (iCCA), when located proximally to the second-order bile ducts within the liver parenchyma, perihilar CCA (pCCA), arising between the second-order bile ducts and the insertion of the cystic duct into the common bile duct, and distal CCA (dCCA), located in the common bile duct below the cystic duct insertion [1].

Several risk factors have been linked to CCA, most of them associated with chronic inflammation of the biliary epithelium and bile stasis. Some recognized risk factors such as obesity, metabolic syndrome, or high alcohol consumption have increased globally over recent decades, which could be contributing to increasing CCA incidence. However, the majority of CCA cases do not present any identifiable risk factors.

In most cases, the diagnosis of CCA is established when the disease is already at advanced stages, which highly compromises access to effective treatment, resulting in a dismal outcome [3,4]. Therefore, prevention and early diagnosis remain the cornerstone for improving the survival of this devasting disease.

Surgical resection is the best therapy for CCA [5,6]. Unfortunately, complete resection is possible in less than 15–35% of cases [4,7], and even in those patients in whom complete tumor removal is achieved, the risk of recurrence is greater than 60% [8,9]. Another radical option is liver transplantation (LT), but its use in CCA is controversial due to the high risk of recurrence and the lower survival benefit compared to other LT indications as well as the limited number of donors [5]. In addition, locoregional therapies have been recently proposed as a reliable treatment alternative for those patients with liver-only, unresectable CCA, but the low level of evidence supporting their efficacy impedes making any robust recommendation [10]. Finally, systemic therapy has rapidly evolved in the last years, and the irruption of targeted therapies and immunotherapy has changed the treatment approach. In this review, we will discuss the controversies in the therapeutical management of CCA.

## 2. Liver Transplantation in Cholangiocarcinoma

Theoretically, LT is an excellent treatment option for primary liver tumors due to (1) its capacity to completely remove the tumor (and the undetected liver micrometastasis), particularly when major resection is needed due to tumor extension/location, (2) its ability to eliminate the underlying chronic liver disease, and (3) the possibility of maximizing the survival benefit compared to alternative therapies. Regrettably, the major problem for the wide application of LT is the shortage of donors, since the number of candidates largely exceeds the available livers to be implanted. Due to the scarce number of donors, it is the usual policy to exclude from transplantation any patients with an expected suboptimal post-transplant survival (with a cutoff arbitrarily established to be at least greater than 50–60% at 5 years) [11,12]. In addition, the LT allocation policy should be adjusted to guarantee real access to LT, preventing drop-out due to tumor progression and withdrawal from the waiting list in an excessive proportion of patients, but at the same time, requiring an observation period which would allow for identifying biologically aggressive tumors that would be associated with a higher risk of unacceptable recurrence.

### *2.1. Liver Transplantation in iCCA*

iCCA was considered a contraindication for LT in most centers worldwide due to very poor initial results, i.e., a reported 2-year survival of around 30% [13,14]. These unacceptable outcomes were directly related to a high prevalence of microvascular invasion and poor tumor differentiation, particularly in patients with an unresectable or locally advanced tumor [15]. However, more recent retrospective studies have demonstrated encouraging results in terms of overall survival (OS) when a thorough selection of the population is performed. The relevance of selection based on tumor burden was demonstrated for the first time in an international multicenter study that included 48 patients who underwent LT and had been diagnosed with incidental iCCA in the explant. A total of 15 patients had "very early" iCCA (single tumor $\leq$ 2 cm) and 33 patients had "advanced" iCCA (single tumor > 2 cm or multifocal disease). After a median follow-up of 35 months, the 1-year, 3-year, and 5-year cumulative risks of recurrence were, 7%, 18%, and 18%, respectively, in the very early iCCA group vs. 30%, 47%, and 61% in the advanced iCCA group ($p < 0.01$). The 1-year, 3-year, and 5-year overall survival rates were 93%, 84%, and 65% in the very early iCCA group vs. 79%, 50%, and 45% in the advanced iCCA group ($p < 0.02$) [16]. Microvascular invasion and poor differentiation were associated with tumor recurrence in the multivariate analysis. Patients in the advanced iCCA group were divided into an intermediate stage ($n = 6$; single tumors 2.1–3 cm, not poorly differentiated) and an advanced stage ($n = 27$; all other tumors). The 1-year, 3-year, and 5-year overall survival rates were, 82%, 61%, and 61%, respectively, in the intermediate stage vs. 55%, 47%, and 42% in the advanced stage ($p < 0.03$) [16]. Ziogas et al. performed a meta-analysis of 18 studies, finding that the 5-year OS for very early iCCA was 71%, versus only 48% for advanced iCCA (single tumor > 2 cm or multiple tumors) [17]. Disappointingly, all these data come from retrospective studies, including mostly LT patients in whom iCCA was an incidental finding. Accordingly, prospective studies with well-defined inclusion and exclusion criteria and a predefined post-LT imaging follow-up are fervently needed. Furthermore, the outcome of LT should be analyzed according to the intention to treat the principle instead of only analyzing those patients finally transplanted. A multicenter, observational study (NCT02878473) aimed to prospectively evaluate the effectiveness of LT for very early iCCA is still ongoing.

### *2.2. Locally Advanced, Unresectable iCCA and Liver Transplantation: Role of Neoadjuvant Therapy*

iCCA patients with a stable, liver-limited disease on neoadjuvant therapy may have a favorable disease biology, with long-term survival after LT. Disappointingly, the evidence is scarce, and most studies are single-center, retrospective, and include a low number of patients and heterogeneous population in terms of tumor stage and neoadjuvant approach. The most relevant study, which is also performed within a study protocol at MD Anderson in Texas, was recently published by McMillan et al. [18]. Patients with CCA liver-only with the absence of vascular or lymph node involvement were considered for LT. Neoadjuvant therapy consisted of the first-line use of gemcitabine plus cisplatin (GemCis), and disease stability was required by radiological evaluation for at least six months. The treatments performed prior to LT in addition to GemCis were heterogeneous, including various types of locoregional therapies, liver resection, and targeted therapies such as inhibitors of isocitrate dehydrogenase 1 (IDH-1), fibroblast growth factor receptor (FGFR), and poly ADP-ribose polymerase (PARP). Over 11 years, 65 patients were evaluated, of whom 28 were denied for listing. Five patients were excluded after being eligible for resection due to tumor regression after neoadjuvant therapy. At the end of the follow-up, 18 out of 32 patients underwent LT and 14 did not (7 were still on the waiting list and 7 were because of tumor progression or death while on the WL). The time range between diagnosis and inclusion in the WL was very wide (74–1054 days), which could implicitly suggest a selection of patients linked to tumor biology. Intent-to-treat (ITT) survival analysis at 1, 3, and 5 years was 90%, 61%, and 49%, respectively. The recurrence-free survival (RFS) at

3 years was 52%, and 7 out of 18 transplant patients (39%) developed tumor recurrence (4 during the first year post-LT). In another retrospective analysis recently published by Ito T. et al. [19], 30 patients who underwent neoadjuvant therapy were finally transplanted. In this series, the neoadjuvant protocol was less defined, and the 5-year overall survival was 49%.

In conclusion, data on LT for unresectable iCCA are scarce and the level of evidence is low (Table 1). The response to neoadjuvant chemotherapy, especially in the context of new personalized target therapies, could identify patients with biologically less aggressive tumors in whom LT may offer long-term results. Prospective studies, with well-characterized and homogeneous populations in terms of baseline tumor burden, with clear-cut multimodal neoadjuvant protocols (locoregional and/or systemic treatment), and relevant outcomes (OS by ITT, RFS, and cancer-related survival), are mandatory to establish the role of LT in patients with locally advanced iCCA.

**Table 1.** Outcomes of LT in patients with unresectable iCCA.

| Study | Design | Number of Patients | Neoadjuvant Therapy | Overall Survival |
|---|---|---|---|---|
| Sapisochin et al. (2016) [16] | Retrospective cohort multicenter. Incidental iCCA by pathological study. | 48 | | 1 year: 93%<br>3 years: 84%<br>5 years: 65% |
| McMillan et al. (2022) [18] | Prospective single-center case series. | 18 | Neoadjuvant chemotherapy (GemCis) and disease stability were required by radiological evaluation for at least six months. Treatments in addition to GemCis were heterogeneous: locoregional therapies, liver resection, and TT (IDH-1, FGFR, and PARP). | 1 year: 100%<br>3 years: 71%<br>5 years: 57% |
| Ito et al. (2022) [19] | Retrospective, single-center, case series. | 30 | Neoadjuvant chemotherapy and or locoregional therapies. | 1 year: 80%<br>3 years: 63%<br>5 years: 49% |

Abbreviations: GemCis, Gemcitabine and Cisplatin; iCCA, intrahepatic cholangiocarcinoma; IDH, isocitrate dehydrogenase; FGFR, fibroblast growth factor receptor; and TT, targeted therapy.

### *2.3. Perihiliar CCA (pCCA) and Liver Transplantation*

The prognosis of pCCA is marked by frequent late diagnosis, which precludes the use of potentially curative treatments. The pCCA could develop in the context of primary sclerosing cholangitis (PSC) or de novo, in the absence of liver disease [1,4].

The indication of LT after neoadjuvant chemoradiation has been established as a therapeutic option with acceptable long-term OS results (>50% at 5 years) in carefully selected patients with early-stage unresectable pCCA and patients with pCCA associated with PSC [20]. Among different neoadjuvant chemoradiation strategies, the Mayo Clinic protocol based on strict criteria for diagnosis and patient selection, aggressive neoadjuvant chemoradiation, and surgical staging before transplantation, has been positioned as the best strategy. Tan et al. recently reported the Mayo Clinic results from 1993 to 2018. A total of 349 patients were initially assessed, but 277 (79%) underwent staging work-up, and in 60% ($n$ = 211) LT was performed. According to ITT analysis (from the start of neoadjuvant therapy, including patients who did not undergo LT), the survival rates at 1, 5, and 10 years were 80%, 51%, and 46%, respectively, while the survival rates in those finally transplanted were 91%, 69%, and 62% [20]. The outcome in pCCA associated with PSC was significantly better than those arising within the healthy liver (5-year survival of 60% vs. 39%, respectively) [20,21], and the center experience positively impacts LT outcomes in patients with pCCA [22]. Other centers have reported poorer survival rates, inferior to 40% at 5 years [23–27], partially explained by the lower proportion of PSC-related pCCA and the lesser center experience. The use of living donor liver transplantation (LDLT) has been postulated as an interesting option since it does not directly impact the principles of allocation justice related to cadaveric LT. Regrettably, robust data comparing both options are scarce, but LDLT for de novo pCCA seems to be associated with higher disease recurrence and slightly worse OS [20].

Patient selection and neoadjuvant chemoradiation protocol are critical, being the Mayo Clinic proposal the most frequently evaluated. The Mayo Clinic protocol requires having a lesion with a radial diameter (perpendicular to the duct) $\leq$3 cm and without extension below the cystic duct. Endoscopic ultrasound-guided aspiration of regional hepatic lymph nodes is routinely performed prior to neoadjuvant therapy, and the presence of lymph node metastases is an exclusion criterion. Diagnostic biopsies, whether transgastric endoscopic or percutaneous transhepatic, are usually dismissed given the potential risk of seeding metastases in the peritoneum [21,28]. Vascular encapsulation and tumor extension along the duct, although not considered contraindications for neoadjuvant treatment, are conditions of a worse prognosis in terms of response [29]. The neoadjuvant therapy includes external beam radiation plus concomitant 5-fluorouracil and brachytherapy, followed by maintenance capecitabine until LT. After completion of neoadjuvant chemoradiation, patients should undergo staging laparoscopy prior to LT, comprising a complete examination of the abdominal cavity, routine biopsy of the regional lymph nodes, and biopsy of any other suspicious lesions. The timing of the staging surgery is a subject of debate, especially in PSC patients with advanced liver disease and complications associated with the presence of portal hypertension. However, the drop-out probability in patients with pCCA-PSC is usually lower than in patients with de novo pCCA (15% vs. 28%) [30]. In addition, the treatment of clinical complications during the neoadjuvant protocol is the cornerstone for LT outcome. Sarcopenia is frequently present in pCCA patients. It has shown an impact on the post-LT outcome and should be actively treated with nutritional support, which in some cases may include nasogastric or nasojejunal tube insertion for feeding [31]. Recurrent cholangitis because of biliary obstruction, biliary stenting, radiation-induced ductal injury, and/or underlying PSC has led to the development of antimicrobial resistance which increases the risk of intra-abdominal infections after LT. Finally, radiotherapy during the neoadjuvant protocol increases the risk of hepatic artery thrombosis and radiation-induced fibrosis [32,33].

Finally, recent publications found that LT in patients who meet the criteria for liver resection had better survival than that observed after resection, even when sub-analysis stratified by PSC was performed [34,35]. However, the survival benefit decreases when analyzed according to the ITT principle, which clearly calls into question the possibility of using donor livers for resectable pCCA [20,36]. An ongoing randomized, ITT multi-center trial in France TRANSPHIL (NCT02232932) comparing neoadjuvant chemoradiation and LT vs. upfront surgical resection will clarify this controversial topic.

## 3. Locoregional Therapies

Locoregional therapies (LRT) applicable to the treatment of iCCA include thermal ablation (TA), chemoembolization (TACE), radioembolization (TARE), chemotherapy hepatic arterial infusion (HAI), and external beam radiotherapy (EBRT). They have been postulated as an alternative to systemic therapy in those patients with liver-only, unresectable CCA, and in some cases, as neoadjuvant therapy prior to surgical resection/transplantation or as rescue therapy after systemic treatment failure.

### 3.1. Thermal Ablation

Although surgical resection is potentially the best therapeutic option in iCCA, some patients have liver-only disease categorized as unresectable due to localization and/or the presence of underlying cirrhosis with clinically significant portal hypertension (CSPH) and/or liver dysfunction that precludes liver resection. In this setting, TA might be a safe and effective treatment option. Unfortunately, the evidence to recommend these treatments is scarce and relies on retrospective studies, and in most cases, they are single-center and include a limited number of patients [37,38]. TA (with radiofrequency and microwave ablation being the most common techniques) is able to achieve local control of small iCCA lesions, focal and unresectable iCCA (either due to inadequate localization or CSPH), although its efficacy in terms of tumor response and survival outcomes are inferior to those obtained in the field of hepatocellular carcinoma (HCC) [39,40]. In a recent systematic review and pooled analysis, which included 15 studies with a total of 645 patients, mostly retrospective and monocentric, TA showed a pooled complete response rate of 93.9%, and a mean OS of 30.2 months (95% CI: 21.8–38.6) [10]. Noticeably, in more than 50% of the cases analyzed, TA was applied after post-resection recurrence, and 30% of the patients had liver cirrhosis. In patients with underlying cirrhosis, TA is a reliable treatment approach, and in those cases with single iCCA < 2 cm, TA obtained a similar survival to that obtained in HCC and comparable to that reported after surgical resection [41]. In summary, TA is feasible, safe, and may be a good alternative in selected unresectable patients.

### *3.2. Transarterial Chemoembolization, Radioembolization, and Chemotherapy Hepatic Arterial Infusion*

Transarterial chemoembolization (TACE) has been evaluated in retrospective studies including a small number of patients with a very heterogeneous clinical profile, which makes it difficult to establish any recommendation. In a recent systematic review and pooled analysis, which included 22 studies with a total of 1145 patients, mostly retrospective and monocentric, TACE was associated with a pooled response rate of 23.4%, a mean PFS of 15 months, and OS of 15.9 months [10]. Moreover, the addition of TACE using irinotecan-loaded drug-eluting microspheres to GemCis vs. GemCis alone was tested in a small ($n = 48$) randomized controlled trial, showing a significant improvement in downsizing to resection (25% vs. 8%, $p < 0.005$) and an improved PFS and OS (33.7 vs. 12.6 months, $p = 0.048$), with an adequate safety profile [42]. Confirmatory, larger studies are needed before supporting this treatment combination.

Transarterial radioembolization (TARE) has also been evaluated in iCCA. However, most studies were single-center, including a small number of patients with heterogeneous inclusion criteria [43–50]. In a recent systematic review and meta-analysis including a total of 921 patients from 21 studies, TARE showed an overall disease-control rate of 82.3%, a median PFS and OS of 7.8 months and 12.7 months, respectively, and in 11% of the cases, patients were downstaged to being surgically resectable [51]. However, the high heterogeneity hampers data reproducibility. In addition, TARE was evaluated as associated with GemCis in a phase 2 study that included 41 naïve patients. The response rate and disease control rate according to RECIST were 41% and 98%, respectively. After a median follow-up of 36 months, the median PFS was 14 months (95% CI, 8–17 months), the median OS was 22 months (95% CI, 14–52 months), and nine patients (22%) could be downstaged to surgical resection, achieving R0 surgical resection in eight cases [52]. In addition, TARE was compared with systemic therapy (GemCis) in patients with locally advanced iCCA with liver-only disease (SIRCCA phase 3 trial, NCT02807181). Disappointingly, the study was prematurely interrupted because of low recruitment and its preliminary results have not yet been reported.

Finally, the efficacy and safety of chemotherapy hepatic arterial infusion (HAI) have been evaluated in small series including a heterogeneous population. In a recent systematic review, 331 patients from 16 studies were identified, many of them with bilobar involvement (75%), multifocality (66%), and a high percentage of macrovascular invasion (~40%). HAI showed a pooled response rate of 41.3%, and a PFS and OS of 10 months and 21.3 months, respectively [10].

### *3.3. External Beam Radiotherapy*

The role of EBRT in the treatment of iCCA is also uncertain. Only one study was reported as prospective [53], but most patients were already treated with chemotherapy. In a systematic review and meta-analysis that included 541 patients, the 2-year local control rate, PFS, and OS were 69.1%, 15.6 months, and 18.9 months, respectively [10]. In addition, a recent registry study from the United States of America reported that the use of high-dose/ablative radiotherapy (>85 Gy) was associated with an improved outcome compared to conventional doses [54], which suggests that ERBT may be effective in some selected cases.

In a summary, locoregional procedures are safe and have shown some signals of efficacy and might be considered an alternative to systemic therapy in selected unresectable patients with iCCA.

## 4. Systemic Therapy

Treatment for patients with locally advanced and/or metastatic disease relies on the use of systemic therapy. Until recently, the only option that demonstrated survival benefit at the first line in advanced patients ineligible for surgical or locoregional options was the combination of GemCis. The pivotal ABC-02 study demonstrated a median overall survival (mOS) of 11.7 months (95% CI, 9.5 to 14.3) for GemCis compared with 8.1 months (95% CI, 7.1 to 8.7) for gemcitabine alone (HR, 0.64; 95% CI, 0.52 to 0.80; $p < 0.001$) [55]. The best results of GemCis were obtained in patients with iCCA, liver-only involvement [56]. Other combinations such as FOLFIRINOX (5-fluorouracil, oxaliplatin, and irinotecan) did not improve PFS compared with GemCis [57], and gemcitabine plus S-1 (an oral combination of the 5-fluorouracil prodrug tegafur with gimeracil and oteracil) demonstrated non-inferiority to gemcitabine and cisplatin in a randomized phase 3 trial [58]. Very recently, an open-label, non-inferiority phase 3 trial showed that capecitabine plus oxaliplatin (XELOX) was not inferior to gemcitabine plus oxaliplatin (GEMOX) in terms of 6-months PFS rate (46.7% [95% CI 41.5–51.8] vs. 44.6% [95% CI 39.7–49.3]) and thus XELOX may be an alternative to GEMOX in first-line setting [59]. Another promising combination is NUC-1031, a phosphoramidate transformation of gemcitabine, combined with cisplatin, which showed a promising objective response rate (ORR) (63.6% in the efficacy-evaluable population) in the ABC-08 phase Ib study [60]. This combination is currently under evaluation in a phase 3 trial in which patients are being randomized to NUC-1031 combined with cisplatin or GemCis (Nutide-121 trial; NCT04163900). Finally, results are also awaited from a phase II/III, multicenter, randomized, placebo-controlled study of GemCis with or without Bintrafusp Alfa (M7824) as the first-line treatment of BTC (NCT 04066491).

Immunotherapy has significantly expanded the scope of cancer treatment in recent years and its role in CCA is extensively revised elsewhere [61]. The most promising results have come from the phase 3 randomized, double-blind, placebo-controlled TOPAZ-01 trial, which demonstrated that durvalumab (PDL-1 antibody) plus GemCis significantly improved survival compared to GemCis plus placebo in advanced biliary tract cancer (BTC) [62]. Patients in the experimental arm received 1500 mg of durvalumab every 3 weeks with GemCis for up to eight cycles, followed by durvalumab 1500 mg every 4 weeks, until disease progression or unacceptable toxicity.

The mOS was 12.8 months vs. 11.5 months (HR, 0.80; 95%CI, 0.66–0.97; $p = 0.021$), the median PFS (mPFS) was 7.2 months vs. 5.7 months (HR, 0.75; 95%CI, 0.64–0.89; $p = 0.001$), and ORR was 26.7% vs. 18.7% in durvalumab plus GemCis vs. GemCis plus placebo, respectively. The combination of durvalumab and GemCis was well tolerated, and grade 3 or 4 treatment-related adverse event rates were similar between both groups (62.7% with durvalumab vs. 64.9% with placebo). Updated OS and safety data after 6.5 months of additional follow-up have been recently reported in ESMO 2022. When compared with placebo, the addition of durvalumab to GemCis resulted in a longer median OS of 12.9 (11.6–14.1) months versus 11.3 (10.1–12.5) months, respectively, HR 0.76 (95% CI 0.64–0.91), together with manageable safety [63]. Based on these positive results, durvalumab plus GemCis has become the new standard first-line systemic therapy option for advanced BTC.

More recently, the addition of tremelimumab (two-dosing regimen) to durvalumab plus GemCis vs. Gemcitabine +/-cisplatin did not add any substantial benefit in a multicenter, German phase 2 trial [64]. Finally, another promising combination is Pembrolizumab plus GemCis (NCT 04003636-KEYNOTE 966), whose results have recently been announced as positive in 1st line treatment.

In the second line setting, the addition of liposomal irinotecan to 5-fluorouracil and leucovorin significantly improved PFS (3.9 months [95% CI, 2.6–4.7] vs. 1.5 months [1.2–1.9]) compared with 5-fluorouracil and leucovorin; HR = 0.38 [0.26–0.54], $p < 0.0001$) in a multicenter, open-label, randomized, phase 2b (NIFTY) study in patients with BTC who progressed on first-line GemCis [65,66]. More recently, the ABC-06 study demonstrated the benefit of leucovorin, 5- fluorouracil, and oxaliplatin (FOLFOX) in the second-line setting compared with the placebo [67]; mOS was significantly longer in the FOLFOX group than in the placebo group (6.2 vs. 5.3 months; HR 0.69, 95%CI 0.50–0.97, $p$ = 0·031). Based on these findings, FOLFOX should become standard-of-care chemotherapy in the second-line treatment for advanced BTC and the reference regimen for further clinical trials. Multiple case reports exist of patients with mismatch repair deficient (dMMR)/microsatellite instability-high (MSI-H) CCA treated with the PD-1 antibody pembrolizumab with promising ORR [68,69]. The phase 2 KEYNOTE-158 basket trial of pembrolizumab for previously treated MSI-H cancer included 22 patients with CCA, for whom complete response (CR) and partial response (PR) were achieved in 3 (13.6%) and 6 (27.3%) patients, respectively, with a median {range} of the duration of response of 30.6 {6.2 to 40.5}, and a mOS of 19.4 months (95% CI 6.5—not reached) [70,71]. Based on those results, the FDA approved Pembrolizumab as a second-line therapy for patients with MSI-H cancers which have progressed through prior therapy.

*Targeted Therapy in CCA*

Relevant progress has been accomplished in the last years on the molecular biology of CCA, and related target therapies. Molecularly, CCA is a highly heterogeneous disease, with genomic differences between intra and extrahepatic CCA [1,2,72,73], with IDH and FGFR pathway alterations predominantly found in intrahepatic cases, along with RAS and ARID1A [74]. The European Society for Medical Oncology (ESMO) scale for the clinical actionability of molecular targets (ESCAT) [75] is present in around 50–60% of cases, including kinases (FGFR1/2/3, PIK3CA, ALK, EGFR, ERBB2, BRAF, and KRAS), other oncogenes (IDH1/2) and tumor-suppressor genes (BRCA1/2) [76]. The main results of targeted therapy in CCA are summarized in Table 2.

The recent discovery of FGFR2 fusions in patients with iCCA has been rapidly translated into a promising therapeutic target [77,78]. Pemigatinib, a selective inhibitor of FGFR1, 2, and 3, was tested in a multicenter, open-label, single-arm, multicohort, phase 2 study (FIGHT-202) in patients with previously treated, locally advanced or metastatic CCA with FGFR2 fusions or rearrangements [79]. A total of 146 patients were enrolled and 107 had FGFR2 fusions or rearrangements. The ORR was achieved in 38 (35.5%) patients (3 patients with complete responses) and the mOS was 21.8 (14.8-not estimated) months. The most frequent adverse events were hyperphosphatemia, alopecia, diarrhea, fatigue, and dysgeusia, but in only 9 % of cases was the treatment interrupted because of toxicity. Another selective, ATP-competitive inhibitor of FGFR, Infigratinib, was tested in a single arm, open-label phase 2 trial including 108 with FGFR2 fusions or rearrangements who were previously treated with at least one gemcitabine-containing regimen. The ORR was 23.1% (one complete response and 24 partial responses) with a safety profile similar to pemigatinib [80] In addition, Futibatinib, a highly selective, irreversible FGFR1–4 inhibitor, showed in an open-label phase 2 (FOENIX-CCA2) trial including 103 patients with an FGFR2 fusion/rearrangement, an ORR of 41.7% and 74% of responses which lasted ≥6 months. After a median follow-up of 25 months, the mature mOS was 20.0 months, with a 12-month OS rate of 73.1% and a similar safety profile [81]. Based on these results, the FDA and EMA granted accelerated approval to pemigatinib, infigratinib, and futibatinib for the treatment of adults with previously treated, unresectable locally advanced or metastatic CCA with FGFR2 fusion or other rearrangements. All these three agents are being currently tested by a large, phase 3 RCT in a first-line setting against GemCis (Pemigatinib: FIGHT 302, NCT: NCT03656536; Futibatinib: FOENIX-CCA3, NCT 04093362; infigratinib: NCT03773302).

**Table 2.** Main trials of targeted therapy in CCA.

| Agent | Trial id and/or Name | Mechanism or Pathway | Phase | Study Population | Arms | Outcomes |
|---|---|---|---|---|---|---|
| | | | **IDH mutations** | | | |
| Ivosidenib | NCT02989857ClarIDHy trial | IDH-1 inhibitor (decreases oncometabolite 2-HG) | 3 | Previously treated, advanced, IDH1-mutant CCA. | Ivosidenib vs. Placebo | mPFS (months): 2.7 (95% CI, 1.6–4.2) vs. 1.4 (1.4–1.6); HR: 0.37; (95% CI, 0.25–0.54) $p < 0.0001$ mOS (months): 10.3 (95% CI, 7.8–12.4) vs. 7.5 (95% CI, 4.8–11.1) |
| | | | **FGFR alterations** | | | |
| Pemigatinib | NCT02924376 FIGHT-202 | FGFR 1, 2, and 3 reversible inhibitors; FGFR fusions or rearrangements | 2 | Advanced, previously treated CCA with and without FGFR2 fusions/rearrangements/ alterations. | FGFR2 rearrangements or fusion CCA Other FGF/FGFR alterations No FGF/FGFR alterations | ORR (%): 37 (95% CI, 27.9 –46.9) mOS (months): 17.5 (95% CI, 14.4–22.9) vs. 6.7 (95% CI, 2.1–10.5) vs. 4.0 (95% CI, 2.0–4.6) |
| Infigratinib (BGJ398) | NCT02150967PROOF-201 | ATP-competitive FGFR 1, 2, and 3 tyrosine kinase reversible inhibitor | 2 | Locally advanced or metastatic CCA with FGFR2 fusions or rearrangements, previously treated with at least one gemcitabine-containing regimen. | Single arm | ORR (%): 23.1 (95% CI, 15.6–32.2) |
| Futibatinib (TAS-120) | NCT02052778FOENIX-CCA2 | Highly selective, irreversible pan-FGFR antagonist | 2 | Advanced, previously treated iCCA with FGFR2 fusions/other rearrangements. | Single arm | ORR (%): 41.7 mPFS (months): 9 mOS (months): 21.7 |
| Erdafitinib | NCT02699606LUC2001 | Pan-FGFR kinase inhibitor | 2 | Patients previously treated, aCCA with FGFR alterations. | Single arm | ORR (%): 50.0 |
| | | | **HER2 alterations** | | | |
| Pertuzumab and trastuzumab | NCT02091141MyPathway | Monoclonal ab targeting HER2 domain II; monoclonal ab binds to domain IV of HER2 | 2 | Previously treated, advanced BTC withHER2 amplification, overexpression, or both. | Single arm | ORR (%): 23 (95% CI, 11–39) |
| Neratinib | NCT01953926(SUMMIT trial) | Pan-HER irreversible TKI, with clinical activity against HER2 | 2 | Previously treated, advanced BTC harboring HER2 somatic mutations. | Single arm | ORR (%): 12 (95% CI, 3–31) mPFS (months): 1.8 (95% CI, 1.1–3.7) |
| | | | **BRAF V600E mutation** | | | |
| Dabrafenib and trametinib | NCT02034110ROAR trial | B-type Raf proto-oncogene, tyrosine kinase in the MAPK pathway | 2 | Previously treated, advanced BTC withBRAF V600E mutation. | Single arm | ORR (%): 47 (95% CI, 31 to 62) |

Abbreviations: aCCA, advanced cholangiocarcinoma; BTC, biliary tract cancer; CCA, cholangiocarcinoma; HER2, human epidermal growth factor receptor 2; iCCA, intrahepatic cholangiocar-cinoma; IDH, isocitrate dehydrogenase; FGFR, fibroblast growth factor receptor; ORR, overall re-sponse rate; mOS, median overall survival; mPFS, median progression-free survival; and TKI, ty-rosine kinase inhibitor.

In addition, IDH1 mutations occur in approximately 15% of patients with iCCA [1]. Ivosidenib (AG-120), is a targeted inhibitor of mutated IDH1 and its efficacy has been shown in a multicenter, randomized, double-blind, placebo-controlled, phase 3 study (CLarIDHy) including patients with previously treated, IDH1-mutant CCA. A total of 185 patients were randomly assigned (2:1) to oral ivosidenib ($n$ = 124) or a matched placebo ($n$ = 61). The placebo to ivosidenib crossover was allowed after radiological progression. PFS was significantly improved with ivosidenib compared with the placebo (2.7 vs. 1.4 months; HR 0.37, 95% CI 0.25–0.54; $p < 0{\cdot}0001$) [82]. Median OS was 10.3 months with ivosidenib vs. 7.5 months with placebo (HR 0.79, 95% CI 0.56–1.12; $p = 0.09$), but when adjusted for crossover, mOS with placebo was 5.1 months (HR 0.49, 95% CI 0.34–0.70; $p < 0.001$) [83]. Based on those results, ivosidenib was recently approved by the FDA for chemotherapy-refractory, IDH1 mutated CCA.

HER2 overexpression or amplification, which is present in 15% of all cases of biliary tract cancer, has been identified as a druggable molecular target. The safety and efficacy of anti-HER2 therapy have been tested in prospective phase I/II studies in first-line associated with GemCis [84] and in second-line in combination with mFOLFOX [85], showing promising activity with acceptable toxicity, warranting further investigation. Finally, the ROAR basket trial evaluated dabrafenib and trametinib in 43 patients with previously treated, $BRAF^{V600E}$-mutated BTC. The independently evaluated ORR was 47% (95% CI 31–62), the mPFS was 9 months (95% CI 5–10), and the mOS was 14 months (95% CI 10–33) [86]. The FDA granted swift approval for the use of dabrafenib and trametinib for patients carrying the $BRAF^{V600E}$ mutation and cancer progression after systemic therapy.

In light of this and other accumulating evidence that advanced BTCs are good candidates for molecular triage, ESMO recently recommended the routine use of next-generation sequencing to be performed in all CCA patients. Therefore, molecular analysis, preferably using whole-gene sequence platforms, should be carried out before or during first-line therapy to evaluate options for second and higher lines of treatments as early as possible in advanced disease [75,87].

## 5. Future Perspectives

Treatment of CCA is rapidly evolving. Figure 1 summarizes the grade of recommendation and level of evidence of the available treatment options in CCA. Despite the fact that CCA has been considered a contraindication for LT, the competitive results in selected patient populations after the application of neoadjuvant therapy, which may allow for the evaluation of tumor biology, forces us to reconsider the potential role that LT may have in these patients. In the future, tumor biology assessment will be a critical determinant for patients' outcomes in the setting of LT, and tools such as liquid biopsy or the integration of genomic and radiological information may help to improve patient selection. Further prospective studies with strict inclusion criteria and homogeneous and clear-cut multimodal neoadjuvant protocols are mandatory to establish the role of LT in patients with CCA.

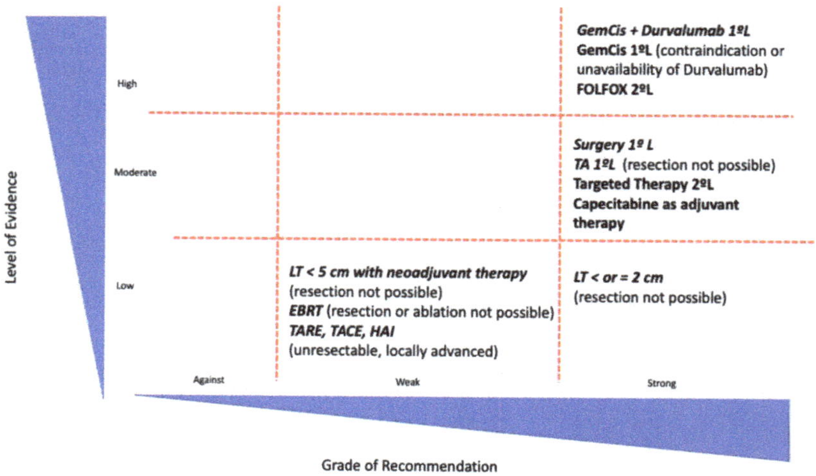


**Figure 1.** Representation of current treatments in CCA according to levels of evidence and strength of recommendation. TA: Thermal ablation; GemCis: Gemcitabine and Cisplatin; LT: Liver transplantation; EBRT: external beam radiotherapy; TACE: Transarterial chemoembolization; TARE: Transarterial radioembolization; and HAI: chemotherapy hepatic arterial infusion.

In addition, locoregional therapies may have a role in unresectable iCCA. In those patients with single tumors smaller than 2 cm, particularly arising in patients with advanced chronic liver disease in whom resection may not be feasible, thermal ablation may become a reliable alternative. Intra-arterial procedures, particularly TARE, are under evaluation in patients with unresectable, liver-only disease as an alternative or in combination with systemic therapy.

Undoubtedly, the greatest advances in the management of CCA have occurred in systemic treatment. Immunotherapy has emerged as an effective treatment associated with chemotherapy, and ongoing trials are evaluating these agents in this orphan disease. The analysis of immune biomarkers will be critical for the selection of patients with a greater benefit from immunotherapy. However, we currently do not have these biomarkers to optimize patient selection. The most hopeful results have been obtained with targeted therapies. In CCA, actionable molecular alterations are found in nearly 50% of cases and several agents directed to those molecular disruptions have shown promising results, with high rates of ORR and durable tumor response and signals of survival benefit. The CCA is a good candidate for molecular triage. The need for adequate tumor tissue for molecular profiling and intra-patient tumor heterogeneity are often limitations when tissue is obtained by percutaneous biopsy or cytology. In this setting, a liquid biopsy may overcome this obstacle and, hopefully, will allow for the monitoring of clonal evolution during selective therapeutic pressure and the detection of residual molecular disease. Cell-free DNA (cfDNA) analysis is an attractive approach as it can provide genomic information when tissue cannot be obtained or is insufficient in quantity or quality, can better capture intra-patient tumor heterogeneity, and can even facilitate the study of the evolution and tumor resistance, with potential predictive and prognostic capacity. These potential benefits might overcome the limitations of the use of tissue for NGS analysis, where, in more than 25% of cases, the isolated DNA does not qualify for a reliable molecular analysis.

## 6. Conclusions

The development of new treatment strategies represents a challenge in the therapeutic approach of CCA. The precision medicine based on molecular profiling is a reality and it will guide the systemic treatment. In addition, the applicability of locoregional therapies or liver transplantation based on an adequate patients' selection will be crucial for improving the outcome. For all these strategies, the multidisciplinary and collaborative management will be essential and further studies are needed to confirm their benefits.

**Author Contributions:** Conceptualization, A.F. and E.M.; methodology A.F., J.F.-F., T.S., A.S., A.C., M.B. and G.I.; writing—original draft preparation, E.M., A.F., J.F.-F., T.S., A.S., A.C., M.B. and G.I.; writing—review and editing, A.F. and E.M.; supervision, A.F. All authors have read and agreed to the published version of the manuscript.

**Funding:** This research received no external funding.

**Acknowledgments:** E. Mauro: Andrew K. Burroughs Short-Term Training Fellowship 2021 from EILF-EASL. G. Iserte Grant: "Beca de enfermería de la AEEH" from Asociación Española para el Estudio del Hígado. Grant "Ajut per la iniciació a la recerca" from Societat Catalana de Digestologia. SCD_INICIACRECE22_02. A. Forner: Grant from Instituto de Salud Carlos III (PI18/00542). CIBERehd is funded by the Instituto de Salud Carlos III. Some of the authors of this article are members of the European Reference Network (ERN) RARE-LIVER. Some of the authors of this article are members of the European Network for the Study of Cholangiocarcinoma (ENS-CCA) and participate in the initiative COST Action EURO-CHOLANGIO-NET granted by the COST Association (CA18122) and European Commission Horizon 2020 program (ESCALON project #825510).

**Conflicts of Interest:** E. Mauro: Received speaker fees from Roche and Sirtex, and travel funding from MSD. J. Ferrer-Fàbrega: Received lecture fees from Bayer and Astrazeneca and consultancy fees from Astrazeneca. T. Sauri: None. A. Soler: None. A. Cobo: None. M. Burrel: None. G. Iserte: Received travel expenses from Bayer. A. Forner: Received lecture fees from Gilead, Boston Scientific, and MSD, as well as consultancy fees from Bayer, AstraZeneca, Roche, SIRTEX, AB Exact Science, and Guerbert.

## References


1. Banales, J.M.; Marin, J.J.G.; Lamarca, A.; Rodrigues, P.M.; Khan, S.A.; Roberts, L.R.; Cardinale, V.; Carpino, G.; Andersen, J.B.; Braconi, C.; et al. Cholangiocarcinoma 2020: The next Horizon in Mechanisms and Management. *Nat. Rev. Gastroenterol. Hepatol.* **2020**, *17*, 557–588. [CrossRef] [PubMed]
2. Brindley, P.; Bachini, M.; Ilyas, S.; Khan, S.A.; Loukas, A.; Sirica, A.E.; Teh, B.; Wongkham, S.; Gores, G.J. Cholangiocarcinoma. *Nat. Rev. Dis. Primers* **2021**, *7*, 65. [CrossRef] [PubMed]
3. Forner, A.; Vidili, G.; Rengo, M.; Bujanda, L.; Ponz-Sarvisé, M.; Lamarca, A. Clinical Presentation, Diagnosis and Staging of Cholangiocarcinoma. *Liver Int.* **2019**, *39*, 98–107. [CrossRef] [PubMed]
4. Izquierdo-Sanchez, L.; Lamarca, A.; la Casta, A.; Buettner, S.; Utpatel, K.; Klümpen, H.-J.; Adeva, J.; Vogel, A.; Lleo, A.; Fabris, L.; et al. Cholangiocarcinoma Landscape in Europe: Diagnostic, Prognostic and Therapeutic Insights from the ENSCCA Registry. *J. Hepatol.* **2022**, *76*, 1109–1121. [CrossRef] [PubMed]
5. Mazzaferro, V.; Gorgen, A.; Roayaie, S.; Droz dit Busset, M.; Sapisochin, G. Liver Resection and Transplantation for Intrahepatic Cholangiocarcinoma. *J. Hepatol.* **2020**, *72*, 364–377. [CrossRef] [PubMed]
6. Molina, V.; Ferrer-Fábrega, J.; Sampson-Dávila, J.; Díaz, A.; Ayuso, C.; Forner, A.; Fondevila, C.; García-Valdecasas, J.C.; Bruix, J.; Fuster, J. Intention-to-Treat Curative Liver Resection in Patients with "Very Early" Intrahepatic Cholangiocarcinoma. *Langenbecks Arch. Surg.* **2020**, *405*, 967–975. [CrossRef]
7. Amini, N.; Ejaz, A.; Spolverato, G.; Kim, Y.; Herman, J.M.; Pawlik, T.M. Temporal Trends in Liver-Directed Therapy of Patients with Intrahepatic Cholangiocarcinoma in the United States: A Population-Based Analysis. *J. Surg. Oncol.* **2014**, *110*, 163–170. [CrossRef]
8. Primrose, J.N.; Fox, R.P.; Palmer, D.H.; Malik, H.Z.; Prasad, R.; Mirza, D.; Anthony, A.; Corrie, P.; Falk, S.; Finch-Jones, M.; et al. Capecitabine Compared with Observation in Resected Biliary Tract Cancer (BILCAP): A Randomised, Controlled, Multicentre, Phase 3 Study. *Lancet Oncol.* **2019**, *20*, 663–673. [CrossRef]
9. Edeline, J.; Hirano, S.; Bertaut, A.; Konishi, M.; Benabdelghani, M.; Uesaka, K.; Watelet, J.; Ohtsuka, M.; Hammel, P.; Kaneoka, Y.; et al. Individual Patient Data Meta-Analysis of Adjuvant Gemcitabine-Based Chemotherapy for Biliary Tract Cancer: Combined Analysis of the BCAT and PRODIGE-12 Studies. *Eur. J. Cancer* **2022**, *164*, 80–87. [CrossRef]
10. Edeline, J.; Lamarca, A.; McNamara, M.G.; Jacobs, T.; Hubner, R.A.; Palmer, D.; Groot Koerkamp, B.; Johnson, P.; Guiu, B.; Valle, J.W. Locoregional Therapies in Patients with Intrahepatic Cholangiocarcinoma: A Systematic Review and Pooled Analysis. *Cancer Treat. Rev.* **2021**, *99*, 102258. [CrossRef]
11. Volk, M.L.; Vijan, S.; Marrero, J.A. A Novel Model Measuring the Harm of Transplanting Hepatocellular Carcinoma Exceeding Milan Criteria. *Am. J. Transpl.* **2008**, *8*, 839–846. [CrossRef] [PubMed]
12. Clavien, P.A.; Lesurtel, M.; Bossuyt, P.M.M.; Gores, G.J.; Langer, B.; Perrier, A. Recommendations for Liver Transplantation for Hepatocellular Carcinoma: An International Consensus Conference Report. *Lancet Oncol.* **2012**, *13*, e11. [CrossRef] [PubMed]
13. Becker, N.S.; Rodriguez, J.A.; Barshes, N.R.; O'Mahony, C.A.; Goss, J.A.; Aloia, T.A. Outcomes Analysis for 280 Patients with Cholangiocarcinoma Treated with Liver Transplantation over an 18-Year Period. *J. Gastrointest. Surg.* **2008**, *12*, 117–122. [CrossRef] [PubMed]
14. Goldstein, R.M.; Stone, M.; Tillery, G.W.; Senzer, N.; Levy, M.; Husberg, B.S.; Gonwa, T.; Klintmalm, G. Is Liver Transplantation Indicated for Cholangiocarcinoma? *Am. J. Surg.* **1993**, *166*, 768–772. [CrossRef]
15. Meyer, C.G.; Penn, I.; James, L. Liver Transplantation for Cholangiocarcinoma: Results in 207 Patients. *Transplantation* **2000**, *69*, 1633–1637. [CrossRef]
16. Sapisochin, G.; Facciuto, M.; Rubbia-Brandt, L.; Marti, J.; Mehta, N.; Yao, F.Y.; Vibert, E.; Cherqui, D.; Grant, D.R.; Hernandez-Alejandro, R.; et al. Liver Transplantation for "Very Early" Intrahepatic Cholangiocarcinoma: International Retrospective Study Supporting a Prospective Assessment. *Hepatology* **2016**, *64*, 1178–1188. [CrossRef]
17. Ziogas, I.A.; Giannis, D.; Economopoulos, K.P.; Hayat, M.H.; Montenovo, M.I.; Matsuoka, L.K.; Alexopoulos, S.P. Liver Transplantation for Intrahepatic Cholangiocarcinoma: A Meta-Analysis and Meta-Regression of Survival Rates. *Transplantation* **2021**, *105*, 2263–2271. [CrossRef]
18. McMillan, R.R.; Javle, M.; Kodali, S.; Saharia, A.; Mobley, C.; Heyne, K.; Hobeika, M.J.; Lunsford, K.E.; Victor, D.W.; Shetty, A.; et al. Survival Following Liver Transplantation for Locally Advanced, Unresectable Intrahepatic Cholangiocarcinoma. *Am. J. Transpl.* **2022**, *22*, 823–832. [CrossRef]
19. Ito, T.; Butler, J.R.; Noguchi, D.; Ha, M.; Aziz, A.; Agopian, V.G.; DiNorcia, J.; Yersiz, H.; Farmer, D.G.; Busuttil, R.W.; et al. A 3-Decade, Single-Center Experience of Liver Transplantation for Cholangiocarcinoma: Impact of Era, Tumor Size, Location, and Neoadjuvant Therapy. *Liver Transpl.* **2022**, *28*, 386–396. [CrossRef]
20. Tan, E.K.; Taner, T.; Heimbach, J.K.; Gores, G.J.; Rosen, C.B. Liver Transplantation for Peri-Hilar Cholangiocarcinoma. *J. Gastrointest. Surg.* **2020**, *24*, 2679–2685. [CrossRef]
21. Bowlus, C.L.; Arrivé, L.; Bergquist, A.; Deneau, M.; Forman, L.; Ilyas, S.I.; Lunsford, K.E.; Martinez, M.; Sapisochin, G.; Shroff, R.; et al. AASLD Practice Guidance on Primary Sclerosing Cholangitis and Cholangiocarcinoma. *Hepatology* **2022**, *77*, 659–702. [CrossRef] [PubMed]
22. Kitajima, T.; Hibi, T.; Moonka, D.; Sapisochin, G.; Abouljoud, M.S.; Nagai, S. Center Experience Affects Liver Transplant Outcomes in Patients with Hilar Cholangiocarcinoma. *Ann. Surg. Oncol.* **2020**, *27*, 5209–5221. [CrossRef] [PubMed]

23. Mantel, H.T.J.; Westerkamp, A.C.; Adam, R.; Bennet, W.F.; Seehofer, D.; Settmacher, U.; Sánchez-Bueno, F.; Prous, J.F.; Boleslawski, E.; Friman, S.; et al. Strict Selection Alone of Patients Undergoing Liver Transplantation for Hilar Cholangiocarcinoma Is Associated with Improved Survival. *PLoS ONE* **2016**, *11*, e0156127. [CrossRef] [PubMed]
24. Loveday, B.P.T.; Knox, J.J.; Dawson, L.A.; Metser, U.; Brade, A.; Horgan, A.M.; Gallinger, S.; Greig, P.D.; Moulton, C.A. Neoadjuvant Hyperfractionated Chemoradiation and Liver Transplantation for Unresectable Perihilar Cholangiocarcinoma in Canada. *J. Surg. Oncol.* **2018**, *117*, 213–219. [CrossRef]
25. Duignan, S.; Maguire, D.; Ravichand, C.S.; Geoghegan, J.; Hoti, E.; Fennelly, D.; Armstrong, J.; Rock, K.; Mohan, H.; Traynor, O. Neoadjuvant Chemoradiotherapy Followed by Liver Transplantation for Unresectable Cholangiocarcinoma: A Single-Centre National Experience. *HPB* **2014**, *16*, 91–98. [CrossRef]
26. Cambridge, W.A.; Fairfield, C.; Powell, J.J.; Harrison, E.M.; Søreide, K.; Wigmore, S.J.; Guest, R.V. Meta-Analysis and Meta-Regression of Survival After Liver Transplantation for Unresectable Perihilar Cholangiocarcinoma. *Ann. Surg.* **2021**, *273*, 240–250. [CrossRef]
27. van Keulen, A.M.; Franssen, S.; van der Geest, L.G.; de Boer, M.T.; Coenraad, M.; van Driel, L.M.J.W.; Erdmann, J.I.; Haj Mohammad, N.; Heij, L.; Klümpen, H.J.; et al. Nationwide Treatment and Outcomes of Perihilar Cholangiocarcinoma. *Liver Int.* **2021**, *41*, 1945–1953. [CrossRef]
28. Heimbach, J.K.; Sanchez, W.; Rosen, C.B.; Gores, G.J. Trans-Peritoneal Fine Needle Aspiration Biopsy of Hilar Cholangiocarcinoma Is Associated with Disease Dissemination. *HPB* **2011**, *13*, 356–360. [CrossRef]
29. Bhat, M.; Hathcock, M.; Kremers, W.K.; Darwish Murad, S.; Schmit, G.; Martenson, J.; Alberts, S.; Rosen, C.B.; Gores, G.J.; Heimbach, J. Portal Vein Encasement Predicts Neoadjuvant Therapy Response in Liver Transplantation for Perihilar Cholangiocarcinoma Protocol. *Transpl. Int.* **2015**, *28*, 1383–1391. [CrossRef]
30. Sio, T.T.; Martenson, J.A.; Haddock, M.G.; Novotny, P.J.; Gores, G.J.; Alberts, S.R.; Miller, R.C.; Heimbach, J.K.; Rosen, C.B. Outcome of Transplant-Fallout Patients With Unresectable Cholangiocarcinoma. *Am. J. Clin. Oncol.* **2016**, *39*, 271–275. [CrossRef]
31. Shin, S.-P.; Koh, D.-H. Clinical Impact of Sarcopenia on Cholangiocarcinoma. *Life* **2022**, *12*, 815. [CrossRef]
32. Mantel, H.T.J.; Rosen, C.B.; Helmbach, J.K.; Nyberg, S.L.; Ishitani, M.B.; Andrews, J.C.; McKusick, M.A.; Haddock, M.G.; Alberts, S.R.; Gores, G.J. Vascular Complications after Orthotopic Liver Transplantation after Neoadjuvant Therapy for Hilar Cholangiocarcinoma. *Liver Transpl.* **2007**, *13*, 1372–1381. [CrossRef]
33. Heimbach, J.K.; Gores, G.J.; Haddock, M.G.; Alberts, S.R.; Pedersen, R.; Kremers, W.; Nyberg, S.L.; Ishitani, M.B.; Rosen, C.B. Predictors of Disease Recurrence Following Neoadjuvant Chemoradiotherapy and Liver Transplantation for Unresectable Perihilar Cholangiocarcinoma. *Transplantation* **2006**, *82*, 1703–1707. [CrossRef] [PubMed]
34. Ethun, C.G.; Lopez-Aguiar, A.G.; Anderson, D.J.; Adams, A.B.; Fields, R.C.; Doyle, M.B.; Chapman, W.C.; Krasnick, B.A.; Weber, S.M.; Mezrich, J.D.; et al. Transplantation Versus Resection for Hilar Cholangiocarcinoma: An Argument for Shifting Treatment Paradigms for Resectable Disease. *Ann. Surg.* **2018**, *267*, 797–805. [CrossRef] [PubMed]
35. Breuer, E.; Mueller, M.; Doyle, M.B.; Yang, L.; Darwish Murad, S.; Anwar, I.J.; Merani, S.; Limkemann, A.; Jeddou, H.; Kim, S.C.; et al. Liver Transplantation as a New Standard of Care in Patients With Perihilar Cholangiocarcinoma? Results From an International Benchmark Study. *Ann. Surg.* **2022**, *276*, 846–853. [CrossRef]
36. Rosen, C.B. Transplantation Versus Resection for Hilar Cholangiocarcinoma: An Argument for Shifting Paradigms for Resectable Disease in Annals of Surgery 2018. *Ann. Surg.* **2018**, *267*, 808–809. [CrossRef] [PubMed]
37. Adeva, J.; Sangro, B.; Salati, M.; Edeline, J.; la Casta, A.; Bittoni, A.; Berardi, R.; Bruix, J.; Valle, J.W. Medical Treatment for Cholangiocarcinoma. *Liver Int.* **2019**, *39*, 123–142. [CrossRef]
38. Valle, J.W.; Borbath, I.; Khan, S.A.; Huguet, F.; Gruenberger, T.; Arnold, D. On behalf of the ESMO Guidelines Committee Biliary Cancer: ESMO Clinical Practice Guidelines for Diagnosis, Treatment and Follow-Up. *Ann. Oncol.* **2016**, *27*, v28–v37. [CrossRef]
39. Kim, J.H.; Won, H.J.; Shin, Y.M.; Kim, K.A.; Kim, P.N. Radiofrequency Ablation for the Treatment of Primary Intrahepatic Cholangiocarcinoma. *AJR Am. J. Roentgenol.* **2011**, *196*, W205–W209. [CrossRef]
40. Fu, Y.; Yang, W.; Wu, W.; Yan, K.; Xing, B.C.; Chen, M.H. Radiofrequency Ablation in the Management of Unresectable Intrahepatic Cholangiocarcinoma. *J. Vasc. Interv. Radiol.* **2012**, *23*, 642–649. [CrossRef]
41. Díaz-González, Á.; Vilana, R.; Bianchi, L.; García-Criado, Á.; Rimola, J.; Rodríguez de Lope, C.; Ferrer, J.; Ayuso, C.; da Fonseca, L.G.; Reig, M.; et al. Thermal Ablation for Intrahepatic Cholangiocarcinoma in Cirrhosis: Safety and Efficacy in Non-Surgical Patients. *J. Vasc. Interv. Radiol.* **2020**, *31*, 710–719. [CrossRef]
42. Martin, R.C.G.; Simo, K.A.; Hansen, P.; Rocha, F.; Philips, P.; McMasters, K.M.; Tatum, C.M.; Kelly, L.R.; Driscoll, M.; Sharma, V.R.; et al. Drug-Eluting Bead, Irinotecan Therapy of Unresectable Intrahepatic Cholangiocarcinoma (DELTIC) with Concomitant Systemic Gemcitabine and Cisplatin. *Ann. Surg. Oncol.* **2022**, *29*, 5462–5473. [CrossRef] [PubMed]
43. Paprottka, K.J.; Galiè, F.; Ingrisch, M.; Geith, T.; Ilhan, H.; Todica, A.; Michl, M.; Nadjiri, J.; Paprottka, P.M. Outcome and Safety after 103 Radioembolizations with Yttrium-90 Resin Microspheres in 73 Patients with Unresectable Intrahepatic Cholangiocarcinoma-An Evaluation of Predictors. *Cancers* **2021**, *13*, 5399. [CrossRef] [PubMed]
44. Gangi, A.; Shah, J.; Hatfield, N.; Smith, J.; Sweeney, J.; Choi, J.; El-Haddad, G.; Biebel, B.; Parikh, N.; Arslan, B.; et al. Intrahepatic Cholangiocarcinoma Treated with Transarterial Yttrium-90 Glass Microsphere Radioembolization: Results of a Single Institution Retrospective Study. *J. Vasc. Interv. Radiol.* **2018**, *29*, 1101–1108. [CrossRef] [PubMed]

45. Buettner, S.; Braat, A.J.A.T.; Margonis, G.A.; Brown, D.B.; Taylor, K.B.; Borgmann, A.J.; Kappadath, S.C.; Mahvash, A.; IJzermans, J.N.M.; Weiss, M.J.; et al. Yttrium-90 Radioembolization in Intrahepatic Cholangiocarcinoma: A Multicenter Retrospective Analysis. *J. Vasc. Interv. Radiol.* **2020**, *31*, 1035–1043. [CrossRef] [PubMed]
46. Mouli, S.; Memon, K.; Baker, T.; Benson, A.B.; Mulcahy, M.F.; Gupta, R.; Ryu, R.K.; Salem, R.; Lewandowski, R.J. Yttrium-90 Radioembolization for Intrahepatic Cholangiocarcinoma: Safety, Response, and Survival Analysis. *J. Vasc. Interv. Radiol.* **2013**, *24*, 1227–1234. [CrossRef]
47. Levillain, H.; Duran Derijckere, I.; Ameye, L.; Guiot, T.; Braat, A.; Meyer, C.; Vanderlinden, B.; Reynaert, N.; Hendlisz, A.; Lam, M.; et al. Personalised Radioembolization Improves Outcomes in Refractory Intra-Hepatic Cholangiocarcinoma: A Multicenter Study. *Eur. J. Nucl. Med. Mol. Imaging* **2019**, *46*, 2270–2279. [CrossRef]
48. Bargellini, I.; Mosconi, C.; Pizzi, G.; Lorenzoni, G.; Vivaldi, C.; Cappelli, A.; Vallati, G.E.; Boni, G.; Cappelli, F.; Paladini, A.; et al. Yttrium-90 Radioembolization in Unresectable Intrahepatic Cholangiocarcinoma: Results of a Multicenter Retrospective Study. *Cardiovasc. Intervent. Radiol.* **2020**, *43*, 1305–1314. [CrossRef]
49. White, J.; Carolan-Rees, G.; Dale, M.; Patrick, H.E.; See, T.C.; Bell, J.K.; Manas, D.M.; Crellin, A.; Slevin, N.J.; Sharma, R.A. Yttrium-90 Transarterial Radioembolization for Chemotherapy-Refractory Intrahepatic Cholangiocarcinoma: A Prospective, Observational Study. *J. Vasc. Interv. Radiol.* **2019**, *30*, 1185–1192. [CrossRef]
50. Köhler, M.; Harders, F.; Lohöfer, F.; Paprottka, P.M.; Schaarschmidt, B.M.; Theysohn, J.; Herrmann, K.; Heindel, W.; Schmidt, H.H.; Pascher, A.; et al. Prognostic Factors for Overall Survival in Advanced Intrahepatic Cholangiocarcinoma Treated with Yttrium-90 Radioembolization. *J. Clin. Med.* **2019**, *9*, 56. [CrossRef]
51. Schartz, D.A.; Porter, M.; Schartz, E.; Kallas, J.; Gupta, A.; Butani, D.; Cantos, A. Transarterial Yttrium-90 Radioembolization for Unresectable Intrahepatic Cholangiocarcinoma: A Systematic Review and Meta-Analysis. *J. Vasc. Interv. Radiol.* **2022**, *33*, 679–686. [CrossRef]
52. Edeline, J.; Touchefeu, Y.; Guiu, B.; Farge, O.; Tougeron, D.; Baumgaertner, I.; Ayav, A.; Campillo-Gimenez, B.; Beuzit, L.; Pracht, M.; et al. Radioembolization Plus Chemotherapy for First-Line Treatment of Locally Advanced Intrahepatic Cholangiocarcinoma: A Phase 2 Clinical Trial. *JAMA Oncol.* **2020**, *6*, 51–59. [CrossRef]
53. Jung, D.H.; Kim, M.S.; Cho, C.K.; Yoo, H.J.; Jang, W.I.; Seo, Y.S.; Paik, E.K.; Kim, K.B.; Han, C.J.; Kim, S.B. Outcomes of Stereotactic Body Radiotherapy for Unresectable Primary or Recurrent Cholangiocarcinoma. *Radiat. Oncol. J.* **2014**, *32*, 163–169. [CrossRef]
54. De, B.; Tran Cao, H.S.; Vauthey, J.N.; Manzar, G.S.; Corrigan, K.L.; Raghav, K.P.S.; Lee, S.S.; Tzeng, C.W.D.; Minsky, B.D.; Smith, G.L.; et al. Ablative Liver Radiotherapy for Unresected Intrahepatic Cholangiocarcinoma: Patterns of Care and Survival in the United States. *Cancer* **2022**, *128*, 2529–2539. [CrossRef] [PubMed]
55. Valle, J.; Wasan, H.; Palmer, D.H.; Cunningham, D.; Anthoney, A.; Maraveyas, A.; Madhusudan, S.; Iveson, T.; Hughes, S.; Pereira, S.P.; et al. Cisplatin plus Gemcitabine versus Gemcitabine for Biliary Tract Cancer. *New Engl. J. Med.* **2010**, *362*, 1273–1281. [CrossRef]
56. Lamarca, A.; Ross, P.; Wasan, H.S.; Hubner, R.A.; McNamara, M.G.; Lopes, A.; Manoharan, P.; Palmer, D.; Bridgewater, J.; Valle, J.W. Advanced Intrahepatic Cholangiocarcinoma: Post Hoc Analysis of the ABC-01, -02, and -03 Clinical Trials. *J. Natl. Cancer Inst.* **2020**, *112*, 200–210. [CrossRef]
57. Phelip, J.-M.; Desrame, J.; Edeline, J.; Barbier, E.; Terrebonne, E.; Michel, P.; Perrier, H.; Dahan, L.; Bourgeois, V.; Akouz, F.K.; et al. Modified FOLFIRINOX Versus CISGEM Chemotherapy for Patients With Advanced Biliary Tract Cancer (PRODIGE 38 AMEBICA): A Randomized Phase II Study. *J. Clin. Oncol.* **2022**, *40*, 262–271. [CrossRef] [PubMed]
58. Morizane, C.; Okusaka, T.; Mizusawa, J.; Katayama, H.; Ueno, M.; Ikeda, M.; Ozaka, M.; Okano, N.; Sugimori, K.; Fukutomi, A.; et al. Combination Gemcitabine plus S-1 versus Gemcitabine plus Cisplatin for Advanced/Recurrent Biliary Tract Cancer: The FUGA-BT (JCOG1113) Randomized Phase III Clinical Trial. *Ann. Oncol.* **2019**, *30*, 1950–1958. [CrossRef] [PubMed]
59. Kim, S.T.; Kang, J.H.; Lee, J.; Lee, H.W.; Oh, S.Y.; Jang, J.S.; Lee, M.A.; Sohn, B.S.; Yoon, S.Y.; Choi, H.J.; et al. Capecitabine plus Oxaliplatin versus Gemcitabine plus Oxaliplatin as First-Line Therapy for Advanced Biliary Tract Cancers: A Multicenter, Open-Label, Randomized, Phase III, Noninferiority Trial. *Ann. Oncol.* **2019**, *30*, 788–795. [CrossRef]
60. McNamara, M.G.; Bridgewater, J.; Palmer, D.H.; Faluyi, O.; Wasan, H.; Patel, A.; Ryder, W.D.; Barber, S.; Gnanaranjan, C.; Ghazaly, E.; et al. A Phase Ib Study of NUC-1031 in Combination with Cisplatin for the First-Line Treatment of Patients with Advanced Biliary Tract Cancer (ABC-08). *Oncologist* **2021**, *26*, e669–e678. [CrossRef]
61. Kang, S.; El-rayes, B.F.; Akce, M. Evolving Role of Immunotherapy in Advanced Biliary Tract Cancers. *Cancers (Basel)* **2022**, *14*, 1748. [CrossRef] [PubMed]
62. Oh, D.-Y.; Ruth, A.; Qin, S.; Chen, L.-T.; Okusaka, T.; Vogel, A.; Kim, J.W.; Suksombooncharoen, T.; Lee, M.A.; Kitano, M.; et al. Durvalumab plus Gemcitabine and Cisplatin in Advanced Biliary Tract Cancer. *NEJM Evid.* **2022**, *7*, 522–532. [CrossRef]
63. Oh, D.-Y. Updated Overall Survival from the Phase 3 TOPAZ-1 Study of Durvalumab or Placebo plus Gemcitabine and Cisplatin in Patients with Advanced Biliary Tract Cancer. (Abstract 56P). *Ann. Oncol.* **2022**, *33*, S19–S26. [CrossRef]
64. Vogel, A.; Boeck, S.; Waidmann, O.; Bitzer, M.; Wenzel, P.; Belle, S.; Springfeld, C.; Schulze, K.; Weinmann, A.; Lindig, U.; et al. A Randomized Phase II Trial of Durvalumab and TremelIMUmab with Gemcitabine or Gemcitabine and Cisplatin Compared to Gemcitabine and Cisplatin in Treatment-Naïve Patients with CHolangio- and GallbladdEr Carcinoma (IMMUCHEC). *Ann. Oncol.* **2022**, *33*, S563. [CrossRef]

65. Yoo, C.; Kim, K.; Jeong, J.H.; Kim, I.; Kang, M.J.; Cheon, J.; Kang, B.W.; Ryu, H.; Lee, J.S.; Kim, K.W.; et al. Liposomal Irinotecan plus Fluorouracil and Leucovorin versus Fluorouracil and Leucovorin for Metastatic Biliary Tract Cancer after Progression on Gemcitabine plus Cisplatin (NIFTY): A Multicentre, Open-Label, Randomised, Phase 2b Study. *Lancet Oncol.* **2021**, *22*, 1560–1572. [CrossRef] [PubMed]
66. Yoo, C.; Kim, K.-P.; Kim, I.; Kang, M.J.; Cheon, J.; Kang, B.W.; Ryu, H.; Jeong, J.; Lee, J.S.; Kim, K.W.; et al. Final Results from the NIFTY Trial, a Phase IIb, Randomized, Open-Label Study of Liposomal Irinotecan (Nal-IRI) plus Fluorouracil (5-FU)/Leucovorin (LV) in Patients (Pts) with Previously Treated Metastatic Biliary Tract Cancer (BTC). *Ann. Oncol.* **2022**, *33*, S565. [CrossRef]
67. Lamarca, A.; Palmer, D.H.; Wasan, H.S.; Ross, P.J.; Ma, Y.T.; Arora, A.; Falk, S.; Gillmore, R.; Wadsley, J.; Patel, K.; et al. Second-Line FOLFOX Chemotherapy versus Active Symptom Control for Advanced Biliary Tract Cancer (ABC-06): A Phase 3, Open-Label, Randomised, Controlled Trial. *Lancet Oncol.* **2021**, *22*, 690–701. [CrossRef]
68. Lee, S.H.; Lee, H.S.; Lee, S.H.; Woo, S.M.; Kim, D.U.; Bang, S. Efficacy and Safety of Pembrolizumab for Gemcitabine/Cisplatin-Refractory Biliary Tract Cancer: A Multicenter Retrospective Study. *J. Clin. Med.* **2020**, *9*, 1769. [CrossRef]
69. Naganuma, A.; Sakuda, T.; Murakami, T.; Aihara, K.; Watanuki, Y.; Suzuki, Y.; Shibasaki, E.; Masuda, T.; Uehara, S.; Yasuoka, H.; et al. Microsatellite Instability-High Intrahepatic Cholangiocarcinoma with Portal Vein Tumor Thrombosis Successfully Treated with Pembrolizumab. *Intern. Med.* **2020**, *59*, 2261–2267. [CrossRef]
70. Marabelle, A.; Le, D.T.; Ascierto, P.A.; di Giacomo, A.M.; de Jesus-Acosta, A.; Delord, J.P.; Geva, R.; Gottfried, M.; Penel, N.; Hansen, A.R.; et al. Efficacy of Pembrolizumab in Patients with Noncolorectal High Microsatellite Instability/Mismatch Repair–Deficient Cancer: Results from the Phase II KEYNOTE-158 Study. *J. Clin. Oncol.* **2020**, *38*, 1–10. [CrossRef]
71. Maio, M.; Ascierto, P.A.; Manzyuk, L.; Motola-Kuba, D.; Penel, N.; Cassier, P.A.; Bariani, G.M.; de Jesus-Acosta, A.; Doi, T.; Longo, F.; et al. Pembrolizumab in Microsatellite Instability High or Mismatch Repair Deficient Cancers: Updated Analysis from the Phase II KEYNOTE-158 Study. *Ann. Oncol.* **2022**, *33*, 929–938. [CrossRef] [PubMed]
72. Montal, R.; Sia, D.; Montironi, C.; Leow, W.Q.; Esteban-Fabró, R.; Pinyol, R.; Torres-Martin, M.; Bassaganyas, L.; Moeini, A.; Peix, J.; et al. Molecular Classification and Therapeutic Targets in Extrahepatic Cholangiocarcinoma. *J. Hepatol.* **2020**, *73*, 315–327. [CrossRef]
73. Sia, D.; Hoshida, Y.; Villanueva, A.; Roayaie, S.; Ferrer, J.; Tabak, B.; Peix, J.; Sole, M.; Tovar, V.; Alsinet, C.; et al. Integrative Molecular Analysis of Intrahepatic Cholangiocarcinoma Reveals 2 Classes That Have Different Outcomes. *Gastroenterology* **2013**, *144*, 829–840. [CrossRef] [PubMed]
74. Nakamura, H.; Arai, Y.; Totoki, Y.; Shirota, T.; Elzawahry, A.; Kato, M.; Hama, N.; Hosoda, F.; Urushidate, T.; Ohashi, S.; et al. Genomic Spectra of Biliary Tract Cancer. *Nat. Genet.* **2015**, *47*, 1003–1010. [CrossRef] [PubMed]
75. Mosele, F.; Remon, J.; Mateo, J.; Westphalen, C.B.; Barlesi, F.; Lolkema, M.P.; Normanno, N.; Scarpa, A.; Robson, M.; Meric-Bernstam, F.; et al. Recommendations for the Use of Next-Generation Sequencing (NGS) for Patients with Metastatic Cancers: A Report from the ESMO Precision Medicine Working Group. *Ann. Oncol.* **2020**, *31*, 1491–1505. [CrossRef] [PubMed]
76. Verdaguer, H.; Saurí, T.; Acosta, D.A.; Guardiola, M.; Sierra, A.; Hernando, J.; Nuciforo, P.; Miquel, J.M.; Molero, C.; Peiró, S.; et al. ESMO Scale for Clinical Actionability of Molecular Targets Driving Targeted Treatment in Patients with Cholangiocarcinoma. *Clin. Cancer Res.* **2022**, *28*, 1662–1671. [CrossRef]
77. Arai, Y.; Totoki, Y.; Hosoda, F.; Shirota, T.; Hama, N.; Nakamura, H.; Ojima, H.; Furuta, K.; Shimada, K.; Okusaka, T.; et al. Fibroblast Growth Factor Receptor 2 Tyrosine Kinase Fusions Define a Unique Molecular Subtype of Cholangiocarcinoma. *Hepatology* **2014**, *59*, 1427–1434. [CrossRef]
78. Sia, D.; Losic, B.; Moeini, A.; Cabellos, L.; Hao, K.; Revill, K.; Bonal, D.; Miltiadous, O.; Zhang, Z.; Hoshida, Y.; et al. Massive Parallel Sequencing Uncovers Actionable FGFR2-PPHLN1 Fusion and ARAF Mutations in Intrahepatic Cholangiocarcinoma. *Nat. Commun.* **2015**, *6*, 6087. [CrossRef]
79. Abou-Alfa, G.K.; Sahai, V.; Hollebecque, A.; Vaccaro, G.; Melisi, D.; Al-Rajabi, R.; Paulson, A.S.; Borad, M.J.; Gallinson, D.; Murphy, A.G.; et al. Pemigatinib for Previously Treated, Locally Advanced or Metastatic Cholangiocarcinoma: A Multicentre, Open-Label, Phase 2 Study. *Lancet Oncol.* **2020**, *21*, 671–684. [CrossRef]
80. Javle, M.; Roychowdhury, S.; Kelley, R.K.; Sadeghi, S.; Macarulla, T.; Weiss, K.H.; Waldschmidt, D.-T.; Goyal, L.; Borbath, I.; El-Khoueiry, A.; et al. Infigratinib (BGJ398) in Previously Treated Patients with Advanced or Metastatic Cholangiocarcinoma with FGFR2 Fusions or Rearrangements: Mature Results from a Multicentre, Open-Label, Single-Arm, Phase 2 Study. *Lancet Gastroenterol. Hepatol.* **2021**, *6*, 803–815. [CrossRef]
81. Goyal, L.; Meric-Bernstam, F.; Hollebecque, A.; Valle, J.W.; Morizane, C.; Karasic, T.B.; Abrams, T.A.; Furuse, J.; Kelley, R.K.; Cassier, P.A.; et al. Futibatinib for FGFR2-Rearranged Intrahepatic Cholangiocarcinoma. *New Engl. J. Med.* **2023**, *388*, 228–239. [CrossRef] [PubMed]
82. Abou-Alfa, G.K.; Macarulla, T.; Javle, M.M.; Kelley, R.K.; Lubner, S.J.; Adeva, J.; Cleary, J.M.; Catenacci, D.V.; Borad, M.J.; Bridgewater, J.; et al. Ivosidenib in IDH1-Mutant, Chemotherapy-Refractory Cholangiocarcinoma (ClarIDHy): A Multicentre, Randomised, Double-Blind, Placebo-Controlled, Phase 3 Study. *Lancet Oncol.* **2020**, *21*, 796–807. [CrossRef] [PubMed]
83. Zhu, A.X.; Macarulla, T.; Javle, M.M.; Kelley, R.K.; Lubner, S.J.; Adeva, J.; Cleary, J.M.; Catenacci, D.V.T.; Borad, M.J.; Bridgewater, J.A.; et al. Final Overall Survival Efficacy Results of Ivosidenib for Patients With Advanced Cholangiocarcinoma With IDH1 Mutation. *JAMA Oncol.* **2021**, *7*, 1669–1677. [CrossRef] [PubMed]

84. Jeong, H.; Jeong, J.H.; Kim, K.P.; Lee, S.S.; Oh, D.W.; Park, D.H.; Song, T.J.; Park, Y.; Hong, S.M.; Ryoo, B.Y.; et al. Feasibility of HER2-Targeted Therapy in Advanced Biliary Tract Cancer: A Prospective Pilot Study of Trastuzumab Biosimilar in Combination with Gemcitabine Plus Cisplatin. *Cancers* **2021**, *13*, 161. [CrossRef] [PubMed]
85. Lee, C.; Chon, H.J.; Cheon, J.; Lee, M.A.; Im, H.-S.; Jang, J.-S.; Kim, M.H.; Park, S.; Kang, B.; Hong, M.; et al. Trastuzumab plus FOLFOX for HER2-Positive Biliary Tract Cancer Refractory to Gemcitabine and Cisplatin: A Multi-Institutional Phase 2 Trial of the Korean Cancer Study Group (KCSG-HB19–14). *Lancet Gastroenterol. Hepatol.* **2023**, *8*, 56–65. [CrossRef]
86. Subbiah, V.; Lassen, U.; Élez, E.; Italiano, A.; Curigliano, G.; Javle, M.; de Braud, F.; Prager, G.W.; Greil, R.; Stein, A.; et al. Dabrafenib plus Trametinib in Patients with BRAFV600E-Mutated Biliary Tract Cancer (ROAR): A Phase 2, Open-Label, Single-Arm, Multicentre Basket Trial. *Lancet Oncol.* **2020**, *21*, 1234–1243. [CrossRef]
87. Vogel, A.; Bridgewater, J.; Edeline, J.; Kelley, R.K.; Klümpen, H.J.; Malka, D.; Primrose, J.N.; Rimassa, L.; Stenzinger, A.; Valle, J.W.; et al. Biliary Tract Cancer: ESMO Clinical Practice Guideline for Diagnosis, Treatment and Follow-Up. *Ann. Oncol.* **2022**, *34*, 127–140. [CrossRef]

cancers

Article

# Stationary Trend in Elevated Serum Alpha-Fetoprotein Level in Hepatocellular Carcinoma Patients

**Yi-Hao Yen [1,*,†], Kwong-Ming Kee [1], Wei-Feng Li [2], Yueh-Wei Liu [2], Chih-Chi Wang [2,*,†], Tsung-Hui Hu [1], Ming-Chao Tsai [1] and Chih-Yun Lin [3]**

1 Division of Hepatogastroenterology, Department of Internal Medicine, Kaohsiung Chang Gung Memorial Hospital and Chang Gung University College of Medicine, Kaohsiung 833, Taiwan
2 Liver Transplantation Center, Department of Surgery, Kaohsiung Chang Gung Memorial Hospital, Kaohsiung 833, Taiwan
3 Biostatistics Center, Kaohsiung Chang Gung Memorial Hospital, Kaohsiung 833, Taiwan
* Correspondence: cassellyen@yahoo.com.tw (Y.-H.Y.); ufel4996@ms26.hinet.net (C.-C.W.); Tel./Fax: +886-7-7317123 (Y.-H.Y.); +886-7-7317123 (C.-C.W.)
† These authors contributed equally to this work.

**Simple Summary:** In this study, we demonstrated that overall 51.2% of patients with hepatocellular carcinoma (HCC) had elevated alpha-fetoprotein (AFP) levels. The proportion of patients with elevated AFP levels was stationary in the period from 2011 to 2020. The proportion of patients with Barcelona Clinic Liver Cancer classification (BCLC) stages 0–A HCC decreased from 2011 to 2020, whereas the proportion of patients with non-HBV- and non-HCV (NBNC)-HCC increased in the same period. Furthermore, the proportion of patients with early-stage HCC (i.e., BCLC stages 0–A) was lower for NBNC-HCC than for HBV- or HCV-related HCC. Advanced tumor stage, severe underlying liver disease, viral etiology, and female gender are associated with elevated AFP levels in HCC patients.

**Abstract:** A recent study from the US showed a decreasing trend in the elevated serum alpha-fetoprotein (AFP) level (i.e., $\geq$20 ng/mL) in hepatocellular carcinoma (HCC) patients at the time of diagnosis. Furthermore, advanced tumor stage and severe underlying liver disease were associated with elevated AFP levels. We aimed to evaluate this issue in an area endemic for hepatitis B virus (HBV). Between 2011 and 2020, 4031 patients were newly diagnosed with HCC at our institution. After excluding 54 patients with unknown AFP data, the remaining 3977 patients were enrolled in this study. Elevated AFP level was defined as $\geq$20 ng/mL. Overall, 51.2% of HCC patients had elevated AFP levels; this proportion remained stationary between 2011 and 2020 (51.8% vs. 51.1%). Multivariate analysis showed that female gender (odds ratio (OR) = 1.462; $p < 0.001$), tumor size per 10 mm increase (OR = 1.155; $p < 0.001$), multiple tumors (OR = 1.406; $p < 0.001$), Barcelona Clinic Liver Cancer stages B–D (OR = 1.247; $p = 0.019$), cirrhosis (OR = 1.288; $p = 0.02$), total bilirubin > 1.4 mg/dL (OR = 1.218; $p = 0.030$), and HBV- or hepatitis C virus (HCV)-positive status (OR = 1.720; $p < 0.001$) were associated with elevated AFP levels. In conclusion, a stationary trend in elevated serum AFP level in HCC patients has been noted in the past 10 years. Advanced tumor stage, severe underlying liver disease, viral etiology, and female gender are associated with elevated AFP levels in HCC patients.

**Keywords:** alpha-fetoprotein; hepatocellular carcinoma; hepatitis B virus; hepatitis C virus

**Citation:** Yen, Y.-H.; Kee, K.-M.; Li, W.-F.; Liu, Y.-W.; Wang, C.-C.; Hu, T.-H.; Tsai, M.-C.; Lin, C.-Y. Stationary Trend in Elevated Serum Alpha-Fetoprotein Level in Hepatocellular Carcinoma Patients. *Cancers* **2023**, *15*, 1222. https://doi.org/10.3390/cancers15041222

Academic Editor: Georgios Germanidis

Received: 2 January 2023
Revised: 5 February 2023
Accepted: 13 February 2023
Published: 15 February 2023



## 1. Introduction

Hepatocellular carcinoma (HCC) is one of the leading causes of cancer-related death worldwide [1]. A meta-analysis showed that HCC surveillance is associated with significant improvements in early-stage tumor detection, the receipt of curative therapy, and survival of cirrhotic patients [2]. The American Association for the Study of Liver Diseases (AASLD) guideline recommends HCC surveillance for high-risk populations. The modality recommended for surveillance is ultrasound with or without an alpha-fetoprotein (AFP) serum assay [3]. Ultrasound with an AFP serum assay is recommended for surveillance because a meta-analysis demonstrated that ultrasound alone had low sensitivity in detecting early-stage tumor in cirrhotic patients. The combination of AFP serum assay and ultrasound significantly increases the sensitivity of tumor detection [4]. Currently, the AASLD guideline recommends diagnostic multiphasic magnetic resonance imaging (MRI)/computed tomography (CT) for further evaluation when the AFP level is ≥20 ng/mL on surveillance [3].

Multiple factors, including advanced tumor stage and viral etiology of chronic liver disease, are associated with elevated AFP levels in HCC patients [5]. A recent study from the US found a downtrend in the percentage of HCC cases with elevated AFP levels at the time of diagnosis from 2010 to 2017 in a large cohort from the National Cancer Database. Elevated AFP was defined as ≥20 ng/mL. Furthermore, advanced tumor stage and severe underlying liver disease were associated with elevated AFP levels. The authors suggested that these changes in AFP values at HCC diagnosis were possibly related to the increasing trend in early-stage tumor detection and the shift from viral (i.e., hepatitis B virus (HBV) or hepatitis C virus (HCV)) to nonviral etiology. However, data on the etiology of liver disease were unavailable in the database analyzed in the study [6].

Approximately 90% of HCC cases are associated with a known underlying etiology [7]. In East Asia, the major risk factor is HBV, whereas, in the Western world, it is HCV [7]. The risk of HCC attributed to HCV infection has largely decreased owing to the eradication of the virus with direct-acting antiviral (DAA) agents [8]. Nonalcoholic fatty liver disease (NAFLD), which is usually associated with obesity, metabolic syndrome, or diabetes mellitus, is becoming the fastest growing etiology of HCC, not only in Western countries [9], but also in Asia [10].

Due to the different etiologies of HCC in the East and the West and viral etiology being associated with elevated AFP levels in HCC patients [5], we aimed to evaluate whether there is a downtrend in the percentages of HCC cases with elevated AFP levels at the time of diagnosis and the factors associated with elevated AFP levels in HCC patients in a country from East Asia, where the leading etiology of HCC is HBV.

## 2. Materials and Methods

The study was conducted according to the guidelines of the Declaration of Helsinki and approved by the Institutional Review Board of Kaohsiung Chang Gung Memorial Hospital (reference number: 202201189B0; date of approval: 8 August 2022).

The Institutional Review Board of Kaohsiung Chang Gung Memorial Hospital waived the need for informed consent due to the retrospective and observational nature of the study design. Data were extracted from Kaohsiung Chang Gung Memorial Hospital's HCC registry database, which holds prospectively collected and annually updated data.

From 2011 to 2020, 4031 patients were newly diagnosed with HCC at the institution. After excluding 54 patients with unknown AFP data, the remaining 3977 patients were enrolled in this study.

### 2.1. Variables of Interest

Patient demographics, tumor size and number, clinical tumor–node–metastasis (TNM) stage (seventh edition of the American Joint Committee on Cancer (AJCC)) [11], Barcelona Clinic Liver Cancer classification (BCLC) stage [12], AFP level, cirrhosis, Child–Pugh class [13], creatinine, bilirubin, international normalized ratio (INR), hepatitis B surface antigen (HBsAg), anti-HCV antibody, alcohol use disorder (AUD), and HCC diagnostic method (i.e., clinical vs. pathological diagnosis) were prospectively collected from the HCC registry data. Infection with HBV was defined as being HBsAg-positive. Infection with HCV was defined as being anti-HCV-antibody-positive, irrespective of viremia. An individual with AUD was defined as a habitual drinker. Demographic information included age, gender, height, and weight. Tumor size was determined according to the results of pathological examination of patients who underwent surgery, whereas it was determined according to the findings of imaging in patients who underwent nonsurgical treatments. Tumor number (solitary vs. multiple) was determined from the findings of imaging. The presence of cirrhosis was indicated by an Ishak score [14] of 5 or 6 in patients who underwent surgery, whereas it was determined according to the findings of imaging in patients who underwent nonsurgical treatments. Cirrhosis was indicated in imaging by small liver size, nodular liver surface, presence of regeneration nodules, left and right lobe liver volume redistribution, etc. [15]. The BCLC stages according to the original version and BCLC stage A were defined within Milan criteria [16].

The raw data for the cohort involved in this study are available via the following digital object identifier: https://www.dropbox.com/scl/fi/6hzj7a3hrqhp5yu4i17qu/afp-trend.xlsx?dl=0&rlkey=srqyaz7t1sqsli6heaoig4bk8 (accessed on 1 January 2023).

### 2.2. Statistical Analysis

Variables are presented as number and percentage or median and interquartile range. The Chi-square test was used to compare categorical variables. Mann–Whitney $U$ test was used to compare continuous variables. Whether there was an increasing or decreasing trend of BCLC stages 0–A or non-HBV- and non-HCV (NBNC)-HCC according to the year of HCC diagnosis was examined for a linear trend using the Chi-square test. Univariate analyses were conducted to explore the association between elevated AFP levels and clinical variables. Variables with $p$-values $\leq 0.1$ in univariate analyses were included in a multivariate logistic regression analysis. To avoid collinearity, we examined the correlation between two independent variables using Spearman's correlation test. If two independent variables had a correlation coefficient above 0.5, then they were determined to be highly correlated with each other and thus, collinear. In this case, we only chose one of the variables for multivariate analysis. In this analysis, we used the following cutoff values for continuous variables: for bilirubin, the upper limit of the normal range (i.e., 1.4 mg/dL); for creatinine, the upper limit of the normal range (i.e., 1.2 mg/dL); and for INR, the upper limit of the normal range (i.e., 1.2). Relative risks are presented as odds ratio (OR) with a 95% confidence interval (CI). To compare with a recent US study [6], we used the same method adopted in that study [6] to interpret the temporal trend of elevated AFP levels. We estimated the percentage of elevated AFP using marginal effects (i.e., the average predicted probability) from a logistic regression model [17]. All statistical analyses were performed using SPSS version 22.0 and R statistical software version 4.0.5. Two-tailed significance values were applied, and the level of statistical significance was defined as $p < 0.05$.

## 3. Results

### 3.1. Trend in Elevated AFP Levels at the Time of HCC Diagnosis

Overall, 2036 (51.2%) patients with HCC had elevated AFP levels. Between 2011 and 2020, the proportion of patients with elevated AFP levels was stationary in the case of all patients (51.8% [95% CI = 49.4–54.1%] in 2011 vs. 51.1% [95% CI = 48.9–53.3%] in 2020); patients in AJCC stage 1 (40.3% [95% CI = 37.5–43.1%] in 2011 vs. 39.7% [95% CI = 37.0–42.3%] in 2020); patients in AJCC stage 2 (45.4% [95% CI = 41.5–49.3%] in 2011 vs. 44.7% [95% CI = 40.9–48.6%] in 2020); patients in AJCC stage 3 (72.6% [95% CI = 68.8–76.3%] in 2011 vs. 71.9% [95% CI = 68.3–75.5%] in 2020); and patients in AJCC stage 4 (79.8% [95% CI = 74.3–85.3%] in 2011 vs. 79.2% [95% CI = 73.7–84.6%] in 2020) (Figure 1).

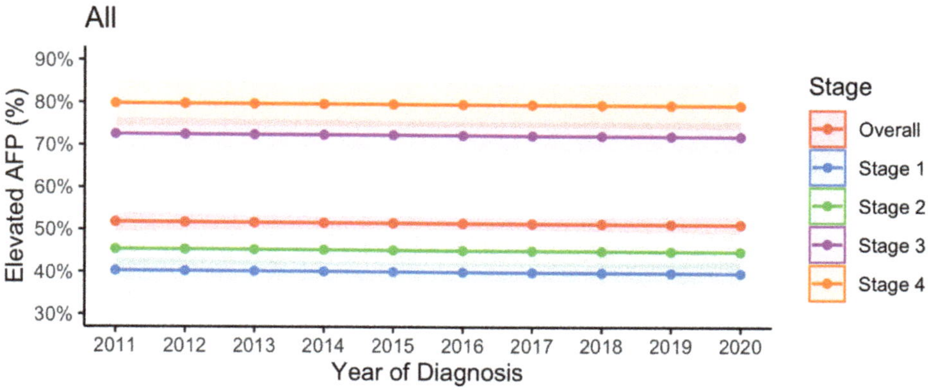


**Figure 1.** Trend in elevated serum alpha-fetoprotein levels (i.e., ≥20 ng/mL) at the time of hepatocellular carcinoma diagnosis in all patients and at different stages (seventh edition American Joint Committee on Cancer).

*3.2. Trends in Early-Stage Tumor Prevalence and Non-Viral Etiology at the Time of HCC Diagnosis*

The proportion of patients with early-stage tumor (i.e., BCLC stages 0–A) decreased in the period from 2011 to 2020 (54.2% vs. 42.5%, $p < 0.001$) (Figure 2). The proportion of patients with non-HBV and non-HCV (NBNC)-HCC increased between the years 2011 and 2020 (from 20.4% to 28.7%, $p < 0.001$) (Figure 3).

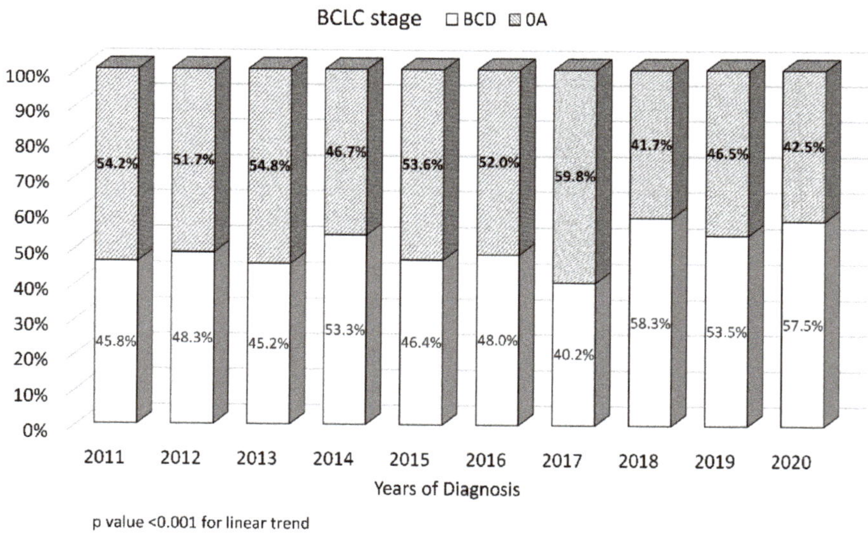


**Figure 2.** Trend in early-stage tumor at the time of hepatocellular carcinoma diagnosis. Early-stage tumor was defined as Barcelona Clinic Liver Cancer classification stages 0–A.

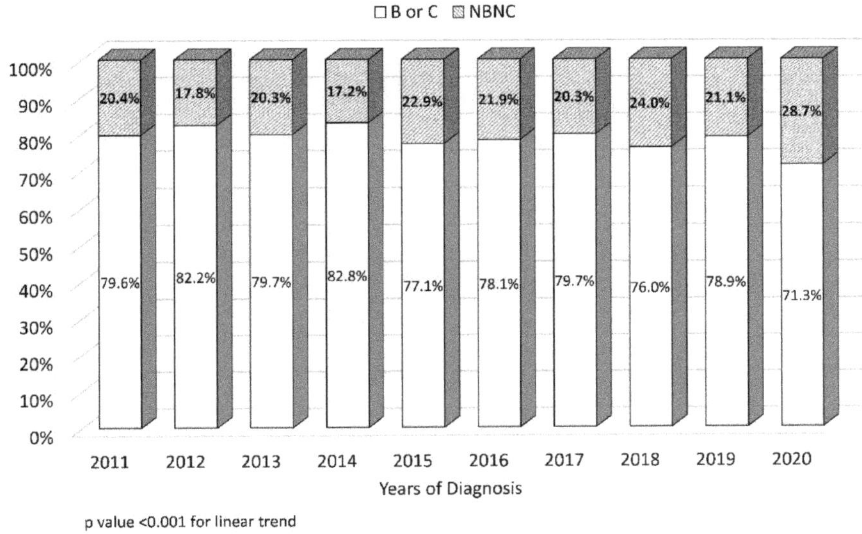


**Figure 3.** Trend in nonviral etiology (i.e., patients negative for both hepatitis B and C viruses) at the time of hepatocellular carcinoma diagnosis.

### 3.3. Patients' Characteristics Categorized by AFP Level

Compared to patients with normal AFP levels, a smaller proportion of patients with elevated AFP were male ($p = 0.02$), had a pathological diagnosis of HCC ($p < 0.001$), were in AJCC stage 1 or 2 HCC ($p < 0.001$), had a solitary tumor ($p < 0.001$), were in BCLC stage 0 or A HCC ($p < 0.001$), were in Child–Pugh class A ($p < 0.001$), and had a low body mass index (BMI) ($p < 0.001$). Furthermore, patients with elevated AFP levels had larger tumors ($p < 0.001$), a higher total bilirubin level ($p < 0.001$), and a higher INR ($p < 0.001$) and a higher proportion of them were cirrhotic ($p = 0.001$) and HBsAg-positive ($p = 0.002$). However, there were no significant differences in age, creatinine level, proportion with AUD, and proportion with anti-HCV-antibody-positive status between the two groups (Table 1).

**Table 1.** Patients' characteristics categorized by alpha-fetoprotein level.

| Characteristic | Elevated AFP (≥20 ng/mL), $n$ = 2036 | Normal AFP (<20 ng/mL), $n$ = 1941 | $p$ |
|---|---|---|---|
| Age (years) | 63 (55–71) | 63 (56–71) | 0.286 |
| Male | 1437 (70.6%) | 1434 (73.9%) | 0.02 |
| BMI (kg/m$^2$) | 24.2 (22.1–26.9) | 24.9 (22.5–27.8) | <0.001 |
| Diagnosis method | | | <0.001 |
| Clinical diagnosis | 896 (44.0%) | 593 (30.6%) | |
| Pathology diagnosis | 1140 (56.0%) | 1348 (69.4%) | |
| AUD | | | 0.460 |
| Yes | 260 (12.8%) | 225 (11.6%) | |
| No | 1762 (86.5%) | 1705 (87.8%) | |
| Not available | 11 (0.6%) | 14 (0.7%) | |
| HBsAg | | | 0.002 |
| Positive | 1006 (49.4%) | 866 (44.6%) | |
| Negative | 1030 (50.6%) | 1075 (55.4%) | |
| Anti-HCV | | | 0.327 |
| Positive | 741(36.4%) | 674 (34.7%) | |
| Negative | 1295 (63.6%) | 1266 (65.2%) | |
| Not available | 0 | 1 (0.1%) | |

**Table 1.** *Cont.*

| Characteristic | Elevated AFP (≥20 ng/mL), $n$ = 2036 | Normal AFP (<20 ng/mL), $n$ = 1941 | $p$ |
|---|---|---|---|
| Cirrhosis | | | 0.001 |
| Yes | 1469 (72.2%) | 1294 (66.7%) | |
| No | 561 (27.6%) | 640 (33.0%) | |
| Not available | 7 (0.4%) | 6 (0.3%) | |
| Child–Pugh class | | | <0.001 |
| A | 1587 (77.9%) | 1627 (83.80%) | |
| B | 345 (16.9%) | 235 (12.1%) | |
| C | 70 (3.4%) | 51 (2.6%) | |
| Not available | 34 (1.7%) | 28 (1.4%) | |
| Creatinine (mg/dL) | 1.0 (0.8–1.3) | 1.0 (0.8–1.3) | 0.731 |
| Total bilirubin (mg/dL) | 1.1 (0.8–1.7) | 1.0 (0.7–1.4) | <0.001 |
| INR | 1.0 (1.0–1.1) | 1.0 (1.0–1.1) | <0.001 |
| Tumor size (mm) | 46 (26–98) | 28 (20–45) | <0.001 |
| Tumor number | | | <0.001 |
| Single | 1064 (52.3%) | 1360 (70.1%) | |
| Multiple | 972 (47.7%) | 581 (29.9%) | |
| 7th edition AJCC stage | | | <0.001 |
| 1 | 788 (38.7%) | 1168 (60.2%) | |
| 2 | 340 (16.7%) | 414 (21.3%) | |
| 3 | 617 (30.3%) | 233 (12.0%) | |
| 4 | 279 (13.7%) | 72 (3.7%) | |
| Not available | 12 (0.6%) | 54 (2.8%) | |
| BCLC stage | | | <0.001 |
| 0 | 206 (10.1%) | 293 (15.1%) | |
| A | 566 (27.3%) | 867 (44.7%) | |
| B | 446 (21.9%) | 362 (18.7%) | |
| C | 676 (33.2%) | 279 (14.4%) | |
| D | 124 (6.1%) | 81(4.2%) | |
| Not available | 28 (1.4%) | 59 (3.0%) | |

AFP, alpha-fetoprotein; BMI, body mass index; AUD, alcohol use disorder; HBsAg, hepatitis B surface antigen; anti-HCV, anti-hepatitis C virus antibody; INR, international normalized ratio; AJCC, American Joint Committee on Cancer; BCLC, Barcelona Clinic Liver Cancer.

### 3.4. Variables Associated with Elevated AFP Level

Univariate analysis showed that the following variables were associated with elevated AFP levels: tumor size per 10 mm increase (OR = 1.167; 95% CI = 1.146–1.189; $p < 0.001$); using AJCC stage 1 as the reference, AJCC stage 2 (OR = 1.202; 95% CI = 1.013–1.426; $p = 0.035$), AJCC stage 3 (OR =3.909; 95% CI = 3.270–4.674; $p < 0.001$), and AJCC stage 4 (OR = 6.071; 95% CI = 4.553–8.096; $p < 0.001$); multiple tumors (OR = 2.076; 95% CI = 1.818–2.371; $p < 0.001$); using BCLC stages 0–A as the reference, BCLC stages B–D (OR = 2.640; 95% CI = 2.317–3.009; $p < 0.001$); cirrhosis (OR = 1.275; 95% CI = 1.110–1.465; $p = 0.001$); Child–Pugh class B or C (OR = 1.525; 95% CI = 1.287–1.806; $p < 0.001$); total bilirubin > 1.4 mg/dL (OR = 1.454; 95% CI = 1.262–1.676; $p < 0.001$); INR > 1.2 (OR = 1.238; 95% CI = 1.002–1.530; $p = 0.048$); and HBV- or HCV-positive status (OR = 1.355; 95% CI = 1.160–1.583; $p < 0.001$). Because of the strong correlation between AJCC and BCLC stages (correlation coefficient = 0.658, $p < 0.001$), we only selected the BCLC stage for multivariate analysis to avoid collinearity (Table 2).

Multivariate analysis showed that the following variables were associated with elevated AFP levels: female gender (OR = 1.462; 95% CI = 1.256–1.701; $p < 0.001$); tumor size per 10 mm increase (OR = 1.155; 95% CI = 1.127–1.183; $p < 0.001$); multiple tumors (OR = 1.406; 95% CI = 1.205–1.641; $p < 0.001$); BCLC stages B–D (OR = 1.247; 95% CI = 1.037–1.500; $p = 0.019$); cirrhosis (OR = 1.288; 95% CI = 1.099–1.509; $p = 0.02$); total bilirubin > 1.4 mg/dL (OR = 1.218; 95% CI = 1.020–1.455; $p = 0.030$); and HBV- or HCV-positive status (OR = 1.720; 95% CI = 1.451–2.038; $p < 0.001$) (Table 2).

**Table 2.** Univariate and multivariate analyses of factors associated with elevated AFP levels.

| Variable | Univariate OR (95% CI) | *p* | Multivariate OR (95% CI) | *p* |
|---|---|---|---|---|
| Age (per 10 years) | 0.955 (0.904–1.010) | 0.105 | | |
| Female vs. male | 1.20 (1.04–1.39) | 0.011 | 1.462 (1.256–1.701) | <0.001 |
| AUD | 1.077 (0.887–1.306) | 0.454 | | |
| HBsAg or anti-HCV-positive | 1.355 (1.160–1.583) | <0.001 | 1.720 (1.451–2.038) | <0.001 |
| Cirrhosis | 1.275 (1.110–1.465) | 0.001 | 1.288 (1.099–1.509) | 0.02 |
| Child–Pugh class B or C vs. A | 1.525(1.287–1.806) | <0.001 | 0.964 (0.770–1.205) | 0.745 |
| Creatinine >1.2 mg/dL | 0.983 (0.849–1.138) | 0.821 | | |
| Total bilirubin >1.4 mg/dL | 1.454 (1.262–1.676) | <0.001 | 1.218 (1.020–1.455) | 0.030 |
| INR >1.2 | 1.238 (1.002–1.530) | 0.048 | 0.897 (0.687–1.172) | 0.425 |
| Tumor size, per 10 mm increase | 1.167 (1.146–1.189) | <0.001 | 1.155 (1.127–1.183) | <0.001 |
| Multiple tumors | 2.076 (1.818–2.371) | <0.001 | 1.406 (1.205–1.641) | <0.001 |
| BCLC stage | | | | |
| (O–A as reference) | | | | |
| B–D | 2.640 (2.317–3.009) | <0.001 | 1.247 (1.037–1.500) | 0.019 |
| 7th edition AJCC | | | | |
| Stage 1 as reference | | | | |
| Stage 2 | 1.202 (1.013–1.426) | 0.035 | | |
| Stage 3 | 3.909 (3.270–4.674) | <0.001 | | |
| Stage 4 | 6.071 (4.553–8.096) | <0.001 | | |

AFP, alpha-fetoprotein; AUD, alcohol use disorder; HBsAg, hepatitis B surface antigen; anti-HCV, anti-hepatitis C virus antibody; INR, international normalized ratio; AJCC, American Joint Committee on Cancer; BCLC, Barcelona Clinic Liver Cancer.

### 3.5. *Proportion of BCLC Stages 0–A Patients in NBNC-HCC vs. HBV- or HCV-Related HCC*

Of the 870 patients with NBNC-HCC, 315 (36.2%) were in BCLC stages 0–A, 529 (60.8%) were in BCLC stages B–D, and 26 (0.3%) were of unknown BCLC stage. Of the 3107 HBV- or HCV-positive patients, 1607 (51.7%) were in BCLC stages 0–A, 1439 (46.3%) were in BCLC stages B–D, and 61 (0.2%) were of unknown BCLC stage. A significantly lower proportion of NBNC-HCC patients were in BCLC stages 0–A compared to HBV- or HCV-related HCC patients ($p < 0.001$).

## 4. Discussion

In this study, we demonstrated that overall 51.2% of patients with HCC had elevated AFP levels. The proportion of patients with an elevated AFP level was stationary in the period from 2011 to 2020. The proportion of patients with BCLC stages 0–A HCC decreased from 2011 to 2020, whereas the proportion of patients with NBNC-HCC increased in the same period. Furthermore, the proportion of patients with early-stage HCC (i.e., BCLC stages 0–A) was lower for NBNC-HCC than for HBV- or HCV-related HCC. Our previous study reported that the HCC cases in our institution accounted for 9.8% of the total cases at the national level [18].

Independent factors associated with elevated AFP levels included female gender, increased tumor size, multiple tumors, BCLC stages B–D, cirrhosis, total bilirubin > 1.4 mg/dL, and viral etiology. Advanced tumor stage and viral etiology were associated with elevated AFP levels. Between 2011 and 2020, the proportion of patients with NBNC-HCC increased (which would have led to a decreasing trend in AFP elevation), which was counterbalanced by the decreased proportion of patients with BCLC stages 0–A HCC (which would have led to an increasing trend in AFP elevation). Ultimately, the proportion of patients with elevated AFP levels was stationary between the years 2011 and 2020 in this study. Furthermore, the decreasing proportion of patients with BCLC stages 0–A HCC during this period may be due to the concurrent increase in the proportion of patients with NBNC-HCC, because the BCLC stages 0–A were less frequently found in NBNC-HCC patients compared to HBV- or HCV-related HCC patients, a result that agrees with the findings of a previous study [19]. In that study, patients with NBNC-HCC presented with larger tumors and at later stages of disease compared to patients with virus-related HCC [19]. This result may be due to the low rate of HCC surveillance in NBNC-HCC patients; furthermore, a significant proportion of NBNC-HCC patients could be NAFLD-related cases [20].

A recent study from the US reported an overall 62.6% of HCC patients with elevated AFP levels (i.e., ≥20 ng/mL) at the time of diagnosis. Between 2010 and 2017, there was a decline in the percentage of HCC patients with elevated AFP levels (68.2% vs. 57.5%). Furthermore, the decline was most evident among patients with early-stage tumors (i.e., seventh edition AJCC stage 1), from 55.7% in 2010 to 40.7% in 2017. However, the authors did not investigate the potential cause of these results. They assumed that these results were likely due to the increasing trend in early-stage tumor detection and the shift from viral to nonviral etiology [6]. In contrast, in this study, an overall 51.2% of patients with HCC had elevated AFP levels. The proportion of patients with elevated AFP levels was stationary in the period from 2011 to 2020 for all patients and across different AJCC stages in the present study. However, it is unclear why there is a discrepancy between this study and the US study [6].

Previous studies have shown that female gender, viral etiology, severe underlying liver disease, and advanced tumor stage are independently associated with elevated levels of AFP [5,6], findings that are compatible with those of the present study. Interestingly, female gender is associated with elevated AFP levels in the present study and previous large-scale studies [5,6]. However, the underlying mechanism is still unknown. It is well known that severe liver disease and advanced tumor stage are associated with elevated AFP levels [5,6,21]. The link between viral etiology and elevated AFP may be due to the former's association with cirrhosis. A previous study demonstrated that NAFLD is the leading cause of non-cirrhotic HCC [22].

In Taiwan, HBV is the leading etiology of HCC [18]. In 1984, Taiwan established a universal HBV vaccination program for newborns [23,24]. HBV vaccination has reduced the incidence of HCC in children and adolescents [23,24]. Antiviral therapy for HBV and HCV has been widely used in Taiwan since 2003. Antiviral therapies reduce the incidence of HBV- and HCV-related HCC [25,26]. Consequently, the incidence of HBV- and HCV-related HCC was expected to decrease. Indeed, of the 3843 HCC patients from five medical centers in Taiwan enrolled by Chang et al. in their study during 2005–2011, only 10.7% had NBNC-HCC [27], in contrast to 20.4% in 2011 and the gradual rise to 28.7% in 2020 in the present study. Therefore, we infer that the universal HBV vaccination program for newborns and the widely employed antiviral therapy for HBV and HCV have resulted in the decreasing incidence of HBV- and HCV-related HCC in Taiwan.

The etiologies of NBNC-HCC may be AUD and NAFLD; other etiologies, such as autoimmune hepatitis, primary biliary cirrhosis, and primary sclerosing cholangitis, related to chronic liver disease are rare [22]. Patients with NBNC-HCC have an increased risk of metabolic comorbidities [28], which implies that a significant proportion of NBNC-HCC cases could be NAFLD-related.

The incidence of HCC attributed to a nonviral etiology (mainly NAFLD) is rising [29,30]; in addition, it is associated with a decline in the percentage of patients with elevated AFP levels. Other tumor markers more specific to NBNC-HCC are needed for HCC surveillance in this population. Another available tumor marker for HCC surveillance is des-γ-carboxyprothrombin (DCP), which is recommended in clinical practice guidelines by the Japan Society of Hepatology [31]. Previous studies have demonstrated that elevated DCP might be a diagnostic marker for NBNC-HCC [32,33].

Due to the obesity pandemic, the challenge is how to screen for HCC in patients with NAFLD. The American Gastroenterological Association (AGA) guidelines recommend that screening for HCC should be considered for cirrhotic patients due to NAFLD. When the quality of ultrasound is suboptimal for HCC screening (e.g., due to obesity), future surveillance should be performed using CT or MRI, with or without determining the AFP level, every 6 months [34]. However, the cost-effectiveness of HCC surveillance, if CT or MRI replaces ultrasound, remains unknown [34].

Modern molecular biology-based technologies (e.g., liquid biopsy) hold considerable promise for early diagnosis of HCC. To date, there are still no US Food and Drug Administration (FDA)-approved liquid biopsy assays for HCC, mainly due to the lack of survival benefit of such assays [35].

The strength of the present study is the use of a large cohort of patients with HCC with prospectively collected data and limited missing data. However, the study has several limitations. First, we did not use the Alcohol Use Disorders Inventory Test (AUDIT) [36] or AUDIT-C [37,38] (which is recommended by the European Association for the Study of the Liver (EASL) guidelines [39]) to screen HCC patients for AUD. In contrast, we reviewed medical records and used habitual drinking to define AUD, which may underestimate the prevalence of AUD in the present study. Second, the study lacked data on antiviral therapy for HBV and HCV, which can affect AFP levels. Previous studies have shown that the cutoff values of AFP for HCC surveillance are lower in HBV-related cirrhosis patients receiving nucleos(t)ide analogue therapy [40] and HCV-related cirrhosis patients treated with DAAs [41]. Third, the study lacked data on HCV ribonucleic acid (RNA). A proportion of patients with anti-HCV-antibody-positive status could have experienced a past episode of resolved HCV infection. Fourth, etiologies other than HBV, HCV, and AUD were unavailable. Furthermore, we also did not have data on hepatic steatosis and metabolic comorbidities [42]. Therefore, we could not define NAFLD in the present study. Finally, this is a retrospective study.

## 5. Conclusions

In the past 10 years, a stationary trend in elevated serum AFP levels in HCC patients was noted in this cohort from an HBV-endemic area. This result may be due to the proportion of patients with early-stage HCC decreasing and the proportion of patients with NBNC-HCC increasing during this period. Furthermore, advanced tumor stage, severe underlying liver disease, female gender, and viral etiology were associated with elevated AFP in HCC patients.

**Author Contributions:** Conceptualization, Y.-H.Y. and C.-C.W.; methodology, Y.-H.Y.; software, C.-Y.L.; validation, all authors; formal analysis, C.-Y.L.; investigation, Y.-H.Y.; resources, C.-C.W.; data curation, all authors; writing—original draft preparation, Y.-H.Y.; writing—review and editing, C.-C.W.; visualization, all authors; supervision, all authors; project administration, Y.-H.Y.; funding acquisition, Y.-H.Y. All authors have read and agreed to the published version of the manuscript.

**Funding:** This study was supported by Grant CMRPG8L0181 from the Kaohsiung Chang Gung Memorial Hospital, Taiwan.

**Institutional Review Board Statement:** The Institutional Review Board of Kaohsiung Chang Gung Memorial Hospital approved this study (Reference number: 202201189B0) and waived the need for informed consent.

**Informed Consent Statement:** The Institutional Review Board of Kaohsiung Chang Gung Memorial Hospital waived the need for informed consent due to the retrospective and observational nature of the study design.

**Data Availability Statement:** The data can be shared up on request.

**Acknowledgments:** The authors thank the Cancer Center, Kaohsiung Chang Gung Memorial Hospital for the provision of HCC registry data. The authors thank Chih-Yun Lin and Nien-Tzu Hsu and the Biostatistics Center, Kaohsiung Chang Gung Memorial Hospital for statistics work.

**Conflicts of Interest:** The authors have no conflict of interest.

## References


1. Bray, F.; Ferlay, J.; Soerjomataram, I.; Siegel, R.L.; Torre, L.A.; Jemal, A. Global cancer statistics 2018: GLOBOCAN estimates of incidence and mortality worldwide for 36 cancers in 185 countries. *CA Cancer J. Clin.* **2018**, *68*, 394–424. [CrossRef] [PubMed]
2. Singal, A.G.; Pillai, A.; Tiro, J. Early detection, curative treatment, and survival rates for hepatocellular carcinoma surveillance in patients with cirrhosis: A meta-analysis. *PLoS Med.* **2014**, *11*, e1001624. [CrossRef] [PubMed]
3. Marrero, J.A.; Kulik, L.M.; Sirlin, C.B.; Zhu, A.X.; Finn, R.S.; Abecassis, M.M.; Roberts, L.R.; Heimbach, J.K. Diagnosis, Staging, and Management of Hepatocellular Carcinoma: 2018 Practice Guidance by the American Association for the Study of Liver Diseases. *Hepatology* **2018**, *68*, 723–750. [CrossRef] [PubMed]
4. Tzartzeva, K.; Obi, J.; Rich, N.E.; Parikh, N.D.; Marrero, J.A.; Yopp, A.; Waljee, A.K.; Singal, A.G. Surveillance Imaging and Alpha Fetoprotein for Early Detection of Hepatocellular Carcinoma in Patients with Cirrhosis: A Meta-analysis. *Gastroenterology* **2018**, *154*, 1706–1718.e1. [CrossRef]
5. Giannini, E.G.; Sammito, G.; Farinati, F.; Ciccarese, F.; Pecorelli, A.; Rapaccini, G.L.; Di Marco, M.; Caturelli, E.; Zoli, M.; Borzio, F.; et al. Italian Liver Cancer (ITA.LI.CA) Group. Determinants of alpha-fetoprotein levels in patients with hepatocellular carcinoma: Implications for its clinical use. *Cancer* **2014**, *120*, 2150–2157. [CrossRef]
6. Vipani, A.; Lauzon, M.; Luu, M.; Roberts, L.R.; Singal, A.G.; Yang, J.D. Decreasing Trend of Serum α-Fetoprotein Level in Hepatocellular Carcinoma. *Clin. Gastroenterol. Hepatol.* **2022**, *20*, 1177–1179.e4. [CrossRef]
7. Global Burden of Disease Liver Cancer Collaboration; Akinyemiju, T.; Abera, S.; Ahmed, M.; Alam, N.; Alemayohu, M.A.; Allen, C.; Al-Raddadi, R.; Alvis-Guzman, N.; Amoako, Y.; et al. The burden of primary liver cancer and underlying etiologies from 1990 to 2015 at the global, regional, and national level. *JAMA Oncol.* **2017**, *3*, 1683–1691.
8. Kanwal, F.; Kramer, J.; Asch, S.M.; Chayanupatkul, M.; Cao, Y.; El-Serag, H.B. Risk of hepatocellular cancer in HCV patients treated with direct-acting antiviral agents. *Gastroenterology* **2017**, *153*, 996–1005.e1. [CrossRef]
9. Estes, C.; Razavi, H.; Loomba, R.; Younossi, Z.; Sanyal, A.J. Modeling the epidemic of nonalcoholic fatty liver disease demonstrates an exponential increase in burden of disease. *Hepatology* **2018**, *67*, 123–133. [CrossRef]
10. Li, J.; Zou, B.; Yeo, Y.H.; Feng, Y.; Xie, X.; Lee, D.H.; Fujii, H.; Wu, Y.; Kam, L.Y.; Ji, F.; et al. Prevalence, incidence, and outcome of non-alcoholic fatty liver disease in Asia, 1999–2019: A systematic review and meta-analysis. *Lancet Gastroenterol. Hepatol.* **2019**, *4*, 389–398. [CrossRef]
11. American Joint Committee on Cancer. *American Joint Committee on Cancer Staging Manual*, 7th ed.; Edge, S.B., Byrd, D.R., Compton, C.C., Fritz, A.G., Greene, F.L., Rotti, A., III, Eds.; Springer: New York, NY, USA, 2010; p. 175.
12. Llovet, J.M.; Bru, C.; Bruix, J. Prognosis of hepatocellular carcinoma: The BCLC staging classification. *Semin. Liver Dis.* **1999**, *19*, 329–338. [CrossRef]
13. Pugh, R.N.; Murray-Lyon, I.M.; Dawson, J.L.; Pietroni, M.C.; Williams, R. Transection of the oesophagus for bleeding oesophageal varices. *Br. J. Surg.* **1973**, *60*, 646–649. [CrossRef] [PubMed]
14. Everhart, J.E.; Wright, E.C.; Goodman, Z.D.; Dienstag, J.L.; Hoefs, J.C.; Kleiner, D.E.; Ghany, M.G.; Mills, A.S.; Nash, S.R.; Govindarajan, S.; et al. Prognostic value of Ishak fibrosis stage: Findings from the hepatitis C antiviral long-term treatment against cirrhosis trial. *Hepatology* **2010**, *51*, 585–594. [CrossRef] [PubMed]
15. Shiha, G.; Ibrahim, A.; Helmy, A.; Sarin, S.K.; Omata, M.; Kumar, A.; Bernstien, D.; Maruyama, H.; Saraswat, V.; Chawla, Y.; et al. Asian-Pacific Association for the Study of the Liver (APASL) consensus guidelines on invasive and non-invasive assessment of hepatic fibrosis: A 2016 update. *Hepatol. Int.* **2017**, *11*, 1–30. [CrossRef]
16. Mazzaferro, V.; Regalia, E.; Doci, R.; Andreola, S.; Pulvirenti, A.; Bozzetti, F.; Montalto, F.; Ammatuna, M.; Morabito, A.; Gennari, L. Liver transplantation for the treatment of small hepatocellular carcinomas in patients with cirrhosis. *N. Engl. J. Med.* **1996**, *334*, 693–699. [CrossRef] [PubMed]
17. Lüdecke, D. sjPlot: Da ta Visualization for Statistics in Social Science. R Package Version 2.8.8. 2021. Available online: https://CRAN.R-project.org/package\protect$\relax\protect{\begingroup1\endgroup\@@over4}$sjPlot (accessed on 1 January 2020).
18. Lin, S.H.; Lin, C.Y.; Hsu, N.T.; Yen, Y.H.; Kee, K.M.; Wang, J.H.; Hu, T.H.; Chen, C.H.; Hung, C.H.; Chen, C.H.; et al. Reappraisal of the roles of alpha-fetoprotein in hepatocellular carcinoma surveillance using large-scale nationwide database and hospital-based information. *J. Formos. Med. Assoc.* **2022**, *121*, 2085–2092. [CrossRef] [PubMed]
19. Jun, T.W.; Yeh, M.L.; Yang, J.D.; Chen, V.L.; Nguyen, P.; Giama, N.H.; Huang, C.F.; Hsing, A.W.; Dai, C.Y.; Huang, J.F.; et al. More advanced disease and worse survival in cryptogenic compared to viral hepatocellular carcinoma. *Liver Int.* **2018**, *38*, 895–902. [CrossRef]
20. Chen, V.L.; Yeh, M.L.; Yang, J.D.; Leong, J.; Huang, D.Q.; Toyoda, H.; Chen, Y.L.; Guy, J.; Maeda, M.; Tsai, P.C.; et al. Effects of Cirrhosis and Diagnosis Scenario in Metabolic-Associated Fatty Liver Disease-Related Hepatocellular Carcinoma. *Hepatol. Commun.* **2020**, *5*, 122–132. [CrossRef]
21. Richardson, P.; Duan, Z.; Kramer, J.; Davila, J.A.; Tyson, G.L.; El-Serag, H.B. Determinants of serum alpha-fetoprotein levels in hepatitis C-infected patients. *Clin. Gastroenterol. Hepatol.* **2012**, *10*, 428–433. [CrossRef]
22. Gawrieh, S.; Dakhoul, L.; Miller, E.; Scanga, A.; deLemos, A.; Kettler, C.; Burney, H.; Liu, H.; Abu-Sbeih, H.; Chalasani, N.; et al. Characteristics, aetiologies and trends of hepatocellular carcinoma in patients without cirrhosis: A United States multicentre study. *Aliment. Pharmacol. Ther.* **2019**, *50*, 809–821. [CrossRef]

23. Chang, M.H.; Chen, C.J.; Lai, M.S.; Hsu, H.M.; Wu, T.C.; Kong, M.S.; Liang, D.C.; Shau, W.Y.; Chen, D.S. Universal hepatitis B vaccination in Taiwan and the incidence of hepatocellular carcinoma in children. Taiwan Childhood Hepatoma Study Group. *N. Engl. J. Med.* **1997**, *336*, 1855–1859. [CrossRef] [PubMed]
24. Chang, M.H.; You, S.L.; Chen, C.J.; Liu, C.J.; Lee, C.M.; Lin, S.M.; Chu, H.C.; Wu, T.C.; Yang, S.S.; Kuo, H.S.; et al. Taiwan Hepatoma Study Group. Decreased incidence of hepatocellular carcinoma in hepatitis B vaccinees: A 20-year follow-up study. *J. Natl. Cancer Inst.* **2009**, *101*, 1348–1355. [CrossRef]
25. Liaw, Y.F.; Sung, J.J.; Chow, W.C.; Farrell, G.; Lee, C.Z.; Yuen, H.; Tanwandee, T.; Tao, Q.M.; Shue, K.; Keene, O.N.; et al. Lamivudine for patients with chronic hepatitis B and advanced liver disease. *N. Engl. J. Med.* **2004**, *351*, 1521–1531. [CrossRef]
26. Beste, L.A.; Green, P.; Berry, K.; Belperio, P.; Ioannou, G.N. Hepatitis C-Related Hepatocellular Carcinoma Incidence in the Veterans Health Administration After Introduction of Direct-Acting Antivirals. *JAMA* **2020**, *324*, 1003–1005. [CrossRef] [PubMed]
27. Chang, I.C.; Huang, S.F.; Chen, P.J.; Chen, C.L.; Chen, C.L.; Wu, C.C.; Tsai, C.C.; Lee, P.H.; Chen, M.F.; Lee, C.M.; et al. The hepatitis viral status in patients with hepatocellular carcinoma: A study of 3843 patients from Taiwan Liver Cancer Network. *Medicine* **2016**, *95*, e3284. [CrossRef] [PubMed]
28. Hsu, P.Y.; Hsu, C.T.; Yeh, M.L.; Huang, C.F.; Huang, C.I.; Liang, P.C.; Lin, Y.H.; Hsieh, M.Y.; Wei, Y.J.; Hsieh, M.H.; et al. Early Fibrosis but Late Tumor Stage and Worse Outcomes in Hepatocellular Carcinoma Patients without Hepatitis B or Hepatitis C. *Dig. Dis. Sci.* **2020**, *65*, 2120–2129. [CrossRef] [PubMed]
29. El-Serag, H.B.; Kanwal, F.; Feng, Z.; Marrero, J.A.; Khaderi, S.; Singal, A.G. Risk factors for cirrhosis in contemporary hepatology practices-findings from the Texas hepatocellular carcinoma Consortium cohort. *Gastroenterology* **2020**, *159*, 376–377. [CrossRef] [PubMed]
30. Huang, D.Q.; El-Serag, H.B.; Loomba, R. Global epidemiology of NAFLD-related HCC: Trends, predictions, risk factors and prevention. *Nat. Rev. Gastroenterol. Hepatol.* **2021**, *18*, 223–238. [CrossRef]
31. Kokudo, N.; Takemura, N.; Hasegawa, K.; Takayama, T.; Kubo, S.; Shimada, M.; Nagano, H.; Hatano, E.; Izumi, N.; Kaneko, S.; et al. Clinical practice guidelines for hepatocellular carcinoma: The Japan Society of Hepatology 2017 (4th JSH-HCC guidelines) 2019 update. *Hepatol. Res.* **2019**, *49*, 1109–1113. [CrossRef]
32. Hayashi, M.; Yamada, S.; Takano, N.; Okamura, Y.; Takami, H.; Inokawa, Y.; Sonohara, F.; Tanaka, N.; Shimizu, D.; Hattori, N.; et al. Different Characteristics of Serum Alfa Fetoprotein and Serum Des-gamma-carboxy Prothrombin in Resected Hepatocellular Carcinoma. *In Vivo* **2021**, *35*, 1749–1760.
33. Taura, N.; Ichikawa, T.; Miyaaki, H.; Ozawa, E.; Tsutsumi, T.; Tsuruta, S.; Kato, Y.; Goto, T.; Kinoshita, N.; Fukushima, M.; et al. Frequency of elevated biomarkers in patients with cryptogenic hepatocellular carcinoma. *Med. Sci. Monit.* **2013**, *19*, 742–750. [CrossRef]
34. Loomba, R.; Lim, J.K.; Patton, H.; El-Serag, H.B. AGA Clinical Practice Update on Screening and Surveillance for Hepatocellular Carcinoma in Patients With Nonalcoholic Fatty Liver Disease: Expert Review. *Gastroenterology* **2020**, *158*, 1822–1830. [CrossRef] [PubMed]
35. Johnson, P.; Zhou, Q.; Dao, D.Y.; Lo, Y.M.D. Circulating biomarkers in the diagnosis and management of hepatocellular carcinoma. *Nat. Rev. Gastroenterol. Hepatol.* **2022**, *19*, 670–681. [CrossRef] [PubMed]
36. Saunders, J.B.; Aasland, O.G.; Babor, T.F.; de la Fuente, J.R.; Grant, M. Development of the alcohol use disorders identification test (AUDIT): WHO collaborative project on early detection of persons with harmful alcohol consumption–II. *Addiction* **1993**, *88*, 791–804. [CrossRef] [PubMed]
37. Bush, K.; Kivlahan, D.R.; McDonell, M.B.; Fihn, S.D.; Bradley, K.A. The AUDIT alcohol consumption questions (AUDIT-C): An effective brief screening test for problem drinking. Ambulatory Care Quality Improvement Project (ACQUIP). Alcohol Use Disorders Identification Test. *Arch. Intern. Med.* **1998**, *158*, 1789–1795. [CrossRef] [PubMed]
38. Gual, A.; Segura, L.; Contel, M.; Heather, N.; Colom, J. Audit-3 and audit-4: Effectiveness of two short forms of the alcohol use disorders identification test. *Alcohol Alcohol.* **2002**, *37*, 591–596. [CrossRef] [PubMed]
39. European Association for the Study of the Liver. EASL Clinical Practice Guidelines: Management of alcohol-related liver disease. *J. Hepatol.* **2018**, *69*, 154–181. [CrossRef]
40. Su, T.H.; Peng, C.Y.; Chang, S.H.; Tseng, T.C.; Liu, C.J.; Chen, C.L.; Liu, C.H.; Yang, H.C.; Chen, P.J.; Kao, J.H. Serum PIVKA-II and alpha-fetoprotein at virological remission predicts hepatocellular carcinoma in chronic hepatitis B related cirrhosis. *J. Formos. Med. Assoc.* **2022**, *121*, 703–711. [CrossRef]
41. Degasperi, E.; Perbellini, R.; D'Ambrosio, R.; Uceda Renteria, S.C.; Ceriotti, F.; Perego, A.; Orsini, C.; Borghi, M.; Iavarone, M.; Bruccoleri, M.; et al. Prothrombin induced by vitamin K absence or antagonist-II and alpha foetoprotein to predict development of hepatocellular carcinoma in Caucasian patients with hepatitis C-related cirrhosis treated with direct-acting antiviral agents. *Aliment. Pharmacol. Ther.* **2022**, *55*, 350–359. [CrossRef] [PubMed]
42. Chalasani, N.; Younossi, Z.; Lavine, J.E.; Charlton, M.; Cusi, K.; Rinella, M.; Harrison, S.A.; Brunt, E.M.; Sanyal, A.J. The diagnosis and management of nonalcoholic fatty liver disease: Practice guidance from the American Association for the Study of Liver Diseases. *Hepatology* **2018**, *67*, 328–357. [CrossRef] [PubMed]

*cancers*

MDPI

*Article*

# Alpha-Fetoprotein Combined with Radiographic Tumor Burden Score to Predict Overall Survival after Liver Resection in Hepatocellular Carcinoma

**Yi-Hao Yen [1,*,†], Yueh-Wei Liu [2], Wei-Feng Li [2], Chih-Chi Wang [2,*,†], Chee-Chien Yong [2], Chih-Che Lin [2] and Chih-Yun Lin [3]**

1 Division of Hepatogastroenterology, Department of Internal Medicine, Kaohsiung Chang Gung Memorial Hospital, College of Medicine, Chang Gung University, 123 Ta Pei Road, Kaohsiung 833401, Taiwan
2 Liver Transplantation Center, Department of Surgery, Kaohsiung Chang Gung Memorial Hospital, 123 Ta Pei Road, Kaohsiung 833401, Taiwan
3 Biostatistics Center of Kaohsiung Chang Gung Memorial Hospital, Kaohsiung 833401, Taiwan
* Correspondence: cassellyen@yahoo.com.tw (Y.-H.Y.); ufel4996@ms26.hinet.net (C.-C.W.); Tel.: +886-7-7317123 (Y.-H.Y. & C.-C.W.); Fax: +886-7-7318762 (Y.-H.Y. & C.-C.W.)
† These authors contributed equally to this work.

**Simple Summary:** The tumor burden score (TBS) is calculated using the Pythagorean theorem based on the largest tumor size and tumor number ($\alpha^2 + \beta^2 = \gamma^2$, where $\alpha$ is the largest tumor size, $\beta$ is the tumor number, and $\gamma$ is the TBS). Patients who underwent liver resection (LR) for Barcelona Clinic Liver Cancer stage 0, A, or B hepatocellular carcinoma (HCC) between 2011 and 2018 were enrolled. Among 743 patients, 193 (26.0%) patients had a low TBS (<2.6), 474 (63.8%) had a moderate TBS (2.6–7.9), and 75 (10.1%) had a high TBS (>7.9). Combining radiographic TBS and alpha-fetoprotein levels could stratify overall survival among HCC patients after LR.

**Abstract:** We evaluated whether combining the radiographic tumor burden score (TBS) and alpha-fetoprotein (AFP) level could be used to stratify overall survival (OS) among hepatocellular carcinoma (HCC) patients after liver resection (LR). Patients who underwent LR for Barcelona Clinic Liver Cancer stage 0, A, or B HCC between 2011 and 2018 were enrolled. TBS scores were calculated using the following equation: $TBS^2$ = (largest tumor size (in cm))$^2$ + (tumor number)$^2$. Among 743 patients, 193 (26.0%) patients had a low TBS (<2.6), 474 (63.8%) had a moderate TBS (2.6–7.9), and 75 (10.1%) had a high TBS (>7.9). Those with a TBS $\leq$ 7.9 and AFP < 400 ng/mL had a significantly better OS than those with a TBS > 7.9 and an AFP < 400 ng/mL ($p$ = 0.003) or $\geq$ 400 ng/mL ($p$ < 0.001). A multivariate analysis using TBS $\leq$ 7.9 and AFP < 400 ng/mL as the reference values showed that a TBS > 7.9 and an AFP < 400 ng/mL (hazard ratio (HR): 2.063; 95% confidence interval [CI]: 1.175–3.623; $p$ = 0.012) or $\geq$ 400 ng/mL (HR: 6.570; 95% CI: 3.684–11.719; $p$ < 0.001) were independent predictors of OS. In conclusion, combining radiographic TBSs and AFP levels could stratify OS among HCC patients undergoing LR.

**Keywords:** tumor burden score; hepatocellular carcinoma; alpha-fetoprotein; liver resection

**Citation:** Yen, Y.-H.; Liu, Y.-W.; Li, W.-F.; Wang, C.-C.; Yong, C.-C.; Lin, C.-C.; Lin, C.-Y. Alpha-Fetoprotein Combined with Radiographic Tumor Burden Score to Predict Overall Survival after Liver Resection in Hepatocellular Carcinoma. *Cancers* **2023**, *15*, 1203. https://doi.org/10.3390/cancers15041203

Academic Editor: Georgios Germanidis

Received: 31 December 2022
Revised: 4 February 2023
Accepted: 9 February 2023
Published: 14 February 2023



## 1. Introduction

Hepatocellular carcinoma (HCC) is one of the leading causes of cancer-related deaths worldwide [1–3]. Liver resection (LR) is a primary treatment modality for HCC [1–3] and has been shown to improve survival across Barcelona Clinic Liver Cancer (BCLC) stages [4,5]. However, LR is also associated with higher risks compared with non-surgical treatment [2,3]. Therefore, preoperative prognostic predictions and the assessment of risk-to-benefit ratios in patients with HCC is critical when determining which patients should undergo LR.

A previous study established the tumor burden score (TBS), which is calculated using the largest tumor size and tumor number in the Pythagorean theorem ($\alpha^2 + \beta^2 = \gamma^2$, where $\alpha$ = largest tumor size, $\beta$ = tumor number, and $\gamma$ = TBS) [6]. The TBS showed a satisfactory ability to stratify prognostic outcomes among patients with HCC who underwent LR [6]. Further, Tsilimigras et al. reported that serum alpha-fetoprotein (AFP) and pathological TBS had synergistic impacts on prognosis following LR for HCC [7]. However, Tsilimigras et al. used pathological TBS, which is not suitable for preoperative prognostic predictions [7]. We aimed to use radiographic TBS combined with AFP evaluation to predict prognosis in patients with BCLC stages 0, A, and B HCC who underwent LR.

## 2. Materials and Methods

The study was conducted according to the guidelines of the Declaration of Helsinki and approved by the Institutional Review Board of Chang Gung Memorial Hospital —Kaohsiung branch (reference number: 202201189B0; approval date: 8 August 2022).

Data were extracted from the Kaohsiung Chang Gung Memorial Hospital HCC registry data. These data are prospectively collected and updated annually. The vital status of every patient is updated annually through a link to the website of the Taiwan Ministry of Health and Welfare (Cancer Screening and Tracing Information Integrated System for Taiwan; https://hosplab.hpa.gov.tw/CSTIIS/index.aspx, accessed on 1 January 2007).

From 2011 to 2018, patients with newly diagnosed HCC who underwent LR at Kaohsiung Chang Gung Memorial Hospital were enrolled in this study. The inclusion criteria were BCLC stage 0, A, and B HCC. The exclusion criteria were unknown preoperative AFP level, unknown tumor differentiation status, age < 18 years, unknown pathological stage, pathological stage N1 or M1, and non-curative LR. Curative LR was defined as the complete resection of all macroscopic tumors with microscopically negative surgical margins. After the application of all inclusion and exclusion criteria, 743 patients who underwent LR during 2011–2018 were enrolled. The raw data for all 743 included patients are available via the following digital object identifier: https://www.dropbox.com/scl/fi/unezavq59ethkc6sprtu9/raw-data.xlsx?dl=0&rlkey=cwu1zq6f9bguxiq0ghqqfvvh3, accessed on 1 January 2023.

Tumor sizes and numbers were assessed using preoperative contrast-enhanced computed tomography or magnetic resonance imaging. Tumor differentiation was assessed using Edmondson and Steiner's classification [8]. Fibrosis was assessed using Ishak scores [9]. Cirrhosis was defined as an Ishak score of 5 or 6. Major resection was defined as the resection of $\geq$3 Couinaud segments.

The Seventh American Joint Committee on Cancer (AJCC)/tumor–node–metastasis (TNM) staging criteria [10] were applied to our HCC registry data from 2011 to 2017, and the Eighth AJCC/TNM staging criteria [11] were applied to our HCC registry data starting in 2018. Therefore, we present the pathological T-stage in this study as stage 1 or 2 versus stage 3 or 4. A T-stage 1 or 2 includes the detection of a single tumor, with or without vascular invasion, or multiple tumors, none of which are >5 cm. T-stages 3 or 4 include the detection of multiple tumors, among which any are >5 cm, or tumors with major vascular invasion, the perforation of the visceral peritoneum, or the direct invasion of adjacent organs other than the gallbladder.

## 3. Outcome Measurement

The primary outcome measure was overall survival (OS), which was defined as the interval between the date of LR and the date of the last follow-up or death.

## 4. Definitions

The detection of a single tumor $\leq$ 2 cm was defined as BCLC stage 0; a single tumor > 2 cm, or 2–3 tumors $\leq$ 3 cm, were defined as BCLC stage A; and 2–3 tumors $\geq$ 3 cm, or $\geq$ 4 tumors, were defined as BCLC stage B [5]. Tumor sizes were defined by the size of the largest tumor if multiple tumors were detected. TBS scores [6] were determined using the following equation: $TBS^2$ = (largest tumor size (in cm))$^2$ + (tumor number)$^2$. The tumor number and the largest tumor size were determined based on the results of imaging studies. Cutoff values for TBS were determined according to OS by using X-tile [12], a bioinformatics tool created by Camp et al.

## 5. Statistical Analysis

The patients' characteristics are presented as the number (percentage) of patients matching the specific characteristics, and they were compared using the Chi-square test. Comparisons of OS between groups were performed using the Kaplan–Meier survival curves and the log-rank test. Covariates in the multivariate model were chosen a priori, based on established clinical relevance. The potential confounders included age, the presence of cirrhosis, TBS, AFP level, and tumor differentiation status [2,3,5,6]. These variables were fully adjusted in the multivariate model. The results are presented as a hazard ratio (HR) with a 95% confidence interval (CI). Statistical analyses were performed using SPSS version 22.0. Two-tailed significance values were calculated, and significance was defined as $p < 0.05$.

## 6. Results

### *6.1. Clinical and Pathological Characteristics of Patients*

A total of 743 patients, who underwent LR for HCC and met the inclusion and exclusion criteria, were enrolled in this study. The patients were divided into three groups according to the TBS: high TBS (>7.9; $n$ = 75, 10.1%), medium TBS (2.6–7.9; $n$ = 474, 63.8%), and low TBS (<2.6; $n$ = 193, 26%) (Figure 1). Overall, 136 patients (18.3%) had BCLC stage 0, 538 (72.4%) had BCLC stage A, and 69 (9.3%) had BCLC stage B. Among all 743 patients, 390 (52.5%) patients were hepatitis B surface antigen (HBsAg)-positive, 251 (33.8%) patients were anti-hepatitis C virus (HCV)-positive, 352 (47.4%) patients had a major resection, 414 (55.7%) patients were < 65 years of age, 572 (77.0%) patients were male, 623 (83.8%) patients had AFP < 400 ng/mL, and 731 (98.4%) patients were Child–Pugh Class A. Pathological examinations showed that 287 (38.6%) patients had cirrhosis, 29 (3.9%) patients had poorly differentiated tumors, and 63 (8.5%) patients had pathological T-stage 3 or 4 (Table 1). After combining TBS and AFP levels, four groups were generated: TBS $\leq$ 7.9/AFP < 400 ng/mL ($n$ = 577, 77.7%); TBS $\leq$ 7.9/AFP $\geq$ 400 ng/mL ($n$ = 90, 12.1%); TBS > 7.9/AFP < 400 ng/mL ($n$ = 46, 6.2%); and TBS > 7.9/AFP $\geq$ 400 ng/mL ($n$ = 30, 4.0%). The proportion of male patients (89.1%) was highest among the TBS > 7.9/AFP < 400 ng/mL group ($p$ = 0.014). The proportion of cirrhotic patients (15.2%) was lowest in the TBS > 7.9/AFP $\leq$ 400 ng/mL group ($p$ = 0.014). The proportion of patients who were anti-HCV-positive (36.4%) was highest in the TBS $\leq$ 7.9/AFP < 400 ng/mL group ($p$ = 0.023). The proportion of patients who underwent major resection (86.7%) was highest in the TBS > 7.9/AFP $\geq$ 400 ng/mL group ($p$ < 0.001). The proportion of patients with poor tumor differentiation (13.3%) was highest in the TBS > 7.9/AFP $\geq$ 400 ng/mL group ($p$ < 0.001). The proportion of patients with pathological stage-T 3 or 4 (60.0%) was highest in the TBS > 7.9/AFP $\geq$ 400 ng/mL group ($p$ < 0.001). No significant differences between groups were observed for the remaining variables (Table 2).

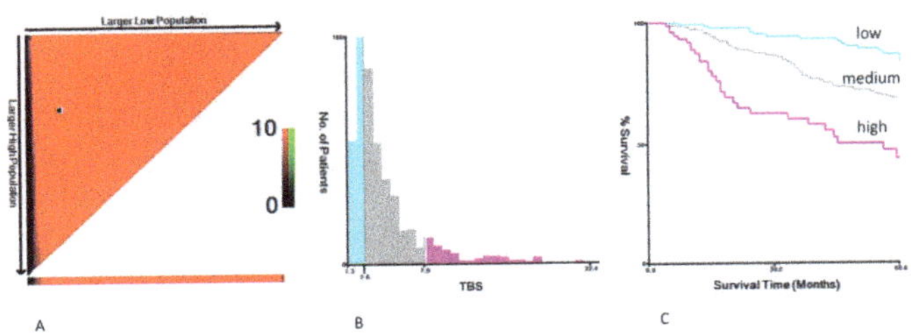


**Figure 1.** Cutoff tumor burden score (TBS) values were determined by overall survival using X-tile, a bioinformatics tool created by Camp et al. (**A**) Data represented graphically in a right-triangular

grid in which each pixel represents the data from a given set of divisions. The vertical axis represents all possible "high" populations, with the size of the high population increasing from top to bottom. Similarly, the horizontal axis represents all possible "low" populations, with the size of the low population increasing from left to right. (**B**) The number of patients in each group for a given set of divisions. (**C**) Kaplan–Meier curves show significant differences in overall survival among patients with a low TBS (i.e., <2.6), moderate TBS (i.e., 2.6–7.9), and high TBS (i.e., >7.9; $p < 0.001$).

**Table 1.** The patients' clinical and pathological characteristics.

| | **Total, *n* = 743** |
|---|---|
| Age (year) | |
| ≤65 | 414 (55.7%) |
| >65 | 329 (44.3%) |
| Sex | |
| Male | 572 (77.0%) |
| Female | 171 (23.0%) |
| Cirrhosis | |
| Presence | 287 (38.6%) |
| Absence | 420 (56.5%) |
| Unknown | 36 (4.8%) |
| HBsAg | |
| Positive | 390 (52.5%) |
| Negative | 353 (47.5%) |
| Anti-HCV | |
| Positive | 251 (33.8%) |
| Negative | 492 (66.2%) |
| AFP (ng/mL) | |
| <400 | 623 (83.8%) |
| ≥400 | 120 (16.2%) |
| Child–Pugh class | |
| A | 731 (98.4%) |
| B | 11 (1.5%) |
| Unknown | 1 (0.1%) |
| Type of resection | |
| Major | 352 (47.4%) |
| Minor | 394 (53.0) |
| Tumor differentiation | |
| Poor | 29 (3.9%) |
| Moderate | 658 (88.6%) |
| Well | 56 (7.5%) |
| Pathology T-stage | |
| 1–2 | 680 (91.5%) |
| 3–4 | 63 (8.5%) |
| Tumor burden score | |
| Low < 2.6 | 193 (26.0%) |
| Medium: 2.6–7.9 | 474 (63.8%) |
| High > 7.9 | 75 (10.1%) |
| BCLC stage | |
| 0 | 136 (18.3) |
| A | 538 (72.4%) |
| B | 69 (9.3%) |
| Radiographic tumor size (cm) | |
| ≤5 | 573 (77.1%) |
| >5 | 170 (22.9%) |
| Radiographic tumor number | |
| 1 | 637 (85.7%) |
| 2 | 78 (10.5%) |
| 3 | 16 (2.2%) |
| 4 | 11 (1.5%) |
| 5 | 1 (0.1%) |

BCLC, Barcelona Clinic Liver Cancer; HBsAg, hepatitis B virus surface antigen; HCV, hepatitis C virus; AFP, α-fetoprotein.

**Table 2.** Clinical and pathological characteristics of patients stratified by AFP and TBS levels.

| | TBS ≤ 7.9 and AFP < 400 ng/mL, $n$ = 577 | TBS ≤ 7.9 and AFP ≥ 400 ng/mL, $n$ = 90 | TBS > 7.9 and AFP < 400 ng/mL, $n$ = 46 | TBS > 7.9 and AFP ≥ 400 ng/mL, $n$ = 30 | $p$ |
|---|---|---|---|---|---|
| Age (year) | | | | | 0.283 |
| ≤65 | 381 (66%) | 58 (64.4%) | 27 (58.7%) | 24 (80%) | |
| >65 | 196 (34%) | 32 (35.6%) | 19 (41.3%) | 6 (20%) | |
| Sex | | | | | 0.014 |
| Male | 488 (77.6%) | 59 (65.6%) | 41 (89.1%) | 23 (76.7%) | |
| Female | 129 (22.4%) | 31 (34.4%) | 5 (10.9%) | 7 (23.3%) | |
| Cirrhosis | | | | | <0.001 |
| Yes | 231 (40%) | 43 (47.8%) | 7 (15.2%) | 6 (20%) | |
| No | 323 (56%) | 43 (47.8%) | 33 (71.7%) | 21 (70%) | |
| Unknown | 23 (4.0%) | 4 (4.4%) | 6 (13.0%) | 3 (10.0%) | |
| HBsAg | | | | | 0.076 |
| Positive | 302 (52.3%) | 52 (57.8%) | 17 (37%) | 19 (36.7%) | |
| Negative | 275 (47.7%) | 38 (42.2%) | 29 (63.0%) | 11 (63.3%) | |
| Anti-HCV | | | | | 0.023 |
| Positive | 210 (36.4%) | 26 (28.9%) | 12 (26.1%) | 4 (13.3%) | |
| Negative | 367 (63.6%) | 64 (71.1%) | 34 (73.9%) | 26 (86.7%) | |
| Child–Pugh class | | | | | 0.168 |
| A | 568 (98.4%) | 90 (100%) | 43 (93.5%) | 30 (100%) | |
| B | 8 (1.4%) | 0 | 3 (6.5%) | 0 | |
| Unknown | 1 (0.2%) | 0 | 0 | 0 | |
| Type of resection | | | | | <0.001 |
| Major | 245 (42.5%) | 41 (42.6%) | 37 (80.4%) | 26 (86.7%) | |
| Minor | 332 (57.5%) | 49 (54.4%) | 9 (19.6%) | 4 (13.3%) | |
| Tumor differentiation | | | | | <0.001 |
| Poor | 14 (2.4%) | 8 (8.9%) | 3 (6.5%) | 4 (13.3%) | |
| Moderate | 510 (88.4%) | 81 (90.0%) | 42 (91.3%) | 24 (83.3%) | |
| Well | 53 (9.2%) | 1 (1.1%) | 1 (2.2%) | 1 (3.3%) | |
| Pathology T-stage | | | | | <0.001 |
| 1–2 | 551 (95.5%) | 84 (93.3%) | 33 (71.7%) | 12 (40%) | |
| 3–4 | 26 (4.5%) | 6 (6.7%) | 13 (28.3%) | 18 (60%) | |

HBsAg, hepatitis B virus surface antigen; HCV, hepatitis C virus; AFP, α-fetoprotein; TBS, tumor burden score.

### 6.2. The Association of TBS and AFP with OS

The median follow-up in this cohort was 19.9 months (interquartile range (IQR): 10.8–58.8 months). Both the TBS and AFP were strong predictors of OS, and the 5-year OS incrementally worsened with higher TBS (low TBS: 86%; medium TBS: 69%; high TBS: 44%; $p < 0.001$) (Figure 1). Patients with high AFP levels (i.e., AFP ≥ 400 ng/mL) had worse 5-year OS compared with patients who had low AFP levels (i.e., AFP < 400 ng/mL; 57% vs. 73%; $p < 0.001$).

When examining different combinations of TBS and AFP relative to OS, TBS and AFP showed a synergistic impact on OS. Patients with TBS ≤ 7.9 and AFP < 400 ng/mL had better OS than those with TBS > 7.9 and AFP < 400 ng/mL ($p = 0.003$) and those with TBS > 7.9 and AFP ≥ 400 ng/mL ($p < 0.001$). Patients with TBS ≤ 7.9 and AFP ≥ 400 ng/mL had better OS than those with TBS > 7.9 and AFP ≥ 400 ng/mL ($p < 0.001$). Patients with TBS > 7.9 and AFP < 400 ng/mL had better OS than those with TBS > 7.9 and AFP ≥ 400 ng/mL ($p = 0.027$). No significant differences in OS were observed between other groups (Figure 2).

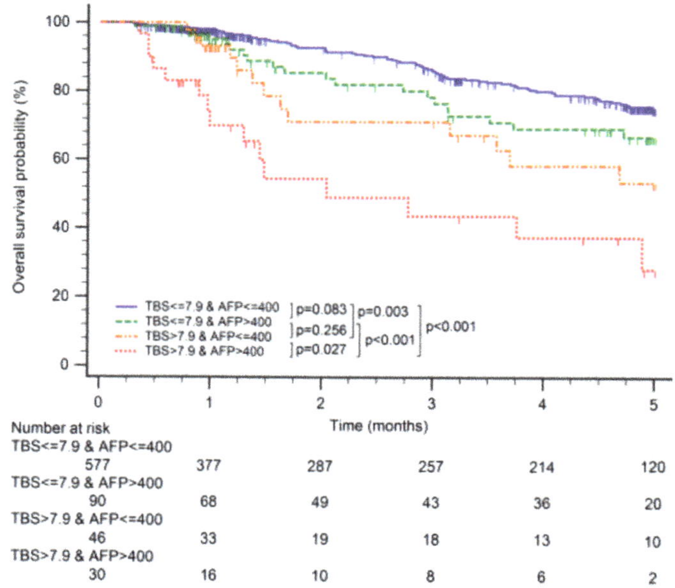


**Figure 2.** Kaplan–Meier curves demonstrating differences in overall survival among patients with tumor burden score (TBS) $\leq$ 7.9/α-fetoprotein (AFP) $\leq$ 400 ng/mL, TBS $\leq$ 7.9/AFP > 400 ng/mL, TBS > 7.9/AFP $\leq$ 400 ng/mL, and TBS > 7.9/AFP > 400 ng/mL.

On multivariate analysis, using TBS $\leq$ 7.9 and AFP < 400 ng/mL as the reference values, and after adjusting for all confounding variables, the combinations of TBS > 7.9/AFP < 400 ng/mL (HR: 2.063; 95% CI: 1.175–3.623; $p = 0.012$) and TBS > 7.9/AFP $\geq$ 400 ng/mL (HR: 6.57; 95% CI: 3.684–11.719; $p < 0.001$) were independent predictors of OS. By contrast, the combination of TBS $\leq$ 7.9/AFP $\geq$ 400 ng/mL was not an independent predictor for OS (HR: 1.394; 95% CI: 0.858–2.266; $p = 0.180$; Table 3).

**Table 3.** Univariate and multivariate Cox regression analysis of factors associated with overall survival.

| | Univariate HR (95% CI) | *p* | Multivariate HR (95% CI) | *p* |
|---|---|---|---|---|
| Age (year) | | | | |
| ≤65 | 1.00 (reference) | | 1.00 (reference) | |
| >65 | 2.083 (1.488–2.915) | <0.001 | 2.146 (1.523–3.024) | <0.001 |
| Cirrhosis | | | | |
| No | 1.00 (reference) | | 1.00 (reference) | |
| Yes | 1.525 (1.081–2.153) | 0.016 | 1.692 (1.190–2.405) | 0.003 |
| Unknown | 1.382 (0.597–3.197) | 0.450 | 0.969 (0.414–2.268) | 0.942 |
| Tumor differentiation | | | | |
| Well or moderate | 1.00 (reference) | | 1.00 (reference) | |
| Poor | 2.895 (1.514–5.535) | 0.001 | 2.027 (1.044–3.939) | 0.037 |
| Group | | | | |
| TBS ≤ 7.9 and AFP < 400 ng/mL | 1.00 (reference) | | 1.00 (reference) | |
| TBS ≤ 7.9 and AFP ≥ 400 ng/mL | 1.560 (0.966–2.520) | 0.069 | 1.394 (0.858–2.266) | 0.180 |
| TBS > 7.9 and AFP < 400 ng/mL | 2.077 (1.193–3.617) | 0.010 | 2.063 (1.175–3.623) | 0.012 |
| TBS > 7.9 and AFP ≥ 400 ng/mL | 5.378 (3.094–9.349) | <0.001 | 6.57 (3.684–11.719) | <0.001 |

AFP, α-fetoprotein; TBS, tumor burden score.

## 7. Discussion

In the current study, TBS combined with AFP levels could stratify OS among patients with BCLC stage 0, A, and B HCC who underwent LR. This model was useful for preoperative prognosis prediction and the assessment of risk-to-benefit ratios. The proportions of cirrhotic patients were lowest in the TBS > 7.9/AFP < 400 ng/mL group, which also contained higher proportions of patients with high tumor burden who required major resection. Major resection can be performed in patients with non-cirrhotic HCC with low rates of major complications and with satisfactory outcomes [13]. The proportion of patients with poor tumor differentiation and pathological T-stage 3 or 4 was highest in the TBS > 7.9/AFP ≥ 400 ng/mL group, indicating a high tumor burden, and a previous study reported that AFP elevation was correlated with vascular invasion and poor tumor differentiation [14].

The BCLC guidelines suggest that LR is indicated for BCLC stages 0–A [5]. However, a meta-analysis reported that LR improved the prognosis of BCLC B HCC patients compared with those who underwent transarterial chemoembolization (TACE) [15]. Furthermore, LR for BCLC B HCC is commonly adopted in daily practice at both Eastern and Western treatment centers [16,17]. We enrolled BCLC stage 0, A, and B patients in the current study.

BCLC stages A and B are heterogeneous. BCLC stage A is defined as multiple tumors within the Milan criteria or a solitary tumor >2 cm, irrespective of the size [5]. However, increasing tumor size is associated with an increased risk of microvascular invasion and micrometastasis, which can lead to a worse prognosis [18–20]. A previous study reported that the prognosis of BCLC stage A patients with a single large HCC > 5 cm is similar to that of patients classified as BCLC stage B HCC among patients who underwent LR [21]. BCLC stage B is well-known for its heterogeneity [3]. The BCLC staging system does not consider AFP in their model [5], despite AFP being a well-known prognostic biomarker for HCC [22]. We, therefore, hypothesized that TBS combined with AFP evaluation could be useful in daily practice for predicting OS among patients with BCLC stage 0, A, and B HCC who undergo LR.

The optimal cutoff values for TBS relative to OS were determined in the current study using the X-tile bioinformatics tool [12] (low TBS was defined as < 2.6, medium TBS was defined as 2.6–7.9, and high TBS was defined as > 7.9); these values are different from those reported by Tsilimigras et al. (low TBS was defined as < 3.36, medium TBS was defined as 3.36–13.74, and high TBS was defined as > 13.74) [7]. The discrepancies between studies could be due to differences in patients' characteristics. Furthermore, Tsilimigras et al. used pathologically defined TBS [7], whereas we used radiographically defined TBS in the current study.

The TBS is simple to use, and numerous studies have demonstrated its predictive value in prognosis in patients with HCC undergoing LR [6,7,23–32]. Among these studies, some studies used pathological TBS, which is not suitable for preoperative prognostic predictions [6,7,23–26]. Two studies only enrolled patients with BCLC stage B [27,28]. Whether the results of these studies could be extrapolated to other BCLC stages is unknown; meanwhile, our study enrolled patients with BCLC stages 0, A, and B. Endo et al. reported that a preoperative model composed of AFP, radiographic TBS, and neutrophil-to-lymphocyte ratio could predict the presence of microvascular invasion [29], whereas the aim of the present study is to predict OS. Fukami et al. enrolled patients with BCLC stages 0, A, and B. The authors used a combination of the Controlling Nutritional Status score and radiographic TBS to predict OS. However, this study only enrolled 96 patients. The case number may be too small to draw any conclusions [30].

Endo et al. enrolled 1676 patients with BCLC stages 0, A, B, and C. The authors used a preoperative model composed of radiographic TBS, AFP, neutrophil-to-lymphocyte ratio, albumin, gamma-glutamyl transpeptidase, and vascular involvement to predict 5-year OS. This model could stratify the OS into three distinct groups. Further, this model outperformed HCC staging systems including the BCLC and the AJCC systems [31].

Lima et al. enrolled 1435 patients with HCC undergoing LR without mentioning their BCLC stage. A risk score that included three variables (i.e., radiographic TBS, AFP, and Child–Pugh class) demonstrated superior predictive value for OS compared with the BCLC stage and further stratified patients within the BCLC stage relative to OS [32].

The advantage of our model, which is composed of radiographic TBS (cutoff value = 7.9) and AFP (cutoff value = 400 ng/mL), is simplicity; meanwhile, Lima et al. used a point system (TBS low/medium/high = 0/1/2; AFP low/high = 0/1; Child–Pugh class A/B = 0/1) to develop a risk score [32] and Endo et al. used an online calculator (https://yutaka-endo.shinyapps.io/PrepoScore_Shiny/, accessed on 1 January 2023) for their complex model [31]. We believe that model simplicity is of paramount importance in clinical application. Although numerous prognostic models have been developed for HCC [33], the BCLC [5] and the AJCC [10,11] staging systems are the most popular, mainly due to their simplicity.

One strength of the current study was the evaluation of radiographic TBS, which is useful for preoperative risk assessment. In addition, the vital status of each patient in the current study was verified through a link to a government website. There were few pieces of missing data identified in the current study. Among the variables included in the multivariate analysis, only 36 (4.8%) patients had unknown cirrhotic status. The limitations of the current study include the retrospective design and the study of a single treatment center, the outcomes of which may not be generalizable to other institutions. Furthermore, the majority of patients in this study were categorized as low tumor burden (i.e., TBS $\leq$ 7.9 and AFP < 400 ng/mL). The case numbers in the remaining groups were limited. Finally, no validation cohort was used to verify the cutoff values for TBS.

## 8. Conclusions

Radiographic TBS combined with AFP levels successfully stratified OS among patients with BCLC stages 0, A, and B in patients who underwent LR, which could be used for the preoperative prognosis prediction and risk-to-benefit ratio assessments among patients with BCLC stage 0, A, and B HCC prior to LR.

**Author Contributions:** Conceptualization, Y.-H.Y. and C.-C.W.; Methodology, Y.-H.Y. and Y.-W.L.; Software, C.-Y.L.; Validation, all authors; Formal Analysis, C.-Y.L.; Investigation, Y.-H.Y. and W.-F.L.; Resources, C.-C.W.; Data Curation, C.-C.Y.; Writing—Original Draft Preparation, Y.-H.Y.; Writing—Review and Editing, C.-C.W.; Visualization, C.-C.L.; Supervision, all authors; Project Administration, Y.-H.Y.; Funding Acquisition, Y.-H.Y. All authors have read and agreed to the published version of the manuscript.

**Funding:** This study was supported by a Grant CMRPG8L0181 from the Chang Gung Memorial Hospital–Kaohsiung Medical Center, Taiwan.

**Institutional Review Board Statement:** The Institutional Review Board of Kaohsiung Chang Gung Memorial Hospital approved this study (reference number: 202201189B0).

**Informed Consent Statement:** Not applicable.

**Data Availability Statement:** The data presented in this study are available via the following digital object identifier: https://www.dropbox.com/scl/fi/unezavq59ethkc6sprtu9/raw-data.xlsx?dl=0&rlkey=cwu1zq6f9bguxiq0ghqqfvvh3, accessed on 1 January 2023.

**Acknowledgments:** The authors thank the Cancer Center, Kaohsiung Chang Gung Memorial Hospital, for the provision of the HCC registry data. The authors thank Chih-Yun Lin, Nien-Tzu Hsu, and the Biostatistics Center, Kaohsiung Chang Gung Memorial Hospital, for statistics work. This study was supported by the Grant CMRPG8L0181 from the Chang Gung Memorial Hospital–Kaohsiung Medical Center, Taiwan.

**Conflicts of Interest:** The authors declare no conflict of interest.

## References


1. Shao, Y.-Y.; Wang, S.-Y.; Lin, S.-M.; Chen, K.-Y.; Tseng, J.-H.; Ho, M.-C.; Lee, R.-C.; Liang, P.-C.; Liao, L.-Y.; Huang, K.-W.; et al. Management consensus guideline for hepatocellular carcinoma: 2020 update on surveillance, diagnosis, and systemic treatment by the Taiwan Liver Cancer Association and the Gastroenterological Society of Taiwan. *J. Formos. Med. Assoc.* **2020**, *120*, 1051–1060. [CrossRef] [PubMed]
2. Marrero, J.A.; Kulik, L.M.; Sirlin, C.B.; Zhu, A.X.; Finn, R.S.; Abecassis, M.M.; Roberts, L.R.; Heimbach, J.K. Diagnosis, Staging, and Management of Hepatocellular Carcinoma: 2018 Practice Guidance by the American Association for the Study of Liver Diseases. *Hepatology* **2018**, *68*, 723–750. [CrossRef] [PubMed]
3. European Association for the Study of the Liver. EASL Clinical Practice Guidelines: Management of hepatocellular carcinoma. *J. Hepatol.* **2018**, *69*, 182–236. [CrossRef] [PubMed]
4. Vitale, A.; Burra, P.; Frigo, A.C.; Trevisani, F.; Farinati, F.; Spolverato, G.; Volk, M.; Giannini, E.G.; Ciccarese, F.; Piscaglia, F.; et al. Survival benefit of liver resection for patients with hepatocellular carcinoma across different Barcelona Clinic Liver Cancer stages: A multicentre study. *J. Hepatol.* **2014**, *62*, 617–624. [CrossRef]
5. Forner, A.; Llovet, J.M.; Bruix, J. Hepatocellular carcinoma. *Lancet* **2012**, *379*, 1245–1255. [CrossRef]
6. Tsilimigras, D.I.; Moris, D.; Hyer, J.M.; Bagante, F.; Sahara, K.; Moro, A.; Paredes, A.Z.; Mehta, R.; Ratti, F.; Marques, H.P.; et al. Hepatocellular carcinoma tumour burden score to stratify prognosis after resection. *Br. J. Surg.* **2020**, *107*, 854–864. [CrossRef]
7. Tsilimigras, D.; Hyer, J.; Diaz, A.; Bagante, F.; Ratti, F.; Marques, H.; Soubrane, O.; Lam, V.; Poultsides, G.; Popescu, I.; et al. Synergistic Impact of Alpha-Fetoprotein and Tumor Burden on Long-Term Outcomes Following Curative-Intent Resection of Hepatocellular Carcinoma. *Cancers* **2021**, *13*, 747. [CrossRef]
8. Edmonson, H.; Steiner, P. Primary carcinoma of the liver: A study of 100 cases among 48,900 necropsies. *Cancer* **1954**, *7*, 462–503. [CrossRef]
9. Everhart, J.E.; Wright, E.C.; Goodman, Z.D.; Dienstag, J.L.; Hoefs, J.C.; Kleiner, D.E.; Ghany, M.G.; Mills, A.S.; Nash, S.R.; Govindarajan, S.; et al. Prognostic value of Ishak fibrosis stage: Findings from the hepatitis C antiviral long-term treatment against cirrhosis trial. *Hepatology* **2009**, *51*, 585–594. [CrossRef]
10. American Joint Committee on Cancer. *American Joint Committee on Cancer Staging Manual*, 7th ed.; Edge, S.B., Byrd, D.R., Compton, C.C., Eds.; Springer: New York, NY, USA, 2010; p. 175.
11. Abou-Alfa, G.K.; Pawlik, T.M.; Shindoh, J. Liver. In *AJCC Cancer Staging Manual*, 8th ed.; Amin, M.B., Ed.; AJCC: Chicago, IL, USA, 2017; p. 287.
12. Camp, R.L.; Dolled-Filhart, M.; Rimm, D.L. X-tile: A new bio-informatics tool for biomarker assessment and outcome-based cut-point optimization. *Clin. Cancer Res.* **2004**, *10*, 7252–7259. [CrossRef]
13. Lang, H.; Sotiropoulos, G.C.; Dömland, M.; Frühauf, N.R.; Paul, A.; Hüsing, J.; Malagó, M.; Broelsch, C.E. Liver resection for hepatocellular carcinoma in non-cirrhotic liver without underlying viral hepatitis. *Br. J. Surg.* **2004**, *92*, 198–202. [CrossRef]
14. Ding, H.-F.; Zhang, X.-F.; Bagante, F.; Ratti, F.; Marques, H.P.; Soubrane, O.; Lam, V.; Poultsides, G.A.; Popescu, I.; Alexandrescu, S.; et al. Prediction of tumor recurrence by α-fetoprotein model after curative resection for hepatocellular carcinoma. *Eur. J. Surg. Oncol. (EJSO)* **2020**, *47*, 660–666. [CrossRef]
15. Hyun, M.H.; Lee, Y.-S.; Kim, J.H.; Lee, C.U.; Jung, Y.K.; Seo, Y.S.; Yim, H.J.; Yeon, J.E.; Byun, K.S. Hepatic resection compared to chemoembolization in intermediate- to advanced-stage hepatocellular carcinoma: A meta-analysis of high-quality studies. *Hepatology* **2018**, *68*, 977–993. [CrossRef]
16. Tsilimigras, D.I.; Bagante, F.; Moris, D.; Ms, J.M.H.; Sahara, K.; Paredes, A.Z.; Mehta, R.; Ratti, F.; Marques, H.P.; Soubrane, O.; et al. Recurrence Patterns and Outcomes after Resection of Hepatocellular Carcinoma within and beyond the Barcelona Clinic Liver Cancer Criteria. *Ann. Surg. Oncol.* **2020**, *27*, 2321–2331. [CrossRef]
17. Liu, Y.; Yong, C.; Lin, C.; Wang, C.; Chen, C.; Cheng, Y.; Wang, J.; Yen, Y.; Chen, C. Liver resection of hepatocellular carcinoma within and beyond the Barcelona Clinic Liver Cancer guideline recommendations: Results from a high-volume liver surgery center in East Asia. *J. Surg. Oncol.* **2020**, *122*, 1587–1594. [CrossRef]
18. Bruix, J.; Reig, M.; Sherman, M. Evidence-Based Diagnosis, Staging, and Treatment of Patients With Hepatocellular Carcinoma. *Gastroenterology* **2016**, *150*, 835–853. [CrossRef]
19. Pawlik, T.M.; Delman, K.A.; Vauthey, J.-N.; Nagorney, D.M.; Ng, I.O.-L.; Ikai, I.; Yamaoka, Y.; Belghiti, J.; Lauwers, G.Y.; Poon, R.T.; et al. Tumor size predicts vascular invasion and histologic grade: Implications for selection of surgical treatment for hepatocellular carcinoma. *Liver Transplant.* **2005**, *11*, 1086–1092. [CrossRef]
20. Cho, Y.; Sinn, D.H.; Yu, S.J.; Gwak, G.Y.; Kim, J.H.; Yoo, Y.J.; Jun, D.W.; Kim, T.Y.; Lee, H.Y.; Cho, E.J.; et al. Survival Analysis of Single Large (>5 cm) Hepatocellular Carcinoma Patients: BCLC A versus B. *PLoS ONE* **2016**, *11*, e0165722. [CrossRef]
21. Tsilimigras, D.I.; Bagante, F.; Sahara, K.; Moris, D.; Hyer, J.M.; Wu, L.; Ratti, F.; Marques, H.P.; Soubrane, O.; Paredes, A.Z.; et al. Prognosis After Resection of Barcelona Clinic Liver Cancer (BCLC) Stage 0, A, and B Hepatocellular Carcinoma: A Comprehensive Assessment of the Current BCLC Classification. *Ann. Surg. Oncol.* **2019**, *26*, 3693–3700. [CrossRef]
22. Galle, P.R.; Foerster, F.; Kudo, M.; Chan, S.L.; Llovet, J.M.; Qin, S.; Schelman, W.R.; Chintharlapalli, S.; Abada, P.B.; Sherman, M.; et al. Biology and significance of alpha-fetoprotein in hepatocellular carcinoma. *Liver Int.* **2019**, *39*, 2214–2229. [CrossRef]

23. Elfadaly, A.N.; Tsilimigras, D.I.; Hyer, J.M.; Paro, A.; Bagante, F.; Ratti, F.; Marques, H.P.; Soubrane, O.; Lam, V.; Poultsides, G.A.; et al. Impact of Tumor Burden Score on Conditional Survival after Curative-Intent Resection for Hepatocellular Carcinoma: A Multi-Institutional Analysis. *World J. Surg.* **2021**, *45*, 3438–3448. [CrossRef] [PubMed]
24. Tsilimigras, D.I.; Mehta, R.; Guglielmi, A.; Ratti, F.; Marques, H.P.; Soubrane, O.; Lam, V.; Poultsides, G.A.; Popescu, I.; Alexandrescu, S.; et al. Recurrence beyond the Milan criteria after curative-intent resection of hepatocellular carcinoma: A novel tumor-burden based prediction model. *J. Surg. Oncol.* **2020**, *122*, 955–963. [CrossRef] [PubMed]
25. Moazzam, Z.; Lima, H.A.; Alaimo, L.; Endo, Y.; Shaikh, C.F.; Ratti, F.; Marques, H.P.; Soubrane, O.; Lam, V.; Poultsides, G.A.; et al. Impact of tumor burden score on timing and patterns of recurrence after curative-intent resection of hepatocellular carcinoma. *Surgery* **2022**, *172*, 1448–1455. [CrossRef] [PubMed]
26. Lima, H.A.; Alaimo, L.; Brown, Z.J.; Endo, Y.; Moazzam, Z.; Tsilimigras, D.I.; Shaikh, C.; Resende, V.; Guglielmi, A.; Ratti, F.; et al. Application of hazard functions to investigate recurrence after curative-intent resection for hepatocellular carcinoma. *HPB* **2022**, *in press*. [CrossRef] [PubMed]
27. Tsilimigras, D.I.; Mehta, R.M.; Paredes, A.Z.M.; Moris, D.M.; Sahara, K.; Bagante, F.; Ratti, F.; Marques, H.P.; Silva, S.; Soubrane, O.; et al. Overall Tumor Burden Dictates Outcomes for Patients Undergoing Resection of Multinodular Hepatocellular Carcinoma Beyond the Milan Criteria. *Ann. Surg.* **2020**, *272*, 574–581. [CrossRef]
28. Lima, H.A.; Endo, Y.; Alaimo, L.; Moazzam, Z.; Munir, M.M.; Shaikh, C.; Resende, V.; Guglielmi, A.; Marques, H.P.; Cauchy, F.; et al. Tumor Burden Score and Serum Alpha-fetoprotein Subclassify Intermediate-Stage Hepatocellular Carcinoma. *J. Gastrointest. Surg.* **2022**, *26*, 2512–2521. [CrossRef]
29. Endo, Y.; Alaimo, L.; Lima, H.A.; Moazzam, Z.; Ratti, F.; Marques, H.P.; Soubrane, O.; Lam, V.; Kitago, M.; Poultsides, G.A.; et al. A Novel Online Calculator to Predict Risk of Microvascular Invasion in the Preoperative Setting for Hepatocellular Carcinoma Patients Undergoing Curative-Intent Surgery. *Ann. Surg. Oncol.* **2022**, *30*, 725–733. [CrossRef]
30. Fukami, Y.; Saito, T.; Osawa, T.; Arikawa, T.; Matsumura, T.; Kurahashi, S.; Komatsu, S.; Kaneko, K.; Sano, T. Preoperative Controlling Nutritional Status plus Tumor Burden Score for the Assessment of Prognosis after Curative Liver Resection for Hepatocellular Carcinoma. *Med. Princ. Pract.* **2020**, *30*, 131–137. [CrossRef]
31. Endo, Y.; Lima, H.A.; Alaimo, L.; Moazzam, Z.; Brown, Z.; Shaikh, C.F.; Ratti, F.; Marques, H.P.; Soubrane, O.; Lam, V.; et al. Preoperative risk score (PreopScore) to predict overall survival after resection for hepatocellular carcinoma. *HPB* 2023, *in press*. [CrossRef]
32. Lima, H.A.; Endo, Y.; Moazzam, Z.; Alaimo, L.; Shaikh, C.; Munir, M.M.; Resende, V.; Guglielmi, A.; Marques, H.P.; Cauchy, F.; et al. TAC score better predicts survival than the BCLC following resection of hepatocellular carcinoma. *J. Surg. Oncol.* **2022**, *127*, 374–384. [CrossRef]
33. Beumer, B.R.; Buettner, S.; Galjart, B.; van Vugt, J.L.; de Man, R.A.; Ijzermans, J.N.; Koerkamp, B.G. Systematic review and meta-analysis of validated prognostic models for resected hepatocellular carcinoma patients. *Eur. J. Surg. Oncol. (EJSO)* **2021**, *48*, 492–499. [CrossRef] [PubMed]

*cancers*

MDPI

*Article*

# Changes in ALBI Score and PIVKA-II within Three Months after Commencing Atezolizumab Plus Bevacizumab Treatment Affect Overall Survival in Patients with Unresectable Hepatocellular Carcinoma

**Shinji Unome [1], Kenji Imai [1,*], Koji Takai [1], Takao Miwa [1], Tatsunori Hanai [1], Yoichi Nishigaki [2], Hideki Hayashi [2], Takahiro Kochi [2], Shogo Shimizu [3], Junji Nagano [3], Soichi Iritani [3], Atsushi Suetsugu [1] and Masahito Shimizu [1]**

1 Department of Gastroenterology/Internal Medicine, Graduate School of Medicine, Gifu University, Gifu 501-1194, Japan
2 Gifu Municipal Hospital, Gifu 500-8513, Japan
3 Gifu Prefectural General Medical Center, Gifu 500-8717, Japan
* Correspondence: ikenji@gifu-u.ac.jp; Tel.: +81-(58)-230-6308; Fax: +81-(58)-230-6310

**Citation:** Unome, S.; Imai, K.; Takai, K.; Miwa, T.; Hanai, T.; Nishigaki, Y.; Hayashi, H.; Kochi, T.; Shimizu, S.; Nagano, J.; et al. Changes in ALBI Score and PIVKA-II within Three Months after Commencing Atezolizumab Plus Bevacizumab Treatment Affect Overall Survival in Patients with Unresectable Hepatocellular Carcinoma. *Cancers* **2022**, *14*, 6089. https://doi.org/10.3390/cancers14246089

Academic Editor: Georgios Germanidis

Received: 24 November 2022
Accepted: 8 December 2022
Published: 10 December 2022



**Simple Summary:** Atezolizumab plus bevacizumab (Atez/Bev) treatment is now recommended as a first-line systemic treatment for unresectable hepatocellular carcinoma. In this study, we evaluated the therapeutic effects and adverse events of Atez/Bev treatment in the real world including patients with Child–Pugh B or non-viral hepatitis and those who received Atez/Bev treatment as a later-line treatment. Furthermore, we analyzed the factors affecting the overall survival among changes in the clinical indicators representing liver function and tumor-related factors within 3 months after the introduction of Atez/Bev treatment. The results of this study may be useful in determining whether to continue or modify Atez/Bev treatment at an early stage after starting this treatment.

**Abstract:** In this study, we aimed to evaluate the efficacy and safety of atezolizumab plus bevacizumab (Atez/Bev) treatment for unresectable hepatocellular carcinoma (HCC) and to analyze the factors affecting overall survival (OS). A total of 69 patients who received Atez/Bev at our institutions for unresectable HCC were enrolled in this study. OS and progression-free survival (PFS) were estimated using the Kaplan–Meier method. Changes in clinical indicators within 3 months were defined as delta ($\Delta$) values, and the Cox proportional hazards model was used to identify which $\Delta$ values affected OS. The median OS, PFS, objective response rate, and disease control rate were 12.5 months, 5.4 months, 23.8%, and 71.4%, respectively. During the observational period, 62 patients (92.5%) experienced AEs (hypertension (33.3%) and general fatigue), and 27 patients (47.4%) experienced grade $\geq$ 3 AEs (hypertension (10.1%) and anemia (7.2%)). There was a significant deterioration in the albumin-bilirubin (ALBI) score ($-2.22$ to $-1.97$; $p < 0.001$), and a reduction in PIVKA-II levels (32,458 to 11,584 mAU/mL; $p = 0.040$) within 3 months after commencing Atez/Bev. Both the worsening $\Delta$ ALBI score ($p = 0.005$) and increasing $\Delta$ PIVKA-II ($p = 0.049$) were significantly associated with the OS of patients.

**Keywords:** hepatocellular carcinoma; atezolizumab; bevacizumab; prognosis factor; ALBI score; PIVKA-II

## 1. Introduction

Hepatocellular carcinoma (HCC) is a prevalent disease worldwide, with approximately 800,000 individuals newly developing and dying from this malignancy each year [1]. HCC is difficult to detect during the early stages, and in most cases is diagnosed only after having progressed to an unresectable state [2]. Approximately 50% of patients with HCC receive systemic therapy [3]. Sorafenib was the first oral active multi-kinase inhibitor confirmed to be effective against unresectable HCC [4], and since its introduction, other multi-kinase inhibitors including lenvatinib, regorafenib, and cabozantinib have similarly been established to be efficacious [5–7].

In recent years, the importance of immune checkpoint inhibitors in HCC treatment has received increasing attention. Among these agents, treatment with atezolizumab, a programmed death-ligand 1 (PD-L1)-targeted antibody, administered in combination with bevacizumab (Atez/Bev), for unresectable HCC was for the first time reported to result in better overall survival (OS) and progression-free survival (PFS) outcomes than sorafenib (IMbrave150) [8]. On the basis of this favorable outcome, Atez/Bev is now recommended as a first-line systemic treatment for unresectable HCC in the recently revised guidelines issued in the United States, Europe, and Japan [9–11].

In the IMbrave150 trial [8], none of the participants had previously received systemic treatment and had a good liver functional reserve (the inclusion criterion was Child–Pugh A). However, in the real world, patients who receive Atez/Bev for unresectable HCC generally receive a range of other treatments including systemic therapy, and some patients have poor liver functional reserve, as seen in Japan, where HCC occurs in elderly patients with reduced hepatic functional reserve. In addition, the incidence of non-viral HCC, which may be less likely to respond to Atez/Bev, is also rapidly increasing [12,13]. Therefore, it is important to evaluate the efficacy and safety of Atez/Bev in clinical settings.

Although Atez/Bev has been established to be an effective treatment for unresectable advanced HCC, there are some patients who do not benefit from this treatment. Indeed, it has been found that only one-third of the patients who receive this treatment show an objective response [8]. In addition, Atez/Bev therapy can cause serious adverse events (AEs) [8]. Consequently, it is essential to identify the factors affecting survival or AEs when deciding whether to continue or discontinue treatment. In this regard, several biomarkers including PD-L1 expression and pre-existing immunity in baseline tumor tissue [14,15] have been identified as having potential utility in predicting the Atez/Bev response and in determining the course of treatment. However, such evaluations are complex and there is a need for more convenient and established biomarkers for use in daily clinical practice.

In this study, we evaluated the therapeutic effects and AEs of Atez/Bev in the treatment of unresectable HCC. Focusing on the factors affecting overall OS after the initiation of this treatment, we found that a deterioration in hepatic functional reserve and an elevation in the levels of protein induced by vitamin K absence-II (PIVKA-II) during the initial 3 months of treatment were associated with a poorer OS in these patients.

## 2. Materials and Methods

### 2.1. Enrolled Patients

A total of 69 patients who had received Atez/Bev for at least 3 months for unresectable HCC at our institutions (Gifu University Hospital, Gifu Municipal Hospital, and Gifu Prefectural General Medical Center) between November 2020 and March 2022 were included in this study. The study design was reviewed and approved by the Ethics Committee of Gifu University School of Medicine on 2 June 2021 (ethical protocol code: 2021–074).

### 2.2. HCC Diagnosis and Therapeutic Strategies

HCC was diagnosed on the basis of a typical hypervascular tumor stain on angiography and typical dynamic computed tomography (CT) or magnetic resonance imaging (MRI) findings of enhanced staining in the early phase and attenuation in the delayed phase [16]. Therapeutic strategies for HCC in this study were determined according to the clinical guidelines for HCC published by the Japan Society of Hepatology [16]. Atez/Bev was administered according to the standard regimen, for which all patients received intravenous atezolizumab (1200 mg) plus bevacizumab (15 mg/kg body weight) every 3 weeks [8]. An alternative treatment was considered when serious AEs, a hyper progressive disease defined as disease progression with a ≥2-fold increase in the first evaluation [17], or progressive disease (PD) for a certain period were observed.

### 2.3. Evaluation of the Efficacy and Safety of Atez/Bev

The therapeutic response of each patient was assessed using dynamic CT or MRI imaging according to the Response Evaluation Criteria in Solid Tumors [18]. OS was defined as the time from the day of commencing Atez/Bev therapy to death or the last visit. PFS was defined as the time from the commencement of Atez/Bev treatment to the observation of clinical disease progression or death. Adverse events were assessed according to the Common Terminology Criteria for Adverse Events (CTCAE), version 5.0.

### 2.4. Determination of Prognostic Factors and Statistical Analyses

Differences in the baseline characteristics within 3 months after the initiation of Atez/Bev therapy were compared using a paired-$t$ test. Changes in clinical indicators representing liver function and tumor-related factors within 3 months after the introduction of the treatment were defined as delta ($\Delta$) values. The Cox proportional hazards model was used to identify which $\Delta$ values affected the OS after the initiation of this treatment.

OS and PFS were estimated using the Kaplan–Meier method, and differences between curves were evaluated using the log-rank test. Maximally selected rank statistics were used to determine the optimal cut-off to maximize the separation of the curves in the two groups [19]. We used the 'maxstat' package (version 0.7-25) in R to conduct these statistical analyses. Statistical significance was set at $p < 0.05$, and all statistical analyses were performed using R (version 4.1.2; R Foundation for Statistical Computing, Vienna, Austria; http://www.R-project.org/, accessed on 26 July 2022).

## 3. Results

### 3.1. Patient Characteristics and HCC Treatment Status

The clinical characteristics of the enrolled patients (55 men with an average age of 74.4 years) immediately prior to the initiation of Atez/Bev treatment are shown in Table 1. With to the underlying liver diseases, 12, 22, 16, 12, and seven patients had hepatitis B virus, hepatitis C virus, non-alcoholic steatohepatitis, alcoholic liver disease, and other diseases, respectively, whereas with respect to liver functional reserve, 37, 24, seven, and one patient had Child–Pugh scores of 5, 6, 7, and 8, respectively.

**Table 1.** Baseline demographic and clinical characteristics of the enrolled patients.

| Variables | ($n$ = 69) |
|---|---|
| Age (years) | 74.4 ± 9.7 |
| Sex (male/female) | 55/14 |
| ECOG PS (0/1/2) | 55/17/2 |
| Etiology (HBV/HCV/NASH/Alcohol/others) | 12/22/16/12/7 |
| BCLC stage (A/B1/B2/C) | 9/8/16/36 |
| Child–Pugh score (5/6/7/8) | 37/24/7/1 |
| ALBI score | −2.22 ± 0.42 |
| ALB (g/dL) | 3.5 ± 0.5 |
| AST (U/L) | 51 ± 40 |
| ALT (U/L) | 37 ± 34 |
| T-Bil (mg/dL) | 0.8 ± 0.4 |
| PT (%) | 96 ± 18 |
| AFP (ng/mL) | 2252 ± 7337 |
| PIVKA-II (mAU/mL) | 32,458 ± 156,378 |

Values are presented as a mean ± standard deviation. ECOG, Eastern Cooperative Oncology Group; PS, performance status; HBV, hepatitis B virus; HCV, hepatitis C virus; NASH, nonalcoholic steatohepatitis; BCLC, Barcelona Clinic Liver Cancer; ALBI score, albumin-bilirubin score; ALB, albumin; AST, aspartate aminotransferase; ALT, alanine aminotransferase; T-Bil, total bilirubin; PT, prothrombin time; AFP, alpha-fetoprotein; PIVKA-II, protein induced by vitamin K absence or antagonists-II.

Among the enrolled patients, 54 (78.2%) had received other treatment for HCC prior to the initiation of Atez/Bev, one (1.4%) had received combination treatment, and 31 (44.9%)

had received other treatments after the Atez/Bev treatment. Details of pre-treatment, combination treatment, and post-treatment are shown in Table 2.

**Table 2.** The pre-, combination, and post-treatment of the patients receiving atezolizumab plus bevacizumab treatment.

| | Pre-Treatment | Combination Treatment | Post-Treatment |
|---|---|---|---|
| Any treatments | 54 (78.2%) | 1 (1.4%) | 31 (44.9%) |
| Hepatectomy | 23 | 0 | 3 |
| RFA | 23 | 0 | 2 |
| TACE | 40 | 1 | 8 |
| Radiation therapy | 9 | 0 | 2 |
| Sorafenib | 6 | 0 | 2 |
| Regorafenib | 1 | 0 | 0 |
| Lenvatinib | 18 | 0 | 21 |
| Ramucirumab | 3 | 0 | 3 |

RFA, radiofrequency ablation; TACE, transcatheter arterial chemo embolization.

### 3.2. *Efficacy and Safety of Atez/Bev for Patients with Unresectable HCC*

The mean observational period for the enrolled patients was 7.8 ± 3.8 months. OS rates at 6 and 12 months and median OS were 77.6%, 50.7%, and 12.5 months, respectively (Figure 1a), whereas the PFS rates at 6 and 12 months and the median PFS were 46.0%, 24.2%, and 5.4 months, respectively (Figure 1b). The therapeutic effects of complete response, partial response, stable disease, and PD were observed in one, 14, 30, and 18 cases, respectively. The objective response rate (ORR) and disease control rate (DCR) were 23.8% and 71.4%, respectively. Fifteen patients had PD within 3 months, and these patients tended to have shorter survival than those who did not ($p$ = 0.067, Figure S1).

(a) (b)

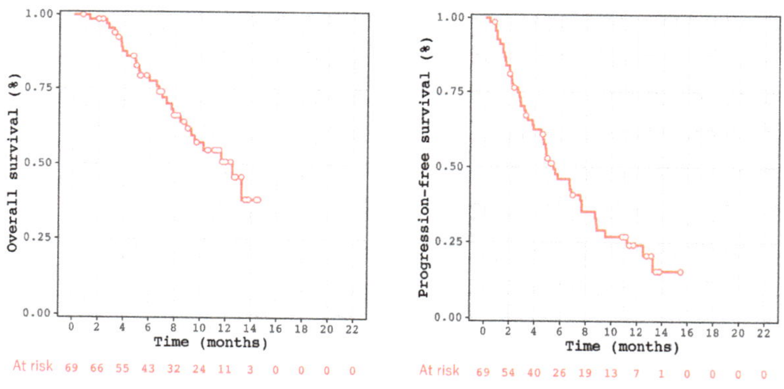


**Figure 1.** Kaplan–Meier curve for overall survival after the introduction of atezolizumab plus bevacizumab treatment for unresectable HCC (**a**) and for progression-free survival (**b**).

Table 3 shows the AEs recorded in response to the Atez/Bev treatment. We found that 62 patients (92.5%) experienced some form of AE, the most frequent of which at any grade was hypertension (33.3%), followed by general fatigue (31.9%), proteinuria (26.1%), liver dysfunction (24.6%), and appetite loss (23.2%). AEs at Grade ≥ 3 were identified in 27 patients (47.4%), the most frequent of which was hypertension (10.1%), followed by anemia (7.2%), appetite loss (5.8%), and hemorrhage (5.8%). With respect to immune-related AEs, three patients experienced interstitial pneumonia, and one experienced myasthenia gravis and rheumatic arthritis. None of the enrolled patients experienced Grade 5 AEs.

**Table 3.** Adverse events during atezolizumab plus bevacizumab treatment.

| | Any Grade (*n* = 69) | Grade 1 | Grade 2 | Grade ≥ 3 |
|---|---|---|---|---|
| Any symptoms | 62 (92.5%) | | | 27 (47.4%) |
| Hypertension | 23 (33.3%) | 8 (11.6%) | 8 (11.6%) | 7 (10.1%) |
| General fatigue | 22 (31.9%) | 16 (23.2%) | 5 (7.2%) | 1 (1.4%) |
| Proteinuria | 18 (26.1%) | 5 (7.2%) | 11 (15.9%) | 2 (2.9%) |
| Liver dysfunction | 17 (24.6%) | 14 (20.3%) | 0 | 3 (4.3%) |
| Appetite loss | 16 (23.2%) | 6 (8.7%) | 6 (8.7%) | 4 (5.8%) |
| Hemorrhage | 11 (15.9%) | 7 (10.1%) | 0 | 4 (5.8%) |
| Platelet count decreased | 10 (14.5%) | 4 (5.8%) | 3 (4.3%) | 3 (4.3%) |
| Anemia | 9 (13.0%) | 2 (2.9%) | 2 (2.9%) | 5 (7.2%) |
| Diarrhea | 7 (10.1%) | 6 (8.7%) | 1 (1.4%) | 0 |
| Hypothyroidism | 5 (7.2%) | 4 (5.8%) | 1 (1.4%) | 0 |
| Skin disorders | 2 (2.9%) | 0 | 0 | 2 (2.9%) |
| Heart failure | 2 (2.9%) | 0 | 2 (2.9%) | 0 |
| Colonic perforation | 1 (1.4%) | 0 | 0 | 1 (1.4%) |
| Interstitial pneumonia | 4 (5.8%) | 3 (4.3%) | 1 (1.4%) | 0 |
| Myasthenia gravis | 1 (1.4%) | 0 | 0 | 1 (1.4%) |
| Rheumatic arthritis | 1 (1.4%) | 0 | 1 (1.4%) | 0 |

### 3.3. *Changes in Clinical Indicators 3 Months after the Induction of Atez/Bev Affecting OS*

Table 4 shows the changes in clinical indicators representing liver functional reserve and tumor markers during the initial 3 months after the initiation of Atez/Bev treatment. Within 3 months after commencing treatment, we detected a significant deterioration in factors representing liver functional reserve including the Child–Pugh score, albumin-bilirubin (ALBI) score [20], serum albumin level, and the presence of ascites ($p < 0.001$). Moreover, there was a significant reduction in the levels of PIVKA-II ($p = 0.040$).

**Table 4.** Changes in clinical indicators 3 months after the introduction of atezolizumab plus bevacizumab treatment.

| Variables | Introduction | After 3 Months | *p* Value |
|---|---|---|---|
| Child-Pugh score | 5.6 ± 0.7 | 6.4 ± 1.4 | <0.001 |
| ALBI score | −2.22 ± 0.42 | −1.97 ± 0.51 | <0.001 |
| ALB (g/dL) | 3.5 ± 0.4 | 3.3 ± 0.5 | <0.001 |
| AST (U/L) | 51.0 ± 40.3 | 49.6 ± 54.5 | 0.253 |
| ALT (U/L) | 37.1 ± 33.7 | 34.5 ± 37.3 | 0.364 |
| T-Bil (mg/dL) | 0.9 ± 0.4 | 1.3 ± 2.3 | 0.143 |
| PT (%) | 96.5 ± 18.3 | 93.1 ± 21.7 | 0.078 |
| Ascites (yes/no) | 0/69 | 15/54 | <0.001 |
| Encephalopathy (yes/no) | 0/69 | 1/68 | 1.000 |
| AFP (ng/mL) | 2252 ± 7337 | 4997 ± 17,802 | 0.079 |
| PIVKA-II (mAU/mL) | 32,458 ± 156,378 | 11,584 ± 28,983 | 0.040 |

Values are compared using the paired-*t* test. ALBI score, albumin-bilirubin score; ALB, albumin; AST, aspartate aminotransferase; ALT, alanine aminotransferase; T-Bil, total bilirubin; PT, prothrombin time; AFP, alpha-fetoprotein; PIVKA-II, protein induced by vitamin K absence or antagonists-II.

When analyzing the Δ values, the Δ Child–Pugh score, Δ ALBI score, Δ albumin, and Δ T-Bil, all representing liver function impairment and Δ PIVKA-II, were selected as prognostic factors in univariate analysis. We analyzed the Δ ALBI score and Δ PIVKA-II in multivariate analysis and identified a deterioration in the ALBI score (hazard ratio (HR): 5.477, 95% confidence interval (CI): 1.656–18.12, $p = 0.005$) and increased PIVKA-II (HR: 1.001, 95%CI: 1.000–1.003, $p = 0.049$) within 3 months after the initiation of Atez/Bev treatment as independent prognostic factors in multivariate analyses (Table 5). However, the AFP and PIVKA-II change ratios, which were defined by the AFP and PIVKA-II values at 3 months after Atez/Bev treatment divided by their values before the treatment, were not associated with OS (Table S1). When limited to the 54 patients who did not have PD within 3 months, increased PIVKA-II (HR: 1.002; 95%CI, 1.001–1.003; $p = 0.033$) was the only independent risk factor for OS (Table S2).

**Table 5.** Univariate and multivariate analyses of possible risk factors for overall survival among the changes of clinical indicators within 3 months by the Cox proportional hazards model.

| Variables | Univariate Analysis | | Multivariate Analysis | |
|---|---|---|---|---|
| | HR (95%CI) | *p* Value | HR (95%CI) | *p* Value |
| Δ Child–Pugh score/3 months | 1.971 (1.420–2.737) | <0.001 | | |
| Δ ALBI score/3 months | 2.951 (1.956–4.453) | <0.001 | 5.477 (1.656–18.12) | 0.005 |
| Δ Albumin (g/dL)/3 months | 0.167 (0.069–0.409) | <0.001 | | |
| Δ AST (U/L)/3 months | 1.001 (0.991–1.011) | 0.846 | | |
| Δ ALT (U/L)/3 months | 0.999 (0.985–1.014) | 0.925 | | |
| Δ T-Bil (mg/dL)/3 months | 1.276 (1.092–1.490) | 0.002 | | |
| Δ PT (%)/3 months) | 0.985 (0.966–1.004) | 0.120 | | |
| Δ AFP (ng/mL)/3 months | 1.002 (0.999–1.004) | 0.129 | | |
| Δ PIVKA-II (mAU/mL)/3 months | 1.002 (1.001–1.003) | 0.003 | 1.001 (1.000–1.003) | 0.049 |

Δ values mean the changes of clinical indicators that represent liver function and tumor markers within 3 months after the introduction of atezolizumab plus bevacizumab treatment. ALBI score, albumin-bilirubin score; ALB, albumin; AST, aspartate aminotransferase; ALT, alanine aminotransferase; T-Bil, total bilirubin; PT, prothrombin time; AFP, alpha-fetoprotein; PIVKA-II, protein induced by vitamin K absence or antagonists-II.

Maximally selected rank statistics revealed that the optimal cutoff values of the Δ ALBI score and Δ PIVKA-II were 0.376 and 672 mAU/mL, respectively (Figure S2). Patients with Δ ALBI scores ≤ 0.376 ($p < 0.001$, Figure 2a) and Δ PIVKA-II ≤ 672 mAU/mL ($p = 0.007$, Figure 2b) had significantly longer survival than those with Δ ALBI scores >0.376 and Δ PIVKA-II > 672 mAU/mL, respectively. Furthermore, the enrolled patients were further divided into three groups based on using the two cutoff values as follows: Group 1, patients with Δ ALBI score ≤0.376 and Δ PIVKA-II ≤ 672 mAU/mL; Group 3, patients with Δ ALBI score > 0.376 and Δ PIVKA-II > 672 mAU/mL; and Group 2, patients with others. Patients in Group 1 had longer survival times than those in Group 2 ($p = 0.012$) and Group 3 ($p < 0.001$) (Figure 2c).

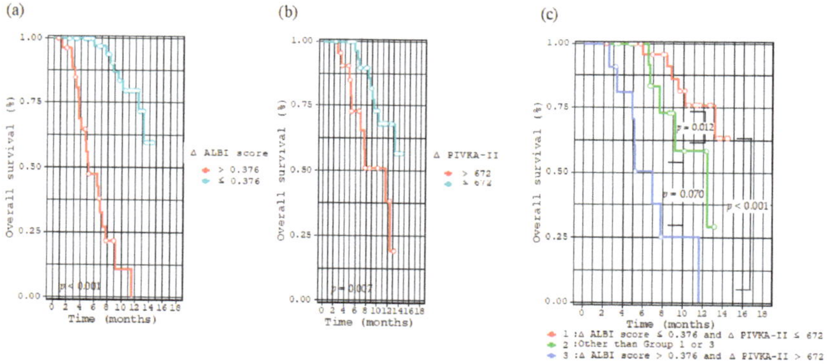


**Figure 2.** Kaplan–Meier curve for overall survival divided by Δ ALBI score of 0.376 (**a**), Δ PIVKA-II of 672 mAU/mL (**b**), and divided into three groups as follows: Group 1, patients with Δ ALBI score ≤ 0.376 and Δ PIVKA-II ≤ 672 mAU/mL; Group 3, patients with Δ ALBI score > 0.376 and Δ PIVKA-II > 672 mAU/mL; and Group 2, patients with others (**c**).

## 4. Discussion

In this study, we describe the clinical outcomes and AEs associated with Atez/Bev therapy for unresectable advanced HCC performed in a clinical setting. Results obtained from the updated IMbrave150 trial revealed median OS and PFS values of 19.2 and 6.9 months, and ORR and DCR values of 30% and 74%, respectively [21]. Compared with these observations, we recorded similar ORR (23.8%) and DCR (71.4%) values in the present study, whereas the median OS (12.5 months) and PFS (5.4 months) values were slightly inferior. These latter differences could be ascribed to the larger number of enrolled patients in our study who had Child–Pugh B, had received pretreatment that included other systemic therapies, or had non-viral hepatitis. In this regard, the findings of some studies have indicated that patients with Child–Pugh B or non-viral hepatitis and those who received Atez/Bev as a later-line treatment had poorer clinical outcomes [13,22,23]. In the present study, we found that patients with Child–Pugh B had significantly poorer survival than those with Child–Pugh A (Figure S3a; $p = 0.027$), whereas there were no significant differences in OS among patients who received Atez/Bev as a first-line and later-line treatment (Figure S3b; $p = 0.472$) or patients with viral and non-viral hepatitis (Figure S3c; $p = 0.178$). Although this study included only a small number of patients, our findings nevertheless tended to indicate that the prognostic benefits of Atez/Bev may be diminished, at least in patients with reduced hepatic functional reserve. Further studies are needed to determine whether the expected effect can be achieved in cases of Atez/Bev post-treatment or in cases of non-viral hepatitis.

In this study, we established that the liver functional reserve, as indicated by the Child–Pugh and ALBI scores, was deteriorated significantly in those patients who received Atez/Bev. Furthermore, we identified an unfavorable change in ALBI score ($\Delta$ ALBI score) as a prognostic factor. Consequently, patients receiving treatment should be aware of the risk of reduced hepatic functional reserve such as deterioration in the Child–Pugh score, ALBI score, albumin levels, and the appearance of ascites, and that maintaining hepatic function reserve may improve patient prognosis. Moreover, we observed a significant reduction in the PIVKA-II levels within 3 months after the commencement of Atez/Bev treatment and identified increasing PIVKA-II ($\Delta$ PIVKA-II) as a poor prognostic factor in these patients. The response of alpha-fetoprotein (AFP), another HCC tumor marker, 6 weeks after initiating Atz/Bev therapy, has been reported to be a potential surrogate biomarker for prognosis in patients with HCC [24]. Moreover, the CRAFITY score, determined by C-reactive protein and AFP levels, has also been reported to be useful for predicting therapeutic outcomes in these patients [25]. In contrast, however, the utility of PIVKA-II assessment for predicting a response to Atz/Bev has hardly been previously reported [26]. Interestingly, even when limiting 54 patients to those who did not have PD within 3 months, $\Delta$ PIVKA-II was the only independent risk factor for OS (Table S2). In clinical practice, it is sometimes difficult to determine whether Atez/Bev treatment should be continued. For patients with deteriorating ALBI score and elevated PIVKA-II, especially those with a $\Delta$ ALBI score >0.376 and $\Delta$ PIVKA-II >672 mAU/mL belonging to Group 3 (Figure 2c), the prognosis is clearly poor, and a change to an alternative treatment should be considered.

The nature and severity of treatment-related AEs observed in this study differed substantially from those previously reported [8,21–23,27–31]. In contrast, we detected significant deterioration in liver functional reserve including albumin levels, Child–Pugh score, ALBI score, and the appearance of ascites within the initial 3 months of treatment. Although the findings of some studies have indicated that ALBI scores tend to decline within the first few weeks of treatment, observations in most previous studies have tended to indicate that these scores do not deteriorate in response to Atez/Bev [23,27,29–31]. The fact that we detected a positive correlation between the $\Delta$ ALBI score and $\Delta$ PIVKA-II in the present study (coefficient of correlation = 0.286, $p = 0.034$; Figure S4) would tend to imply that a deterioration in the ALBI score is associated with the progression of HCC itself, rather than with the AEs of this treatment. Clearly, in patients with a low hepatic functional reserve, Atz/Bev may promote a further deterioration of function. In addition, it is important to understand that when a tumor is not controlled by Atz/Bev, the liver functional reserve may deteriorate during the early stages of treatment.

This study did, however, have certain limitations, notably the fact that this was a retrospective study with a small sample size. Furthermore, the observational period was short and a substantial number of enrolled patients were censored at the end of this study. Additionally, the Δ ALBI score and Δ PIVKA-II, which were selected as independent risk factors for OS in this study, showed a modest positive correlation. This may have affected the reliability of the results of this study. Prospective studies involving a larger number of patients and a more extended observational period should be conducted in the future to overcome these limitations.

## 5. Conclusions

We observed a significant deterioration in ALBI score and a significant reduction in PIVKA-II levels within 3 months after initiating Atez/Bev therapy for unresectable HCC. Furthermore, a deterioration in the ALBI score and elevation of PIVKA-II within 3 months were both independent prognostic factors of the treatment. Evaluation of these factors may be useful in determining whether to continue or modify Atez/Bev treatment.

**Supplementary Materials:** The following supporting information can be downloaded at: https://www.mdpi.com/article/10.3390/cancers14246089/s1, Figure S1: Kaplan–Meier curve for overall survival divided into two groups who had progression disease within three months and who did not; Figure S2: The result of the optimal cutoff values of the Δ ALBI score and Δ PIVKA-II according to the maximally selected rank statistics; Figure S3: Kaplan–Meier curve for overall survival divided by Child–Pugh A and B (**a**), first-line and later-line (**b**), and viral hepatitis and non-viral hepatitis (**c**); Figure S4: A correlation between the Δ ALBI score and Δ PIVKA-II; Table S1: Univariate and mutivariate analyses of possible risk factors for overall survival among the changes of clinical indicators within 3 months by Cox proportional hazards model using AFP and PIVKA-II change ratio instead of Δ AFP and Δ PIVKA-II; Table S2: Univariate and mutivariate analyses of possible risk factors for overall survival in 54 patients who did not have progression disease at 3 months.

**Author Contributions:** S.U., K.I., K.T., T.M., T.H., Y.N., H.H., T.K., S.S., J.N., S.I., A.S. and M.S. designed the study. K.I. analyzed the data. S.U. drafted the manuscript. K.T. supervised the participants' treatment. K.I., K.T., S.U., T.M., T.H., Y.N., H.H., T.K., S.S., J.N., S.I., A.S. and M.S. contributed to the selection of the participants and collected the data. K.I., K.T., S.U., T.M., T.H., Y.N., H.H., T.K., S.S., J.N., S.I. and A.S. revised the manuscript, and M.S. mainly reviewed and amended the manuscript. All authors have read and agreed to the published version of the manuscript.

**Funding:** This research received no external funding.

**Institutional Review Board Statement:** The study design including the consent procedure was approved by the ethics committee of the Gifu University School of Medicine (ethical protocol code: 2021–074).

**Informed Consent Statement:** We were unable to obtain written informed consent in advance due to the retrospective design of our study. Instead, by disclosing the details of the study, we provided the study participants with an opportunity to opt-out.

**Data Availability Statement:** The data presented in this study are available upon request from the corresponding author.

**Conflicts of Interest:** The authors declare no conflict of interest.

## References

1. Bray, F.; Ferlay, J.; Soerjomataram, I.; Siegel, R.L.; Torre, L.A.; Jemal, A. Global Cancer Statistics 2018: GLOBOCAN Estimates of Incidence and Mortality Worldwide for 36 Cancers in 185 Countries. *CA Cancer J. Clin.* **2018**, *68*, 394–424. [CrossRef]
2. Njei, B.; Rotman, Y.; Ditah, I.; Lim, J.K. Emerging Trends in Hepatocellular Carcinoma Incidence and Mortality. *Hepatology* **2015**, *61*, 191–199. [CrossRef] [PubMed]

3. Llovet, J.M.; Castet, F.; Heikenwalder, M.; Maini, M.K.; Mazzaferro, V.; Pinato, D.J.; Pikarsky, E.; Zhu, A.X.; Finn, R.S. Immunotherapies for Hepatocellular Carcinoma. *Nat. Rev. Clin. Oncol.* **2022**, *19*, 151–172. [CrossRef] [PubMed]
4. Llovet, J.M.; Ricci, S.; Mazzaferro, V.; Hilgard, P.; Gane, E.; Blanc, J.-F.F.; de Oliveira, A.C.; Santoro, A.; Raoul, J.-L.L.; Forner, A.; et al. Sorafenib in Advanced Hepatocellular Carcinoma. *N. Engl. J. Med.* **2008**, *359*, 378–390. [CrossRef] [PubMed]
5. Bruix, J.; Qin, S.; Merle, P.; Granito, A.; Huang, Y.H.; Bodoky, G.; Pracht, M.; Yokosuka, O.; Rosmorduc, O.; Breder, V.; et al. Regorafenib for Patients with Hepatocellular Carcinoma Who Progressed on Sorafenib Treatment (RESORCE): A Randomised, Double-Blind, Placebo-Controlled, Phase 3 Trial. *Lancet* **2017**, *389*, 56–66. [CrossRef] [PubMed]
6. Kudo, M.; Finn, R.S.; Qin, S.; Han, K.H.; Ikeda, K.; Piscaglia, F.; Baron, A.; Park, J.W.; Han, G.; Jassem, J.; et al. Lenvatinib versus Sorafenib in First-Line Treatment of Patients with Unresectable Hepatocellular Carcinoma: A Randomised Phase 3 Non-Inferiority Trial. *Lancet* **2018**, *391*, 1163–1173. [CrossRef]
7. Abou-Alfa, G.K.; Meyer, T.; Cheng, A.-L.; El-Khoueiry, A.B.; Rimassa, L.; Ryoo, B.-Y.; Cicin, I.; Merle, P.; Chen, Y.; Park, J.-W.; et al. Cabozantinib in Patients with Advanced and Progressing Hepatocellular Carcinoma. *N. Engl. J. Med.* **2018**, *379*, 54–63. [CrossRef]
8. Finn, R.S.; Qin, S.; Ikeda, M.; Galle, P.R.; Ducreux, M.; Kim, T.-Y.; Kudo, M.; Breder, V.; Merle, P.; Kaseb, A.O.; et al. Atezolizumab plus Bevacizumab in Unresectable Hepatocellular Carcinoma. *N. Engl. J. Med.* **2020**, *382*, 1894–1905. [CrossRef]
9. Bruix, J.; Chan, S.L.; Galle, P.R.; Rimassa, L.; Sangro, B. Systemic Treatment of Hepatocellular Carcinoma: An EASL Position Paper. *J. Hepatol.* **2021**, *75*, 960–974. [CrossRef]
10. Kudo, M.; Kawamura, Y.; Hasegawa, K.; Tateishi, R.; Kariyama, K.; Shiina, S.; Toyoda, H.; Imai, Y.; Hiraoka, A.; Ikeda, M.; et al. Management of Hepatocellular Carcinoma in Japan: JSH Consensus Statements and Recommendations 2021 Update. *Liver Cancer* **2021**, *10*, 181–223. [CrossRef]
11. Gordan, J.D.; Kennedy, E.B.; Abou-Alfa, G.K.; Beg, M.S.; Brower, S.T.; Gade, T.P.; Goff, L.; Gupta, S.; Guy, J.; Harris, W.P.; et al. Systemic Therapy for Advanced Hepatocellular Carcinoma: ASCO Guideline. *J. Clin. Oncol.* **2020**, *38*, 4317–4345. [CrossRef] [PubMed]
12. Tateishi, R.; Uchino, K.; Fujiwara, N.; Takehara, T.; Okanoue, T.; Seike, M.; Yoshiji, H.; Yatsuhashi, H.; Shimizu, M.; Torimura, T.; et al. A Nationwide Survey on Non-B, Non-C Hepatocellular Carcinoma in Japan: 2011–2015 Update. *J. Gastroenterol.* **2019**, *54*, 367–376. [CrossRef] [PubMed]
13. Pfister, D.; Núñez, N.G.; Pinyol, R.; Govaere, O.; Pinter, M.; Szydlowska, M.; Gupta, R.; Qiu, M.; Deczkowska, A.; Weiner, A.; et al. NASH Limits Anti-Tumour Surveillance in Immunotherapy-Treated HCC. *Nature* **2021**, *592*, 450–456. [CrossRef] [PubMed]
14. Zhu, A.X.; Abbas, A.R.; de Galarreta, M.R.; Guan, Y.; Lu, S.; Koeppen, H.; Zhang, W.; Hsu, C.-H.; He, A.R.; Ryoo, B.-Y.; et al. Molecular correlates of clinical response and resistance to atezolizumab in combination with bevacizumab in advanced hepatocellular carcinoma. *Nat. Med.* **2022**, *28*, 1599–1611. [CrossRef]
15. Li, X.S.; Li, J.W.; Li, H.; Jiang, T. Prognostic Value of Programmed Cell Death Ligand 1 (PD-L1) for Hepatocellular Carcinoma: A Meta-Analysis. *Biosci. Rep.* **2020**, *40*, BSR20200459. [CrossRef]
16. Kokudo, N.; Takemura, N.; Hasegawa, K.; Takayama, T.; Kubo, S.; Shimada, M.; Nagano, H.; Hatano, E.; Izumi, N.; Kaneko, S.; et al. Clinical Practice Guidelines for Hepatocellular Carcinoma: The Japan Society of Hepatology 2017 (4th JSH-HCC Guidelines) 2019 Update. *Hepatol. Res.* **2019**, *49*, 1109–1113. [CrossRef]
17. Champiat, S.; Dercle, L.; Ammari, S.; Massard, C.; Hollebecque, A.; Postel-Vinay, S.; Chaput, N.; Eggermont, A.; Marabelle, A.; Soria, J.C.; et al. Hyperprogressive Disease Is a New Pattern of Progression in Cancer Patients Treated by Anti-PD-1/PD-L1. *Clin. Cancer Res.* **2017**, *23*, 1920–1928. [CrossRef]
18. Schwartz, L.H.; Seymour, L.; Litière, S.; Ford, R.; Gwyther, S.; Mandrekar, S.; Shankar, L.; Bogaerts, J.; Chen, A.; Dancey, J.; et al. RECIST 1.1—Standardisation and Disease-Specific Adaptations: Perspectives from the RECIST Working Group. *Eur. J. Cancer* **2016**, *62*, 138–145. [CrossRef]
19. Hothorn, T.; Zeileis, A. Generalized Maximally Selected Statistics. *Biometrics* **2008**, *64*, 1263–1269. [CrossRef]
20. Johnson, P.J.; Berhane, S.; Kagebayashi, C.; Satomura, S.; Teng, M.; Reeves, H.L.; O'Beirne, J.; Fox, R.; Skowronska, A.; Palmer, D.; et al. A Nssessment of Liver Function in Patients with Hepatocellular Carcinoma: A New Evidence-Based Approach—The Albi Grade. *J. Clin. Oncol.* **2015**, *33*, 550–558. [CrossRef]
21. Cheng, A.L.; Qin, S.; Ikeda, M.; Galle, P.R.; Ducreux, M.; Kim, T.Y.; Lim, H.Y.; Kudo, M.; Breder, V.; Merle, P.; et al. Updated Efficacy and Safety Data from IMbrave150: Atezolizumab plus Bevacizumab vs. Sorafenib for Unresectable Hepatocellular Carcinoma. *J. Hepatol.* **2022**, *76*, 862–873. [CrossRef] [PubMed]
22. Chuma, M.; Uojima, H.; Hattori, N.; Arase, Y.; Fukushima, T.; Hirose, S.; Kobayashi, S.; Ueno, M.; Tezuka, S.; Iwasaki, S.; et al. Safety and Efficacy of Atezolizumab plus Bevacizumab in Patients with Unresectable Hepatocellular Carcinoma in Early Clinical Practice: A Multicenter Analysis. *Hepatol. Res.* **2022**, *52*, 269–280. [CrossRef] [PubMed]
23. Tanaka, T.; Hiraoka, A.; Tada, T.; Hirooka, M.; Kariyama, K.; Tani, J.; Atsukawa, M.; Takaguchi, K.; Itobayashi, E.; Fukunishi, S.; et al. Therapeutic efficacy of atezolizumab plus bevacizumab treatment for unresectable hepatocellular carcinoma in patients with Child-Pugh class A or B liver function in real-world clinical practice. *Hepatol. Res.* **2022**, *52*, 773–783. [CrossRef] [PubMed]
24. Zhu, A.X.; Dayyani, F.; Yen, C.-J.; Ren, Z.; Bai, Y.; Meng, Z.; Pan, H.; Dillon, P.; Mhatre, S.K.; Gaillard, V.E.; et al. Alpha-Fetoprotein as a Potential Surrogate Biomarker for Atezolizumab + Bevacizumab Treatment of Hepatocellular Carcinoma. *Clin. Cancer Res.* **2022**, *28*, 3537–3545. [CrossRef] [PubMed]

25. Scheiner, B.; Pomej, K.; Kirstein, M.M.; Hucke, F.; Finkelmeier, F.; Waidmann, O.; Himmelsbach, V.; Schulze, K.; von Felden, J.; Fründt, T.W.; et al. Prognosis of Patients with Hepatocellular Carcinoma Treated with Immunotherapy—Development and Validation of the CRAFITY Score. *J. Hepatol.* **2022**, *76*, 353–363. [CrossRef] [PubMed]
26. Kaneko, S.; Kurosaki, M.; Tsuchiya, K.; Yasui, Y.; Hayakawa, Y.; Inada, K.; Tanaka, Y.; Ishido, S.; Kirino, S.; Yamashita, K.; et al. Clinical evaluation of Elecsys PIVKA-II for patients with advanced hepatocellular carcinoma. *PLoS ONE* **2022**, *17*, e0265235. [CrossRef]
27. Kuzuya, T.; Kawabe, N.; Hashimoto, S.; Miyahara, R.; Nakano, T.; Nakaoka, K.; Tanaka, H.; Miyachi, Y.; Mii, A.; Tanahashi, Y.; et al. Initial Experience of Atezolizumab Plus Bevacizumab for Advanced Hepatocellular Carcinoma in Clinical Practice. *Cancer Diagn. Progn.* **2021**, *1*, 83–88. [CrossRef] [PubMed]
28. Hiraoka, A.; Kumada, T.; Tada, T.; Hirooka, M.; Kariyama, K.; Tani, J.; Atsukawa, M.; Takaguchi, K.; Itobayashi, E.; Fukunishi, S.; et al. Atezolizumab plus bevacizumab treatment for unresectable hepatocellular carcinoma: Early clinical experience. *Cancer Rep.* **2021**, *5*, e1464. [CrossRef]
29. Maesaka, K.; Sakamori, R.; Yamada, R.; Doi, A.; Tahata, Y.; Miyazaki, M.; Ohkawa, K.; Mita, E.; Iio, S.; Nozaki, Y.; et al. Comparison of Atezolizumab plus Bevacizumab and Lenvatinib in Terms of Efficacy and Safety as Primary Systemic Chemotherapy for Hepatocellular Carcinoma. *Hepatol. Res.* **2022**, *52*, 630–640. [CrossRef]
30. Hayakawa, Y.; Tsuchiya, K.; Kurosaki, M.; Yasui, Y.; Kaneko, S.; Tanaka, Y.; Ishido, S.; Inada, K.; Kirino, S.; Yamashita, K.; et al. Early experience of atezolizumab plus bevacizumab therapy in Japanese patients with unresectable hepatocellular carcinoma in real-world practice. *Investig. New Drugs* **2022**, *40*, 392–402. [CrossRef]
31. Ando, Y.; Kawaoka, T.; Kosaka, M.; Shirane, Y.; Johira, Y.; Miura, R.; Murakami, S.; Yano, S.; Amioka, K.; Naruto, K.; et al. Early Tumor Response and Safety of Atezolizumab plus Bevacizumab for Patients with Unresectable Hepatocellular Carcinoma in Real-World Practice. *Cancers* **2021**, *13*, 3958. [CrossRef] [PubMed]

cancers

MDPI

*Article*

# Baseline Splenic Volume Outweighs Immuno-Modulated Size Changes with Regard to Survival Outcome in Patients with Hepatocellular Carcinoma under Immunotherapy

**Lukas Müller [1], Simon Johannes Gairing [2], Roman Kloeckner [3], Friedrich Foerster [2], Arndt Weinmann [2], Jens Mittler [4], Fabian Stoehr [1], Tilman Emrich [1,5,6], Christoph Düber [1], Peter Robert Galle [2] and Felix Hahn [1,*]**

1 Department of Diagnostic and Interventional Radiology, University Medical Center of the Johannes Gutenberg University Mainz, 55131 Mainz, Germany; lukas.mueller@unimedizin-mainz.de (L.M.); fabian.stoehr@unimedizin-mainz.de (F.S.); tilman.emrich@unimedizin-mainz.de (T.E.); christoph.dueber@unimedizin-mainz.de (C.D.)

2 Department of Internal Medicine I, University Medical Center of the Johannes Gutenberg University Mainz, 55131 Mainz, Germany; simonjohannes.gairing@unimedizin-mainz.de (S.J.G.); friedrich.foerster@unimedizin-mainz.de (F.F.); arndt.weinmann@unimedizin-mainz.de (A.W.); peter.galle@unimedizin-mainz.de (P.R.G.)

3 Department of Interventional Radiology, University Hospital Schleswig-Holstein–Campus Luebeck, 23562 Luebeck, Germany; roman.kloeckner@uksh.de

4 Department of General, Visceral and Transplant Surgery, University Medical Center of the Johannes Gutenberg University Mainz, 55131 Mainz, Germany; jens.mittler@unimedizin-mainz.de

5 German Center for Cardiovascular Research (DZHK), Partner-Site Rhine-Main, 55131 Mainz, Germany

6 Division of Cardiovascular Imaging, Department of Radiology and Radiological Science, Medical University of South Carolina, Charleston, SC 29425, USA

* Correspondence: felix.hahn@unimedizin-mainz.de

**Citation:** Müller, L.; Gairing, S.J.; Kloeckner, R.; Foerster, F.; Weinmann, A.; Mittler, J.; Stoehr, F.; Emrich, T.; Düber, C.; Galle, P.R.; et al. Baseline Splenic Volume Outweighs Immuno-Modulated Size Changes with Regard to Survival Outcome in Patients with Hepatocellular Carcinoma under Immunotherapy. *Cancers* **2022**, *14*, 3574. https://doi.org/10.3390/cancers14153574

Academic Editor: Georgios Germanidis

Received: 6 July 2022
Accepted: 20 July 2022
Published: 22 July 2022



**Simple Summary:** Splenic volume (SV) has been identified as a highly predictive parameter for prognosis in patients with hepatocellular carcinoma (HCC). Moreover, an association between immunotherapy and an increase in SV has been described for various types of cancer. In our cohort of patients with HCC under immunotherapy, SV was a highly predictive factor for overall survival at baseline and initial follow-up. Although a large proportion of patients (76%) showed an SV increase after the initiation of immunotherapy, this additional immuno-modulated SV change was negligible compared to long-standing changes in the splanchnic circulation in our patient cohort.

**Abstract:** *Background*: An association between immunotherapy and an increase in splenic volume (SV) has been described for various types of cancer. SV is also highly predictive of overall survival (OS) in patients with hepatocellular carcinoma (HCC). We evaluated SV and its changes with regard to their prognostic influence in patients with HCC undergoing immunotherapy. *Methods*: All patients with HCC who received immunotherapy in first or subsequent lines at our tertiary care center between 2016 and 2021 were screened for eligibility. SV was assessed at baseline and follow-up using an AI-based tool for spleen segmentation. Patients were dichotomized into high and low SV based on the median value. *Results*: Fifty patients were included in the analysis. The median SV prior to treatment was 532 mL. The median OS of patients with high and low SV was 5.1 months and 18.1 months, respectively ($p = 0.01$). An increase in SV between treatment initiation and the first follow-up was observed in 28/37 (75.7%) patients with follow-up imaging available. This increase in itself was not prognostic for median OS (7.0 vs. 8.5 months, $p = 0.73$). However, patients with high absolute SV at the first follow-up continued to have impaired survival (4.0 months vs. 30.7 months, $p = 0.004$). *Conclusion*: High SV prior to and during treatment was a significant prognostic factor for impaired outcome. Although a large proportion of patients showed an SV increase after the initiation of immunotherapy, this additional immuno-modulated SV change was negligible compared to long-standing changes in the splanchnic circulation in patients with HCC.

**Keywords:** carcinoma; hepatocellular; immunotherapy; diagnostic imaging; treatment outcome; spleen volume; prognosis

## 1. Introduction

Hepatocellular carcinoma (HCC) is the most common primary liver cancer and one of the leading causes of cancer-related deaths worldwide [1]. Patients suffering from HCC tend to have two underlying diseases that influence their prognosis and treatment outcome; in more than 80% of Western patients, HCC developed in existing liver cirrhosis [2]. Thus, in addition to the tumor burden, survival is heavily influenced by the remaining liver function. Liver cirrhosis itself leads to the development of portal hypertension [3]. Portal hypertension, in turn, is a factor influencing the risk of hepatic decompensation during HCC treatment and is furthermore a prognostic factor for overall survival (OS) [4–6]. The reference standard for measuring portal hypertension is direct measurement of the hepatic vein pressure gradient (HVPG) through a transjugular approach [2,4]. However, due to its invasive nature and high effort, HVPG measurement is not routinely performed in the diagnostic evaluation of patients with HCC. Consequently, other clinical parameters, such as low platelet count, the presence of esophageal/gastric varices, and ascites, are considered surrogates in the identification of patients with clinically relevant portal hypertension (CRPH) [7–9].

Splenic volume (SV) at baseline and during treatment has also been identified as a surrogate for CRPH in patients with HCC [10]. Furthermore, it is highly relevant for predicting the prognosis in patients with HCC undergoing curative and palliative treatment [11–15]. Novel AI-based methods enable a fully automated assessment of the SV using computed tomography (CT) data [15,16]. Thus, it can be considered a promising imaging biomarker with the potential for full integration into the routine radiology workflow.

In recent years, the results of the IMbrave150 trial led to changes in the treatment paradigm: Immunotherapy has become a first-line systemic treatment option for patients with advanced HCC and for patients in whom other treatment options have failed [17–21]. Furthermore, several ongoing trials are currently investigating other potential immunotherapeutic agents in various tumor stages [19,22,23]. However, immunotherapy has been linked to systemic reactions and shown to influence several organ systems besides the target [24]. One organ that is affected is the spleen. A change in SV during treatment has been previously reported for patients with melanoma and lung cancer [25,26]. Furthermore, SV has been identified as a risk factor for survival outcomes [26].

No study has yet investigated the role of SV and changes in SV in patients with HCC receiving immunotherapy. Given the high coincidence of concomitant liver cirrhosis and increased SV prior to treatment, the present study aimed to investigate whether additional immuno-modulated changes in SV occur and have a detrimental effect in HCC patients undergoing immunotherapy.

## 2. Materials and Methods

The Ethics Committee of the Medical Association of Rhineland Palatinate, Mainz, Germany, approved this study (permit number 837.199.10). The requirement for informed consent was waived due to the retrospective nature of the study. This report followed the guidelines for reporting observational studies (STROBE) [27].

### 2.1. Patients

This retrospective study included all patients with HCC who presented in our dedicated HCC outpatient clinic between May 2016 and October 2021 for the initiation of immunotherapy. Inclusion criteria were age > 18 years, histological or image-derived HCC diagnosis based on the EASL criteria, immunotherapy as systemic treatment, CT images available prior to immunotherapy, and demographic, clinical, and laboratory data available at initiation of the immunotherapy and during follow-up. Of the scanned 64 patients, 50 (78.1%) patients fulfilled all inclusion criteria (Figure 1).

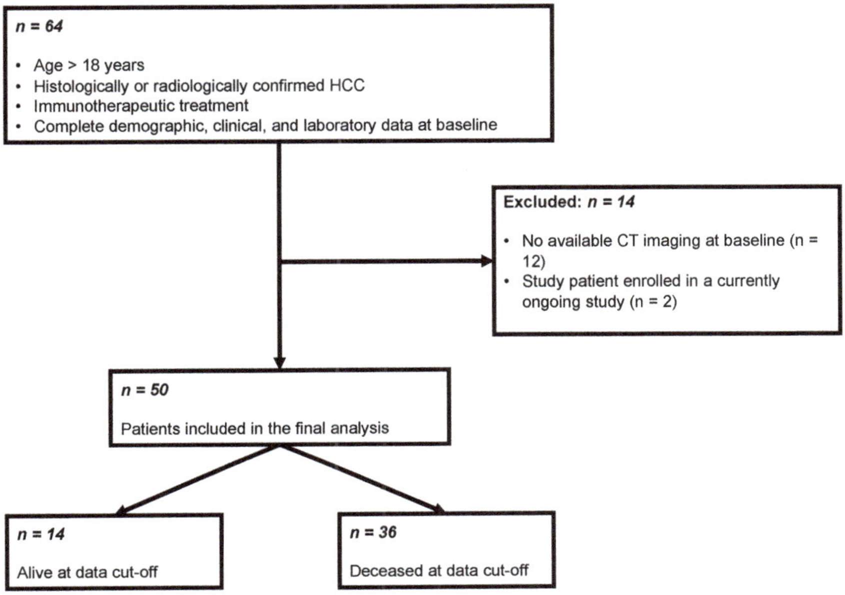


**Figure 1.** Flowchart of the patient selection process for this study.

*2.2. Diagnosis, Treatment, and Follow-Up*

As previously reported, histological or image-derived EASL criteria were used for the diagnosis of HCC [2,28]. The decision to initiate immunotherapy was made by an interdisciplinary tumor board. The board consisted of hepatologists/oncologists, diagnostic and interventional radiologists, visceral surgeons, pathologists, and radiation therapists, who discussed each case prior to the treatment decision. All patients received contrast-enhanced multiphasic CT imaging prior to treatment initiation. Follow-up consisted of clinical examination, blood sampling, and cross-sectional imaging, which was typically repeated every 6 to 12 weeks.

*2.3. Splenic Volume Assessment*

SV was assessed using an established tool for fully automated segmentation and volumetry of the spleen installed at our institution [15]. This algorithm employs the open-source MIScnn library, a convolutional neural network with a U-Net architecture, and has previously been trained for spleen segmentation in patients with HCC undergoing transarterial chemoembolization (TACE) [29]. Detailed information on the features of the network, the settings for training and validation, and the model's performance can be found in the original publication [15]. The output of the network consisted of graphic overlays, which were reviewed by two independent readers. The quality of the graphic overlays was rated as perfect, acceptable, or poor. Consensus reading was performed in the case of discrepancies ($n$ = 2 (4.0%)). Patients with perfect or acceptable SVs were included in the statistical analyses ($n$ = 48); patients with a poor grade ($n$ = 2) were manually re-segmented to obtain the proper SV for further analyses as reported previously [15]. For manual segmentation, the freely available LIFEx software was used (www.lifexsoft.org) [30]. In a second step, SV was normalized to the body surface area (BSA), which was calculated using the patient's height and weight.

### 2.4. Statistical Analysis

Statistical analyses and graphic design were performed in R 4.0.3 (A Language and Environment for Statistical Computing, R Foundation for Statistical Computing, http://www.R-project.org; accessed on 31 May 2022). Data distribution of the continuous variables was assessed for normality using the Shapiro–Wilk test. Normally distributed variables were expressed as mean and standard deviation (SD), whereas non-normally distributed variables were expressed as median and interquartile range (IQR). Categorical and binary baseline parameters were reported as absolute numbers and percentages. Categorical parameters were compared using Fisher's exact test and continuous parameters using the Student's t-test in case of normal distribution and the Mann–Whitney test in case of non-normal distribution. Survival analyses and creation of the Kaplan–Meier curves were performed using the packages "survminer" and "survival" (https://cran.r-project.org/package=survminer, https://CRAN.R-project.org/package=survival, accessed on 31 May 2022). For all patients, OS and progression-free survival (PFS) were calculated from the initiation of treatment. In addition, for patients with available follow-up imaging, OS was calculated from the first follow-up. Log-rank testing was used to compare survival times. Cox proportional hazards regression models assessing hazard ratios (HRs) and corresponding 95% confidence intervals (CIs) were used to determine the effect of the risk stratification. Significance was set at $p < 0.05$.

## 3. Results

### 3.1. Baseline Characteristics

A total of 50 patients, 40 males (80.0%) and 10 females (20.0%), with a median age of 68 years (IQR 62–73 years), were included in the final analysis. For the 37 (74.0%) patients with follow-up CT available, the median time between treatment initiation and follow-up imaging was 85 days (range 68–100 days). Baseline characteristics are provided in Table 1.

**Table 1.** Baseline characteristics.

| Parameter | All Patients ($n$ = 50) |
|---|---|
| **Age, years** * | 67.2 (9.0) |
| **Sex** *** | |
| Female | 10 (20.0) |
| Male | 40 (80.0) |
| **Etiology of cirrhosis** *** | |
| Alcohol | 19 (38.0) |
| Viral | 7 (14.0) |
| Other | 11 (22.0) |
| No cirrhosis | 13 (26.0) |
| **Child–Pugh stage** *** | |
| A | 25 (50.0) |
| B | 10 (20.0) |
| C | 2 (4.0) |
| No cirrhosis | 13 (26.0) |
| **ECOG** *** | |
| ≤1 | 47 (94.0) |
| 2 | 3 (6.0) |
| **BCLC stage** *** | |
| B | 5 (10.0) |
| C | 42 (84.0) |
| D | 3 (6.0) |
| **Portal vein invasion** *** | |
| Yes | 26 (52.0) |
| No | 24 (48.0) |

**Table 1.** *Cont.*

| Parameter | All Patients ($n$ = 50) |
|---|---|
| **Distant metastasis ***** | |
| Yes | 25 (50.0) |
| No | 25 (50.0) |
| **Focality of the liver lesions ***** | |
| Unifocal | 11 (22.0) |
| Multifocal | 39 (78.0) |
| **Sum of the target lesion sizes, mm **** | 83 (51–135) |
| **AFP, ng/mL **** | 277 (16–4485) |
| **Albumin, g/L *** | 30.4 (5.4) |
| **Bilirubin, mg/dL **** | 1.5 (0.7–2.3) |
| **INR **** | 1.2 (1.1–1.3) |
| **Creatinine, mg/dL **** | 0.9 (0.7–1.1) |
| **Thrombocytes, per nL **** | 139 (94–260) |
| **Immunotherapy agent ***** | |
| Atezolizumab + bevazizumab | 29 (58.0) |
| Pembrolizumab | 11 (22.0) |
| Nivolumab | 10 (20.0) |
| **Line of systemic treatment ***** | |
| First | 29 (58.0) |
| Second | 11 (22.0) |
| Third | 10 (20.0) |
| **Previous therapy ***** | |
| Yes | 42 (84.0) |
| No | 8 (16.0) |
| **Subsequent therapy ***** | |
| Yes | 13 (26.0) |
| No | 37 (64.0) |

Values are given as * mean (SD), ** median (IQR) or *** n (%). AFP, alpha-fetoprotein; INR, International Normalized Ratio.

### 3.2. *Increase in Splenic Volume after Initiation of Immunotherapy*

The median SV for all patients was 531.8 mL (IQR 270.4–784.4 mL) and the SV to BSA ratio was 261.9 mL/m$^2$ (IQR 148.1–397.8 mL/m$^2$). For the 37 (74.0%) patients with CT follow-up imaging available, the median SV at baseline was 524.8 mL (IQR 268.7–784.8 mL) and the SV to BSA ratio was 273.0 mL/m$^2$ (IQR 163.3–414.8 mL/m$^2$). The median SV at the first follow-up was 576.9 mL (IQR 307.6–860.7 mL) for these patients ($p$ = 0.37; Figure 2A). An increase in the SV was observed in 28 (75.7%) patients, whereas 9 (24.3%) patients had a decrease in SV during early treatment (Figure 2B). The median change in SV was 17.8% (IQR 2.2–27.3%; range-36.1–141.7%).

For the following analyses, patients were dichotomized into high and low SV based on the median SV to BSA ratio of the patient cohort. According to this stratification, among initial and follow-up imaging, a change from the low to high SV group was observed in only 2 (5.4%) patients, whereas 35 (94.6%) patients remained in their initial group.

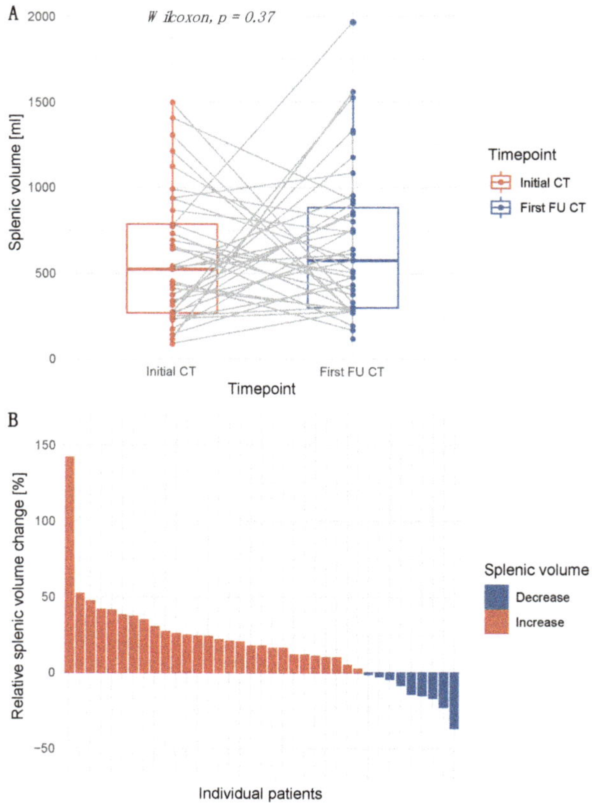


**Figure 2.** Splenic volume (SV) at baseline and during treatment with immunotherapy agents. (**A**) Boxplots of the SV at baseline and during follow-up. (**B**) Relative individual changes in SV between baseline and first follow-up.

### 3.3. *Correlation of Splenic Volume with Parameters of Liver Function, but Not with Tumor Burden*

Patients with high SV had significantly lower albumin levels, higher bilirubin levels, and fewer thrombocytes. No significant differences were observed regarding the INR, the sum of the target lesions, the presence of portal vein infiltration, and the presence of distant metastasis (Table 2).

**Table 2.** Comparison of liver function- and tumor burden-related parameters in patients with low and high splenic volume (SV).

| Parameter | Low SV ($n$ = 25) | High SV ($n$ = 25) | $p$-Value |
|---|---|---|---|
| Liver function | | | |
| Albumin, g/L * | 32.2 (5.9) | 28.5 (4.21) | 0.014 |
| Bilirubin, mg/dL ** | 0.8 (0.6–1.6) | 2.1 (1.5–2.7) | <0.001 |
| Thrombocytes, per nL ** | 224 (138–315) | 101 (75–139) | <0.001 |
| INR ** | 1.1 (1.1–1.3) | 1.2 (1.1–1.4) | 0.190 |
| Tumor burden | | | |
| Sum of the target lesions, mm ** | 76 (50–122) | 88 (51–156) | 0.663 |
| Presence of portal vein infiltration *** | 13 (52.0) | 13 (52.0) | 1.000 |
| Presence of distant metastasis *** | 16 (64.0) | 9 (36.0) | 0.089 |

Values are given as * mean (SD), ** median (IQR) or *** n (%).

### 3.4. Independence of Splenic Volume and Radiological Response

For patients with available follow-up imaging, radiological response was assessed according to mRECIST. The baseline SV of patients with a partial response, stable disease, and progressive disease was 324 mL (IQR 280–581 mL), 670 mL (IQR 471–938 mL), and 412 mL (IQR 249–733 mL), respectively. The follow-up SV in patients with a partial response, stable disease, and progressive disease was 407 mL (IQR 345–666 mL), 744 mL (IQR 531–923 mL), and 479 mL (IQR 293–758 mL), respectively. The median relative change in SV in patients with a partial response, stable disease, and progressive disease between initial imaging and follow-up was 19.1% (IQR 13.3–23.1%), 4.4% (IQR −7.2–24.0%), and 21.9% (IQR 11.9–37.3%), respectively (Figure 3).

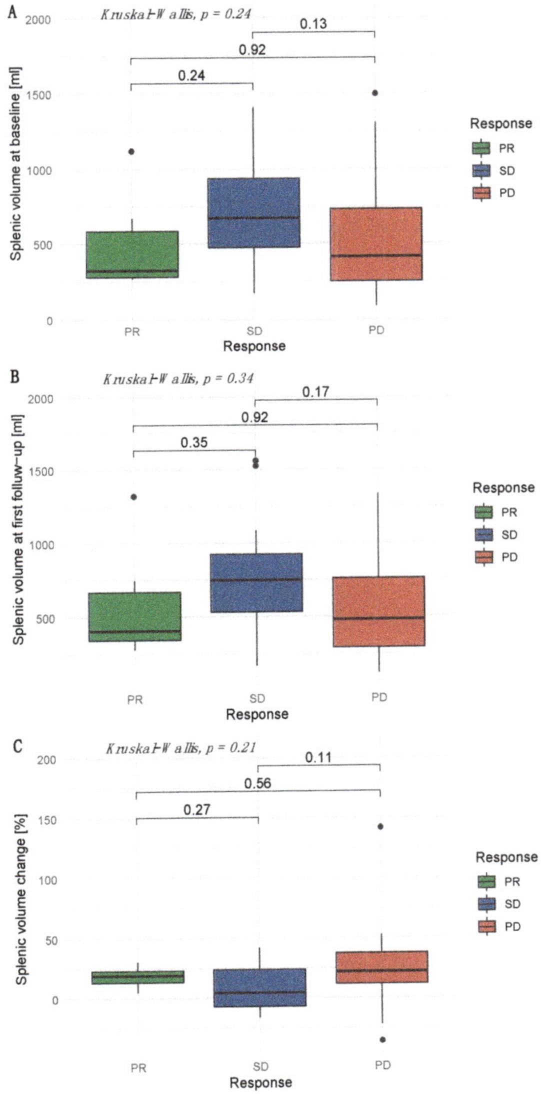


**Figure 3.** Splenic volume among the various response categories. (**A**) Baseline, (**B**) follow-up, and (**C**) relative change.

### 3.5. Significant Impact of High Splenic Volume at Treatment Initiation and during Follow-Up on Overall Survival

The median OS of patients with high SV at baseline was 5.1 months, whereas patients with a low SV had a median OS of 18.1 months ($p = 0.013$; Figure 4A). The PFS in patients with high SV at baseline was 4.6 months, whereas patients with a low SV had a median PFS of 5.3 months ($p = 0.410$; Figure 4B).

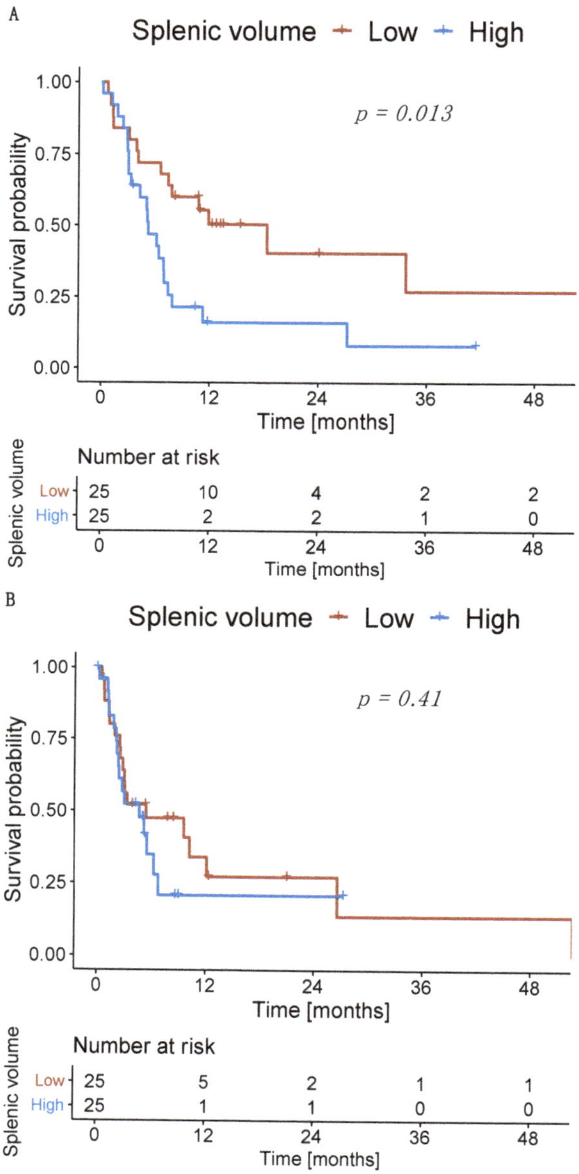


**Figure 4.** Kaplan–Meier curves for patients with low and high splenic volume. (**A**) Overall survival and (**B**) progression-free survival.

Subsequently, we investigated the survival of patients with high and low SV at the first follow-up. Patients with high and low SV at the first follow-up had a median OS of 4.0 months and 30.7 months ($p = 0.004$), respectively (Figure 5A). Patients with an increase in SV from baseline to first follow-up had a median OS of 7.0 months, whereas patients with a decrease in SV had a median OS of 8.5 months ($p = 0.730$; Figure 5B).

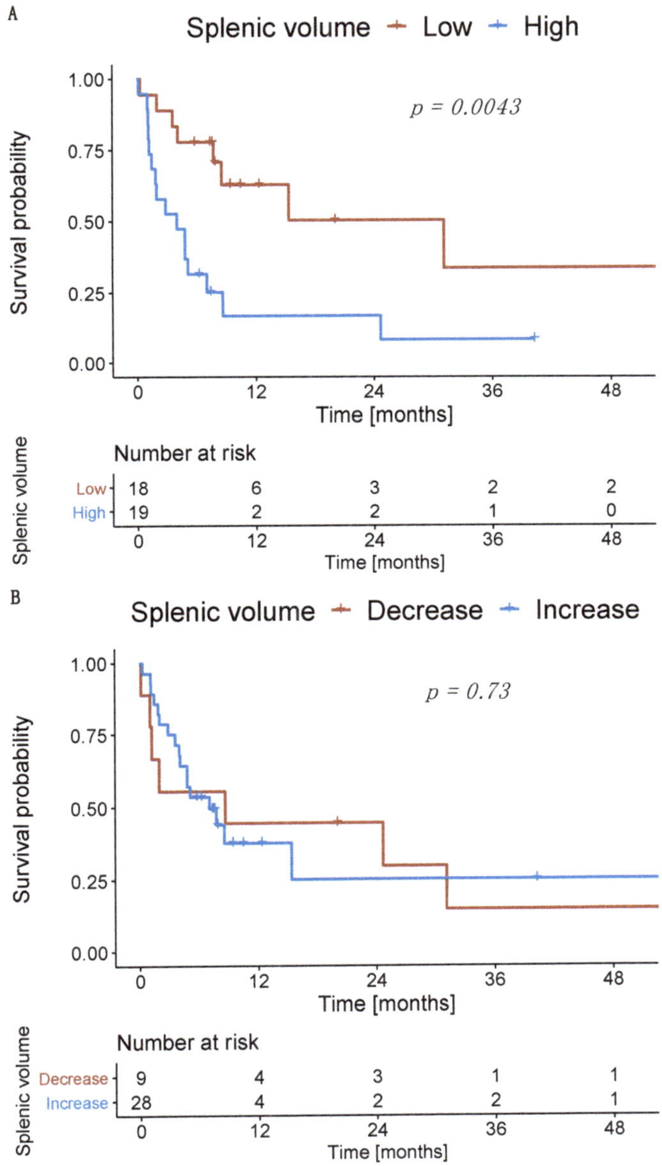


**Figure 5.** Kaplan–Meier curves for overall survival. (**A**) Patients stratified according to the splenic volume at first follow-up and (**B**) according to the relative change compared to baseline.

## 4. Discussion

In this study, we investigated the role of SV and changes in SV with regard to survival outcomes after the initiation of immunotherapy in patients with HCC. Baseline SV was a significant prognostic factor for OS. During early follow-up, the majority of patients had an increase in SV after the initiation of treatment. However, only the absolute SV at the first follow-up remained a significant prognostic factor, and there was no significant survival difference in patients with an increase or decrease in SV.

Our results are in line with previous reports on the changes in SV in patients treated with immunotherapy for other cancer entities [25,26]. Susok et al. investigated the changes in SV during treatment initiation in 49 patients with stage III and IV melanoma [25]. The authors reported a significant increase in the SV after 3 months of follow-up and particularly with the use of anti-CTLA-4 and anti-CTLA-4/anti-PD-1 regimens [25]. However, they did not identify a significant relationship with other clinical parameters. In our study, approximately three-fourths of the patients showed an increase in SV during follow-up, and the median SV increased from 525 to 577 mL, though this increase was not significant.

The median SV change of approximately 18% in our cohort was higher than previously reported for patients with non-small-cell lung cancer undergoing immunotherapy [26]. In their study, Galland et al. reported an increase in 63.5% of patients and a median change of 4.4%. Similar to our results, PFS was not associated with the SV, and the authors reported a significant influence of the baseline SV and the SV during treatment on OS. Unfortunately, the authors did not provide the median absolute SV at treatment initiation and during follow-up. However, the cut-offs used for patient stratification indicate a large difference in the median SV in our patients [26] due to the high proportion of patients with chronic liver disease in our cohort and the associated increase in SV due to increased pressure in the splanchnic circulation [3]. In contrast to our results, Galland et al. postulated that the increase in SV during treatment was significantly associated with impaired survival [26]. In our study, log-rank testing did not show a significant difference in the survival distribution of patients with an increase or decrease in SV during treatment. Moreover, the change in SV under immunotherapy resulted in a change from the low to high SV group in only two (5%) patients. Therefore, the short-term immuno-modulated increase in SV seems to be less important than the pre-existing increase in SV induced by long-standing changes to the splanchnic circulation. Thus, the etiology of changes in SV seems to play a role in investigating correlations between SV and patient outcomes. This is underlined by the significant association between SV and liver function in our study.

In patients with HCC, the baseline SV has been identified as a relevant prognostic factor in various treatment modalities [11–15]. Our results confirm the importance of SV during initial patient evaluation. However, manual spleen segmentation is time-consuming and has a high risk of inter-rater variance [31]. Thus, AI-based solutions for automated SV assessment have the potential to facilitate and standardize this task and enable easy integration into radiological routine. The feasibility of such concepts was reported previously for patients with liver cirrhosis and HCC [15,16]. In this study, we used an algorithm that we had previously trained for patients with HCC undergoing TACE and showed high accuracy in both training and validation [15]. In our study, the algorithm showed sufficient segmentation in 96%, confirming the results of the original study [15]. The present study highlights the easy integration of SV assessment into the routine workflow, together with the high prognostic importance of SV for patients with HCC undergoing immunotherapy. Thus, SV assessment should be contemplated in the diagnostic work-up and for estimating the prognosis in these patients prior to initiating treatment.

The results of this study must be considered in light of several limitations. First, this study was conducted in a retrospective manner and included a limited number of patients. However, this dataset was well-investigated and only patients with complete clinical, laboratory, and imaging data were included. No imputation of missing values was performed. Second, we decided to include patients treated with various immunotherapeutic agents to validate the role of SV in a real-life clinical setting. We did not perform subgroup analysis on each immunotherapy agent due to the small number of patients in each subgroup. However, future studies should validate SV as a novel prognostic factor for various immunotherapy agents and treatment lines.

## 5. Conclusions

In patients with HCC undergoing immunotherapy, high SV prior to and during treatment was a significant prognostic factor for impaired survival. Although a large proportion of HCC patients in our cohort had an SV increase after the initiation of immunotherapy, this increase during treatment did not negatively affect OS per se. Thus, additional immuno-modulated changes in SV were negligible compared to long-standing changes in the splanchnic circulation in patients with HCC.

**Author Contributions:** L.M., S.J.G., R.K., F.F., A.W., J.M., F.S., T.E., C.D., P.R.G. and F.H. devised the study, assisted in data collection, participated in the interpretation of the data, and helped draft the manuscript. L.M., S.J.G., R.K., F.F., A.W. and F.H. carried out the data collection. J.M., F.S., T.E., C.D. and P.R.G. supported the data collection efforts. L.M., R.K. and F.H. created all of the figures and participated in the interpretation of data. L.M., R.K. and F.H. performed the statistical analysis. All authors have read and agreed to the published version of the manuscript.

**Funding:** This research received no external funding.

**Institutional Review Board Statement:** The study was conducted in accordance with the Declaration of Helsinki and approved by the Ethics Committee of the Medical Association of Rhineland Palatinate, Mainz, Germany (permit number 837.199.10).

**Informed Consent Statement:** The requirement for informed consent was waived due to the retrospective nature of the study by the responsible Ethics Committee.

**Data Availability Statement:** Data cannot be shared publicly because of institutional and national data policy restrictions imposed by the Ethics Committee of the Medical Association of Rhineland Palatinate, Mainz, Germany, since the data contain potentially identifying patient information. Data are available upon request for researchers who meet the criteria for access to confidential data.

**Conflicts of Interest:** L.M., F.S. and S.J.G. are supported by the Clinician Scientist Fellowship "Else Kröner Research College: 2018_Kolleg.05". A.W. has received speaker fees and travel grants from Bayer. R.K. has received consultancy fees from Boston Scientific, Bristol-Myers Squibb, Guerbet, Roche, and SIRTEX and lectures fees from BTG, EISAI, Guerbet, Ipsen, Roche, Siemens, SIRTEX, and MSD Sharp & Dohme. FF reports receiving consulting and lectures fees from Roche; lectures fees from Lilly and Pfizer. PRG reports receiving consulting and lectures fees from Adaptimmune, AstraZeneca, Bayer, BMS, Eisai, Ipsen, Lilly, MSD, Roche, and Sirtex. The funders had no role in the design of the study; in the collection, analyses, or interpretation of data; in the writing of the manuscript, or in the decision to publish the results.

## References

1. Bray, F.; Ferlay, J.; Soerjomataram, I.; Siegel, R.L.; Torre, L.A.; Jemal, A. Global cancer statistics 2018: GLOBOCAN estimates of incidence and mortality worldwide for 36 cancers in 185 countries. *CA Cancer J. Clin.* **2018**, *68*, 394–424. [CrossRef] [PubMed]
2. Galle, P.R.; Forner, A.; Llovet, J.M.; Mazzaferro, V.; Piscaglia, F.; Raoul, J.-L.; Schirmacher, P.; Vilgrain, V. EASL Clinical Practice Guidelines: Management of hepatocellular carcinoma. *J. Hepatol.* **2018**, *69*, 182–236. [CrossRef] [PubMed]
3. Iwakiri, Y. Pathophysiology of Portal hypertension. *Clin. Liver Dis.* **2014**, *18*, 281–291. [CrossRef] [PubMed]
4. Bosch, J.; Abraldes, J.G.; Berzigotti, A.; García-Pagan, J.C. The clinical use of HVPG measurements in chronic liver disease. *Nat. Rev. Gastroenterol. Hepatol.* **2009**, *6*, 573. [CrossRef]
5. Berzigotti, A.; Reig, M.; Abraldes, J.G.; Bosch, J.; Bruix, J. Portal hypertension and the outcome of surgery for hepatocellular carcinoma in compensated cirrhosis: A systematic review and meta-analysis. *Hepatology* **2015**, *61*, 526–536. [CrossRef]
6. Müller, L.; Hahn, F.; Mähringer-Kunz, A.; Stoehr, F.; Gairing, S.J.; Foerster, F.; Weinmann, A.; Galle, P.R.; Mittler, J.; Pinto dos Santos, D. Prevalence and clinical significance of clinically evident portal hypertension in patients with hepatocellular carcinoma undergoing transarterial chemoembolization. *United Eur. Gastroenterol. J.* **2022**, *10*, 41–53. [CrossRef] [PubMed]
7. Kim, N.H.; Lee, T.; Cho, Y.K.; Kim, B.I.; Kim, H.J. Impact of clinically evident portal hypertension on clinical outcome of patients with hepatocellular carcinoma treated by transarterial chemoembolization. *J. Gastroenterol. Hepatol.* **2018**, *33*, 1397–1406. [CrossRef]
8. European Association For The Study Of The Liver. EASL–EORTC clinical practice guidelines: Management of hepatocellular carcinoma. *J. Hepatol.* **2012**, *56*, 908–943.

9. Choi, J.W.; Chung, J.W.; Lee, D.H.; Kim, H.-C.; Hur, S.; Lee, M.; Jae, H.J. Portal hypertension is associated with poor outcome of transarterial chemoembolization in patients with hepatocellular carcinoma. *Eur. Radiol.* **2018**, *28*, 2184–2193. [CrossRef]
10. Iranmanesh, P.; Vazquez, O.; Terraz, S.; Majno, P.; Spahr, L.; Poncet, A.; Morel, P.; Mentha, G.; Toso, C. Accurate computed tomography-based portal pressure assessment in patients with hepatocellular carcinoma. *J. Hepatol.* **2014**, *60*, 969–974. [CrossRef]
11. Takeishi, K.; Kawanaka, H.; Itoh, S.; Harimoto, N.; Ikegami, T.; Yoshizumi, T.; Shirabe, K.; Maehara, Y. Impact of splenic volume and splenectomy on prognosis of hepatocellular carcinoma within Milan criteria after curative hepatectomy. *World J. Surg.* **2018**, *42*, 1120–1128. [CrossRef] [PubMed]
12. Bae, J.S.; Lee, D.H.; Yoo, J.; Yi, N.-J.; Lee, K.-W.; Suh, K.-S.; Kim, H.; Lee, K.B. Association between spleen volume and the post-hepatectomy liver failure and overall survival of patients with hepatocellular carcinoma after resection. *Eur. Radiol.* **2021**, *31*, 2461–2471. [CrossRef] [PubMed]
13. Ha, Y.; Kim, D.; Han, S.; Chon, Y.E.; Lee, Y.; Bin, Y.L.; Kim, M.N.; Lee, J.H.; Park, H.; Rim, K.S.; et al. Sarcopenia predicts prognosis in patients with newly diagnosed hepatocellular carcinoma, independent of tumor stage and liver function. *Cancer Res. Treat. Off. J. Korean Cancer Assoc.* **2018**, *50*, 843.
14. Wu, W.-C.; Chiou, Y.-Y.; Hung, H.-H.; Kao, W.-Y.; Chou, Y.-H.; Su, C.-W.; Wu, J.-C.; Huo, T.-I.; Huang, Y.-H.; Lee, K.-C. Prognostic significance of computed tomography scan-derived splenic volume in hepatocellular carcinoma treated with radiofrequency ablation. *J. Clin. Gastroenterol.* **2012**, *46*, 789–795. [CrossRef]
15. Müller, L.; Kloeckner, R.; Mähringer-Kunz, A.; Stoehr, F.; Düber, C.; Arnhold, G.; Gairing, S.J.; Foerster, F.; Weinmann, A.; Galle, P.R.; et al. Fully automated AI-based splenic segmentation for predicting survival and estimating the risk of hepatic decompensation in TACE patients with HCC. *Eur. Radiol.* **2022**. [CrossRef]
16. Lee, C.; Lee, S.S.; Choi, W.-M.; Kim, K.M.; Sung, Y.S.; Lee, S.; Lee, S.J.; Yoon, J.S.; Suk, H.-I. An index based on deep learning–measured spleen volume on CT for the assessment of high-risk varix in B-viral compensated cirrhosis. *Eur. Radiol.* **2021**, *31*, 3355–3365. [CrossRef]
17. Reig, M.; Forner, A.; Rimola, J.; Ferrer-Fábrega, J.; Burrel, M.; Garcia-Criado, A.; Kelley, R.K.; Galle, P.R.; Mazzaferro, V.; Salem, R. BCLC strategy for prognosis prediction and treatment recommendation: The 2022 update. *J. Hepatol.* **2022**, *76*, 681–693. [CrossRef]
18. Galle, P.R.; Dufour, J.-F.; Peck-Radosavljevic, M.; Trojan, J.; Vogel, A. Systemic therapy of advanced hepatocellular carcinoma. *Futur. Oncol.* **2021**, *17*, 1237–1251. [CrossRef]
19. Foerster, F.; Gairing, S.J.; Ilyas, S.I.; Galle, P.R. Emerging Immunotherapy for Hepatocellular Carcinoma: A Guide for Hepatologists. *Hepatology* **2022**, *75*, 1604–1626. [CrossRef]
20. Finn, R.S.; Qin, S.; Ikeda, M.; Galle, P.R.; Ducreux, M.; Kim, T.-Y.; Kudo, M.; Breder, V.; Merle, P.; Kaseb, A.O.; et al. Atezolizumab plus Bevacizumab in Unresectable Hepatocellular Carcinoma. *N. Engl. J. Med.* **2020**, *382*, 1894–1905. [CrossRef]
21. Cheng, A.-L.; Qin, S.; Ikeda, M.; Galle, P.R.; Ducreux, M.; Kim, T.-Y.; Lim, H.Y.; Kudo, M.; Breder, V.; Merle, P. Updated efficacy and safety data from IMbrave150: Atezolizumab plus bevacizumab vs. sorafenib for unresectable hepatocellular carcinoma. *J. Hepatol.* **2022**, *76*, 862–873. [CrossRef] [PubMed]
22. Foerster, F.; Galle, P.R. The current landscape of clinical trials for systemic treatment of hcc. *Cancers* **2021**, *13*, 1962. [CrossRef] [PubMed]
23. Llovet, J.M.; Castet, F.; Heikenwalder, M.; Maini, M.K.; Mazzaferro, V.; Pinato, D.J.; Pikarsky, E.; Zhu, A.X.; Finn, R.S. Immunotherapies for hepatocellular carcinoma. *Nat. Rev. Clin. Oncol.* **2022**, *19*, 151–172. [CrossRef] [PubMed]
24. Ramos-Casals, M.; Brahmer, J.R.; Callahan, M.K.; Flores-Chávez, A.; Keegan, N.; Khamashta, M.A.; Lambotte, O.; Mariette, X.; Prat, A.; Suárez-Almazor, M.E. Immune-related adverse events of checkpoint inhibitors. *Nat. Rev. Dis. Prim.* **2020**, *6*, 1–21. [CrossRef]
25. Susok, L.; Reinert, D.; Lukas, C.; Stockfleth, E.; Gambichler, T. Volume increase of spleen in melanoma patients undergoing immune checkpoint blockade. *Immunotherapy* **2021**, *13*, 885–891. [CrossRef]
26. Galland, L.; Lecuelle, J.; Favier, L.; Fraisse, C.; Lagrange, A.; Kaderbhai, C.; Truntzer, C.; Ghiringhelli, F. Splenic Volume as a Surrogate Marker of Immune Checkpoint Inhibitor Efficacy in Metastatic Non Small Cell Lung Cancer. *Cancers* **2021**, *13*, 3020. [CrossRef]
27. von Elm, E.; Altman, D.G.; Egger, M.; Pocock, S.J.; Gøtzsche, P.C.; Vandenbroucke, J.P. The Strengthening the Reporting of Observational Studies in Epidemiology (STROBE) statement: Guidelines for reporting observational studies. *Lancet* **2007**, *370*, 1453–1457. [CrossRef]
28. Müller, L.; Hahn, F.; Mähringer-Kunz, A.; Stoehr, F.; Gairing, S.J.; Foerster, F.; Weinmann, A.; Galle, P.R.; Mittler, J.; Pinto dos Santos, D. Immunonutritive Scoring in Patients With Hepatocellular Carcinoma Undergoing Transarterial Chemoembolization: Prognostic Nutritional Index or Controlling Nutritional Status Score? *Front. Oncol.* **2021**, *11*, 2205. [CrossRef]
29. Müller, D.; Kramer, F. MIScnn: A framework for medical image segmentation with convolutional neural networks and deep learning. *BMC Med. Imaging* **2021**, *21*, 1–11. [CrossRef]
30. Nioche, C.; Orlhac, F.; Boughdad, S.; Reuzé, S.; Goya-Outi, J.; Robert, C.; Pellot-Barakat, C.; Soussan, M.; Frouin, F.; Buvat, I. LIFEx: A freeware for radiomic feature calculation in multimodality imaging to accelerate advances in the characterization of tumor heterogeneity. *Cancer Res.* **2018**, *78*, 4786–4789. [CrossRef]
31. Nuffer, Z.; Marini, T.; Rupasov, A.; Kwak, S.; Bhatt, S. The best single measurement for assessing splenomegaly in patients with cirrhotic liver morphology. *Acad. Radiol.* **2017**, *24*, 1510–1516. [CrossRef] [PubMed]

*cancers*

MDPI

*Review*

# Salvage versus Primary Liver Transplantation for Hepatocellular Carcinoma: A Twenty-Year Experience Meta-Analysis

**Gian Piero Guerrini *, Giuseppe Esposito, Tiziana Olivieri, Paolo Magistri, Roberto Ballarin, Stefano Di Sandro and Fabrizio Di Benedetto**

Hepato-Pancreato-Biliary Surgery and Liver Transplantation Unit, Policlinico Modena Hospital, Azienda Ospedaliero Universitaria di Modena, Via del Pozzo 71, 41125 Modena, Italy; giuseppe.esposito@unimore.it (G.E.); olivieri.tiziana@aou.mo.it (T.O.); paolo.magistri@unimore.it (P.M.); ballarin.roberto@aou.mo.it (R.B.); sdisandro@unimore.it (S.D.S.); fabrizio.dibenedetto@unimore.it (F.D.B.)
* Correspondence: guerrini.gianpiero@aou.mo.it; Tel.: +39-0594223664

**Simple Summary:** Primary liver transplantation (PLT) for HCC represents the ideal treatment. However, since organ shortage increases the risk of drop-out from the waiting list for tumor progression, a new surgical strategy has been developed: Salvage Liver Transplantation (SLT) can be offered as an additional curative strategy for HCC recurrence after liver resection. The aim of this updated meta-analysis is to compare surgical and long-term outcomes of SLT versus PLT for HCC. The findings of our analysis reveal that SLT offers comparable surgical outcomes but slightly poorer oncological long-term outcomes with respect to PLT.

**Abstract:** (1) Background: Primary liver transplantation (PLT) for HCC represents the ideal treatment. However, since organ shortage increases the risk of drop-out from the waiting list for tumor progression, a new surgical strategy has been developed: Salvage Liver Transplantation (SLT) can be offered as an additional curative strategy for HCC recurrence after liver resection. The aim of this updated meta-analysis is to compare surgical and long-term outcomes of SLT versus PLT for HCC. (2) Materials and Methods: A systematic review and meta-analysis was conducted using the published papers comparing SLT and PLT up to January 2022. (3) Results: 25 studies describing 11,275 patients met the inclusion criteria. The meta-analysis revealed no statistical difference in intraoperative blood loss, overall vascular complications, retransplantation rate, and hospital stay in the SLT group compared with the PLT group. However, the SLT group showed a slightly significant lower 5-year OS rate and 5-year disease-free survival rate. (4) Conclusion: meta-analysis advocates the relative safety and feasibility of both Salvage LT and Primary LT strategies. Specifically, SLT seems to have comparable surgical outcomes but slightly poorer long-term survival than PLT.

**Keywords:** HCC; salvage liver transplantation; rescue liver transplantation; liver transplantation; liver resection

**Citation:** Guerrini, G.P.; Esposito, G.; Olivieri, T.; Magistri, P.; Ballarin, R.; Di Sandro, S.; Di Benedetto, F. Salvage versus Primary Liver Transplantation for Hepatocellular Carcinoma: A Twenty-Year Experience Meta-Analysis. *Cancers* **2022**, *14*, 3465. https://doi.org/10.3390/cancers14143465

Academic Editor: Georgios Germanidis

Received: 18 May 2022
Accepted: 12 July 2022
Published: 16 July 2022



## 1. Introduction

Hepatocellular carcinoma (HCC) is a major contributor to the world's cancer burden and is currently the third leading cause of cancer-related death, with incidences increasing continuously in recent years [1]. Locoregional treatments (mainly radiofrequency ablation and transarterial chemoembolization), liver resection (LR), and liver transplantation (LT) are well-defined and widely accepted treatments for hepatocellular carcinoma [2–4]. However, the best therapy for HCC is still an open and controversial oncological challenge. LT is considered the gold standard therapy for early HCC within liver cirrhosis since it radically removes the cancer and any dysplastic foci and it treats liver disease-related complications (e.g., portal hypertension) [5–7]. The oncological benefits of LT for HCC in terms of 5-year overall survival (OS) and 5-year disease-free survival (DFS) are well-documented: 75% and 90%, respectively [8]. However, organ shortage and the risk of drop-out from the waiting list for tumor progression and deterioration of liver function represent the main limitations for LT [9]. Nowadays, liver surgery for HCC has been demonstrated to be feasible and safe with very low postoperative morbidity and almost zero perioperative mortality [10,11]. Studies on minimally invasive liver surgery have also strongly confirmed these findings [12]. Therefore, primary LR for early HCC with preserved liver function and mild portal hypertension is considered the first-choice treatment [13,14]. Nevertheless, most published data showed a 5-year survival rate and a 5-year DSF after LR for HCC due to cancer relapses of 60% and 30%, respectively [15]. Salvage Liver Transplantation (SLT) is an alternative and promising curative strategy for HCC recurrence or deterioration of liver function after primary liver resection [16]. Moreover, some authors recently described "de principe" Salvage LT (pre-emptive transplantation before tumor recurrence) for a subgroup of patients who present poor histological features and aggressive biological tumor behavior on the final pathology of the resected specimen [17]. Previous studies comparing SLT with primary liver transplantation (PLT) have reported conflicting results in terms of surgical complication and risk of HCC recurrence [18–21]. However, with the advancement of surgical techniques, recent papers have shown SLT to be an effective and feasible treatment for patients with HCC recurrence after primary liver resection with a good long-term survival rate [22]. The purpose of this meta-analysis is to investigate the technical, postoperative, oncological and survival outcomes of PLT compared with SLT.

## 2. Materials and Methods

### 2.1. Study Design

Our meta-analysis was designed according to the Preferred Reporting Items for Systematic Reviews and Meta-Analyses (PRISMA) statement [23], while the authors predetermined the eligibility criteria for the study. Two investigators (E.G. and G.G.P.) independently searched the literature. All retrospective clinical studies that compared Salvage LT with Primary LT for HCC were included in the present systematic review. No prospective studies have been published so far. Case reports, reviews, letters, and animal studies were excluded. All discrepancies during the data collection, synthesis, and analysis were resolved by the consensus of two authors (E.G. and G.G.).

### 2.2. Literature Search and Data Collection

We systematically searched the literature using the PubMed, MEDLINE, and Cochrane library databases for articles published up to January 2022; querying three databases maximizes the probability of capturing articles, as recently demonstrated by Goossen et al. [24]. Our search included the words "HCC", "salvage liver transplantation", "rescue liver transplantation", and "salvage liver transplantation or liver transplantation and liver resection". The search strategy was confined to English language papers and is described in Supplementary File S1 [23] and Supplementary File S2.

### 2.3. Quality Assessment

The quality of the included articles was estimated using the Methodological Index for Non-Randomized Studies (MINORS) [25].

### 2.4. Statistical Analysis

Meta-analysis was realized using the software Review Manager (RevMan) [Version 5.1. Copenhagen: The Nordic Cochrane Centre, The Cochrane Collaboration, 2011). Dichotomous outcomes are displayed as odds ratios (OR) with a 95% confidence interval (CI) by using the Mantel–Haenszel method and continuous variables are displayed as Mean difference (MD) with a 95% CI by utilizing the generic inverse variance method. Mean and standard deviation (SD) for continuous data, if not reported, were estimated using the method illustrated by Hozo et al. [26]. However, for continuous data provided as median and interquartile range (IQR), mean and SD were estimated by employing the method described by Luo et al. [27] and Wan et al. [28], respectively. The cut-off for statistical significance was set at $p \leq 0.05$. Heterogeneities between the studies were evaluated using Q statistics and total variation was computed by $I^2$. A random-effects model (REM) was always adopted due to the conceptual heterogeneity of clinical studies. Publication bias of the included papers is illustrated in Supplementary File S3.

## 3. Results

### 3.1. Studies and Patient Characteristics

Our search strategy disclosed 857 publications concerning Salvage LT. Twenty-nine full papers were examined; however, five studies were not included in the analysis because they did not meet the inclusion criteria. Finally, 25 articles and a total of 11,275 patients were included in the meta-analysis; 9645 patients were offered a Primary LT for HCC, whereas 1630 underwent Salvage LT for HCC recurrence or impaired liver function after primary liver resection. No randomized trials have been published so far. The flow diagram in Figure 1 shows the search process. The baseline characteristics of the two groups are presented in Tables 1 and 2. Technical and postoperative outcomes and oncological and survival features are tabulated in Table 3. The two groups were similar as regards etiology, HBV, and/or HCV infection rates and maximum tumor diameter pre-LT and on post-LT pathology. The number of patients in each study ranged from a minimum of 42 to up to 6975. The MINORS scale assessed a low-quality heterogeneity between studies, providing a mean score of 21.8 (SD: 0.85) and a median score of 22 (range 20–23) (Table 1).

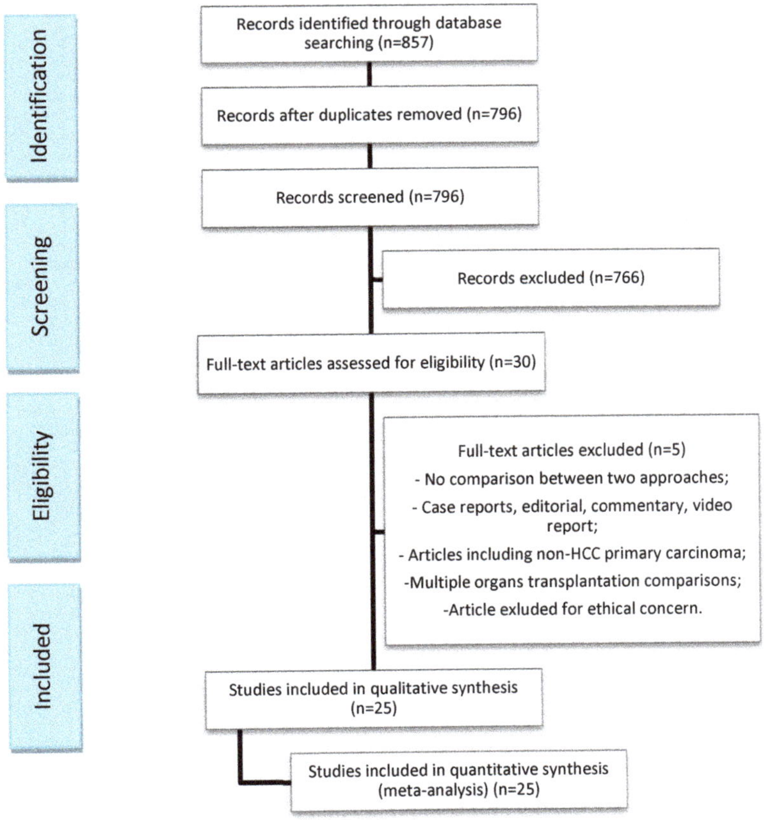


**Figure 1.** Search flow diagram.

**Table 1.** Summary of studies included in the Meta-analysis.

| n. | Author | Region | Year | Study Period | Study Design | Sample Size | | Follow-Up (mo) | | LDLT/DDLT | MINORS (Quality) |
|---|---|---|---|---|---|---|---|---|---|---|---|
| | | | | | | SLT | PLT | SLT | PLT | | |
| 1 | Adam [29] | France | 2003 | 1984–2000 | OCS (R) | 17 | 195 | 49 | 51 | DDLT | 21 |
| 2 | Belghiti [30] | France | 2003 | 1991–2001 | OCS (R) | 18 | 70 | 56.2 | 56.2 | DDLT | 21 |
| 3 | Margarit [31] | Spain | 2005 | 1988–2002 | OCS (P) | 6 | 36 | NA | NA | NA | 20 |
| 4 | Hwang [32] | Korea | 2007 | 1997–2006 | OCS (R) | 17 | 200 | 30.7 | 40.1 | LDLT | 22 |
| 5 | Vennarecci [33] | Italy | 2007 | 2001–2006 | OCS (P) | 9 | 37 | 26.3 | 26.3 | NA | 23 |
| 6 | Del Gadio [34] | Italy | 2008 | 1996–2005 | OCS (R) | 16 | 147 | 26.2 | 36 | DDLT | 23 |
| 7 | Kim [35] | Korea | 2008 | 2005–2007 | OCS (NA) | 15 | 31 | 18.3 | 18.7 | DDLT + LDLT | 20 |
| 8 | Shao [36] | China | 2008 | 2003–2005 | OCS (P) | 15 | 62 | 18 | 22.4 | DDLT | 22 |
| 9 | Cherqui [37] | France | 2009 | 1990–2007 | OCS (R) | 18 | 136 | 57.6 | 576 | DDLT | 21 |
| 10 | Sapisochin [38] | Spain | 2010 | 1990–2007 | OCS (P) | 17 | 34 | 70 | 70 | NA | 22 |
| 11 | Hu [39] | China | 2012 | 1999–2009 | OCS (R) | 888 | 6087 | 15.2 | 15 | DDLT + LDLT | 22 |
| 12 | Kaido [40] | Japan | 2012 | 1999–2009 | OCS (R) | 19 | 48 | 77 | 77 | LDLT | 22 |
| 13 | Liu [41] | China | 2012 | 2001–2011 | OCS (R) | 39 | 180 | 30 | 33 | DDLT + LDLT | 22 |
| 14 | Moon [42] | Korea | 2012 | 1996–2008 | OCS (R) | 17 | 169 | 27.3 | 39 | LDLT | 21 |
| 15 | De Carlis [43] | Italy | 2013 | 2000–2009 | OCS (R) | 26 | 153 | NA | NA | NA | 22 |
| 16 | Guerrini [44] | Italy | 2014 | 2000–2011 | OCS (P) | 28 | 198 | 44.2 | 44.2 | DDLT + LDLT | 22 |
| 17 | Abe [45] | Japan | 2015 | 2001–2011 | OCS (R) | 15 | 45 | 66.3 | 73.2 | LDLT | 22 |
| 18 | Bhangui [46] | France | 2015 | 1990–2012 | OCS (P) | 31 | 340 | 62 | 62 | DDLT | 23 |
| 19 | Vasavada [47] | China | 2015 | 2002–2012 | OCS (R) | 18 | 91 | NA | NA | LDLT | 22 |
| 20 | Whang [48] | China | 2016 | 2001–2011 | OCS (P) | 76 | 295 | 32.4 | 32.4 | DDLT | 23 |
| 21 | Shan [49] | China | 2017 | 2006–2015 | OCS (R) | 28 | 211 | 35 | 35 | DDLT + LDLT | 21 |
| 22 | Yong [50] | Taiwan | 2018 | 2000–2015 | OCS (R) | 100 | 100 | NA | NA | LDLT | 22 |
| 23 | Chan [51] | Taiwan | 2019 | 2001–2018 | OCS (R) | 58 | 245 | NA | NA | LDLT | 22 |
| 24 | Guo [52] | Singapore | 2019 | 2006–2017 | OCS (P) | 14 | 35 | 43.9 | 43.9 | DDLT + LDLT | 22 |
| 25 | Hwan [53] | Korea | 2020 | 2007–2018 | OCS (R) | 125 | 500 | NA | NA | LDLT | 23 |

**Table 2.** General and Patients characteristics.

| | SLT | PLT | Patient (Studies) |
|---|---|---|---|
| **Total patients included** | **1630** | **9645** | **11,275 (25)** |
| Follow-up (months) | 41.3 | 43.8 | 19 |
| HBV infection (%) | 1166/1399 (83.3) | 7157/8652 (82.7) | 16 |
| HCV infection (%) | 103/1240 (8.3) | 786/7842 (10) | 10 |
| MELD score | 11 | 14 | 12 |
| AFP (ng/dl) pre-LT | 184.2 | 208.4 | 11 |
| MILAN in pre-Lt (%) | 264/419 (63) | 1683/2391 (70.4) | 15 |
| MILAN IN on explant (%) | 183/268 (68.2) | 702/948 (74) | 4 |
| Pre-LT Locoregional Treatments (%) | 812/1221 (66.5) | 2901/7600 (38.2) | 11 |
| Waiting list time (months) | 9.6 | 7.2 | 6 |
| Maximum tumor diameter pre LT (cm) | 2.6 | 2.6 | 4 |
| Maximum tumor diameter on explant (cm) | 2.6 | 2.9 | 12 |
| Number of HCC nodule pre LT | 2 | 1.6 | 4 |
| Number of HCC nodule on explant | 3.3 | 2 | 9 |
| Sum of tumor size on explant (cm) | 3.1 | 3.8 | 4 |
| Microvascular invasion (%) | 145/491 (29.5) | 394/1860 (21.2) | 13 |

**Table 3.** Technical and postoperative outcomes; Oncological and survival outcomes.

| Technical and postoperative outcomes | | | | |
|---|---|---|---|---|
| Surgical outcome | Type of surgery | Observations (n) | Mean or % | Studies included (n) |
| Operating time (min) | SLT | 1348 | 600.44 | 16 |
| | PLT | 7971 | 547.12 | |
| Blood loss (ml) | SLT | 1146 | 3174.55 | 6 |
| | PLT | 6722 | 2342.02 | |
| RBC transfusion | SLT | 155 | 7.8 | 8 |
| | PLT | 899 | 6.5 | |
| FFP transfusion | SLT | 126 | 9 | 6 |
| | PLT | 669 | 8 | |
| Reoperation rate | SLT | 48/283 | 16.9% | 9 |
| | PLT | 103/1090 | 9.4% | |
| Mortality rate | SLT | 32/507 | 6.3% | 18 |
| | PLT | 100/2235 | 4.5% | |
| Re-transplantation rate | SLT | 8/131 | 6.1% | 7 |
| | PLT | 70/969 | 7.2% | |
| Postoperative bleeding | SLT | 88/1066 | 8.25% | 10 |
| | PLT | 411/7165 | 5.73% | |
| ICU stay (days) | SLT | 1100 | 8.34 | 8 |
| | PLT | 6574 | 5.44 | |
| Hospital stay (days) | SLT | 1034 | 33.01 | 9 |
| | PLT | 6801 | 26.44 | |
| Vascular complication | SLT | 55/1176 | 4.68% | 12 |
| | PLT | 258/7404 | 3.48% | |
| Arterial thrombosis | SLT | 12/216 | 5.56% | 8 |
| | PLT | 22/790 | 2.78% | |

**Table 3.** *Cont.*

| | | | | |
|---|---|---|---|---|
| Biliary complication | SLT | 162/1191 | 13.6% | 13 |
| | PLT | 838/7449 | 11.2% | |
| Infection and sepsis | SLT | 299/1059 | 28.2% | 10 |
| | PLT | 1826/7149 | 25.5% | |
| | | Oncological and survival outcomes | | |
| Oncological outcome | Type of surgery | Observations (n) | % | Studies included (n) |
| 1-yr OS | SLT | 1072/1375 | 77.9% | 13 |
| | PLT | 6801/8666 | 78.5% | |
| 3-yr OS | SLT | 837/1410 | 59.3% | 15 |
| | PLT | 5508/8950 | 61.9% | |
| 5-yr OS | SLT | 810/1503 | 53.9% | 20 |
| | PLT | 5327/9424 | 56.5% | |
| HCC recurrence | SLT | 37/240 | 15.4% | 10 |
| | PLT | 98/896 | 10.9% | |
| 1-yr DFS | SLT | 967/1358 | 71.2% | 12 |
| | PLT | 5855/8218 | 71.2% | |
| 3-yr DFS | SLT | 763/1393 | 54.8% | 14 |
| | PLT | 4821/8457 | 57% | |
| 5-yr DFS | SLT | 721/1468 | 49.1% | 18 |
| | PLT | 4538/8840 | 51.3% | |

## 3.2. Technical Outcomes

### 3.2.1. Duration of Surgery

The mean operating time was 600.44 min in the SLT group and 547.12 min in the PLT group; sixteen articles reported this item. Operating time was shorter in the Primary LT group, and the meta-analysis showed a statistically significant difference (MD 33.30, (95% CI 17.60, 49.00) $p < 0.0001$), as shown in Figure 2.

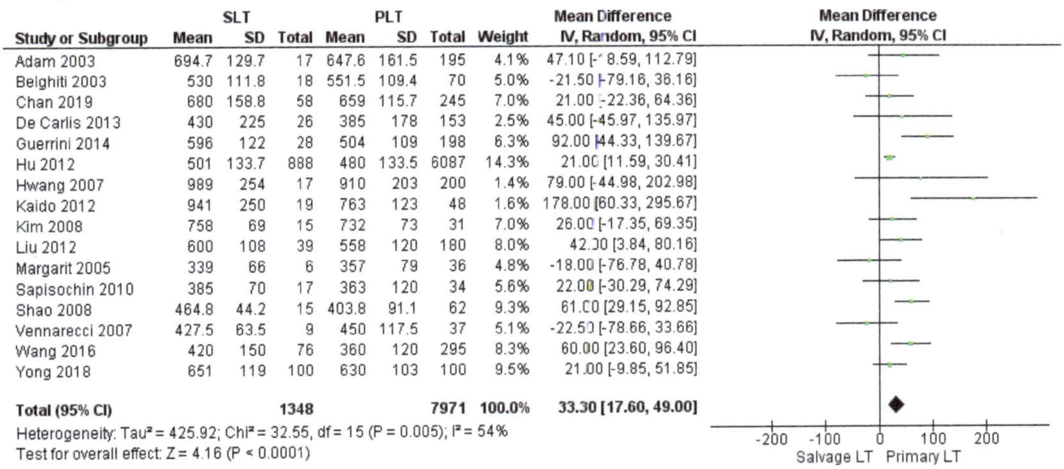


**Figure 2.** Operating time.

### 3.2.2. Intraoperative Blood Loss, Intraoperative Red Blood Cell (RBC), and Fresh Frozen Plasma (FFP) Transfusion

The meta-analysis showed no statistically significant increased intraoperative blood loss in the Salvage LT group when compared with the Primary one (MD 290.35, (95% CI −82.63, 663.32) $p = 0.13$), as shown in Figure 3. The mean intraoperative blood loss in the SLT and PLT groups was 3174.55 cc and 2342.02 cc, respectively. The mean of intraoperative RBC and FFP transfusion was 7.8 RBC units and 9 FFP units in the Salvage LT group, and 6.5 RBC units and 8 FFP units in the Primary LT group. However, our analysis revealed no statistically significant differences between the two approaches: (MD 0.92, (95% CI −0.48, 2.32) $p = 0.07$) and (MD 0.34, (95% CI −0.69, 1.36) $p = 0.52$), respectively, as shown in Figures 4 and 5.

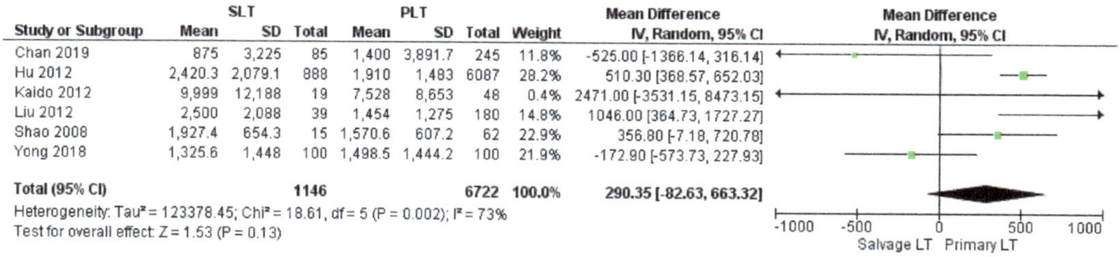


**Figure 3.** Intraoperative blood loss.

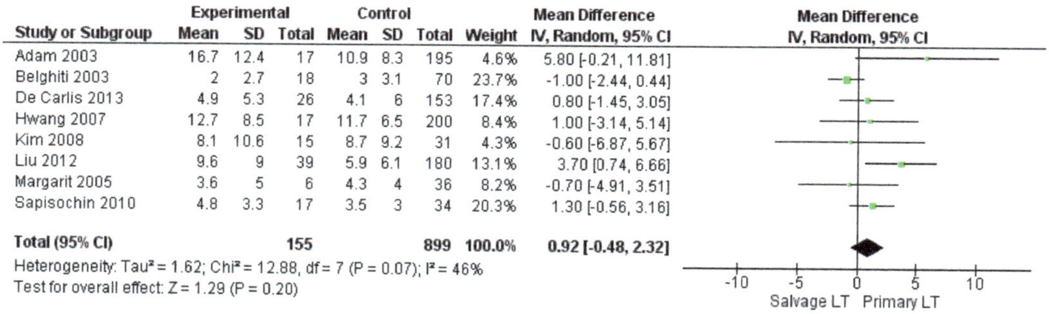


**Figure 4.** Intraoperative Red Blood Cell (RBC) transfusion.

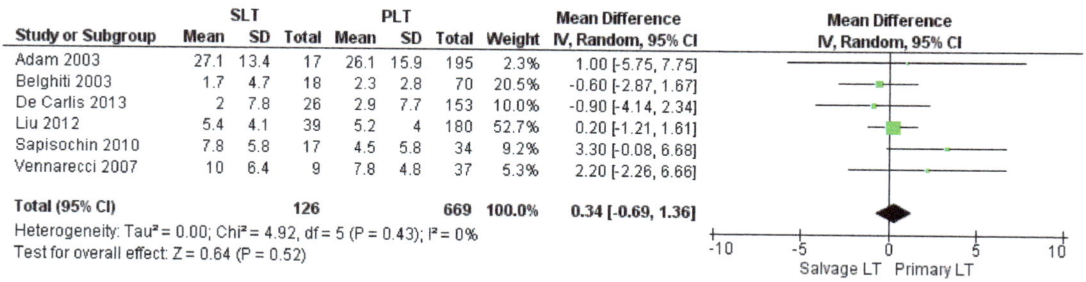


**Figure 5.** Fresh frozen plasma (FFP) transfusion.

### 3.2.3. Reoperation Rate

Reoperation rate was 16.96% (48/283) in the SLT group and 9.45% (103/1090) in the PLT group. The meta-analysis showed a statistically significant difference in the rate of reoperation between the two groups, higher in the SLT than in the PLT group (OR 2.34, (95% CI 1.53, 3.59) $p < 0.0001$), as shown in Figure 6.

### 3.2.4. Perioperative Mortality Rate

Perioperative mortality rate was 6.31% (32/507) in the SLT group and 4.47% (100/2235) in the PLT group; slightly higher in the former group. The meta-analysis of the 18 trials showed a statistically significant difference in the rate of perioperative mortality between the two groups (OR 1.83, (95% CI 1.18, 2.84) $p = 0.007$), as shown in Figure 7.

| Study or Subgroup | SLT Events | SLT Total | PLT Events | PLT Total | Weight | Odds Ratio M-H, Random, 95% CI |
|---|---|---|---|---|---|---|
| Adam 2003 | 0 | 17 | 7 | 195 | 2.1% | 0.72 [0.04, 13.11] |
| Belghiti 2003 | 7 | 18 | 9 | 70 | 13.0% | 4.31 [1.33, 14.01] |
| De Carlis 2013 | 6 | 26 | 20 | 153 | 17.2% | 2.00 [0.71, 5.57] |
| Guo 2019 | 6 | 14 | 7 | 35 | 10.0% | 3.00 [0.78, 11.50] |
| Margarit 2005 | 0 | 6 | 5 | 36 | 2.0% | 0.44 [0.02, 8.99] |
| Moon 2012 | 4 | 17 | 33 | 169 | 12.9% | 1.27 [0.39, 4.14] |
| Vennarecci 2007 | 0 | 9 | 1 | 37 | 1.7% | 1.28 [0.05, 34.01] |
| Wang 2016 | 9 | 76 | 16 | 295 | 24.5% | 2.34 [0.99, 5.53] |
| Yong 2018 | 16 | 100 | 5 | 100 | 16.5% | 3.62 [1.27, 10.30] |
| **Total (95% CI)** | | **283** | | **1090** | **100.0%** | **2.34 [1.53, 3.59]** |
| Total events | 48 | | 103 | | | |

Heterogeneity: Tau² = 0.00; Chi² = 4.90, df = 8 (P = 0.77); I² = 0%
Test for overall effect: Z = 3.93 (P < 0.0001)

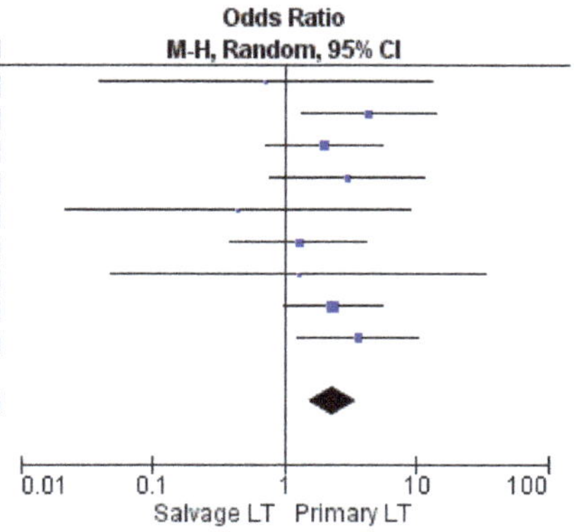


**Figure 6.** Reoperation rate.

| Study or Subgroup | SLT Events | SLT Total | PLT Events | PLT Total | Weight | Odds Ratio M-H, Random, 95% CI |
|---|---|---|---|---|---|---|
| Adam 2003 | 4 | 17 | 4 | 195 | 8.6% | 14.69 [3.29, 65.55] |
| Belghiti 2003 | 1 | 18 | 4 | 70 | 3.8% | 0.97 [0.10, 9.26] |
| Chan 2019 | 3 | 58 | 14 | 245 | 11.7% | 0.90 [0.25, 3.24] |
| De Carlis 2013 | 1 | 26 | 6 | 153 | 4.1% | 0.98 [0.11, 8.49] |
| Del Gaudio 2008 | 0 | 16 | 8 | 147 | 2.3% | 0.50 [0.03, 9.02] |
| Guerrini 2014 | 2 | 28 | 7 | 198 | 7.3% | 2.10 [0.41, 10.65] |
| Guo 2019 | 1 | 14 | 2 | 35 | 3.1% | 1.27 [0.11, 15.23] |
| Hwang 2007 | 1 | 17 | 14 | 200 | 4.4% | 0.83 [0.10, 6.73] |
| Kaido 2012 | 1 | 19 | 4 | 48 | 3.8% | 0.61 [0.06, 5.85] |
| Kim 2008 | 0 | 15 | 1 | 31 | 1.8% | 0.66 [0.03, 17.06] |
| Liu 2012 | 2 | 39 | 8 | 180 | 7.6% | 1.16 [0.24, 5.70] |
| Margarit 2005 | 0 | 6 | 2 | 36 | 1.9% | 1.06 [0.05, 24.77] |
| Moon 2012 | 1 | 17 | 6 | 169 | 4.1% | 1.70 [0.19, 15.00] |
| Sapisochin 2010 | 1 | 17 | 1 | 34 | 2.4% | 2.06 [0.12, 35.14] |
| Shao 2008 | 1 | 15 | 1 | 62 | 2.4% | 4.36 [0.26, 73.98] |
| Vennarecci 2007 | 1 | 9 | 4 | 37 | 3.6% | 1.03 [0.10, 10.53] |
| Wang 2016 | 9 | 76 | 14 | 295 | 24.9% | 2.70 [1.12, 6.49] |
| Yong 2018 | 3 | 100 | 0 | 100 | 2.2% | 7.22 [0.37, 141.52] |
| **Total (95% CI)** | | **507** | | **2235** | **100.0%** | **1.83 [1.18, 2.84]** |
| Total events | 32 | | 100 | | | |

Heterogeneity: Tau² = 0.00; Chi² = 14.76, df = 17 (P = 0.61); I² = 0%
Test for overall effect: Z = 2.71 (P = 0.007)

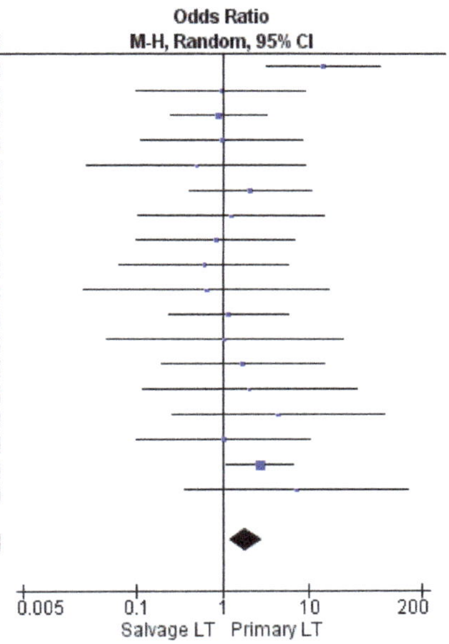


**Figure 7.** Perioperative mortality rate.

#### 3.2.5. Retransplantation Rate

Seven studies reported the retransplantation rate. The retransplantation rate was 6.11% (8/131) in the Salvage LT group and 7.22% (70/969) in the PLT sample. However, the different rates were not statistically significant between the two treatment strategies (OR 1.07, (95% CI 0.51, 2.24) $p = 0.86$), as shown in Figure 8.

### 3.3. *Postoperative Outcomes*

#### 3.3.1. Postoperative Bleeding

Ten studies reported the postoperative bleeding rate. The Salvage LT group's postoperative bleeding rate was considerably higher than the Primary LT group: 8.25% (88/1066) and 5.73% (411/7165), respectively. The difference in bleeding rates was statistically significant (OR 2.19, (95% CI 1.25, 3.81) $p = 0.006$), as shown in Figure 9.

| Study or Subgroup | SLT Events | SLT Total | PLT Events | PLT Total | Weight | Odds Ratio M-H, Random, 95% CI |
|---|---|---|---|---|---|---|
| Adam 2003 | 1 | 17 | 21 | 195 | 12.8% | 0.52 [0.07, 4.11] |
| Belghiti 2003 | 2 | 18 | 8 | 70 | 20.4% | 0.97 [0.19, 5.01] |
| De Carlis 2013 | 1 | 26 | 11 | 153 | 12.6% | 0.52 [0.06, 4.18] |
| Del Gaudio 2008 | 0 | 16 | 12 | 147 | 6.7% | 0.33 [0.02, 5.81] |
| Guerrini 2014 | 2 | 28 | 9 | 198 | 21.9% | 1.62 [0.33, 7.89] |
| Moon 2012 | 2 | 17 | 7 | 169 | 20.0% | 3.09 [0.59, 16.20] |
| Vennarecci 2007 | 0 | 9 | 2 | 37 | 5.7% | 0.75 [0.03, 16.92] |
| **Total (95% CI)** | | **131** | | **969** | **100.0%** | **1.07 [0.51, 2.24]** |
| Total events | 8 | | 70 | | | |

Heterogeneity: Tau² = 0.00; Chi² = 3.65, df = 6 (P = 0.72); I² = 0%
Test for overall effect: Z = 0.17 (P = 0.86)

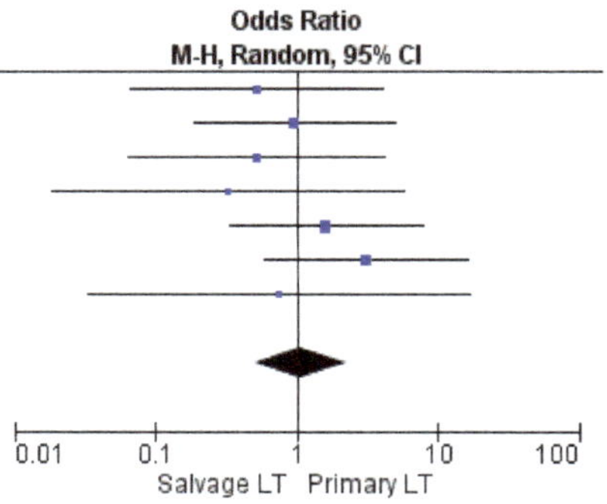


**Figure 8.** Retransplantation rate.

| Study or Subgroup | SLT Events | SLT Total | PLT Events | PLT Total | Weight | Odds Ratio M-H, Random, 95% CI |
|---|---|---|---|---|---|---|
| Abe 2015 | 5 | 15 | 3 | 45 | 8.4% | 7.00 [1.43, 34.29] |
| Adam 2003 | 1 | 17 | 7 | 195 | 5.3% | 1.68 [0.19, 14.51] |
| Belghiti 2003 | 4 | 18 | 2 | 70 | 7.1% | 9.71 [1.62, 58.31] |
| De Carlis 2013 | 4 | 26 | 6 | 153 | 10.4% | 4.45 [1.16, 17.05] |
| Guo 2019 | 2 | 14 | 5 | 35 | 7.2% | 1.00 [0.17, 5.88] |
| Hu 2012 | 59 | 888 | 338 | 6087 | 25.0% | 1.21 [0.91, 1.61] |
| Hwang 2007 | 5 | 17 | 20 | 200 | 12.6% | 3.75 [1.20, 11.74] |
| Kim 2008 | 5 | 15 | 6 | 31 | 10.0% | 2.08 [0.52, 8.41] |
| Liu 2012 | 1 | 39 | 9 | 180 | 5.6% | 0.50 [0.06, 4.07] |
| Moon 2012 | 2 | 17 | 15 | 169 | 8.5% | 1.37 [0.29, 6.57] |
| **Total (95% CI)** | | **1066** | | **7165** | **100.0%** | **2.19 [1.25, 3.81]** |
| Total events | 88 | | 411 | | | |

Heterogeneity: Tau² = 0.30; Chi² = 16.44, df = 9 (P = 0.06); I² = 45%
Test for overall effect: Z = 2.76 (P = 0.006)

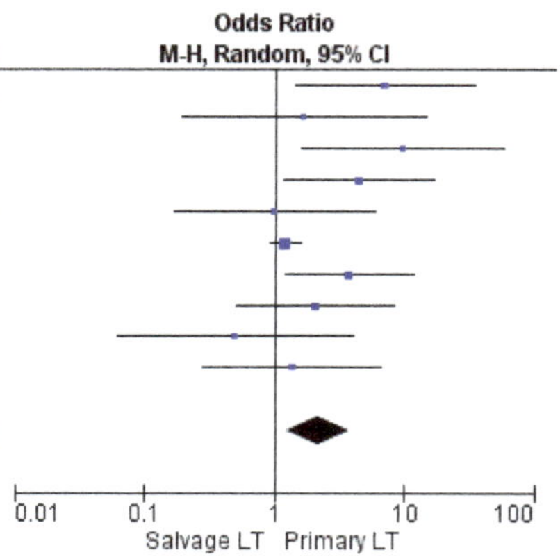


**Figure 9.** Postoperative bleeding.

3.3.2. Intensive Care Unit Stay

The mean Intensive Care Unit (ICU) stay was 8.34 days in the SLT group and 5.44 days in the PLT group. No statistically significant mean difference was recorded (MD −0.12, (95% CI −0.88, 0.63) $p = 0.75$), although a higher mean ICU stay was displayed in the SLT group, as shown in Figure 10.

| Study or Subgroup | SLT Mean | SLT SD | SLT Total | PLT Mean | PLT SD | PLT Total | Weight | Mean Difference IV, Random, 95% CI |
|---|---|---|---|---|---|---|---|---|
| Belghiti 2003 | 9 | 9 | 18 | 10 | 8 | 70 | 2.5% | -1.00 [-5.56, 3.56] |
| Guo 2019 | 5 | 2.6 | 14 | 2 | 9.8 | 35 | 4.0% | 3.00 [-0.52, 6.52] |
| Hu 2012 | 4.9 | 3.7 | 888 | 4.9 | 3 | 6087 | 31.2% | 0.00 [-0.25, 0.25] |
| Kim 2008 | 4.1 | 2.1 | 15 | 4.6 | 2.6 | 31 | 15.2% | -0.50 [-1.90, 0.90] |
| Liu 2012 | 10 | 8.5 | 39 | 10 | 3.5 | 180 | 6.2% | 0.00 [-2.72, 2.72] |
| Sapisochin 2010 | 13.8 | 11.3 | 17 | 5 | 12.3 | 34 | 1.2% | 8.80 [2.02, 15.58] |
| Vennarecci 2007 | 4.5 | 0.9 | 9 | 5.6 | 2 | 37 | 22.5% | -1.10 [-1.97, -0.23] |
| Yong 2018 | 15.4 | 4.5 | 100 | 15.4 | 4.5 | 100 | 17.1% | 0.00 [-1.25, 1.25] |
| **Total (95% CI)** | | | **1100** | | | **6574** | **100.0%** | **-0.12 [-0.88, 0.63]** |

Heterogeneity: Tau² = 0.46; Chi² = 15.68, df = 7 (P = 0.03); I² = 55%
Test for overall effect: Z = 0.32 (P = 0.75)

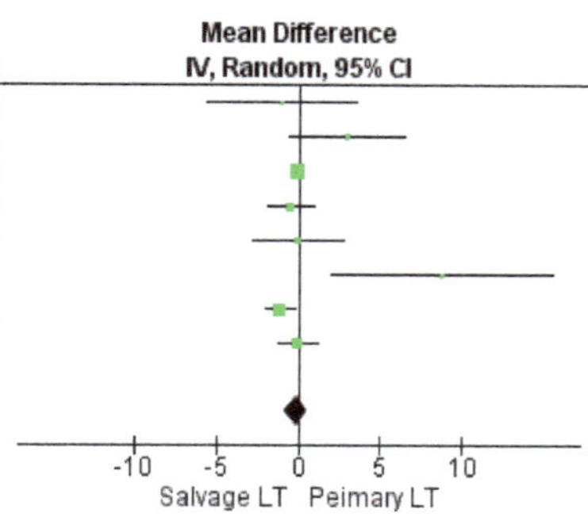


**Figure 10.** Intensive care unit stay.

3.3.3. Length of Hospitalization

The mean hospital stay was 33.01 days in the SLT group and 26.44 in the PLT group; nine articles described this variable. The meta-analysis reported that the mean hospitalization was shorter in the PLT group than in the Salvage LT group, although this imbalance was not significant (MD 0.49, (95% CI −2.13, 3.11) $p$ = 0.71), as shown in Figure 11.

| | SLT | | | PLT | | | | Mean Difference |
|---|---|---|---|---|---|---|---|---|
| Study or Subgroup | Mean | SD | Total | Mean | SD | Total | Weight | IV, Random, 95% CI |
| Belghiti 2003 | 31.7 | 15.6 | 18 | 32.7 | 16.5 | 70 | 7.9% | -1.00 [-9.18, 7.18] |
| De Carlis 2013 | 19 | 13 | 26 | 17 | 15 | 153 | 13.6% | 2.00 [-3.53, 7.53] |
| Guo 2019 | 26.5 | 16.5 | 14 | 19 | 19.5 | 35 | 5.0% | 7.50 [-3.29, 18.29] |
| Hu 2012 | 33.5 | 18.6 | 888 | 33.8 | 15.6 | 6087 | 31.9% | -0.30 [-1.58, 0.98] |
| Liu 2012 | 35 | 18.3 | 39 | 37 | 12.8 | 180 | 12.2% | -2.00 [-8.04, 4.04] |
| Margarit 2005 | 18.8 | 10 | 6 | 29 | 41 | 36 | 2.6% | -10.20 [-25.80, 5.40] |
| Moon 2012 | 97.3 | 78.3 | 17 | 37 | 43.3 | 169 | 0.5% | 60.30 [22.51, 98.09] |
| Sapisochin 2010 | 20.5 | 10 | 17 | 18.5 | 47.5 | 34 | 2.3% | 2.00 [-14.66, 18.66] |
| Vennarecci 2007 | 14.8 | 1.2 | 9 | 14 | 8.8 | 37 | 24.1% | 0.80 [-2.14, 3.74] |
| **Total (95% CI)** | | | **1034** | | | **6801** | **100.0%** | **0.49 [-2.13, 3.11]** |

Heterogeneity: Tau² = 5.16; Chi² = 14.81, df = 8 (P = 0.06); I² = 46%
Test for overall effect: Z = 0.37 (P = 0.71)

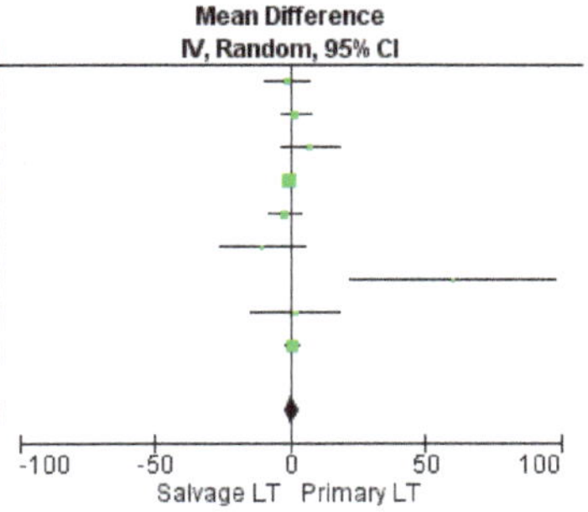


**Figure 11.** Length of hospital stay.

### 3.3.4. Overall Vascular Complication

The rate of vascular complications was evaluated by 12 studies. The vascular complications rate was similar between SLT and PLT: 4.68% (55/1176) and 3.48% (258/7404), respectively. The meta-analysis revealed a statistically significant difference (OR 1.37, (95% CI 1.01, 1.86) $p$ = 0.04), as shown in Figure 12.

| | SLT | | PLT | | | Odds Ratio |
|---|---|---|---|---|---|---|
| Study or Subgroup | Events | Total | Events | Total | Weight | M-H, Random, 95% CI |
| Adam 2003 | 1 | 17 | 3 | 195 | 1.7% | 4.00 [0.39, 40.70] |
| Belghiti 2003 | 1 | 18 | 3 | 70 | 1.7% | 1.31 [0.13, 13.43] |
| De Carlis 2013 | 1 | 26 | 5 | 153 | 2.0% | 1.18 [0.13, 10.56] |
| Del Gaudio 2008 | 1 | 16 | 13 | 147 | 2.1% | 0.69 [0.08, 5.63] |
| Guo 2019 | 2 | 14 | 1 | 35 | 1.5% | 5.67 [0.47, 68.28] |
| Hu 2012 | 37 | 888 | 202 | 6087 | 73.0% | 1.27 [0.89, 1.81] |
| Hwang 2007 | 2 | 17 | 11 | 200 | 3.7% | 2.29 [0.46, 11.30] |
| Kim 2008 | 1 | 15 | 1 | 31 | 1.2% | 2.14 [0.12, 36.80] |
| Liu 2012 | 3 | 39 | 10 | 180 | 5.2% | 1.42 [0.37, 5.41] |
| Moon 2012 | 1 | 17 | 3 | 169 | 1.7% | 3.46 [0.34, 35.21] |
| Vennarecci 2007 | 1 | 9 | 2 | 37 | 1.5% | 2.19 [0.18, 27.20] |
| Yong 2018 | 4 | 100 | 4 | 100 | 4.7% | 1.00 [0.24, 4.11] |
| **Total (95% CI)** | | **1176** | | **7404** | **100.0%** | **1.37 [1.01, 1.86]** |
| Total events | 55 | | 258 | | | |

Heterogeneity: Tau² = 0.00; Chi² = 4.13, df = 11 (P = 0.97); I² = 0%
Test for overall effect: Z = 2.01 (P = 0.04)

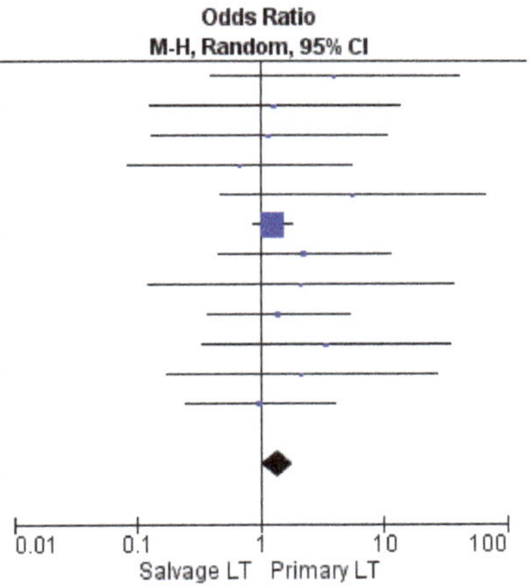


**Figure 12.** Overall vascular complication.

### 3.3.5. Arterial Thrombosis

A total of 34 patients developed arterial thrombosis in twelve studies. The arterial thrombosis rate in the SLT group was higher than within the PLT group: 5.56% (12/216) and 2.78% (22/790), respectively. However, a statistically significant difference in these rates was not recognized between the two approaches (OR 1.87, (95% CI 0.87, 4.03) $p$ = 0.11), as shown in Figure 13.

### 3.3.6. Biliary Complications

Thirteen papers analyzed the frequency of biliary complications (stenosis, leakage, and fistula). The biliary complication rate was significantly higher in the SLT group than the PLT group: 13.6% (162/1191) and 11.2% (838/7449), (OR 1.22, (95% CI 1.01, 1.47) $p$ = 0.04), as shown in Figure 14.

| Study or Subgroup | SLT Events | SLT Total | PLT Events | PLT Total | Weight | Odds Ratio M-H, Random, 95% CI |
|---|---|---|---|---|---|---|
| Adam 2003 | 1 | 17 | 3 | 195 | 10.9% | 4.00 [0.39, 40.70] |
| Belghiti 2003 | 1 | 18 | 3 | 70 | 10.8% | 1.31 [0.13, 13.43] |
| De Carlis 2013 | 1 | 26 | 5 | 153 | 12.2% | 1.18 [0.13, 10.56] |
| Guo 2019 | 2 | 14 | 1 | 35 | 9.5% | 5.67 [0.47, 68.28] |
| Kim 2008 | 1 | 15 | 1 | 31 | 7.2% | 2.14 [0.12, 36.80] |
| Moon 2012 | 1 | 17 | 3 | 169 | 10.9% | 3.46 [0.34, 35.21] |
| Vennarecci 2007 | 1 | 9 | 2 | 37 | 9.2% | 2.19 [0.18, 27.20] |
| Yong 2018 | 4 | 100 | 4 | 100 | 29.3% | 1.00 [0.24, 4.11] |
| **Total (95% CI)** | | **216** | | **790** | **100.0%** | **1.87 [0.87, 4.03]** |
| Total events | 12 | | 22 | | | |

Heterogeneity: Tau² = 0.00; Chi² = 2.50, df = 7 (P = 0.93); I² = 0%
Test for overall effect: Z = 1.61 (P = 0.11)

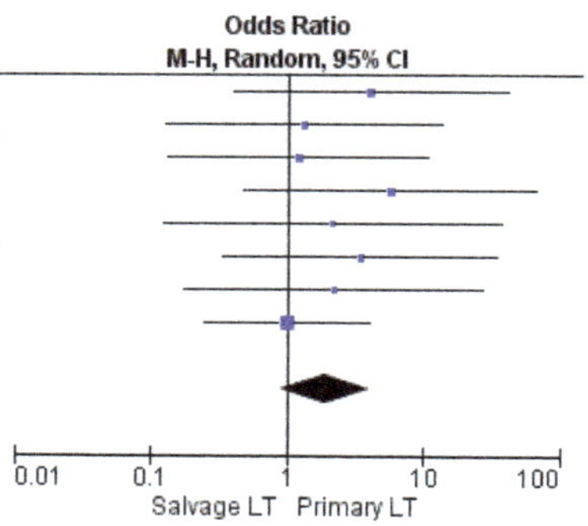


**Figure 13.** Arterial thrombosis.

| Study or Subgroup | SLT Events | SLT Total | PLT Events | PLT Total | Weight | Odds Ratio M-H, Random, 95% CI |
|---|---|---|---|---|---|---|
| Abe 2015 | 5 | 15 | 15 | 45 | 2.3% | 1.00 [0.29, 3.45] |
| Adam 2003 | 0 | 17 | 6 | 195 | 0.4% | 0.83 [0.05, 15.41] |
| Belghiti 2003 | 1 | 18 | 3 | 70 | 0.6% | 1.31 [0.13, 13.43] |
| De Carlis 2013 | 1 | 26 | 6 | 153 | 0.8% | 0.98 [0.11, 8.49] |
| Del Gaudio 2008 | 5 | 16 | 38 | 147 | 2.8% | 1.30 [0.43, 4.00] |
| Guo 2019 | 2 | 14 | 2 | 35 | 0.8% | 2.75 [0.35, 21.76] |
| Hu 2012 | 113 | 888 | 681 | 6087 | 77.5% | 1.16 [0.94, 1.43] |
| Hwang 2007 | 4 | 17 | 36 | 200 | 2.5% | 1.40 [0.43, 4.55] |
| Kim 2008 | 4 | 15 | 6 | 31 | 1.7% | 1.52 [0.36, 6.46] |
| Liu 2012 | 3 | 39 | 5 | 180 | 1.6% | 2.92 [0.67, 12.76] |
| Moon 2012 | 3 | 17 | 17 | 169 | 1.9% | 1.92 [0.50, 7.35] |
| Vennarecci 2007 | 0 | 9 | 8 | 37 | 0.4% | 0.18 [0.01, 3.47] |
| Yong 2018 | 21 | 100 | 15 | 100 | 6.6% | 1.51 [0.73, 3.13] |
| **Total (95% CI)** | | **1191** | | **7449** | **100.0%** | **1.22 [1.01, 1.47]** |
| Total events | 162 | | 838 | | | |

Heterogeneity: Tau² = 0.00; Chi² = 4.89, df = 12 (P = 0.96); I² = 0%
Test for overall effect: Z = 2.04 (P = 0.04)

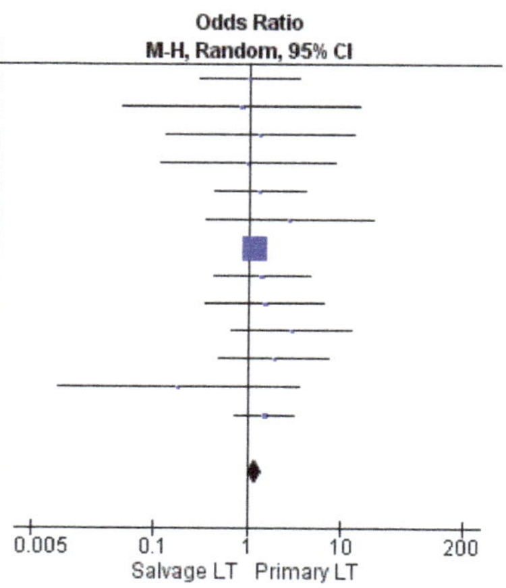


**Figure 14.** Biliary complication.

### 3.3.7. Infection and Sepsis

Ten studies retrospectively assessed overall infection and sepsis rate. Infection rate of the SLT group was slightly higher than the PLT group: 28.2% (299/1059) and 25.5% (1826/7149), respectively. Nevertheless, the meta-analysis stated that the result was not significant (OR 1.14, (95% CI 0.98, 1.32) $p = 0.08$), as shown in Figure 15.

| Study or Subgroup | SLT Events | SLT Total | PLT Events | PLT Total | Weight | Odds Ratio M-H, Random, 95% CI |
|---|---|---|---|---|---|---|
| Abe 2015 | 3 | 15 | 4 | 45 | 0.8% | 2.56 [0.50, 13.07] |
| Adam 2003 | 1 | 17 | 11 | 195 | 0.5% | 1.05 [0.13, 8.62] |
| Belghiti 2003 | 7 | 18 | 29 | 70 | 1.9% | 0.90 [0.31, 2.60] |
| De Carlis 2013 | 5 | 26 | 24 | 153 | 1.9% | 1.28 [0.44, 3.72] |
| Del Gaudio 2008 | 3 | 16 | 31 | 147 | 1.2% | 0.86 [0.23, 3.22] |
| Guo 2019 | 2 | 14 | 5 | 35 | 0.7% | 1.00 [0.17, 5.88] |
| Hu 2012 | 268 | 888 | 1678 | 6087 | 89.4% | 1.14 [0.97, 1.32] |
| Hwang 2007 | 1 | 17 | 9 | 200 | 0.5% | 1.33 [0.16, 11.14] |
| Liu 2012 | 8 | 39 | 32 | 180 | 2.8% | 1.19 [0.50, 2.84] |
| Vennarecci 2007 | 1 | 9 | 3 | 37 | 0.4% | 1.42 [0.13, 15.47] |
| **Total (95% CI)** | | **1059** | | **7149** | **100.0%** | **1.14 [0.98, 1.32]** |
| Total events | 299 | | 1826 | | | |

Heterogeneity: Tau² = 0.00; Chi² = 1.45, df = 9 (P = 1.00); I² = 0%
Test for overall effect: Z = 1.75 (P = 0.08)

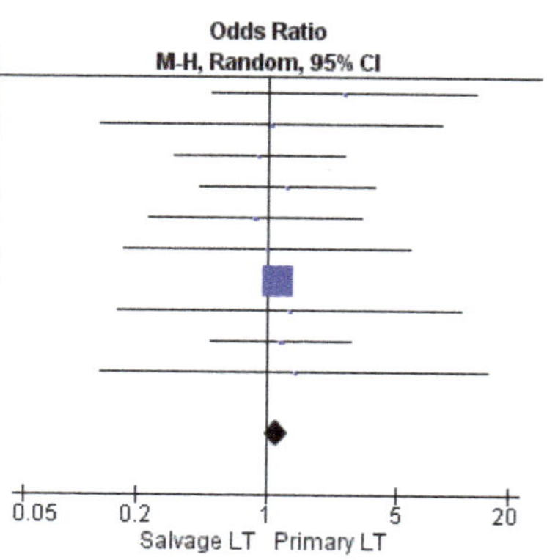


**Figure 15.** Infection and sepsis.

## *3.4. Oncological and Survival Outcomes*

### 3.4.1. Overall Survival Rates

Thirteen, fifteen, and twenty studies reported the 1-year, 3-year, and 5-year overall survival (OS) rate, respectively. Our meta-analysis revealed a similar 1-year OS rate of 77.9% (1072/1375) in the SLT group and 78.5% (6801/8666) in the PLT group, although this evidence was not statistically significant (OR 0.80, (95% CI 0.62, 1.03) $p = 0.08$), as shown in Figure 16. On the other hand, the meta-analysis showed a statistically significant difference in the 3-year and 5-year OS rates between the two groups with a slightly lower OS rate in the SLT group: SLT 59.3% (837/1410) and PLT 61.9% (5508/8905) (OR 0.72, (95% CI 0.60, 0.86) $p = 0.0002$), as shown in Figure 17; and SLT 53.9% (810/1503) and PLT 56.5% (5327/9424) (OR 0.68, (95% CI 0.56, 0.82) $p < 0.0001$), as shown in Figure 18, respectively.

| Study or Subgroup | SLT Events | SLT Total | PLT Events | PLT Total | Weight | Odds Ratio M-H, Random, 95% CI |
|---|---|---|---|---|---|---|
| Adam 2003 | 12 | 17 | 156 | 195 | 4.8% | 0.60 [0.20, 1.80] |
| Bhangui 2015 | 25 | 31 | 279 | 340 | 6.5% | 0.91 [0.36, 2.32] |
| Chan 2019 | 50 | 58 | 207 | 245 | 8.0% | 1.15 [0.50, 2.61] |
| Guerrini 2014 | 24 | 28 | 161 | 198 | 4.7% | 1.38 [0.45, 4.21] |
| Hu 2012 | 648 | 888 | 4534 | 6087 | 41.8% | 0.92 [0.79, 1.08] |
| Hwang 2007 | 15 | 17 | 175 | 200 | 2.6% | 1.07 [0.23, 4.97] |
| Hwang 2020 | 121 | 125 | 478 | 500 | 5.0% | 1.39 [0.47, 4.12] |
| Liu 2012 | 34 | 39 | 162 | 180 | 5.2% | 0.76 [0.26, 2.18] |
| Sapisochin 2010 | 10 | 17 | 29 | 34 | 3.3% | 0.25 [0.06, 0.95] |
| Shan 2017 | 18 | 28 | 169 | 211 | 7.7% | 0.45 [0.19, 1.04] |
| Vasavada 2015 | 14 | 18 | 84 | 91 | 3.3% | 0.29 [0.08, 1.13] |
| Vennarecci 2007 | 8 | 9 | 29 | 37 | 1.3% | 2.21 [0.24, 20.35] |
| Yong 2018 | 93 | 100 | 338 | 348 | 5.8% | 0.39 [0.15, 1.06] |
| **Total (95% CI)** | | **1375** | | **8666** | **100.0%** | **0.80 [0.62, 1.03]** |
| Total events | 1072 | | 6801 | | | |

Heterogeneity: Tau² = 0.03; Chi² = 14.33, df = 12 (P = 0.28); I² = 16%
Test for overall effect: Z = 1.74 (P = 0.08)

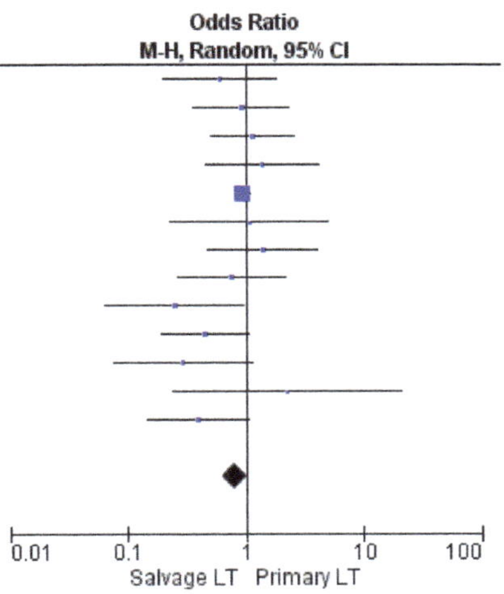


**Figure 16.** 1-year overall survival rates.

| Study or Subgroup | SLT Events | SLT Total | PLT Events | PLT Total | Weight | Odds Ratio M-H, Random, 95% CI |
|---|---|---|---|---|---|---|
| Adam 2003 | 9 | 17 | 133 | 195 | 2.9% | 0.52 [0.19, 1.42] |
| Belghiti 2003 | 15 | 18 | 57 | 70 | 1.6% | 1.14 [0.29, 4.52] |
| Bhangui 2015 | 13 | 31 | 182 | 340 | 5.1% | 0.63 [0.30, 1.32] |
| Chan 2019 | 39 | 58 | 192 | 245 | 7.0% | 0.57 [0.30, 1.06] |
| Guerrini 2014 | 19 | 28 | 138 | 198 | 4.0% | 0.92 [0.39, 2.15] |
| Hu 2012 | 460 | 888 | 3354 | 6087 | 44.7% | 0.88 [0.76, 1.01] |
| Hwang 2007 | 11 | 17 | 156 | 200 | 2.7% | 0.52 [0.18, 1.48] |
| Hwang 2020 | 98 | 125 | 425 | 500 | 10.6% | 0.64 [0.39, 1.05] |
| Liu 2012 | 30 | 39 | 146 | 180 | 4.1% | 0.78 [0.34, 1.79] |
| Moon 2012 | 10 | 17 | 140 | 169 | 2.7% | 0.30 [0.10, 0.84] |
| Sapisochin 2010 | 9 | 17 | 26 | 34 | 1.9% | 0.35 [0.10, 1.19] |
| Shan 2017 | 15 | 28 | 135 | 211 | 4.5% | 0.65 [0.29, 1.44] |
| Vasavada 2015 | 14 | 18 | 77 | 91 | 1.9% | 0.64 [0.18, 2.22] |
| Vennarecci 2007 | 8 | 9 | 23 | 37 | 0.6% | 4.87 [0.55, 43.18] |
| Yong 2018 | 87 | 100 | 324 | 348 | 5.5% | 0.50 [0.24, 1.01] |
| **Total (95% CI)** | | **1410** | | **8905** | **100.0%** | **0.72 [0.60, 0.86]** |
| Total events | 837 | | 5508 | | | |

Heterogeneity: Tau² = 0.01; Chi² = 15.63, df = 14 (P = 0.34); I² = 10%
Test for overall effect: Z = 3.67 (P = 0.0002)

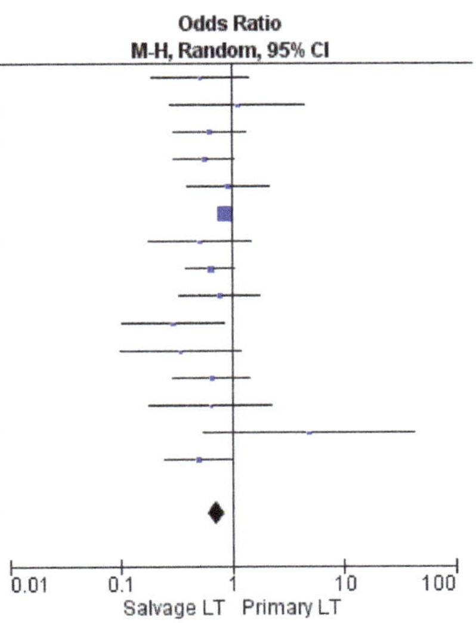


**Figure 17.** 3-year overall survival rates.

| Study or Subgroup | SLT Events | SLT Total | PLT Events | PLT Total | Weight | Odds Ratio M-H, Random, 95% CI |
|---|---|---|---|---|---|---|
| Adam 2003 | 7 | 17 | 119 | 195 | 3.2% | 0.45 [0.16, 1.22] |
| Belghiti 2003 | 11 | 18 | 41 | 70 | 2.9% | 1.11 [0.38, 3.21] |
| Bhangui 2015 | 10 | 31 | 135 | 340 | 4.9% | 0.72 [0.33, 1.58] |
| Chan 2019 | 36 | 58 | 176 | 245 | 7.4% | 0.64 [0.35, 1.17] |
| Cherqui 2009 | 13 | 18 | 101 | 136 | 2.7% | 0.90 [0.30, 2.71] |
| De Carlis 2013 | 21 | 26 | 115 | 153 | 3.0% | 1.39 [0.49, 3.93] |
| Del Gaudio 2008 | 10 | 16 | 107 | 147 | 2.8% | 0.62 [0.21, 1.83] |
| Guerrini 2014 | 14 | 28 | 129 | 198 | 4.8% | 0.53 [0.24, 1.19] |
| Guo 2019 | 9 | 14 | 26 | 35 | 1.9% | 0.62 [0.16, 2.36] |
| Hu 2012 | 407 | 888 | 2971 | 6087 | 24.2% | 0.89 [0.77, 1.02] |
| Hwang 2007 | 9 | 17 | 144 | 200 | 3.2% | 0.44 [0.16, 1.19] |
| Hwang 2020 | 94 | 125 | 396 | 500 | 10.7% | 0.80 [0.50, 1.26] |
| Kaido 2012 | 15 | 19 | 36 | 48 | 2.1% | 1.25 [0.35, 4.50] |
| Liu 2012 | 24 | 39 | 130 | 180 | 5.6% | 0.62 [0.30, 1.27] |
| Moon 2012 | 10 | 17 | 135 | 169 | 3.0% | 0.36 [0.13, 1.01] |
| Sapisochin 2010 | 9 | 17 | 22 | 34 | 2.4% | 0.61 [0.19, 2.00] |
| Shan 2017 | 12 | 28 | 125 | 211 | 4.8% | 0.52 [0.23, 1.15] |
| Vasavada 2015 | 12 | 18 | 77 | 91 | 2.6% | 0.36 [0.12, 1.13] |
| Vennarecci 2007 | 8 | 9 | 23 | 37 | 0.7% | 4.87 [0.55, 43.18] |
| Yong 2018 | 79 | 100 | 319 | 348 | 7.2% | 0.34 [0.19, 0.63] |
| **Total (95% CI)** | | **1503** | | **9424** | **100.0%** | **0.68 [0.56, 0.82]** |
| Total events | 810 | | 5327 | | | |

Heterogeneity: Tau² = 0.03; Chi² = 24.59, df = 19 (P = 0.17); I² = 23%
Test for overall effect: Z = 3.95 (P < 0.0001)

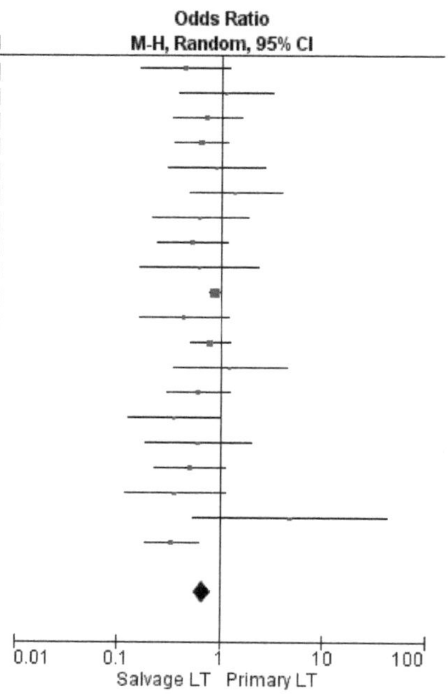


**Figure 18.** 5-year overall survival rates.

### 3.4.2. HCC Recurrence Rate

Types of HCC recurrence after LT were locoregional and/or systemic. Ten studies assessed tumor recurrence rate. Disease recurrence rate was 15.4% (37/240) in the SLT group and 10.9% (98/896) in the Primary LT group. The meta-analysis showed a statistically significant difference in the rate of disease recurrence between the two groups with a lower rate in the PLT group (OR 1.93, (95% CI 1.23, 3.04) $p = 0.004$), as shown in Figure 19.

| Study or Subgroup | SLT Events | SLT Total | PLT Events | PLT Total | Weight | Odds Ratio M-H, Random, 95% CI |
|---|---|---|---|---|---|---|
| Adam 2003 | 7 | 13 | 34 | 191 | 15.5% | 5.39 [1.70, 17.04] |
| Belghiti 2003 | 1 | 18 | 3 | 70 | 3.8% | 1.31 [0.13, 13.43] |
| Del Gaudio 2008 | 3 | 16 | 13 | 147 | 10.8% | 2.38 [0.60, 9.44] |
| Guerrini 2014 | 4 | 28 | 17 | 198 | 15.0% | 1.77 [0.55, 5.71] |
| Guo 2019 | 2 | 14 | 4 | 35 | 6.2% | 1.29 [0.21, 8.00] |
| Kaido 2012 | 4 | 19 | 4 | 48 | 9.1% | 2.93 [0.65, 13.21] |
| Margarit 2005 | 1 | 6 | 5 | 36 | 3.7% | 1.24 [0.12, 12.95] |
| Sapisochin 2010 | 5 | 17 | 4 | 34 | 9.4% | 3.13 [0.71, 13.67] |
| Vennarecci 2007 | 0 | 9 | 4 | 37 | 2.3% | 0.39 [0.02, 7.94] |
| Yong 2018 | 10 | 100 | 10 | 100 | 24.1% | 1.00 [0.40, 2.52] |
| **Total (95% CI)** | | **240** | | **896** | **100.0%** | **1.93 [1.23, 3.04]** |
| Total events | 37 | | 98 | | | |

Heterogeneity: Tau² = 0.00; Chi² = 7.37, df = 9 (P = 0.60); I² = 0%
Test for overall effect: Z = 2.85 (P = 0.004)

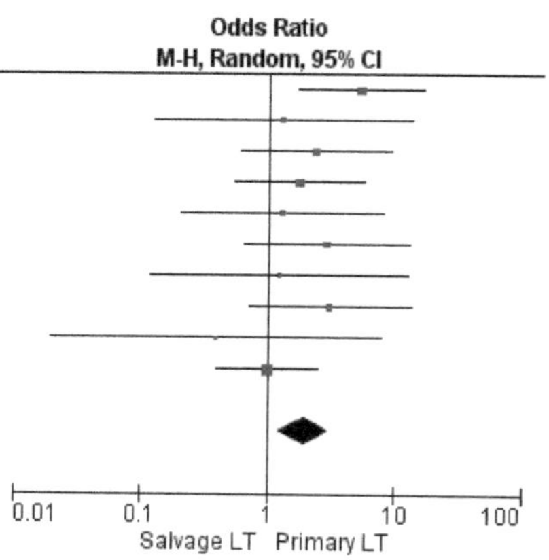


**Figure 19.** HCC recurrence rate.

#### 3.4.3. Disease-Free Survival Rates

Twelve, fourteen, and eighteen papers retrospectively assessed the 1-year, 3-year, and 5-year disease-free survival (DFS) rates, respectively. The meta-analysis showed statistically significant differences in the DFS rate of HCC between the two groups with the same 1-year DFS rate in the SLT group and PLT group, with 71.2% (967/1358) and 71.2% (5855/8218), respectively (OR 0.66, (95% CI 0.47, 0.92) $p = 0.01$), as shown in Figure 20. However, the 3-year and 5-year DFS rates were lower in the SLT group than the PLT group: SLT 54.8% (763/1393) and PLT 57% (4821/8457) (OR 0.59, (95% CI 0.44, 0.88) $p = 0.007$), as shown in Figure 21; and SLT 49.1% (721/1468) and PLT 51.3% (4538/8840), (OR 0.65, (95% CI 0.52, 0.82) $p = 0.0002$), as shown in Figure 22, respectively.

| Study or Subgroup | SLT Events | SLT Total | PLT Events | PLT Total | Weight | Odds Ratio M-H, Random, 95% CI |
|---|---|---|---|---|---|---|
| Adam 2003 | 8 | 17 | 148 | 195 | 7.7% | 0.28 [0.10, 0.77] |
| Bhangui 2015 | 24 | 31 | 274 | 340 | 9.2% | 0.83 [0.34, 2.00] |
| Chan 2019 | 50 | 58 | 235 | 245 | 8.0% | 0.27 [0.10, 0.71] |
| Guerrini 2014 | 27 | 28 | 186 | 198 | 2.4% | 1.74 [0.22, 13.94] |
| Hu 2012 | 575 | 888 | 4041 | 6087 | 24.2% | 0.93 [0.80, 1.08] |
| Hwang 2020 | 108 | 125 | 428 | 500 | 14.7% | 1.07 [0.60, 1.89] |
| Liu 2012 | 34 | 39 | 160 | 180 | 7.3% | 0.85 [0.30, 2.42] |
| Sapisochin 2010 | 15 | 17 | 33 | 34 | 1.7% | 0.23 [0.02, 2.71] |
| Shan 2017 | 14 | 28 | 141 | 211 | 10.5% | 0.50 [0.22, 1.10] |
| Vasavada 2015 | 13 | 18 | 83 | 91 | 5.5% | 0.25 [0.07, 0.88] |
| Vennarecci 2007 | 9 | 9 | 33 | 37 | 1.2% | 2.55 [0.13, 51.74] |
| Yong 2018 | 90 | 100 | 93 | 100 | 7.7% | 0.68 [0.25, 1.86] |
| **Total (95% CI)** | | **1358** | | **8218** | **100.0%** | **0.66 [0.47, 0.92]** |
| Total events | 967 | | 5855 | | | |

Heterogeneity: Tau² = 0.12; Chi² = 19.67, df = 11 (P = 0.05); I² = 44%
Test for overall effect: Z = 2.47 (P = 0.01)

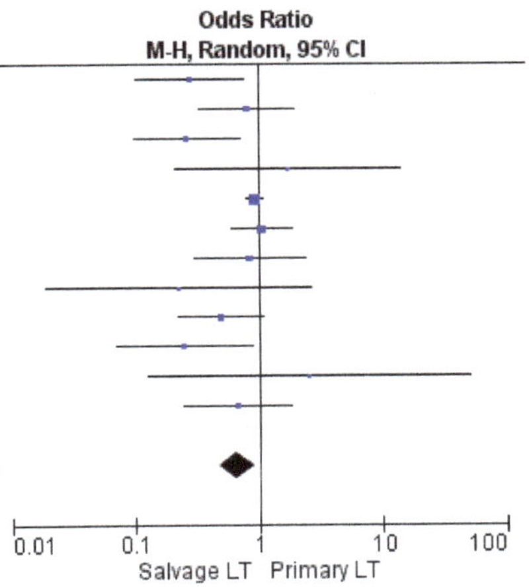


**Figure 20.** 1-year disease free survival rates.

| Study or Subgroup | SLT Events | SLT Total | PLT Events | PLT Total | Weight | Odds Ratio M-H, Random, 95% CI |
|---|---|---|---|---|---|---|
| Adam 2003 | 5 | 17 | 125 | 195 | 5.5% | 0.23 [0.08, 0.69] |
| Belghiti 2003 | 14 | 18 | 54 | 70 | 4.5% | 1.04 [0.30, 3.60] |
| Bhangui 2015 | 12 | 31 | 176 | 340 | 8.5% | 0.59 [0.28, 1.25] |
| Chan 2019 | 42 | 58 | 223 | 245 | 8.9% | 0.26 [0.13, 0.53] |
| Guerrini 2014 | 23 | 28 | 178 | 198 | 5.6% | 0.52 [0.18, 1.51] |
| Hu 2012 | 405 | 888 | 3067 | 6087 | 17.2% | 0.83 [0.72, 0.95] |
| Hwang 2020 | 93 | 125 | 383 | 500 | 12.8% | 0.89 [0.57, 1.40] |
| Liu 2012 | 30 | 39 | 135 | 180 | 7.8% | 1.11 [0.49, 2.52] |
| Moon 2012 | 11 | 17 | 143 | 169 | 5.5% | 0.33 [0.11, 0.98] |
| Sapisochin 2010 | 12 | 17 | 32 | 34 | 2.5% | 0.15 [0.03, 0.88] |
| Shan 2017 | 9 | 28 | 122 | 211 | 7.6% | 0.35 [0.15, 0.80] |
| Vasavada 2015 | 13 | 18 | 66 | 91 | 5.2% | 0.98 [0.32, 3.05] |
| Vennarecci 2007 | 9 | 9 | 27 | 37 | 1.0% | 7.25 [0.39, 136.04] |
| Yong 2018 | 85 | 100 | 90 | 100 | 7.4% | 0.63 [0.27, 1.48] |
| **Total (95% CI)** | | **1393** | | **8457** | **100.0%** | **0.59 [0.44, 0.80]** |
| Total events | 763 | | 4821 | | | |

Heterogeneity: Tau² = 0.14; Chi² = 28.39, df = 13 (P = 0.008); I² = 54%
Test for overall effect: Z = 3.37 (P = 0.0007)

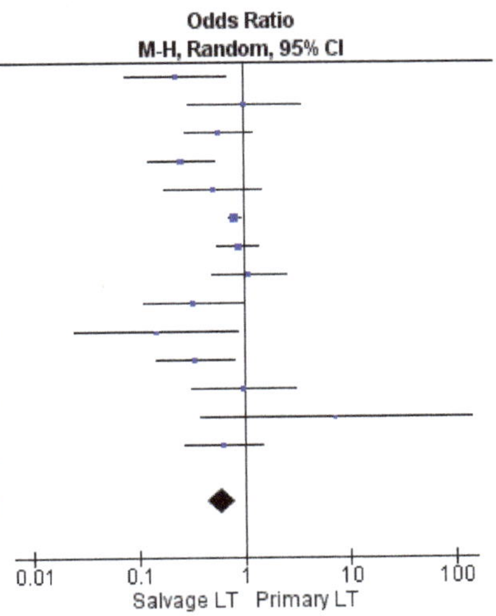


**Figure 21.** 3-year disease free survival rates.

| Study or Subgroup | SLT Events | SLT Total | PLT Events | PLT Total | Weight | Odds Ratio M-H, Random, 95% CI |
|---|---|---|---|---|---|---|
| Adam 2003 | 5 | 17 | 113 | 195 | 3.7% | 0.30 [0.10, 0.89] |
| Belghiti 2003 | 10 | 18 | 39 | 70 | 3.9% | 0.99 [0.35, 2.82] |
| Bhangui 2015 | 9 | 31 | 129 | 340 | 5.9% | 0.67 [0.30, 1.50] |
| Chan 2019 | 42 | 58 | 217 | 245 | 7.2% | 0.34 [0.17, 0.68] |
| De Carlis 2013 | 21 | 26 | 120 | 153 | 3.9% | 1.16 [0.40, 3.30] |
| Del Gaudio 2008 | 8 | 16 | 104 | 147 | 3.9% | 0.41 [0.15, 1.17] |
| Guerrini 2014 | 23 | 28 | 178 | 198 | 3.7% | 0.52 [0.18, 1.51] |
| Guo 2019 | 8 | 14 | 25 | 35 | 2.7% | 0.53 [0.15, 1.93] |
| Hu 2012 | 336 | 888 | 2648 | 6087 | 21.1% | 0.79 [0.68, 0.91] |
| Hwang 2020 | 91 | 125 | 360 | 500 | 12.3% | 1.04 [0.67, 1.62] |
| Kaido 2012 | 15 | 19 | 37 | 48 | 2.7% | 1.11 [0.31, 4.06] |
| Liu 2012 | 26 | 39 | 124 | 180 | 6.7% | 0.90 [0.43, 1.89] |
| Moon 2012 | 11 | 17 | 140 | 169 | 3.7% | 0.38 [0.13, 1.11] |
| Sapisochin 2010 | 10 | 17 | 30 | 34 | 2.3% | 0.19 [0.05, 0.79] |
| Shan 2017 | 9 | 28 | 116 | 211 | 5.5% | 0.39 [0.17, 0.90] |
| Vasavada 2015 | 8 | 18 | 42 | 91 | 4.1% | 0.93 [0.34, 2.58] |
| Vennarecci 2007 | 9 | 9 | 27 | 37 | 0.6% | 7.25 [0.39, 136.04] |
| Yong 2018 | 80 | 100 | 89 | 100 | 6.0% | 0.49 [0.22, 1.10] |
| **Total (95% CI)** | | **1468** | | **8840** | **100.0%** | **0.65 [0.52, 0.82]** |
| Total events | 721 | | 4538 | | | |

Heterogeneity: Tau² = 0.06; Chi² = 24.97, df = 17 (P = 0.10); I² = 32%
Test for overall effect: Z = 3.69 (P = 0.0002)

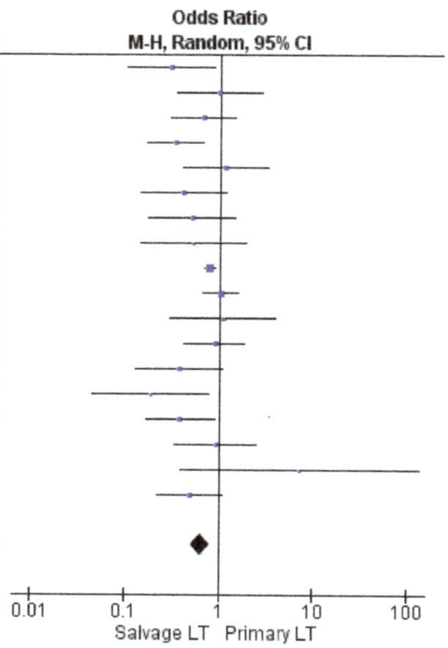


**Figure 22.** 5-year disease free survival rates.

## 4. Discussion

Liver transplantation (LT) represents the ideal treatment option for patients with HCC since it achieves radical tumor clearance and eradicates the underlying liver diseases. However, several patients on the waiting list for LT are faced with tumor progression, the loss of chance for transplantation, or even death due to severe organ shortage and long waiting list times [54]. Thus, in order to overcome the gap between the numbers of donors and recipients, salvage liver transplantation has been proposed in the last decade as an attractive and feasible strategy that combines liver resection and subsequent LT in the case of HCC recurrence [55–59].

This meta-analysis includes the highest number of articles comparing the findings of primary and salvage liver transplantation for HCC and also demonstrates completely new results compared to other studies on the same topic, bringing different and innovative concepts to the strategy of salvage transplantation for recurrence of HCC after liver resection [19].

Operating time and intraoperative blood loss are some of the surgical variables in terms of safety and feasibility most taken into consideration when Salvage LT is compared with Primary LT. Several studies reported a longer mean operating time for SLT than PLT. Although considerable differences exist in terms of duration of surgery and blood loss among the included articles, the duration of the operation and the extent of bleeding are necessarily affected by some technical and anatomical issues. SLT increases the difficulty of surgery due to severe adhesion in the abdominal operation area and due to abnormal anatomical structures as a consequence of previous hepatic resection [22,60]. Our meta-analysis revealed a significantly longer duration of surgery for SLT than Primary LT and also disclosed that intraoperative bleeding was slightly higher in the Salvage LT strategy, but this finding was not statistically significant. Moreover, differences between the two surgical approaches in terms of the mean need for intraoperative RBC and FFP transfusion were not statistically significant. Several papers showed that innovations in surgical techniques and accumulation of surgical experience have gradually decreased the risk of perioperative bleeding for SLT. It has been shown that reducing intraoperative bleeding and blood transfusions rate leads to a better postoperative recovery.

Our study showed that the reoperation rate was significantly higher in the SLT group than in the PLT group, and the perioperative mortality was slightly higher for the Salvage

LT approach. Multiple preoperative bridging and downstaging treatments and the liver resection before salvage LT led to the formation of dense adhesions, portal collateral circulations due to hypertension, and coagulopathy [61–63]. Therefore, these factors increase bleeding after SLT, likely accounting for the higher re-exploration rate. Surgery for salvage liver transplantation is technically demanding, and this could explain the slightly higher perioperative mortality in patients undergoing Salvage rather than Primary LT.

In terms of intensive care unit stay and length of hospital stay, our data showed a longer recovery in the SLT group, although these findings were not statistically significant. On the other hand, overall vascular complication rate, overall infection, and sepsis rate were statistically similar between the two groups.

Several studies have found that the outcome of patients with HCC was similar between liver resection and liver transplantation [15,64–66]. Therefore, liver resection and LT are not opposing alternatives, but, rather, represent the components of a combined strategy for the management of HCC: liver resection can potentially improve the survival of patients listed for LT by decreasing the risk of dropout [67]. Moreover, minimally invasive liver resection (MILR) has a minor technical impact on a subsequent liver transplantation and seems to be associated with shorter operation time, reduced blood loss, and transfusion requirement during Salvage LT [68,69]. Therefore, MILR (laparoscopic or robotic) may become the gold standard for "early" HCC in patient cirrhosis and mild portal hypertension. In 2008, Felli et al. [70] introduced the concept of liver resection as a selection tool for LT. In fact, some pathological characteristics of the resected specimen can identify a subgroup of patients with favorable histological factors (small and well-differentiated HCC, without satellite nodules or microvascular invasion) who could avoid upfront LT because the risk of recurrence appears to be relatively low and if it should occur, then transplantation remains a salvage option at a later date [42,71–73]. On the other hand, patients showing negative prognostic histological features on the resected specimen (e.g., microvascular infiltration, high grade of differentiation) could undergo liver transplantation prior to tumor recurrence: so-called "de principe" SLT [17,74]. Indeed, the French allocation system recently integrated the SLT strategy within its algorithm, although no priority is given to patients at a high risk of HCC recurrence. These results and future research would clarify the role of the molecular and biological pattern of HCC in order to stratify patients with a high risk of recurrence and then arrive at defining the best personalized treatment [75].

A clear definition of "transplantability criteria in SLT", that is, criteria that identify the group of patients who benefit most from transplantation for HCC recurrence after liver resection, has not yet been established [46,76]. Most authors agree that the criteria of patients with a limited recurrence within the Milan criteria is acceptable in order to achieve a good survival post-SLT [77]. Recently, Liu et al. observed the efficacy of SLT for patients with recurrent HCC after liver resection within the University of California San Francisco (UCSF) criteria, since in that study there was no significant difference in OS and DFS rates between the SLT and PLT groups [41].

Recurrence of HCC after transplantation is still a devastating event as no surgical or pharmacological therapy has shown significant prolongation of these patients' survival [78–81]. Some authors have observed that the strategy of the salvage liver transplantation may increase the risk of recurrence of post-transplant patients, thus limiting their survival [51].

In our meta-analysis, the HCC recurrence rate was 15.4% in the SLT group and 10.9% in the PLT group. However, between the different studies taken into account by our meta-analysis, contrasting results can be observed with regard to tumor recurrence. Adam et al. [29]. reported that SLT had an increased risk of recurrence and poorer survival compared with primary transplantation. By contrast, in the same period, Belghiti et al. [30] showed that recurrence rate, operative mortality, and long-term survival were comparable between the two groups.

Important end points of this meta-analysis were overall survival (OS) rate and disease free-survival rate between SLT and PLT. Our meta-analysis showed statistically significant lower 1-, 3-, and 5-year DFS rates for SLT compared to PLT. However, DFS as a long-term outcome indicator could be misleading because it is a composite end point influenced by two events: death and tumor recurrence. However, to better determine long-term outcomes, future studies should match patients based on histological features (tumor size and nodule number) at explant pathology which clearly influence tumor recurrence and mortality [72].

The 1-year OS rate presented no significant difference between SLT and PLT, whereas 3- and 5-year overall survival rates were significantly slightly lower in SLT than after PLT. However, previous studies disclosed that the 5-year survival rates did not differ significantly for patients with SLT and for those with PLT (69% vs 73%; $p = 0.34$) [39].

Our results on survival post-SLT appear to be in contrast to the recent meta-analyses published on the subject [19–21]. However, this is not surprising because most of the studies included in the meta-analysis show a lower survival in the SLT group than in the PLT groups [39]. On the other hand, while survival differences often do not reach a statistically significant difference within an individual study, this difference in survival becomes statistically significant in the meta-analysis, which represents a statistical tool of great relevance and precision (since it "weights" the result in individual studies according to its precision) [82].

Despite the relatively high quality of the included articles, there are several limitations concerning this meta-analysis. The included studies were retrospective and not randomized, so the variables analyzed exhibited heterogeneity. However, the heterogeneity within the studies was treated and resolved by applying the random effect model on all the variables in the study [83]. Moreover, some studies included heterogeneous patient populations with transplantation for HCC recurrence and those who underwent SLT due to liver failure, although this latter indication represents less than 5% of the SLT. Therefore, because of the inherent risk of bias in the considered articles, it is desirable that further well-designed studies are conducted.

Nevertheless, our systematic review summarizes most of the available evidence in comparing outcomes of SLT and PLT. To our knowledge, it is the largest and most recent meta-analysis that makes these comparisons. It introduces completely new results that can form the scientific basis on which to develop further studies on the topic of liver transplantation as an integrated therapy in the treatment of HCC.

## 5. Conclusions

Our meta-analysis advocates the relative safety and feasibility of SLT over the PLT approach for patients with HCC. Specifically, the results of our study confirm that SLT offers comparable technical outcomes but slightly lower survival outcomes with respect to PLT.

**Supplementary Materials:** The following supporting information can be downloaded at: https://www.mdpi.com/article/10.3390/cancers14143465/s1, Files S1 and S2: Search strategy, and File S3: Publication bias.

**Author Contributions:** Study conception and design: G.P.G., G.E. and P.M.; acquisition of data: G.E., R.B. and T.O.; analysis and interpretation of data: G.P.G., G.E. and P.M.; drafting of manuscript: G.P.G., S.D.S. and R.B.; critical revision: G.P.G., S.D.S. and F.D.B. All authors have read and agreed to the published version of the manuscript.

**Funding:** This research received no external funding. The authors of this study declare no financial support.

**Institutional Review Board Statement:** The study did not require institutional review board as is a review of published studies.

**Informed Consent Statement:** Not applicable.

**Data Availability Statement:** Not applicable.

**Conflicts of Interest:** The authors declare no conflict of interest.

## References


1. Sung, H.; Ferlay, J.; Siegel, R.L.; Laversanne, M.; Soerjomataram, I.; Jemal, A.; Bray, F. Global Cancer Statistics 2020: GLOBOCAN Estimates of Incidence and Mortality Worldwide for 36 Cancers in 185 Countries. *CA Cancer J. Clin.* **2021**, *71*, 209–249. [CrossRef] [PubMed]
2. Guerrini, G.P.; Pleguezuelo, M.; Maimone, S.; Calvaruso, V.; Xirouchakis, E.; Patch, D.; Rolando, N.; Davidson, B.; Rolles, K.; Burroughs, A. Impact of tips preliver transplantation for the outcome posttransplantation. *Am. J. Transplant.* **2009**, *9*, 192–200. [CrossRef] [PubMed]
3. DuBray, B.J., Jr.; Chapman, W.C.; Anderson, C.D. Hepatocellular carcinoma: A review of the surgical approaches to management. *Mo. Med.* **2011**, *108*, 195–198.
4. Abrams, P.; Marsh, J.W. Current approach to hepatocellular carcinoma. *Surg. Clin. North Am.* **2010**, *90*, 803–816. [CrossRef]
5. Di Benedetto, F.; Tarantino, G.; De Ruvo, N.; Cautero, N.; Montalti, R.; Guerrini, G.P.; Ballarin, R.; Spaggiari, M.; Smerieri, N.; Serra, V.; et al. University of Modena Experience in HIV-Positive Patients Undergoing Liver Transplantation. *Transplant. Proc.* **2011**, *43*, 1114–1118. [CrossRef]
6. Lewin, S.M.; Mehta, N.; Kelley, R.K.; Roberts, J.P.; Yao, F.Y.; Brandman, D. Liver Transplant (LT) recipients with Nonalcoholic Steatohepatitis (NASH) Have Lower Risk Hepatocellular Carcinoma (HCC). *Liver Transpl.* **2017**, *23*, 1015–1022. [CrossRef] [PubMed]
7. Mazzaferro, V. Squaring the circle of selection and allocation in liver transplantation for HCC: An adaptive approach. *Hepatology* **2016**, *63*, 1707–1717. [CrossRef]
8. Mazzaferro, V.; Lencioni, R.; Majno, P. Early hepatocellular carcinoma on the procrustean bed of ablation, resection, and transplantation. *Semin. Liver Dis.* **2014**, *34*, 415–426.
9. Alver, S.K.; Lorenz, D.J.; Washburn, K.; Marvin, M.R.; Brock, G.N. Comparison of two equivalent MELD scores for hepatocellular carcinoma patients using data from the United Network for Organ Sharing liver transplant waiting list registry. *Transpl. Int. Off. J. Eur. Soc. Organ Transplant.* **2017**, *30*, 1098.
10. Akamatsu, N.; Cillo, U.; Cucchetti, A.; Donadon, M.; Pinna, A.D.; Torzilli, G.; Kokudo, N. Surgery and Hepatocellular Carcinoma. *Liver Cancer* **2016**, *6*, 44–50. [CrossRef]
11. Bruix, J.; Fuster, J. A Snapshot of the Effective Indications and Results of Surgery for Hepatocellular Carcinoma in Tertiary Referral Centers: Is It Adherent to the EASL/AASLD Recommendations? An Observational Study of the HCC East-West Study Group. *Ann. Surg.* **2015**, *262*, e30. [CrossRef]
12. Laurent, A.; Tayar, C.; Andréoletti, M.; Lauzet, J.Y.; Merle, J.C.; Cherqui, D. Laparoscopic liver resection facilitates salvage liver transplantation for hepatocellular carcinoma. *J. Hepatobiliary Pancreat. Surg.* **2009**, *16*, 310–314. [CrossRef] [PubMed]
13. Aube, C.; Oberti, F.; Lonjon, J.; Pageaux, G.; Seror, O.; N'Kontchou, G.; Rode, A.; Radenne, S.; Cassinotto, C.; Vergniol, J.; et al. EASL and AASLD recommendations for the diagnosis of HCC to the test of daily practice. *Liver Int.* **2017**, *37*, 1515–1525. [CrossRef] [PubMed]
14. Scatton, O.; Goumard, C.; Cauchy, F.; Fartoux, L.; Perdigao, F.; Conti, F.; Calmus, Y.; Boelle, P.Y.; Belghiti, J.; Rosmorduc, O.; et al. Early and resectable HCC: Definition and validation of a subgroup of patients who could avoid liver transplantation. *J. Surg. Oncol.* **2015**, *111*, 1007–1015. [CrossRef] [PubMed]
15. Facciuto, M.E.; Rochon, C.; Pandey, M.; Rodriguez-Davalos, M.; Samaniego, S.; Wolf, D.C.; Kim-Schluger, L.; Rozenblit, G.; Sheiner, P.A. Surgical dilemma: Liver resection or liver transplantation for hepatocellular carcinoma and cirrhosis. Intention-to-treat analysis in patients within and outwith Milan criteria. *HPB* **2009**, *11*, 398–404. [CrossRef] [PubMed]
16. Majno, P.E.; Sarasin, F.P.; Mentha, G.; Hadengue, A. Primary liver resection and salvage transplantation or primary liver transplantation in patients with single, small hepatocellular carcinoma and preserved liver function: An outcome-oriented decision analysis. *Hepatology* **2000**, *31*, 899–906. [CrossRef]
17. Goumard, C.; Scatton, O. Resectable HCC: Should salvage liver transplantation for HCC be discussed de principe? *Clin. Res. Hepatol. Gastroenterol.* **2020**, *44*, 117–118. [CrossRef]
18. Muaddi, H.; Al-Adra, D.P.; Beecroft, R.; Ghanekar, A.; Moulton, C.A.; Doyle, A.; Selzner, M.; Wei, A.; McGilvary, I.D.; Gallinger, S.; et al. Liver Transplantation is Equally Effective as a Salvage Therapy for Patients with Hepatocellular Carcinoma Recurrence Following Radiofrequency Ablation or Liver Resection with Curative Intent. *Ann. Surg. Oncol.* **2018**, *25*, 991–999. [CrossRef]
19. Zhu, Y.; Dong, J.; Wang, W.L.; Li, M.X.; Lu, Y. Short- and long-term outcomes after salvage liver transplantation versus primary liver transplantation for hepatocellular carcinoma: A meta-analysis. *Transplant. Proc.* **2013**, *45*, 3329–3342. [CrossRef] [PubMed]
20. Yadav, D.K.; Chen, W.; Bai, X.; Singh, A.; Li, G.; Ma, T.; Yu, X.; Xiao, Z.; Huang, B.; Liang, T. Salvage Liver Transplant versus Primary Liver Transplant for Patients with Hepatocellular Carcinoma. *Ann. Transplant.* **2018**, *23*, 524–545. [CrossRef]
21. Xiong, Q.; Geng, T.T.; He, L.; Gao, H. Harm and Benefits of Salvage Transplantation for Hepatocellular Carcinoma: An Updated Meta-analysis. *Transplant. Proc.* **2016**, *48*, 3336–3347. [CrossRef] [PubMed]
22. Zheng, S.; Xie, Q.; Cheng, J. Salvage liver transplant for hepatocellular carcinoma: Rescues and benefits. *Transl. Gastroenterol. Hepatol.* **2018**, *3*, 65. [CrossRef] [PubMed]
23. Page, M.J.; McKenzie, J.E.; Bossuyt, P.M.; Boutron, I.; Hoffmann, T.C.; Mulrow, C.D.; Shamseer, L.; Tetzlaff, J.M.; Akl, E.A.; Brennan, S.E.; et al. The PRISMA 2020 statement: An updated guideline for reporting systematic reviews. *BMJ* **2021**, *88*, 105906. [CrossRef]

24. Goossen, K.; Tenckhoff, S.; Probst, P.; Grummich, K.; Mihaljevic, A.L.; Büchler, M.W.; Diener, M.K. Optimal literature search for systematic reviews in surgery. *Langenbecks Arch. Surg.* **2018**, *403*, 119–129. [CrossRef] [PubMed]
25. Slim, K.; Nini, E.; Forestier, D.; Kwiatkowski, F.; Panis, Y.; Chipponi, J. Methodological index for non-randomized studies (minors): Development and validation of a new instrument. *ANZ J. Surg.* **2003**, *73*, 712–716. [CrossRef]
26. Hozo, S.P.; Djulbegovic, B.; Hozo, I. Estimating the mean and variance from the median, range, and the size of a sample. *BMC Med. Res. Methodol.* **2005**, *5*, 13. [CrossRef]
27. Luo, D.; Wan, X.; Liu, J.; Tong, T. Optimally estimating the sample mean from the sample size, median, mid-range, and/or mid-quartile range. *Stat. Methods Med. Res.* **2018**, *27*, 1785–1805. [CrossRef]
28. Wan, X.; Wang, W.; Liu, J.; Tong, T. Estimating the sample mean and standard deviation from the sample size, median, range and/or interquartile range. *BMC Med. Res. Methodol.* **2014**, *14*, 135. [CrossRef]
29. Adam, R.; Azoulay, D.; Castaing, D.; Eshkenazy, R.; Pascal, G.; Hashizume, K.; Samuel, D.; Bismuth, H. Liver resection as a bridge to transplantation for hepatocellular carcinoma on cirrhosis: A reasonable strategy? *Ann. Surg.* **2003**, *238*, 508–518; discussion 18–19. [CrossRef]
30. Belghiti, J.; Cortes, A.; Abdalla, E.K.; Régimbeau, J.M.; Prakash, K.; Durand, F.; Sommacale, D.; Dondero, F.; Lesurtel, M.; Sauvanet, A.; et al. Resection prior to liver transplantation for hepatocellular carcinoma. *Ann. Surg.* **2003**, *238*, 885–893; discussion 92–93. [CrossRef]
31. Margarit, C.; Escartín, A.; Castells, L.; Vargas, V.; Allende, E.; Bilbao, I. Resection for hepatocellular carcinoma is a good option in Child-Turcotte-Pugh class A patients with cirrhosis who are eligible for liver transplantation. *Liver Transplant.* **2005**, *11*, 1242–1251. [CrossRef] [PubMed]
32. Hwang, S.; Lee, S.-G.; Moon, D.-B.; Ahn, C.-S.; Kim, K.-H.; Lee, Y.-J.; Ha, T.-Y.; Song, G.-W. Salvage living donor liver transplantation after prior liver resection for hepatocellular carcinoma. *Liver Transplant.* **2007**, *13*, 741–746. [CrossRef] [PubMed]
33. Vennarecci, G.; Ettorre, G.M.; Antonini, M.; Santoro, R.; Maritti, M.; Tacconi, G.; Spoletini, D.; Tessitore, L.; Perracchio, L.; Visco, G.; et al. First-line liver resection and salvage liver transplantation are increasing therapeutic strategies for patients with hepatocellular carcinoma and child a cirrhosis. *Transplant. Proc.* **2007**, *39*, 1857–1860. [CrossRef] [PubMed]
34. Del Gaudio, M.; Ercolani, G.; Ravaioli, M.; Cescon, M.; Lauro, A.; Vivarelli, M.; Zanello, M.; Cuccheti, A.; Vetrone, G.; Tuci, F.; et al. Liver transplantation for recurrent hepatocellular carcinoma on cirrhosis after liver resection: University of Bologna experience. *Am. J. Transplant.* **2008**, *8*, 1177–1185. [CrossRef]
35. Kim, B.W.; Park, Y.K.; Kim, Y.B.; Wang, H.J.; Kim, M.W. Salvage liver transplantation for recurrent hepatocellular carcinoma after liver resection: Feasibility of the Milan criteria and operative risk. *Transplant. Proc.* **2008**, *40*, 3558–3561. [CrossRef]
36. Shao, Z.; Lopez, R.; Shen, B.; Yang, G.S. Orthotopic liver transplantation as a rescue operation for recurrent hepatocellular carcinoma after partial hepatectomy. *World J. Gastroenterol.* **2008**, *14*, 4370–4376. [CrossRef]
37. Cherqui, D.; Laurent, A.; Mocellin, N.; Tayar, C.; Luciani, A.; Van Nhieu, J.T.; Decaens, T.; Hurtova, M.; Memeo, R.; Mallat, A.; et al. Liver resection for transplantable hepatocellular carcinoma: Long-term survival and role of secondary liver transplantation. *Ann. Surg.* **2009**, *250*, 738–746. [CrossRef]
38. Sapisochin, G.; Bilbao, I.; Balsells, J.; Dopazo, C.; Caralt, M.; Lázaro, J.L.; Castells, L.; Allende, H.; Charco, R. Optimization of liver transplantation as a treatment of intrahepatic hepatocellular carcinoma recurrence after partial liver resection: Experience of a single European series. *World J. Surg.* **2010**, *34*, 2146–2154. [CrossRef]
39. Hu, Z.; Zhou, J.; Xu, X.; Li, Z.; Zhou, L.; Wu, J.; Zhang, M.; Zheng, S. Salvage liver transplantation is a reasonable option for selected patients who have recurrent hepatocellular carcinoma after liver resection. *PLoS ONE* **2012**, *7*, e36587. [CrossRef]
40. Kaido, T.; Mori, A.; Ogura, Y.; Hata, K.; Yoshizawa, A.; Iida, T.; Yagi, S.; Uemoto, S. Living donor liver transplantation for recurrent hepatocellular carcinoma after liver resection. *Surgery* **2012**, *151*, 55–60. [CrossRef]
41. Liu, F.; Wei, Y.; Wang, W.; Chen, K.; Yan, L.; Wen, T.; Zhao, J.; Xu, M.; Li, B. Salvage liver transplantation for recurrent hepatocellular carcinoma within UCSF criteria after liver resection. *PLoS ONE* **2012**, *7*, e48932. [CrossRef]
42. Moon, J.I.; Kwon, C.H.; Joh, J.W.; Choi, G.S.; Jung, G.O.; Kim, J.M.; Shin, M.; Choi, S.; Kim, S.; Lee, S.-K. Primary versus salvage living donor liver transplantation for patients with hepatocellular carcinoma: Impact of microvascular invasion on survival. *Transplant. Proc.* **2012**, *44*, 487–493. [CrossRef] [PubMed]
43. De Carlis, L.; Di Sandro, S.; Giacomoni, A.; Mangoni, I.; Lauterio, A.; Mihaylov, P.; Cusumano, C.; Rampoldi, A. Liver transplantation for hepatocellular carcinoma recurrence after liver resection: Why deny this chance of cure? *J. Clin. Gastroenterol.* **2013**, *47*, 352–358. [CrossRef] [PubMed]
44. Guerrini, G.P.; Gerunda, G.E.; Montalti, R.; Ballarin, R.; Cautero, N.; De Ruvo, N.; Spaggiari, M.; Di Benedetto, F. Results of salvage liver transplantation. *Liver Int.* **2014**, *34*, e96–e104. [CrossRef]
45. Abe, T.; Tashiro, H.; Teraoka, Y.; Hattori, M.; Tanimine, N.; Kuroda, S.; Tahara, H.; Ohira, M.; Tanaka, Y.; Kobayashi, T.; et al. Efficacy and Feasibility of Salvage Living Donor Liver Transplantation after Initial Liver Resection in Patients with Hepatocellular Carcinoma. *Dig. Surg.* **2015**, *33*, 8–14. [CrossRef]
46. Bhangui, P.; Allard, M.A.; Vibert, E.; Cherqui, D.; Pelletier, G.; Cunha, A.S.; Guettier, C.; Valle, J.-C.D.; Saliba, F.; Bismuth, H.; et al. Salvage Versus Primary Liver Transplantation for Early Hepatocellular Carcinoma: Do Both Strategies Yield Similar Outcomes? *Ann. Surg.* **2016**, *264*, 155–163. [CrossRef]
47. Bhavin Bhupendra Vasavada C-LC. Salvage transplantation for post-resection recurrence in hepatocellular carcinoma associated with hepatitis C virus. etiology: A feasible strategy? *Hepatoma Res.* **2015**, *1*, 36–40. [CrossRef]

48. Wang, P.; Pu, Y.; Li, H.; Shi, B.; Zheng, S.; Zhong, L. Prognosis for recipients with hepatocellular carcinoma of salvage liver transplantation versus those of primary liver transplantation: A retrospective single-center study. *Springer Plus* **2016**, *5*, 1809. [CrossRef]
49. Shan, Y.; Huang, L.; Xia, Q. Salvage Liver Transplantation Leads to Poorer Outcome in Hepatocellular Carcinoma Compared with Primary Liver Transplantation. *Sci. Rep.* **2017**, *7*, 44652. [CrossRef]
50. Yong, C.C.; Elsarawy, A.M.; Wang, S.H.; Lin, T.S.; Wang, C.C.; Li, W.F.; Lin, T.-L.; Kuo, F.-Y.; Cheng, Y.-F.; Chen, C.-L.; et al. The surgical challenges of salvage living donor liver transplantation for Hepatocellular carcinoma; The cumulative experience of 100 cases-A retrospective cohort study and a propensity score analysis. *Int. J. Surg.* **2018**, *54*, 187–192. [CrossRef]
51. Chan, K.M.; Cheng, C.H.; Wu, T.H.; Lee, C.F.; Wu, T.J.; Chou, H.S.; Lee, W.-C. Salvage living donor liver transplantation for posthepatectomy recurrence: A higher incidence of recurrence but promising strategy for long-term survival. *Cancer Manag. Res.* **2019**, *11*, 7295–7305. [CrossRef] [PubMed]
52. Guo, Y.; Tan, E.K.; Krishnamoorthy, T.L.; Tan, C.K.; Tan, B.H.; Tan, T.T.; Lee, S.-Y.; Chan, C.-Y.; Cheow, P.-C.; Chung, A.Y.F.; et al. Outcomes of salvage liver transplant for recurrent hepatocellular carcinoma: A comparison with primary liver transplant. *Ann. Hepatobiliary Pancreat. Surg.* **2019**, *23*, 1–7. [CrossRef] [PubMed]
53. Hwang, S.; Song, G.W.; Ahn, C.S.; Kim, K.H.; Moon, D.B.; Ha, T.Y.; Jung, D.; Park, G.; Yoon, Y.; Lee, S. Salvage living donor liver transplantation for hepatocellular carcinoma recurrence after hepatectomy: Quantitative prediction using ADV score. *J. Hepatobiliary Pancreat. Sci.* **2021**, *28*, 1000–1013. [CrossRef] [PubMed]
54. Toniutto, P.; Zanetto, A.; Ferrarese, A.; Burra, P. Current challenges and future directions for liver transplantation. *Liver Int.* **2016**, *37*, 317–327. [CrossRef]
55. Lee, S.; Ahn, C.; Ha, T.; Moon, D.; Choi, K.; Song, G.; Chung, D.; Park, G.; Yu, Y.; Choi, N.; et al. Liver transplantation for hepatocellular carcinoma: Korean experience. *J. Hepatobiliary Pancreat. Sci.* **2010**, *17*, 539–547. [CrossRef] [PubMed]
56. Azzam, A.Z. Liver transplantation as a management of hepatocellular carcinoma. *World J. Hepatol.* **2015**, *7*, 1347–1354. [CrossRef]
57. Cucchetti, A.; Cescon, M.; Bertuzzo, V.; Bigonzi, E.; Ercolani, G.; Morelli, M.C.; Ravaioli, M.; Pinna, A.D. Can the dropout risk of candidates with hepatocellular carcinoma predict survival after liver transplantation? *Am. J. Transplant.* **2011**, *11*, 1696–1704. [CrossRef] [PubMed]
58. Di Benedetto, F.; Tarantino, G.; De Ruvo, N.; Cautero, N.; Montalti, R.; Guerrini, G.P.; Ballarin, R.; Spaggiari, M.; Serra, V.; Guaraldi, G.; et al. Liver Transplantation for Hepatocellular Carcinoma in HIV Co-Infected Patients: A Single Centre Experience. *Liver Transplant.* **2011**, *17*, S273–S274.
59. Guerrini, G.P.; Berretta, M.; Guaraldi, G.; Magistri, P.; Esposito, G.; Ballarin, R.; Serra, V.; Di Sandro, S.; Di Benedetto, F. Liver Transplantation for HCC in HIV-Infected Patients: Long-Term Single-Center Experience. *Cancers* **2021**, *13*, 4727. [CrossRef]
60. Xu, D.W.; Wan, P.; Xia, Q. Liver transplantation for hepatocellular carcinoma beyond the Milan criteria: A review. *World J. Gastroenterol.* **2016**, *22*, 3325–3334. [CrossRef]
61. Di Benedetto, F.; Mimmo, A.; De Ruvo, N.; Montalti, R.; Cautero, N.; Guerrini, G.P.; Gerunda, G.E. Liver Transplantation Due to TIPS Complications. *Liver Transplant.* **2010**, *16*, S161.
62. Parikh, N.D.; Waljee, A.K.; Singal, A.G. Downstaging hepatocellular carcinoma: A systematic review and pooled analysis. *Liver Transplant.* **2015**, *21*, 1142–1152. [CrossRef] [PubMed]
63. Jarnagin, W.R. Management of small hepatocellular carcinoma: A review of transplantation, resection, and ablation. *Ann. Surg. Oncol.* **2010**, *17*, 1226–1233. [CrossRef]
64. Morise, Z.; Kawabe, N.; Tomishige, H.; Nagata, H.; Kawase, J.; Arakawa, S.; Yoshida, R.; Isetani, M. Recent advances in liver resection for hepatocellular carcinoma. *Front. Surg.* **2014**, *1*, 21. [CrossRef]
65. Hanish, S.I.; Knechtle, S.J. Liver transplantation for the treatment of hepatocellular carcinoma. *Oncology* **2011**, *25*, 752–757.
66. Llovet, J.M.; Fuster, J.; Bruix, J. Intention-to-treat analysis of surgical treatment for early hepatocellular carcinoma: Resection versus transplantation. *Hepatology* **1999**, *30*, 1434–1440. [CrossRef]
67. Colombo, M.; Sangiovanni, A. Treatment of hepatocellular carcinoma: Beyond international guidelines. *Liver Int.* **2015**, *35* (Suppl. S1), 129–138. [CrossRef]
68. Magistri, P.; Olivieri, T.; Assirati, G.; Guerrini, G.P.; Ballarin, R.; Tarantino, G.; Di Benedetto, F. Robotic Liver Resection Expands the Opportunities of Bridging Before Liver Transplantation. *Liver Transpl.* **2019**, *25*, 1110–1112. [CrossRef] [PubMed]
69. Kim, S.H.; Kim, K.H.; Ha, T.Y.; Jung, D.H.; Park, G.C.; Lee, S.G. Salvage living donor liver transplantation for recurrent hepatocellular carcinoma after prior laparoscopic hepatectomy. *Hepatobiliary Pancreat Dis Int.* **2018**, *17*, 473–476. [CrossRef] [PubMed]
70. Felli, E.; Baumert, T.; Pessaux, P. Is minimally invasive true anatomical HCC resection a future way to improve results in bridge or salvage liver transplantation? *Clin. Res. Hepatol. Gastroenterol.* **2021**, *45*, 101396. [CrossRef]
71. Qu, W.; Zhu, Z.J.; Sun, L.Y.; Wei, L.; Liu, Y.; Zeng, Z.G. Salvage liver transplantation for hepatocellular carcinoma recurrence after primary liver resection. *Clin. Res. Hepatol. Gastroenterol.* **2015**, *39*, 93–97. [CrossRef] [PubMed]
72. Pawlik, T.M.; Delman, K.A.; Vauthey, J.N.; Nagorney, D.M.; Ng, I.O.; Ikai, I.; Yamaoka, Y.; Belghiti, J.; Lauwers, G.Y.; Poon, R.T.; et al. Tumor size predicts vascular invasion and histologic grade: Implications for selection of surgical treatment for hepatocellular carcinoma. *Liver Transplant.* **2005**, *11*, 1086–1092. [CrossRef] [PubMed]

73. Guerrini, G.P.; Pinelli, D.; Di Benedetto, F.; Marini, E.; Corno, V.; Guizzetti, M.; Aluffi, A.; Zambelli, M.; Fagiuoli, S.; Lucà, M.G.; et al. Predictive value of nodule size and differentiation in HCC recurrence after liver transplantation. *Surg. Oncol.* **2015**, *25*, 419–428. [CrossRef] [PubMed]
74. Ferrer-Fàbrega, J.; Forner, A.; Liccioni, A.; Miquel, R.; Molina, V.; Navasa, M.; Fondevila, C.; García-Valdecasas, J.C.; Bruix, J.; Fuster, J. Prospective validation of ab initio liver transplantation in hepatocellular carcinoma upon detection of risk factors for recurrence after resection. *Hepatology* **2016**, *63*, 839–849. [CrossRef]
75. Sala, M.; Fuster, J.; Llovet, J.M.; Navasa, M.; Solé, M.; Varela, M.; Pons, F.; Rimola, A.; García-Valdecasas, J.C.; Brú, C.; et al. High pathological risk of recurrence after surgical resection for hepatocellular carcinoma: An indication for salvage liver transplantation. *Liver Transplant.* **2004**, *10*, 1294–1300. [CrossRef]
76. Lee, S.G. Salvage living-donor liver transplantation to previously hepatectomized hepatocellular carcinoma patients: Is it a reasonable strategy? *Hepatobiliary Pancreat. Dis. Int.* **2013**, *12*, 10–11. [CrossRef]
77. Fuks, D.; Dokmak, S.; Paradis, V.; Diouf, M.; Durand, F.; Belghiti, J. Benefit of initial resection of hepatocellular carcinoma followed by transplantation in case of recurrence: An intention-to-treat analysis. *Hepatology* **2012**, *55*, 132–140. [CrossRef]
78. Chok, K. Management of recurrent hepatocellular carcinoma after liver transplant. *World J. Hepatol.* **2015**, *7*, 1142–1148. [CrossRef]
79. Tarantino, G.; Magistri, P.; Ballarin, R.; Di Francia, R.; Berretta, M.; Di Benedetto, F. Oncological Impact of M-Tor Inhibitor Immunosuppressive Therapy after Liver Transplantation for Hepatocellular Carcinoma: Review of the Literature. *Front. Pharmacol.* **2016**, *7*, 387. [CrossRef]
80. Duvoux, C.; Toso, C. mTOR inhibitor therapy: Does it prevent HCC recurrence after liver transplantation? *Transplant. Rev.* **2015**, *29*, 168–174. [CrossRef]
81. Perricone, G.; Mancuso, A.; Belli, L.S.; Mazzarelli, C.; Zavaglia, C. Sorafenib for the treatment of recurrent hepatocellular carcinoma after liver transplantation: Does mTOR inhibitors association augment toxicity? *Eur. J. Gastroenterol. Hepatol.* **2014**, *26*, 577–578. [CrossRef] [PubMed]
82. Jackson, D.; Turner, R. Power analysis for random-effects meta-analysis. *Res. Synth. Methods* **2017**, *8*, 290–302. [CrossRef] [PubMed]
83. Higgins, J.; Thompson, S.; Deeks, J.; Altman, D. Statistical heterogeneity in systematic reviews of clinical trials: A critical appraisal of guidelines and practice. *J. Health Serv. Res. Policy* **2002**, *7*, 51–61. [CrossRef] [PubMed]

*cancers*

*Review*

# Could We Predict the Response of Immune Checkpoint Inhibitor Treatment in Hepatocellular Carcinoma?

**Choong-kun Lee [1], Stephen L. Chan [2,*] and Hong Jae Chon [3,*]**

1 Division of Medical Oncology, Department of Internal Medicine, Yonsei University College of Medicine, Seoul 03722, Korea; cklee512@yuhs.ac
2 State Key Laboratory of Translational Oncology, Department of Clinical Oncology, Sir YK Pao Centre for Cancer, Prince of Wales Hospital, The Chinese University of Hong Kong, Hong Kong, China
3 Medical Oncology, Department of Internal Medicine, CHA Bundang Medical Center, CHA University, Seongnam 13496, Korea
* Correspondence: chanlam_stephen@cuhk.edu.hk (S.L.C.); minidoctor@cha.ac.kr (H.J.C.); Tel.: +85-23-505-2166 (S.L.C.); +82-31-780-7590 (H.J.C.)

**Simple Summary:** The use of anti-programmed cell-death protein (ligand)-1 (PD-[L]1) is now a standard of care for treating hepatocellular carcinoma (HCC). However, the treatment only benefits 10–20% of patients when used as a monotherapy. The unique environments of hepatitis and/or cirrhosis, which continuously interact with the hosts' immune systems, make it difficult to find appropriate biomarkers to predict the response or lack of response of anti-PD-1/PD-L1 treatment in HCC. The current review aimed to present both clinical and translational biomarkers for anti-PD-1/PD-L1 treatment in HCC.

**Abstract:** The use of anti-programmed cell-death protein (ligand)-1 (PD-[L]1) is an important strategy for treating hepatocellular carcinoma (HCC). However, the treatment only benefits 10–20% of patients when used as a monotherapy. Therefore, the selection of patients for anti-PD-1/PD-L1 treatment is crucial for both patients and clinicians. This review aimed to explore the existing literature on tissue or circulating markers for the identification of responders or non-responders to anti-PD-1/PD-L1 in HCC. For the clinically available markers, both etiological factors (viral versus non-viral) and disease extent (intra-hepatic vs. extrahepatic) impact the responses to anti-PD-1/PD-L1, warranting further studies. Preliminary data suggested that inflammatory indices (e.g., neutrophil-lymphocyte ratio) may be associated with clinical outcomes of HCC during the anti-PD-1/PD-L1 treatment. Finally, although PD-L1 expression in tumor tissues is a predictive marker for multiple cancer types, its clinical application is less clear in HCC due to the lack of a clear-cut association with responders to anti-PD-1/PD-L1 treatment. Although all translational markers are not routinely measured in HCC, recent data suggest their potential roles in selecting patients for anti-PD-1/PD-L1 treatment. Such markers, including the immune classification of HCC, selected signaling pathways, tumor-infiltrating lymphocytes, and auto-antibodies, were discussed in this review.

**Keywords:** hepatocellular carcinoma; anti-programmed cell-death protein (ligand)-1; immune checkpoint inhibitor; predictive biomarker; clinical biomarker; translational biomarker

**Citation:** Lee, C.-k.; Chan, S.L.; Chon, H.J. Could We Predict the Response of Immune Checkpoint Inhibitor Treatment in Hepatocellular Carcinoma? *Cancers* **2022**, *14*, 3213. https://doi.org/10.3390/cancers14133213

Academic Editor: Georgios Germanidis

Received: 1 June 2022
Accepted: 27 June 2022
Published: 30 June 2022



## 1. Introduction

Hepatocellular carcinoma (HCC) accounts for over 80% of primary liver cancer [1,2]. Typically, more than 80% of HCCs occur in the background of cirrhotic liver, which is characterized by long-standing inflammation due to viral hepatitis or metabolic or chemical injury [3,4]. HCC is highly lethal due to its delayed presentation, resistance to drug treatment, and underlying hepatic decompensation [5,6]. The mainstay of systemic therapy for HCC has been muti-targeted tyrosine kinase inhibitors (TKIs), such as sorafenib, lenvatinib, regorafenib, and cabozantinib [7–10]. TKIs typically lead to disease control or result in modest response for a period of time; however, resistance to TKIs is inevitable in most patients after a few months of treatment.

Recent advances in immune checkpoint inhibitors (ICIs) have, however, changed the above scenario. ICIs, particularly the anti-programmed cell death protein (PD)-1/ligand (PD-L1) antibodies, can potentially reverse the immune-exhausted microenvironment of HCC and induce cytotoxic T cell-mediated destruction of HCC [11]. In clinical trials, monotherapy using anti-PD1 was associated with a radiological response rate of 10 to 20% in patients with HCC, including both complete and partial responses [12,13]. The responses can potentially be durable in some patients, thus explaining the observation of a plateau, frequently known as the tail, in the Kaplan–Meier survival curves of clinical trials. However, the initial high expectations from ICIs were disappointed by the failures of phase III clinical trials on anti-PD-1 monotherapy to reach the primary objectives of improved overall survival (OS) compared to sorafenib [14,15]. The negative results of clinical trials on monotherapy using anti-PD-1 could be explained in multiple ways, including the use of subsequent therapy, heterogeneity of patients with HCC, the lack of useful predictive biomarkers for patient selection, and accelerated progression and neutralizing auto-antibodies [16–18].

Strategically, there are two approaches to improve the outcome of anti-PD1/PDL-1 treatment in HCC. First, combinational treatment of anti-PD-1/PD-L1 with other ICIs or targeted agents could be synergistic, thereby significantly enhancing the treatment outcomes in patients. The approval of the atezolizumab–bevacizumab combination as the first-line treatment in HCC is the first notable example [19]. Recently, another combinational regimen of tremelimumab–durvalumab was shown to improve the median OS over that of sorafenib treatment in a phase III clinical trial [20]. The second approach is to develop methodologies to select patients who are more likely to derive benefits from the anti-PD-1/PDL-1 treatment. Experiences with other cancer types suggested that patients could be enriched by clinical biomarkers to improve the outcomes of anti-PD-1/PD-L1 treatment. For example, the phenotype of deficit mismatch repair (d-MMR) is a tumor-agnostic marker that predicts high responses and prolonged survival in response to ICI treatment in different cancers [21]. In lung cancer, the high immunohistochemical staining of PD-L1 in tumor tissues is known to be associated with clinical benefits in anti-PD-1 treatment [22]. For HCC, robust studies were conducted by different groups to identify markers predictive of benefits or resistance to anti-PD-1/anti-PD-L1 treatment. However, the overall picture is more complex in HCC than in other solid tumors due to the unique environment of hepatitis and/or cirrhosis, which continuously interacts with the hosts' immune systems. The current review aimed to present both clinical (Table 1) and translational (Table 2) biomarkers for anti-PD-1/PD-L1 treatment in HCC.

## 2. Clinical Biomarkers

### 2.1. Etiology

Chronic infection with hepatitis B virus (HBV) and hepatitis C virus (HCV) is a traditional major risk factor associated with HCC [23]. The virus-associated mechanisms that cause liver cancer are complex, and HCC develops mostly in cirrhotic liver (about in 90% of the cases), whereas HCC development in the normal liver is a rare event (less than 10% of the cases) [24]. HBV infections account for 75–80% of virus-associated HCCs, and the integration of genetic material of HBV into the human genome leads to p53 inactivation, inflammation, or activation of various oncogenic pathways, including PI3K/Akt/STAT3 pathway and Wnt/b-catenin (induction of oxidative stress), which induce hepatocarcinogenesis [25–28]. Unlike that in HBV infection, the genetic material of HCV is not integrated into the host's genome; rather, the HCV proteins induce chronic inflammation, which leads to the development of HCC [28,29]. In addition to viral causes, fatty liver disease, especially non-alcoholic fatty liver disease (NAFLD) [30], which includes non-alcoholic steatohepatitis (NASH), is the fastest-growing etiology due to lifestyle changes in the western dietary pattern, increased obesity, and improved antiviral therapy [31]. Multiple mechanisms, including steatosis-induced necroinflammation, the release of inflammatory cytokines, and immune microenvironment alterations, are the key driving forces in NAFLD-associated HCC [32,33].

A recent meta-analysis [16] evaluated the effect of etiology in terms of efficacy across three large randomized controlled phase III trials of immunotherapies for HCC, namely anti-PD-L1 in combination with anti-vascular endothelial growth factor (VEGF) (IMbrave150 [19]), or anti-PD-1 monotherapy (CheckMate 459 [14]) compared to sorafenib, or second-line anti-PD-1 monotherapy compared to placebo (Keynote-240 [15])-treated patients. In this large meta-analysis (total $n$ = 1656), patients with HBV-related HCC and HCV-related HCC showed superior survival benefits from immunotherapy than the control, although it was not so in patients with non-viral HCC. Among the additional validation cohort with HCC patients treated with anti-PD/PD-L1, NAFLD was independently associated with shortened survival of patients with HCC after anti-PD-1/PD-L1 treatment. Preclinical evidence showed that NASH progression is associated with increased activated $CD8^{+}PD1^{+}$T cells; anti-PD-1 treatment did not lead to tumor regression, indicating that tumor immune surveillance was impaired. A recent preclinical study suggested that an anti-PD1 and CXCR2 inhibitor combination selectively reprograms tumor-associated neutrophils from a pro-tumor to an anti-tumor phenotype that can overcome the resistance of NASH-HCC to anti-PD1 therapy [34]. In the recent HIMALAYA [35] phase III trial, testing the combination of anti-CTLA4 and anti-PD-L1 inhibitors for first-line treatment of advanced HCC, patients with HBV-related or non-viral-etiology HCC were benefitted in terms of OS, compared to those receiving sorafenib, although it was not so in cases of HCV-related HCC. Further investigation would be required for such contradictory results.

### 2.2. Disease Extent

Treatment options for HCC are dependent on the stage (Barcelona clinic liver cancer, BCLC, staging system [36]) of the disease. Patients in the intermediate stage (BCLC-B) and advanced stage (BCLC-C) are candidates for systemic treatment. In recent first-line phase III trials for advanced HCC (IMbrave150, HIMALAYA), atezolizumab and bevacizumab or tremelimumab and durvalumab were reported to be superior to sorafenib in terms of OS of patients with BCLC-C, although not for those with BCLC-B [19,35]. However, in a Chinese phase III trial conducted mostly for patients with B-viral HCC, those with BCLC-B or BCLC-C also benefitted from anti-PD-1 and anti-VEGF treatments relative to that from sorafenib treatment [37]. Patients in the BCLC-C stage presented with vascular invasion or extrahepatic spread. In IMbrave150 and HIMALAYA trials, patients with extrahepatic spread achieved OS benefit from the first-line atezolizumab and bevacizumab or tremelimumab and durvalumab treatment than from sorafenib treatment. Anti-tumor immune response to ICIs differs in an organ-specific manner [38], and liver metastasis is associated with poor response to immunotherapy monotherapy. Accordingly, intrahepatic tumors of HCC were reported to possibly be less responsive to immunotherapy monotherapy than extrahepatic lesions [39,40]. Preclinical evidence also supported that liver tumors show reduced peripheral T cell numbers and diminished tumoral T cell diversity and function, creating an immune desert. Yu et al. showed that in mouse models, liver-directed radiotherapy could eliminate immunosuppressive hepatic macrophages, enhancing the anti-tumor effect of immunotherapy [41]. Further strategies would be required to enhance the anti-tumor effect of ICIs in intrahepatic lesions of patients with advanced HCC, along with the combination of local control.

Tumoral macrovascular invasion (MVI) of hepatic and/or portal vein branches is a common phenomenon in advanced HCC and is usually associated with a poorer prognosis than HCC without MVI. Patients with HCC and MVI, including Vp4 (presence of a tumor thrombus in the main trunk and/or contralateral portal vein), show superior survival when treated with atezolizumab and bevacizumab (anti-VEGF) than with sorafenib [19,42]; however, there was no additional survival benefit in durvalumab and tremelimumab treatment compared to that in sorafenib treatment in a subgroup analysis of the HIMALAYA trial [35]. Further studies (translational and clinical) would be required to investigate whether anti-VEGF treatment in combination with immunotherapy has any additional benefit in HCC with MVI.

### 2.3. Laboratory Tests

In daily practice, we performed laboratory tests to examine patient status; some features from laboratory (blood) tests can be used as biomarkers to predict immunotherapeutic efficacy in patients with advanced HCC and low invasiveness.

Elevated tumor markers, especially $\alpha$-fetoprotein (AFP), are considered prognostic markers for poor clinical outcomes among patients with HCC. Recent randomized phase III studies showed contradictory results in terms of the benefit of immunotherapy compared to that in the control. In the first-line phase III CheckMate 459 [14] and HIMALAYA [35] trials, patients with high baseline AFP levels ($\geq$400 ng/mL) achieved longer OS when treated with immunotherapy rather than sorafenib. However, results of the IMbrave150 study showed that patients with low baseline AFP levels (AFP < 400 ng/mL) were associated with longer OS and PFS when treated with immunotherapy rather than sorafenib [19]. Since AFP level is related to the tumor or patient characteristics, interpretation should be performed with caution. As generally seen in other treatments, a decline in post-treatment tumor marker level is associated with better efficacy of immunotherapy in advanced HCC; the AFP response at 6 weeks after atezolizumab plus bevacizumab initiation especially seemed to be a potential surrogate biomarker for prognosis [43–45].

The usage of circulating immune cells as predictive biomarkers for immunotherapy was extensively investigated. Contrary to specific immune cells that require additional experiments to obtain, we can inexpensively and reproducibly obtain information about complete blood cell differential counts from the patients in daily laboratory tests. A neutrophil-to-lymphocyte ratio (NLR), defined by the ratio of an absolute number of neutrophils to that of lymphocytes, is an especially well-known marker for selecting patients that are benefitted from immunotherapy in various tumor types [46]. A correlation was reported between circulating neutrophils and neutrophils in the tumor microenvironment, and low circulating lymphocyte levels were associated with low levels of tumor-infiltrating lymphocytes (TILs), thereby resulting in reduced anti-tumor T-cell responses [47,48]. In addition to NLR, platelet-to-lymphocyte ratio (PLR) is regarded as a biomarker of immunotherapy response since platelets are also part of an inflammatory process [49]. In the CheckMate 040 study, OS benefit was observed in patients with low NLR or PLR tertile due to nivolumab treatment than in others. Other studies also reported the predictive role of NLR or PLR in immunotherapy of advanced HCC [44,50]. Kim et al. reported that elevated NLR could predict the occurrence of hyper-progressive disease and inferior survival rate after anti-PD-1 blockade [51]. However, since NLR is also an independent prognostic factor for patients with HCC treated with sorafenib [52,53], further studies would be required to confirm the role of NLR or PLR in patients with HCC, to clarify whether it is a prognostic biomarker for the general population or whether there is any specific role by which it can identify patients with maximum possible benefit from immunotherapy than from tyrosine kinase inhibitor therapy.

### 2.4. PD-L1 Expression

PD-L1 is widely expressed on the surface of tumor cells, and its high expression in the tumor microenvironment is generally regarded as a biomarker for anti-PD-1/PD-L1 immunotherapy in various tumors, especially in NSCLC [22,54,55]. In HCC, PD-L1 expression was reported to be approximately 10 to 20% in tumor cells [14,56], and PD-L1 expression in HCC tumor cells is considered to be associated with tumor aggressiveness and poor survival [57]. Several clinical trials evaluated whether PD-L1 expression has predictive value as a biomarker for immune checkpoint inhibitor efficacy in patients with HCC. However, the types of PD-L1 antibodies for immunohistochemistry (28-8, 22C3,

SP142, SP263) and the way of interpretation vary across trials, and the determination of their roles has been challenging. Among the patients treated with anti-PD-1 monotherapy, PD-L1 positive HCC, whether in tumor cells or tumor and immune cells combined, seem to respond better than those with negative PD-L1 expression [12,13,58]. Genomic analyses for the phase I trial of atezolizumab and bevacizumab in patients with HCC reported that high expression of PD-L1, as per RNA-seq, is related to better response and longer PFS [59]. In recent phase III trials, atezolizumab plus bevacizumab showed benefit over sorafenib in terms of PFS for a tumor or immune cells in PD-L1-positive patients [60]. However, since the recent phase III HIMALAYA trial showed the benefit of doublet immunotherapy over that of sorafenib, regardless of PD-L1 expression [35], further investigation is warranted in this regard.

**Table 1.** Clinical biomarkers.

| Factor | Detail | Outcome | Regimen | Line of Treatment | Trial (Phase) | Ref. |
|---|---|---|---|---|---|---|
| Etiology | Hepatitis B | OS (HR 0.51) and PFS (HR 0.47) benefit | Atezolizumab + Bevacizumab vs. Sorafenib | 1st | IMbrae150 (III) | Finn et al. NEJM 2020 [19] |
| | Hepatitis C | OS (HR 0.43) benefit | Atezolizumab + Bevacizumab vs. Sorafenib | 1st | IMbrae150 (III) | Finn et al. NEJM 2020 [19] |
| | Hepatitis B | OS (HR 0.64) benefit | Durvalumab + Tremelimumab vs. Sorafenib | 1st | Himalaya (III) | Abou-alfa et al. NEJM Evidence 2022 [20] |
| | Non-viral | OS (HR 0.74) benefit | Durvalumab + Tremelimumab vs. Sorafenib | 1st | Himalaya (III) | Abou-alfa et al. NEJM Evidence 2022 [20] |
| | HBV | OS (HR 0.57) benefit | Pembrolizumab vs. Placebo | 2nd | KEYNOTE-240 (III) | Finn et al. JCO 2019 [15] |
| | HBV | OS benefit | Nivolumab, Atezolizumab + Bevacizumab, Pembrolizumab (Meta-analysis) | 1st–2nd | CheckMate-459, IMbrave150 and KEYNOTE-240 (III) | Pfister et al. Nature 2021 [16] |
| | HCV | OS benefit | Nivolumab, Atezolizumab + Bevacizumab, Pembrolizumab (Meta-analysis) | 1st–2nd | CheckMate-459, IMbrave150 and KEYNOTE-240 (III) | Pfister et al. Nature 2021 [16] |
| | NAFLD | Worst survival | Nivolumab, Atezolizumab + Bevacizumab, Pembrolizumab (Meta-analysis) | 1st–2nd | CheckMate-459, IMbrave150 and KEYNOTE-240 (III) | Pfister et al. Nature 2021 [16] |
| BCLC stage | BCLC C (no benefit for BCLC B) | OS (HR 0.58) and PFS (HR 0.58) benefit | Atezolizumab + Bevacizumab vs. Sorafenib | 1st | IMbrae150 (III) | Finn et al. NEJM 2020 [19] |
| | BCLC C (no benefit for BCLC B) | OS (HR 0.76) benefit | Durvalumab + Tremelimumab vs. Sorafenib | 1st | Himalaya (III) | Abou-alfa et al. NEJM Evidence 2022 [20] |
| | BCLC B and C | OS and PFS benefit | Sintilimab + bevacizumab biosimilar vs. Sorafenib | 1st | ORIENT-32 (III) | Ren et al. Lancet Oncol. 2021 [35] |
| Extrahepatic Spread | Extrahepatic spread | OS (HR 0.5) benefit | Atezolizumab + Bevacizumab vs. Sorafenib | 1st | IMbrae150 (III) | Finn et al. NEJM 2020 [19] |
| | Extrahepatic spread | OS (HR 0.67) benefit | Durvalumab + Tremelimumab vs. Sorafenib | 1st | Himalaya (III) | Abou-alfa et al. NEJM Evidence 2022 [20] |
| Macrovascular invasion | Macrovascular invasion | OS (HR 0.58) benefit | Atezolizumab + Bevacizumab vs. Sorafenib | 1st | IMbrae150 (III) | Finn et al. NEJM 2020 [19] |
| | No macrovascular invasion | OS (HR 0.77) benefit | Durvalumab + Tremelimumab vs. Sorafenib | 1st | Himalaya (III) | Abou-alfa et al. NEJM Evidence 2022 [20] |
| Tumor Marker | AFP < 400 ng/mL | OS (HR 0.52) and PFS (0.49) benefit | Atezolizumab + Bevacizumab vs. Sorafenib | 1st | IMbrae150 (III) | Finn et al. NEJM 2020 [19] |
| | AFP ≥ 400 ng/mL | OS (HR 0.64) benefit | Durvalumab + Tremelimumab vs. Sorafenib | 1st | Himalaya (III) | Abou-alfa et al. NEJM Evidence 2022 [20] |
| | AFP ≥ 400 ng/mL | OS benefit | Nivolumab vs. Sorafenib | 1st | CheckMate-459 (III) | Yau et al. Lancet Oncol. 2022 [14] |
| | AFP < 400 ng/mL | OS benefit | Nivolumab | 1st–2nd | CheckMate-040 (I/II) | Sangro et al. J. Hep. 2020 [61] |
| | AFP < 200 ng/mL | OS (HR 0.68) and PFS (HR 0.64) benefit | Pembrolizumab vs. Placebo | 2nd | KEYNOTE-240 (III) | Finn et al. JCO 2019 [15] |

**Table 1.** *Cont.*

| Factor | Detail | Outcome | Regimen | Line of Treatment | Trial (Phase) | Ref. |
|---|---|---|---|---|---|---|
| Other laboratory tests | Neutrophil-to-lymphocyte ratio | OS benefit for pts with low tertile | Nivolumab | 1st–2nd | CheckMate-040 (I/II) | Sangro et al. J. Hep. 2020 [61] |
| | Platelet-to-lymphocyte ratio | OS benefit for pts with low tertile | Nivolumab | 1st–2nd | CheckMate-040 (I/II) | Sangro et al. J. Hep. 2020 [61] |
| PD-L1 IHC | PD-L1 TC (28-8) $\geq$ 1% | No significant benefit | Nivolumab vs. Sorafenib | 1st | CheckMate-459 (III) | Yau et al. Lancet Oncol. 2022 [14] |
| | PD-L1 TC (28-8) $\geq$ 1% | ORR (28% vs. 16%) and OS (28.1 vs. 16.6 months, $p = 0.032$) benefit | Nivolumab | 1st–2nd | CheckMate-040 (I/II) | El-Khoueiry et al. Lancet 2017 [12] |
| | PD-L1 CPS (22C3) $\geq$ 1% | ORR (32% vs. 20%, $p = 0.021$) benefit | Pembrolizumab | 2nd | KEYNOTE-224 (II) | Zhu et al. Lancet Oncol. 2018 [13] |
| | PD-L1 TPS (SP142) $\geq$ 1% | ORR 36% vs. 11% | Camrelizumab | 2nd | NCT02989922 (II) | Qin et al. Lancet Oncol. 2020 [14] |
| | PD-L1 TC or IC (SP263) $\geq$ 1% | PFS (OR 2.69) benefit | Atezolizumab + Bevacizumab vs. Sorafenib | 1st | IMbrae150 (III) | Cheng et al. J. Hepatol. 2022 [58] |

## 3. Translational Biomarkers

### 3.1. Immune-Specific Class of HCC

Daniela et al., previously characterized patients with high immune infiltration and molecular features resembling melanoma who responded to ICIs, as the immune class of HCC (approximately 25% of patients) [62]. Recently, Carla et al. further dichotomized the immunogenomic classification of HCC into inflamed and non-inflamed tumors [63]. However, their analyses were not based on NGS data of advanced patients who had received ICI treatment but rather on the results of pathology and immunohistochemical analyses to evaluate the correlation between expression patterns and the presence of both immune cell infiltrates and immune regulatory molecules. Therefore, the predictive capacity of such classification would need further investigation in patients receiving immunotherapy.

### 3.2. Tumor-Infiltrating Lymphocytes (TILs) and T-Cell Inflamed Gene Expression Profiles (GEP)

Tumor-infiltrating lymphocyte density and phenotypes are good predictive indicators of better responses to immunotherapy [61,64–66]. In the exploratory analysis of the CheckMate 040 trial [67], improved OS of patients with HCC who were being treated with nivolumab correlated with higher densities of $CD3^+$ or $CD8^+$ TILs. Gene expression, known to be related to immune cytolytic activity, was also demonstrated to be associated with the clinical outcome of certain tumors after checkpoint blockade treatment [68,69]. Recently, a T cell-inflamed gene expression profile (GEP) was presented as a predictive indicator of response to anti-PD-1-based therapy [70]. In the CheckMate 040 trial [67], patients receiving nivolumab and having HCC tumor tissues with inflammatory signature GEP consisting of CD274 (PD-L1), CD8α, LAG3, and STAT1, had improved objective response rate (ORR) and OS, suggesting the possibility of a relationship between underlying inflammation within the tumor environments and improved clinical outcomes. Exploratory analysis of the GO30140 study demonstrated that T-effector gene (GZM, PRF1, and CXCL9) signatures were associated with improved responses and longer PFS in patients treated with atezolizumab and bevacizumab [59].

### 3.3. Tumor Mutational Burden and High Microsatellite Instability

Tumor mutational burden (TMB) and microsatellite instability (MSI) are indirect indices of tumor antigenicity resulting from somatic tumor mutations, and these were most extensively studied for their role as predictive biomarkers in anti-PD-1 therapy. Based on KEYNOTE-158, the US FDA granted accelerated approval to pembrolizumab for the treatment of unresectable or metastatic tumor mutational burden-high (TMB-H) ($\geq$10 mutations/megabase (mut/Mb)) solid tumors in adult and pediatric patients [71]. However, patients with HCC were not included in this study, and TMB was not high in

HCC compared to that in melanoma or lung cancer [72]; moreover, TMB was not proven to be very predictive of ICI response in HCC [73]. Exploratory analysis of the GO30140 study demonstrated that TMB is unable to predict the response or PFS in patients with HCC treated with atezolizumab and bevacizumab [59]. Moreover, the phenotype MSI-high or d-MMR is very rare in HCC, with an incidence of approximately 1% [73,74]. In addition, studies have shown that it is mainly found in the early stage rather than the late stage. Therefore, as of now, routine MSI test is not considered informative in HCC.

### 3.4. WNT/β-Catenin

Mutations in *CTNNB1*, the gene responsible for encoding beta-catenin, and other alterations that affect the Wnt/beta-catenin signaling pathway are commonly found in HCC [75–78]; they are detected in approximately one-third of HCC tumors. Studies suggested that *CTNNB1* (β-catenin) mutations and consequent activation of the Wnt/β-catenin pathway could be responsible for the scarcity of immune cells in the tumor microenvironment and hence, poor clinical response to ICI [79,80]. In a genetically engineered mouse model of melanoma with constitutively active β-catenin, the latter was shown to reduce CCL4 expression, which is important for recruiting dendritic cells and, consequently, T-cells into the tumor microenvironment (TME) [79]. The mechanism by which β-catenin reduces CCL4 expression is associated with the induction of transcriptional repressor ATF3 and its binding to the CCL4 promoter [79,81,82]. The immune evasion mechanism was reproduced in an engineered HCC mouse model in which β-catenin was constantly activated; aberrant β-catenin activation resulted in increased resistance to anti-PD-1 therapy [83]. Harding et al. reported that alterations in WNT/β-catenin signaling are associated with lower disease control rate (DCR), shorter median progression-free survival (PFS), and shorter median OS in patients with advanced HCC treated with ICI [84]. Hong et al. also showed that only non-responders to pembrolizumab exhibited somatic mutations in *CTNNB1* [85]. Haber et al., on the other hand, reported that there was no association between the overall immune infiltrate or *CTNNB1* mutations and response [86]. According to the immune-specific class of HCC defined by Montironi et al., one-third was classified as inflamed tumor with Wnt/β-catenin pathway activation, and the remaining were classified as non-inflamed tumors [63]. The discordant results of Wnt/β-catenin pathway activation on its predictive potential in HCC suggest the need for further analysis.

### 3.5. Other Gene Signatures Associated with Adverse Clinical Outcomes

The biomarker study [87] with tislelizumab, an anti-PD-1 monoclonal antibody, was given to patients with advanced HCC previously treated with sorafenib (NCT02407990 and NCT04068519), and it was demonstrated that non-responders had elevated expression of genes related to angiogenesis (TEK, KDR, HGF, and EGR1), immune exhaustion (CD274, CTLA-4, TIGIT, and CD96), and cell cycle (E2F7, FOXA1, and FANCD2), compared to responders. Exploratory analysis of the GO30140 study demonstrated that gene expression related to Notch pathway activation (i.e., high expression of HES1) was associated with a lack of response and shorter PFS in patients treated with atezolizumab and bevacizumab [59].

### 3.6. Circulating Biomarkers

Unlike that in lung cancer or melanoma, studies on circulating biomarkers for immunotherapy in HCC are limited. Feun et al. reported that, among the 11 cytokines and chemokines that were tested in 24 patients with unresectable HCC and receiving pembrolizumab, only baseline TGF-β cytokine level in peripheral blood was significantly higher in non-responders than in responders [88]. Winogrand et al. reported the relevance between the presence of PD-L1$^{+}$ circulating tumor cells (CTCs) and favorable immunotherapy outcome ($n$ = 10); however, it was also a negative prognostic biomarker and an overall survival predictor ($n$ = 87) [89]. Additional verification would still be required to support the small-scale studies before their incorporation as biomarkers in immunotherapy.

### 3.7. Anti-Drug Antibody against Atezolizumab

Humanized antibodies could be immunogenic and induce undesirable anti-drug antibody (ADA) responses upon administration [18,90,91]. ADAs are known to interfere with the action of a therapeutic antibody by affecting drug clearance and serum concentration [91,92] or by neutralization. ICIs were also shown to generate ADA responses in patients with cancer [91,93–95]. Among the various ICI antibodies, atezolizumab has the highest incidence rate of ADA (29.8%) compared to others (around 5% to 10%) [18,90,91,96]. The results of the IMbrave 150 study showed that the incidence of atezolizumab ADA reached 29.6% in patients with HCC at one or more timepoints following atezolizumab–bevacizumab treatment [97]. Although ADA-negative patients had improved OS, ADA-positive ones showed a similar OS with atezolizumab plus bevacizumab vs. sorafenib treatment (HR of ADA-positive patients vs. those of sorafenib was 0.96 (95% CI, 0.621–1.4184)). To date, however, there is no available method to predict which drug may induce ADAs, and there is no FDA-approved commercial test yet to identify the patients who may develop ADA after atezolizumab treatment [98]. Data to guide treatment decisions in patients who develop ADAs are still unavailable. Therefore, a study that can evaluate the overall effect of ADA would be appropriate in the future.

**Table 2.** Translational biomarkers.

| Marker | Assay | Treatment | N | Findings Associated with Clinical Response | Reference |
|---|---|---|---|---|---|
| | | | **TIL Based Biomarkers** | | |
| Baseline $CD3^+$ or $CD8^+$ TILs | IHC | Nivolumab | 189 (CD3)<br>192 (CD8) | $CD3^+$ or $CD8^+$ TILs exhibited a trend towards improved OS | Sangro et al. J. Hep. 2020 [65] |
| $CD3^+$ or $CD8^+$ TILs after Treatment | IHC | Tremelimumab with RFA or TACE | 9 | Responder had higher $CD3^+$ or $CD8^+$ TILs than non-responder | Duffy et al. J. Hep. 2017 [61] |
| | | | **Sequencing based biomarkers** | | |
| T-effector signature (*GZM, PRF1, CXCL9*) | RNA seq | Atezolizumab–Bevacizumab | 90 | Associated with response and longer PFS | Zhu et al. Cancer Res. 2020 [69] |
| Baseline inflammation signature of tumor | RNA seq | Nivolumab | 37 | Inflammatory signature consisting of CD274 (PD-L1), CD8A, LAG3, and STAT1 was associated with both improved objective response rate and OS. | Sangro et al. J. Hep. 2020 [65] |
| WNT/β-catenin | NGS | Immune checkpoint inhibitors | 31 | Activating alteration of WNT/β-catenin signaling was associated with lower DCR, shorter median PFS, and shorter median OS | Harding et al. Clin. Cancer Res. 2019 [82] |
| WNT/β-catenin | NGS | Pembrolizumab | 60 | Somatic mutations in CTNNB1 were found only in non-responders | Hong et al. Genome Med. 2022 [83] |
| Angiogenesis, Immune exhaustion, cell-cycle gene signatures | NGS | Tislelizumab | 41 | Non-responders had elevated angiogenesis, immune exhaustion, and cell-cycle gene signature than responders | Hou et al. J. ImmunoTher. Cancer. 2020 [85] |
| TCR signaling | RNA seq | Pembrolizumab | 60 | Responders demonstrated T cell receptor (TCR) signaling activation with expressions of MHC genes | Hong et al. Genome Med. 2022 [83] |
| Notch pathway activation genes | RNA seq | Atezolizumab–Bevacizumab | 90 | Associated with lack of response and shorter PFS | Zhu et al. Cancer Res. 2020 [69] |
| TMB | WES | Atezolizumab–Bevacizumab | 73 | Not associated with response or PFS | Zhu et al. Cancer Res. 2020 [69] |
| | | | **Circulating biomarkers** | | |
| plasma TGF-β levels | ELISA | Pembrolizumab | 24 | High baseline plasma TGF-β levels (≥200 pg/mL) significantly associated with unfavorable outcomes | Feun et al. Cancer 2019 [86] |
| Anti-drug antibody (ADA) | ELISA | Atezolizumab–Bevacizumab | 336 | While patients with ADA− had an improved OS, those with ADA+ had a similar OS with Ate/Bev vs. sorafenib | Galle et al. Cancer Res. 2021 [95] |
| PD-$L1^+$CTCs | Immunocytochemistry | PD-1 blockade | 10 | PD-$L1^+$CTCs were associated with favorable immunotherapy outcome | Winogrand et al. Hepatol. Commun. 2020 [87] |

## 4. Conclusions

We reviewed the data on ICI biomarkers obtained from recent pivotal studies on HCC (Figure 1). Although several potential candidates were evaluated for predicting response to ICI treatment, there is currently no standard biomarker for ICI-treated patients with HCC. Since tissue biopsy is not mandatory for the diagnosis of HCC, the discovery of predictive biomarkers by tumor tissue analyses is limited compared to that in other solid cancers. Although PD-L1 expression in tumor tissues is known to be a predictive marker for multiple cancer types, its clinical use is less clear in HCC due to the less clear-cut association between PD-L1 expression and responders to anti-PD-1/PD-L1. In HCC, the overall picture is more complex than in other solid tumors due to the unique environment involving hepatitis and/or cirrhosis, which constantly interacts with the host's immune system. Considering the complexity of predicting ICI treatment response in HCC, an integrative multi-parameter approach combining histopathology, imaging, and immune features would need to be applied as a novel strategy. Since immunotherapy has become the new standard of care in HCC, and various biomarker studies are being conducted in parallel, personalized therapy through a biomarker-based approach is expected to improve patient survival outcomes in the future.

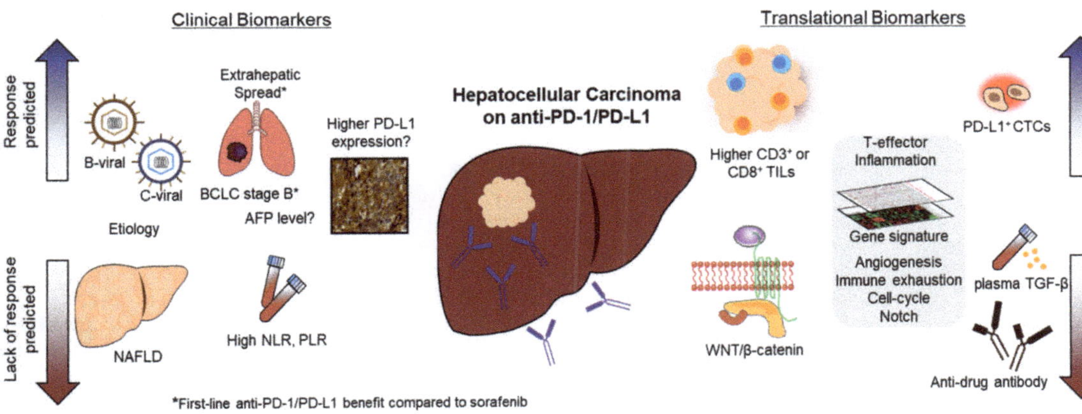


**Figure 1.** Clinical and translational biomarkers to predict the response and lack of response of immune checkpoint inhibitor treatment in hepatocellular carcinoma.

**Author Contributions:** Conceptualization, C.-k.L., S.L.C. and H.J.C.; writing—original draft preparation, C.-k.L. and H.J.C.; writing—review and editing, S.L.C. and H.J.C.; visualization, C.-k.L.; supervision, S.L.C. and H.J.C.; funding acquisition, C.-k.L. and H.J.C. Study design, writing—original draft preparation, review, and editing, all authors. All authors have read and agreed to the published version of the manuscript.

**Funding:** This work was supported by the National Research Foundation of Korea grant funded by the Korean government [MSIT] [NRF-2020R1C1C1004461 to C.-k.L.; NRF-2020R1C1C1010722 to H.J.C.].

**Conflicts of Interest:** Hong Jae Chon has been a consultant/advisor to Roche, ONO, BMS, Eisai, Bayer, MSD, and AstraZeneca, received research grants from Roche, and received lecture fees from Roche, Bayer, and BMS. Stephen L. Chan has been an advisor to Astrazeneca, Autem, Eisai, Ipsen, MSD, and Novartis, received a research grant from Bayer, Ipsen, and SIRTEX, and received a lecture fee from Astrazeneca, Bayer, Eisai, Roche, and MSD. The authors declare no conflict of interest.

## References


1. Llovet, J.M.; Kelley, R.K.; Villanueva, A.; Singal, A.G.; Pikarsky, E.; Roayaie, S.; Lencioni, R.; Koike, K.; Zucman-Rossi, J.; Finn, R.S. Hepatocellular Carcinoma. *Nat. Rev. Dis. Primers* **2021**, *7*, 6. [CrossRef] [PubMed]
2. Center, M.M.; Jemal, A. International Trends in Liver Cancer Incidence Rates. *Cancer Epidemiol. Biomark. Prev.* **2011**, *20*, 2362–2368. [CrossRef] [PubMed]
3. Singal, A.G.; Lampertico, P.; Nahon, P. Epidemiology and Surveillance for Hepatocellular Carcinoma: New Trends. *J. Hepatol.* **2020**, *72*, 250–261. [CrossRef] [PubMed]
4. Bray, F.; Ferlay, J.; Soerjomataram, I.; Siegel, R.L.; Torre, L.A.; Jemal, A. Global Cancer Statistics 2018: GLOBOCAN Estimates of Incidence and Mortality Worldwide for 36 Cancers in 185 Countries. *CA Cancer J. Clin.* **2018**, *68*, 394–424. [CrossRef]
5. Bertuccio, P.; Turati, F.; Carioli, G.; Rodriguez, T.; La Vecchia, C.; Malvezzi, M.; Negri, E. Global Trends and Predictions in Hepatocellular Carcinoma Mortality. *J. Hepatol.* **2017**, *67*, 302–309. [CrossRef]
6. Chan, S.L.; Wong, V.W.S.; Qin, S.; Chan, H.L.Y. Infection and Cancer: The Case of Hepatitis B. *J. Clin. Oncol.* **2016**, *34*, 83–90. [CrossRef]
7. Cheng, A.-L.; Kang, Y.-K.; Chen, Z.; Tsao, C.-J.; Qin, S.; Kim, J.S.; Luo, R.; Feng, J.; Ye, S.; Yang, T.-S.; et al. Efficacy and Safety of Sorafenib in Patients in the Asia-Pacific Region with Advanced Hepatocellular Carcinoma: A Phase III Randomised, Double-Blind, Placebo-Controlled Trial. *Lancet Oncol.* **2009**, *10*, 25–34. [CrossRef]
8. Kudo, M.; Finn, R.S.; Qin, S.; Han, K.-H.; Ikeda, K.; Piscaglia, F.; Baron, A.; Park, J.-W.; Han, G.; Jassem, J.; et al. Lenvatinib versus Sorafenib in First-Line Treatment of Patients with Unresectable Hepatocellular Carcinoma: A Randomised Phase 3 Non-Inferiority Trial. *Lancet* **2018**, *391*, 1163–1173. [CrossRef]
9. Bruix, J.; Qin, S.; Merle, P.; Granito, A.; Huang, Y.-H.; Bodoky, G.; Pracht, M.; Yokosuka, O.; Rosmorduc, O.; Breder, V.; et al. Regorafenib for Patients with Hepatocellular Carcinoma Who Progressed on Sorafenib Treatment (RESORCE): A Randomised, Double-Blind, Placebo-Controlled, Phase 3 Trial. *Lancet* **2017**, *389*, 56–66. [CrossRef]
10. Abou-Alfa, G.K.; Meyer, T.; Cheng, A.-L.; El-Khoueiry, A.B.; Rimassa, L.; Ryoo, B.-Y.; Cicin, I.; Merle, P.; Chen, Y.; Park, J.-W.; et al. Cabozantinib in Patients with Advanced and Progressing Hepatocellular Carcinoma. *N. Engl. J. Med.* **2018**, *379*, 54–63. [CrossRef]
11. Sangro, B.; Sarobe, P.; Hervás-Stubbs, S.; Melero, I. Advances in Immunotherapy for Hepatocellular Carcinoma. *Nat. Rev. Gastroenterol. Hepatol.* **2021**, *18*, 525–543. [CrossRef]
12. El-Khoueiry, A.B.; Sangro, B.; Yau, T.; Crocenzi, T.S.; Kudo, M.; Hsu, C.; Kim, T.-Y.; Choo, S.-P.; Trojan, J.; Welling, T.H., 3rd; et al. Nivolumab in Patients with Advanced Hepatocellular Carcinoma (CheckMate 040): An Open-Label, Non-Comparative, Phase 1/2 Dose Escalation and Expansion Trial. *Lancet* **2017**, *389*, 2492–2502. [CrossRef]
13. Zhu, A.X.; Finn, R.S.; Edeline, J.; Cattan, S.; Ogasawara, S.; Palmer, D.; Verslype, C.; Zagonel, V.; Fartoux, L.; Vogel, A.; et al. Pembrolizumab in Patients with Advanced Hepatocellular Carcinoma Previously Treated with Sorafenib (KEYNOTE-224): A Non-Randomised, Open-Label Phase 2 Trial. *Lancet Oncol.* **2018**, *19*, 940–952. [CrossRef]
14. Yau, T.; Park, J.-W.; Finn, R.S.; Cheng, A.-L.; Mathurin, P.; Edeline, J.; Kudo, M.; Harding, J.J.; Merle, P.; Rosmorduc, O.; et al. Nivolumab versus Sorafenib in Advanced Hepatocellular Carcinoma (CheckMate 459): A Randomised, Multicentre, Open-Label, Phase 3 Trial. *Lancet Oncol.* **2022**, *23*, 77–90. [CrossRef]
15. Finn, R.S.; Ryoo, B.-Y.; Merle, P.; Kudo, M.; Bouattour, M.; Lim, H.Y.; Breder, V.; Edeline, J.; Chao, Y.; Ogasawara, S.; et al. Pembrolizumab as Second-Line Therapy in Patients with Advanced Hepatocellular Carcinoma in KEYNOTE-240: A Randomized, Double-Blind, Phase III Trial. *J. Clin. Oncol.* **2020**, *38*, 193–202. [CrossRef]
16. Pfister, D.; Núñez, N.G.; Pinyol, R.; Govaere, O.; Pinter, M.; Szydlowska, M.; Gupta, R.; Qiu, M.; Deczkowska, A.; Weiner, A.; et al. NASH Limits Anti-Tumour Surveillance in Immunotherapy-Treated HCC. *Nature* **2021**, *592*, 450–456. [CrossRef]
17. Chan, S.L. Hyperprogression in Hepatocellular Carcinoma: Illusion or Reality? *J. Hepatol.* **2021**, *74*, 269–271. [CrossRef]
18. Enrico, D.; Paci, A.; Chaput, N.; Karamouza, E.; Besse, B. Antidrug Antibodies Against Immune Checkpoint Blockers: Impairment of Drug Efficacy or Indication of Immune Activation? *Clin. Cancer Res.* **2020**, *26*, 787–792. [CrossRef]
19. Finn, R.S.; Qin, S.; Ikeda, M.; Galle, P.R.; Ducreux, M.; Kim, T.-Y.; Kudo, M.; Breder, V.; Merle, P.; Kaseb, A.O.; et al. Atezolizumab plus Bevacizumab in Unresectable Hepatocellular Carcinoma. *N. Engl. J. Med.* **2020**, *382*, 1894–1905. [CrossRef]
20. Abou-Alfa, G.K.; Lau, G.; Kudo, M.; Chan, S.L.; Kelley, R.K.; Furuse, J.; Sukeepaisarnjaroen, W.; Kang, Y.-K.; Van Dao, T.; De Toni, E.N.; et al. Tremelimumab plus Durvalumab in Unresectable Hepatocellular Carcinoma. *NEJM Evid.* **2022**. [CrossRef]
21. Marabelle, A.; Le, D.T.; Ascierto, P.A.; Di Giacomo, A.M.; De Jesus-Acosta, A.; Delord, J.-P.; Geva, R.; Gottfried, M.; Penel, N.; Hansen, A.R.; et al. Efficacy of Pembrolizumab in Patients with Noncolorectal High Microsatellite Instability/Mismatch Repair-Deficient Cancer: Results from the Phase II KEYNOTE-158 Study. *J. Clin. Oncol.* **2020**, *38*, 1–10. [CrossRef]
22. Reck, M.; Rodríguez-Abreu, D.; Robinson, A.G.; Hui, R.; Csőszi, T.; Fülöp, A.; Gottfried, M.; Peled, N.; Tafreshi, A.; Cuffe, S.; et al. Pembrolizumab versus Chemotherapy for PD-L1–Positive Non–Small-Cell Lung Cancer. *N. Engl. J. Med.* **2016**, *375*, 1823–1833. [CrossRef]
23. Yang, J.D.; Hainaut, P.; Gores, G.J.; Amadou, A.; Plymoth, A.; Roberts, L.R. A Global View of Hepatocellular Carcinoma: Trends, Risk, Prevention and Management. *Nat. Rev. Gastroenterol. Hepatol.* **2019**, *16*, 589–604. [CrossRef]
24. Kanda, T.; Goto, T.; Hirotsu, Y.; Moriyama, M.; Omata, M. Molecular Mechanisms Driving Progression of Liver Cirrhosis towards Hepatocellular Carcinoma in Chronic Hepatitis B and C Infections: A Review. *Int. J. Mol. Sci.* **2019**, *20*, 1358. [CrossRef]

25. Jiang, Z.; Jhunjhunwala, S.; Liu, J.; Haverty, P.M.; Kennemer, M.I.; Guan, Y.; Lee, W.; Carnevali, P.; Stinson, J.; Johnson, S.; et al. The Effects of Hepatitis B Virus Integration into the Genomes of Hepatocellular Carcinoma Patients. *Genome Res.* **2012**, *22*, 593–601. [CrossRef]
26. Jia, L.; Gao, Y.; He, Y.; Hooper, J.D.; Yang, P. HBV Induced Hepatocellular Carcinoma and Related Potential Immunotherapy. *Pharmacol. Res.* **2020**, *159*, 104992. [CrossRef]
27. Neuveut, C.; Wei, Y.; Buendia, M.A. Mechanisms of HBV-Related Hepatocarcinogenesis. *J. Hepatol.* **2010**, *52*, 594–604. [CrossRef]
28. Choudhari, S.R.; Khan, M.A.; Harris, G.; Picker, D.; Jacob, G.S.; Block, T.; Shailubhai, K. Deactivation of Akt and STAT3 Signaling Promotes Apoptosis, Inhibits Proliferation, and Enhances the Sensitivity of Hepatocellular Carcinoma Cells to an Anticancer Agent, Atiprimod. *Mol. Cancer Ther.* **2007**, *6*, 112–121. [CrossRef]
29. Banerjee, A.; Ray, R.B.; Ray, R. Oncogenic Potential of Hepatitis C Virus Proteins. *Viruses* **2010**, *2*, 2108–2133. [CrossRef]
30. Wong, R.J.; Aguilar, M.; Cheung, R.; Perumpail, R.B.; Harrison, S.A.; Younossi, Z.M.; Ahmed, A. Nonalcoholic Steatohepatitis Is the Second Leading Etiology of Liver Disease among Adults Awaiting Liver Transplantation in the United States. *Gastroenterology* **2015**, *148*, 547–555. [CrossRef]
31. Colombo, M.; Lleo, A. The Impact of Antiviral Therapy on Hepatocellular Carcinoma Epidemiology. *Hepat. Oncol.* **2018**, *5*, HEP03. [CrossRef] [PubMed]
32. Wolf, M.J.; Adili, A.; Piotrowitz, K.; Abdullah, Z.; Boege, Y.; Stemmer, K.; Ringelhan, M.; Simonavicius, N.; Egger, M.; Wohlleber, D.; et al. Metabolic Activation of Intrahepatic CD8+ T Cells and NKT Cells Causes Nonalcoholic Steatohepatitis and Liver Cancer via Cross-Talk with Hepatocytes. *Cancer Cell* **2014**, *26*, 549–564. [CrossRef] [PubMed]
33. Ma, C.; Kesarwala, A.H.; Eggert, T.; Medina-Echeverz, J.; Kleiner, D.E.; Jin, P.; Stroncek, D.F.; Terabe, M.; Kapoor, V.; ElGindi, M.; et al. NAFLD Causes Selective CD4+ T Lymphocyte Loss and Promotes Hepatocarcinogenesis. *Nature* **2016**, *531*, 253–257. [CrossRef] [PubMed]
34. Leslie, J.; Mackey, J.B.G.; Jamieson, T.; Ramon-Gil, E.; Drake, T.M.; Fercoq, F.; Clark, W.; Gilroy, K.; Hedley, A.; Nixon, C.; et al. CXCR2 Inhibition Enables NASH-HCC Immunotherapy. *Gut* **2022**. [CrossRef]
35. Abou-Alfa, G.K.; Chan, S.L.; Kudo, M.; Lau, G.; Kelley, R.K.; Furuse, J.; Sukeepaisarnjaroen, W.; Kang, Y.-K.; Dao, T.V.; De Toni, E.N.; et al. Phase 3 Randomized, Open-Label, Multicenter Study of Tremelimumab (T) and Durvalumab (D) as First-Line Therapy in Patients (Pts) with Unresectable Hepatocellular Carcinoma (UHCC): HIMALAYA. *J. Clin. Oncol.* **2022**, *40*, 379. [CrossRef]
36. Reig, M.; Forner, A.; Rimola, J.; Ferrer-Fàbrega, J.; Burrel, M.; Garcia-Criado, Á.; Kelley, R.K.; Galle, P.R.; Mazzaferro, V.; Salem, R.; et al. BCLC Strategy for Prognosis Prediction and Treatment Recommendation: The 2022 Update. *J. Hepatol.* **2022**, *76*, 681–693. [CrossRef]
37. Ren, Z.; Xu, J.; Bai, Y.; Xu, A.; Cang, S.; Du, C.; Li, Q.; Lu, Y.; Chen, Y.; Guo, Y.; et al. Sintilimab plus a Bevacizumab Biosimilar (IBI305) versus Sorafenib in Unresectable Hepatocellular Carcinoma (ORIENT-32): A Randomised, Open-Label, Phase 2–3 Study. *Lancet Oncol.* **2021**, *22*, 977–990. [CrossRef]
38. Osorio, J.C.; Arbour, K.C.; Le, D.T.; Durham, J.N.; Plodkowski, A.J.; Halpenny, D.F.; Ginsberg, M.S.; Sawan, P.; Crompton, J.G.; Yu, H.A.; et al. Lesion-Level Response Dynamics to Programmed Cell Death Protein (PD-1) Blockade. *J. Clin. Oncol.* **2019**, *37*, 3546–3555. [CrossRef]
39. Lu, L.-C.; Hsu, C.; Shao, Y.-Y.; Chao, Y.; Yen, C.-J.; Shih, I.-L.; Hung, Y.-P.; Chang, C.-J.; Shen, Y.-C.; Guo, J.-C.; et al. Differential Organ-Specific Tumor Response to Immune Checkpoint Inhibitors in Hepatocellular Carcinoma. *Liver Cancer* **2019**, *8*, 480–490. [CrossRef]
40. Kim, H.S.; Hong, J.Y.; Cheon, J.; Kim, I.; Kim, C.G.; Kang, B.; Kim, D.J.; Kim, C.; Chon, H.; Choi, H.J.; et al. Different Organ-Specific Response to Nivolumab to Determine the Survival Outcome of Patients with Advanced Hepatocellular Carcinoma (AHCC). *J. Clin. Oncol.* **2020**, *38*, 4584. [CrossRef]
41. Yu, J.; Green, M.D.; Li, S.; Sun, Y.; Journey, S.N.; Choi, J.E.; Rizvi, S.M.; Qin, A.; Waninger, J.J.; Lang, X.; et al. Liver Metastasis Restrains Immunotherapy Efficacy via Macrophage-Mediated T Cell Elimination. *Nat. Med.* **2021**, *27*, 152–164. [CrossRef]
42. Breder, V.V.; Vogel, A.; Merle, P.; Finn, R.S.; Galle, P.R.; Zhu, A.X.; Cheng, A.-L.; Feng, Y.-H.; Li, D.; Gaillard, V.E.; et al. IMbrave150: Exploratory Efficacy and Safety Results of Hepatocellular Carcinoma (HCC) Patients (Pts) with Main Trunk and/or Contralateral Portal Vein Invasion (Vp4) Treated with Atezolizumab (Atezo) + Bevacizumab (Bev) versus Sorafenib (Sor) in a Global Ph III Study. *J. Clin. Oncol.* **2021**, *39*, 4073.
43. Lee, P.-C.; Chao, Y.; Chen, M.-H.; Lan, K.-H.; Lee, C.-J.; Lee, I.-C.; Chen, S.-C.; Hou, M.-C.; Huang, Y.-H. Predictors of Response and Survival in Immune Checkpoint Inhibitor-Treated Unresectable Hepatocellular Carcinoma. *Cancers* **2020**, *12*, 182. [CrossRef]
44. Cheon, J.; Yoo, C.; Hong, J.Y.; Kim, H.S.; Lee, D.-W.; Lee, M.A.; Kim, J.W.; Kim, I.; Oh, S.-B.; Hwang, J.-E.; et al. Efficacy and Safety of Atezolizumab plus Bevacizumab in Korean Patients with Advanced Hepatocellular Carcinoma. *Liver Int.* **2022**, *42*, 674–681. [CrossRef]
45. Zhu, A.X.; Dayyani, F.; Yen, C.-J.; Ren, Z.; Bai, Y.; Meng, Z.; Pan, H.; Dillon, P.; Mhatre, S.K.; Gaillard, V.E.; et al. Alpha-Fetoprotein as a Potential Surrogate Biomarker for Atezolizumab + Bevacizumab Treatment of Hepatocellular Carcinoma. *Clin. Cancer Res.* **2022**. [CrossRef]
46. Valero, C.; Lee, M.; Hoen, D.; Weiss, K.; Kelly, D.W.; Adusumilli, P.S.; Paik, P.K.; Plitas, G.; Ladanyi, M.; Postow, M.A.; et al. Pretreatment Neutrophil-to-Lymphocyte Ratio and Mutational Burden as Biomarkers of Tumor Response to Immune Checkpoint Inhibitors. *Nat. Commun.* **2021**, *12*, 729. [CrossRef]

47. Ohki, S.; Shibata, M.; Gonda, K.; Machida, T.; Shimura, T.; Nakamura, I.; Ohtake, T.; Koyama, Y.; Suzuki, S.; Ohto, H.; et al. Circulating Myeloid-Derived Suppressor Cells Are Increased and Correlate to Immune Suppression, Inflammation and Hypoproteinemia in Patients with Cancer. *Oncol. Rep.* **2012**, *28*, 453–458. [CrossRef]
48. Gonzalez, H.; Hagerling, C.; Werb, Z. Roles of the Immune System in Cancer: From Tumor Initiation to Metastatic Progression. *Genes Dev.* **2018**, *32*, 1267–1284. [CrossRef]
49. Stone, R.L.; Nick, A.M.; McNeish, I.A.; Balkwill, F.; Han, H.D.; Bottsford-Miller, J.; Rupairmoole, R.; Armaiz-Pena, G.N.; Pecot, C.V.; Coward, J.; et al. Paraneoplastic Thrombocytosis in Ovarian Cancer. *N. Engl. J. Med.* **2012**, *366*, 610–618. [CrossRef]
50. Dharmapuri, S.; Özbek, U.; Lin, J.-Y.; Sung, M.; Schwartz, M.; Branch, A.D.; Ang, C. Predictive Value of Neutrophil to Lymphocyte Ratio and Platelet to Lymphocyte Ratio in Advanced Hepatocellular Carcinoma Patients Treated with Anti-PD-1 Therapy. *Cancer Med.* **2020**, *9*, 4962–4970. [CrossRef]
51. Kim, C.G.; Kim, C.; Yoon, S.E.; Kim, K.H.; Choi, S.J.; Kang, B.; Kim, H.R.; Park, S.-H.; Shin, E.-C.; Kim, Y.-Y.; et al. Hyperprogressive Disease during PD-1 Blockade in Patients with Advanced Hepatocellular Carcinoma. *J. Hepatol.* **2021**, *74*, 350–359. [CrossRef]
52. Hong, Y.M.; Yoon, K.T.; Hwang, T.H.; Heo, J.; Woo, H.Y.; Cho, M. Changes in the Neutrophil-to-Lymphocyte Ratio Predict the Prognosis of Patients with Advanced Hepatocellular Carcinoma Treated with Sorafenib. *Eur. J. Gastroenterol. Hepatol.* **2019**, *31*, 1250–1255. [CrossRef]
53. Johnson, P.J.; Dhanaraj, S.; Berhane, S.; Bonnett, L.; Ma, Y.T. The Prognostic and Diagnostic Significance of the Neutrophil-to-Lymphocyte Ratio in Hepatocellular Carcinoma: A Prospective Controlled Study. *Br. J. Cancer* **2021**, *125*, 714–716. [CrossRef]
54. Lopes, G.; Wu, Y.-L.; Kudaba, I.; Kowalski, D.; Cho, B.C.; Castro, G.; Srimuninnimit, V.; Bondarenko, I.; Kubota, K.; Lubiniecki, G.M.; et al. Pembrolizumab (Pembro) versus Platinum-Based Chemotherapy (Chemo) as First-Line Therapy for Advanced/Metastatic NSCLC with a PD-L1 Tumor Proportion Score (TPS) $\geq$ 1%: Open-Label, Phase 3 KEYNOTE-042 Study. *J. Clin. Oncol.* **2018**, *36*, LBA4. [CrossRef]
55. Wu, Y.; Chen, W.; Xu, Z.P.; Gu, W. PD-L1 Distribution and Perspective for Cancer Immunotherapy-Blockade, Knockdown, or Inhibition. *Front. Immunol.* **2019**, *10*, 2022. [CrossRef]
56. Pinato, D.J.; Mauri, F.A.; Spina, P.; Cain, O.; Siddique, A.; Goldin, R.; Victor, S.; Pizio, C.; Akarca, A.U.; Boldorini, R.L.; et al. Clinical Implications of Heterogeneity in PD-L1 Immunohistochemical Detection in Hepatocellular Carcinoma: The Blueprint-HCC Study. *Br. J. Cancer* **2019**, *120*, 1033–1036. [CrossRef]
57. Gao, Q.; Wang, X.-Y.; Qiu, S.-J.; Yamato, I.; Sho, M.; Nakajima, Y.; Zhou, J.; Li, B.-Z.; Shi, Y.-H.; Xiao, Y.-S.; et al. Overexpression of PD-L1 Significantly Associates with Tumor Aggressiveness and Postoperative Recurrence in Human Hepatocellular Carcinoma. *Clin. Cancer Res.* **2009**, *15*, 971–979. [CrossRef]
58. Qin, S.; Ren, Z.; Meng, Z.; Chen, Z.; Chai, X.; Xiong, J.; Bai, Y.; Yang, L.; Zhu, H.; Fang, W.; et al. Camrelizumab in Patients with Previously Treated Advanced Hepatocellular Carcinoma: A Multicentre, Open-Label, Parallel-Group, Randomised, Phase 2 Trial. *Lancet Oncol.* **2020**, *21*, 571–580. [CrossRef]
59. Zhu, A.X.; Guan, Y.; Abbas, A.R.; Koeppen, H.; Lu, S.; Hsu, C.-H.; Lee, K.-H.; Lee, M.S.; He, A.R.; Mahipal, A.; et al. Abstract CT044: Genomic correlates of clinical benefits from atezolizumab combined with bevacizumab vs. atezolizumab alone in patients with advanced hepatocellular carcinoma (HCC). In Proceedings of the Annual Meeting of the American Association for Cancer Research, Philadelphia, PA, USA, 15 August 2020.
60. Cheng, A.-L.; Qin, S.; Ikeda, M.; Galle, P.R.; Ducreux, M.; Kim, T.-Y.; Lim, H.Y.; Kudo, M.; Breder, V.; Merle, P.; et al. Updated Efficacy and Safety Data from IMbrave150: Atezolizumab plus Bevacizumab vs. Sorafenib for Unresectable Hepatocellular Carcinoma. *J. Hepatol.* **2022**, *76*, 862–873. [CrossRef]
61. Harlin, H. Chemokine Expression in Melanoma Metastases Associated with CD81 T-Cell Recruitment. *Cancer Res.* **2009**, *69*, 3077–3085. [CrossRef]
62. Sia, D.; Jiao, Y.; Martinez-Quetglas, I.; Kuchuk, O.; Villacorta-Martin, C.; Castro de Moura, M.; Putra, J.; Camprecios, G.; Bassaganyas, L.; Akers, N.; et al. Identification of an Immune-Specific Class of Hepatocellular Carcinoma, Based on Molecular Features. *Gastroenterology* **2017**, *153*, 812–826. [CrossRef] [PubMed]
63. Montironi, C.; Castet, F.; Haber, P.K.; Pinyol, R.; Torres-Martin, M.; Torrens, L.; Mesropian, A.; Wang, H.; Puigvehi, M.; Maeda, M.; et al. Inflamed and Non-Inflamed Classes of HCC: A Revised Immunogenomic Classification. *Gut* **2022**. [CrossRef] [PubMed]
64. Galon, J.; Costes, A.; Sanchez-Cabo, F.; Kirilovsky, A.; Mlecnik, B.; Lagorce-Pagès, C.; Tosolini, M.; Camus, M.; Berger, A.; Wind, P.; et al. Type, Density, and Location of Immune Cells within Human Colorectal Tumors Predict Clinical Outcome. *Science* **2006**, *313*, 1960–1964. [CrossRef] [PubMed]
65. Cipponi, A.; Wieers, G.; van Baren, N.; Coulie, P.G. Tumor-Infiltrating Lymphocytes: Apparently Good for Melanoma Patients. But Why? *Cancer Immunol. Immunother.* **2011**, *60*, 1153–1160. [CrossRef] [PubMed]
66. Duffy, A.G.; Ulahannan, S.V.; Makorova-Rusher, O.; Rahma, O.; Wedemeyer, H.; Pratt, D.; Davis, J.L.; Hughes, M.S.; Heller, T.; ElGindi, M.; et al. Tremelimumab in Combination with Ablation in Patients with Advanced Hepatocellular Carcinoma. *J. Hepatol.* **2017**, *66*, 545–551. [CrossRef]
67. Sangro, B.; Melero, I.; Wadhawan, S.; Finn, R.S.; Abou-Alfa, G.K.; Cheng, A.-L.; Yau, T.; Furuse, J.; Park, J.-W.; Boyd, Z.; et al. Association of Inflammatory Biomarkers with Clinical Outcomes in Nivolumab-Treated Patients with Advanced Hepatocellular Carcinoma. *J. Hepatol.* **2020**, *73*, 1460–1469. [CrossRef]
68. Tumeh, P.C.; Harview, C.L.; Yearley, J.H.; Shintaku, I.P.; Taylor, E.J.M.; Robert, L.; Chmielowski, B.; Spasic, M.; Henry, G.; Ciobanu, V.; et al. PD-1 Blockade Induces Responses by Inhibiting Adaptive Immune Resistance. *Nature* **2014**, *515*, 568–571. [CrossRef]

69. Rooney, M.S.; Shukla, S.A.; Wu, C.J.; Getz, G.; Hacohen, N. Molecular and Genetic Properties of Tumors Associated with Local Immune Cytolytic Activity. *Cell* **2015**, *160*, 48–61. [CrossRef]
70. Ayers, M.; Lunceford, J.; Nebozhyn, M.; Murphy, E.; Loboda, A.; Kaufman, D.R.; Albright, A.; Cheng, J.D.; Kang, S.P.; Shankaran, V.; et al. IFN-γ-Related MRNA Profile Predicts Clinical Response to PD-1 Blockade. *J. Clin. Investig.* **2017**, *127*, 2930–2940. [CrossRef]
71. Marabelle, A.; Fakih, M.; Lopez, J.; Shah, M.; Shapira-Frommer, R.; Nakagawa, K.; Chung, H.C.; Kindler, H.L.; Lopez-Martin, J.A.; Miller, W.H., Jr.; et al. Association of Tumour Mutational Burden with Outcomes in Patients with Advanced Solid Tumours Treated with Pembrolizumab: Prospective Biomarker Analysis of the Multicohort, Open-Label, Phase 2 KEYNOTE-158 Study. *Lancet Oncol.* **2020**, *21*, 1353–1365. [CrossRef]
72. Alexandrov, L.B.; Australian Pancreatic Cancer Genome Initiative; Nik-Zainal, S.; Wedge, D.C.; Aparicio, S.A.J.R.; Behjati, S.; Biankin, A.V.; Bignell, G.R.; Bolli, N.; Borg, A.; et al. Signatures of Mutational Processes in Human Cancer. *Nature* **2013**, *500*, 415–421. [CrossRef]
73. Ang, C.; Klempner, S.J.; Ali, S.M.; Madison, R.; Ross, J.S.; Severson, E.A.; Fabrizio, D.; Goodman, A.; Kurzrock, R.; Suh, J.; et al. Prevalence of Established and Emerging Biomarkers of Immune Checkpoint Inhibitor Response in Advanced Hepatocellular Carcinoma. *Oncotarget* **2019**, *10*, 4018–4025. [CrossRef]
74. Goumard, C.; Desbois-Mouthon, C.; Wendum, D.; Calmel, C.; Merabtene, F.; Scatton, O.; Praz, F. Low Levels of Microsatellite Instability at Simple Repeated Sequences Commonly Occur in Human Hepatocellular Carcinoma. *Cancer Genom. Proteom.* **2017**, *14*, 329–339.
75. Zucman-Rossi, J.; Villanueva, A.; Nault, J.-C.; Llovet, J.M. Genetic Landscape and Biomarkers of Hepatocellular Carcinoma. *Gastroenterology* **2015**, *149*, 1226–1239.e4. [CrossRef]
76. Schulze, K.; Imbeaud, S.; Letouzé, E.; Alexandrov, L.B.; Calderaro, J.; Rebouissou, S.; Couchy, G.; Meiller, C.; Shinde, J.; Soysouvanh, F.; et al. Exome Sequencing of Hepatocellular Carcinomas Identifies New Mutational Signatures and Potential Therapeutic Targets. *Nat. Genet.* **2015**, *47*, 505–511. [CrossRef]
77. Gao, C.; Wang, Y.; Broaddus, R.; Sun, L.; Xue, F.; Zhang, W. Exon 3 Mutations of CTNNB1 Drive Tumorigenesis: A Review. *Oncotarget* **2018**, *9*, 5492–5508. [CrossRef]
78. Li, W.; Wang, H.; Ma, Z.; Zhang, J.; Ou-Yang, W.; Qi, Y.; Liu, J. Multi-Omics Analysis of Microenvironment Characteristics and Immune Escape Mechanisms of Hepatocellular Carcinoma. *Front. Oncol.* **2019**, *9*, 1019. [CrossRef]
79. Spranger, S.; Bao, R.; Gajewski, T.F. Melanoma-Intrinsic β-Catenin Signalling Prevents Anti-Tumour Immunity. *Nature* **2015**, *523*, 231–235. [CrossRef]
80. Luke, J.J.; Bao, R.; Sweis, R.F.; Spranger, S.; Gajewski, T.F. WNT/β-Catenin Pathway Activation Correlates with Immune Exclusion across Human Cancers. *Clin. Cancer Res.* **2019**, *25*, 3074–3083. [CrossRef]
81. Spranger, S.; Gajewski, T.F. A New Paradigm for Tumor Immune Escape: β-Catenin-Driven Immune Exclusion. *J. Immunother. Cancer* **2015**, *3*, 43. [CrossRef]
82. Spranger, S.; Dai, D.; Horton, B.; Gajewski, T.F. Tumor-Residing Batf3 Dendritic Cells Are Required for Effector T Cell Trafficking and Adoptive T Cell Therapy. *Cancer Cell* **2017**, *31*, 711–723.e4. [CrossRef]
83. Ruiz de Galarreta, M.; Bresnahan, E.; Molina-Sánchez, P.; Lindblad, K.E.; Maier, B.; Sia, D.; Puigvehi, M.; Miguela, V.; Casanova-Acebes, M.; Dhainaut, M.; et al. Β-Catenin Activation Promotes Immune Escape and Resistance to Anti-PD-1 Therapy in Hepatocellular Carcinoma. *Cancer Discov.* **2019**, *9*, 1124–1141. [CrossRef]
84. Harding, J.J.; Nandakumar, S.; Armenia, J.; Khalil, D.N.; Albano, M.; Ly, M.; Shia, J.; Hechtman, J.F.; Kundra, R.; El Dika, I.; et al. Prospective Genotyping of Hepatocellular Carcinoma: Clinical Implications of next-Generation Sequencing for Matching Patients to Targeted and Immune Therapies. *Clin. Cancer Res.* **2019**, *25*, 2116–2126. [CrossRef]
85. Hong, J.Y.; Cho, H.J.; Sa, J.K.; Liu, X.; Ha, S.Y.; Lee, T.; Kim, H.; Kang, W.; Sinn, D.H.; Gwak, G.-Y.; et al. Hepatocellular Carcinoma Patients with High Circulating Cytotoxic T Cells and Intra-Tumoral Immune Signature Benefit from Pembrolizumab: Results from a Single-Arm Phase 2 Trial. *Genome Med.* **2022**, *14*, 1. [CrossRef]
86. Haber, P.K.; Torres-Martin, M.; Dufour, J.-F.; Verslype, C.; Marquardt, J.; Galle, P.R.; Vogel, A.; Meyer, T.; Labgaa, I.; Roberts, L.R.; et al. Molecular Markers of Response to Anti-PD1 Therapy in Advanced Hepatocellular Carcinoma. *J. Clin. Oncol.* **2021**, *39*, 4100. [CrossRef]
87. Hou, M.-M.; Rau, K.-M.; Kang, Y.-K.; Lee, J.-S.; Pan, H.; Yuan, Y.; Yu, C.; Zhang, Y.; Ma, X.; Wu, X.; et al. 77 Association between programmed death-ligand 1 (PD-L1) expression and gene signatures of response or resistance to tislelizumab monotherapy in hepatocellular carcinoma (HCC). *J. ImmunoTher. Cancer* **2020**, *8*.
88. Feun, L.G.; Li, Y.-Y.; Wu, C.; Wangpaichitr, M.; Jones, P.D.; Richman, S.P.; Madrazo, B.; Kwon, D.; Garcia-Buitrago, M.; Martin, P.; et al. Phase 2 Study of Pembrolizumab and Circulating Biomarkers to Predict Anticancer Response in Advanced, Unresectable Hepatocellular Carcinoma. *Cancer* **2019**, *125*, 3603–3614. [CrossRef]
89. Winograd, P.; Hou, S.; Court, C.M.; Lee, Y.-T.; Chen, P.-J.; Zhu, Y.; Sadeghi, S.; Finn, R.S.; Teng, P.-C.; Wang, J.J.; et al. Hepatocellular Carcinoma-Circulating Tumor Cells Expressing PD-L1 Are Prognostic and Potentially Associated with Response to Checkpoint Inhibitors. *Hepatol. Commun.* **2020**, *4*, 1527–1540. [CrossRef]
90. Davda, J.; Declerck, P.; Hu-Lieskovan, S.; Hickling, T.P.; Jacobs, I.A.; Chou, J.; Salek-Ardakani, S.; Kraynov, E. Immunogenicity of Immunomodulatory, Antibody-Based, Oncology Therapeutics. *J. Immunother. Cancer* **2019**, *7*, 105. [CrossRef]

91. Vaisman-Mentesh, A.; Gutierrez-Gonzalez, M.; DeKosky, B.J.; Wine, Y. The Molecular Mechanisms That Underlie the Immune Biology of Anti-Drug Antibody Formation Following Treatment with Monoclonal Antibodies. *Front. Immunol.* **2020**, *11*, 1951. [CrossRef]
92. Pratt, K.P. Anti-Drug Antibodies: Emerging Approaches to Predict, Reduce or Reverse Biotherapeutic Immunogenicity. *Antibodies* **2018**, *7*, 19. [CrossRef] [PubMed]
93. Agrawal, S.; Statkevich, P.; Bajaj, G.; Feng, Y.; Saeger, S.; Desai, D.D.; Park, J.-S.; Waxman, I.M.; Roy, A.; Gupta, M. Evaluation of Immunogenicity of Nivolumab Monotherapy and Its Clinical Relevance in Patients with Metastatic Solid Tumors. *J. Clin. Pharmacol.* **2017**, *57*, 394–400. [CrossRef] [PubMed]
94. McDermott, D.F.; Huseni, M.A.; Atkins, M.B.; Motzer, R.J.; Rini, B.I.; Escudier, B.; Fong, L.; Joseph, R.W.; Pal, S.K.; Reeves, J.A.; et al. Clinical Activity and Molecular Correlates of Response to Atezolizumab Alone or in Combination with Bevacizumab versus Sunitinib in Renal Cell Carcinoma. *Nat. Med.* **2018**, *24*, 749–757. [CrossRef] [PubMed]
95. Kverneland, A.H.; Enevold, C.; Donia, M.; Bastholt, L.; Svane, I.M.; Nielsen, C.H. Development of Anti-Drug Antibodies Is Associated with Shortened Survival in Patients with Metastatic Melanoma Treated with Ipilimumab. *Oncoimmunology* **2018**, *7*, e1424674. [CrossRef]
96. Hammer, C.; Ruppel, J.; Hunkapiller, J.; Mellman, I.; Quarmby, V. Allelic Variation in HLA-DRB1 Is Associated with Development of Anti-Drug Antibodies in Cancer Patients Treated with Atezolizumab That Are Neutralizing in Vitro. *bioRxiv* **2021**. [CrossRef]
97. Galle, P.R.; Finn, R.S.; Cheng, A.-L.; Bernaards, C.; Shemesh, C.S.; Vilimovskij, A.; Verret, W.J.; Stanzel, S.F.; Ma, N.; Ducreux, M.; et al. Abstract CT185: Assessment of the Impact of Anti-Drug Antibodies on PK and Clinical Outcomes with Atezolizumab + Bevacizumab in HCC. *Cancer Res.* **2021**, *81*, CT185. [CrossRef]
98. Casak, S.J.; Donoghue, M.; Fashoyin-Aje, L.; Jiang, X.; Rodriguez, L.; Shen, Y.L.; Xu, Y.; Jiang, X.; Liu, J.; Zhao, H.; et al. FDA Approval Summary: Atezolizumab Plus Bevacizumab for the Treatment of Patients with Advanced Unresectable or Metastatic Hepatocellular Carcinoma. *Clin. Cancer Res.* **2021**, *27*, 1836–1841. [CrossRef]

cancers

MDPI

*Article*

# Cytological Comparison between Hepatocellular Carcinoma and Intrahepatic Cholangiocarcinoma by Image Analysis Software Using Touch Smear Samples of Surgically Resected Specimens

**Sho Kitamura [1], Keita Kai [1,*], Mitsuo Nakamura [1], Tomokazu Tanaka [2], Takao Ide [2], Hirokazu Noshiro [2], Eisaburo Sueoka [3] and Shinich Aishima [1,4]**

[1] Department of Pathology, Saga University Hospital, Saga 849-8501, Japan; sn5388@cc.saga-u.ac.jp (S.K.); nakamuml@cc.saga-u.ac.jp (M.N.); saish@cc.saga-u.ac.jp (S.A.)
[2] Department of Surgery, Saga University Faculty of Medicine, Saga 849-8501, Japan; f8642@cc.saga-u.ac.jp (T.T.); idetaka@cc.saga-u.ac.jp (T.I.); noshiro@cc.saga-u.ac.jp (H.N.)
[3] Department of Clinical Laboratory Medicine, Saga University Faculty of Medicine, Saga 849-8501, Japan; sueokae@cc.saga-u.ac.jp
[4] Department of Pathology and Microbiology, Saga University Faculty of Medicine, Saga 849-8501, Japan
* Correspondence: kaikeit@cc.saga-u.ac.jp; Tel.: +81-952-34-3264

**Simple Summary:** This study cytologically compared hepatocellular carcinoma (HCC) and intrahepatic cholangiocarcinoma (ICC) using image analyzing software. The results indicated that the major/minor axis ratio of ICC was significantly larger than in HCC in Papanicolaou staining. This difference was consistently observed in clinical samples of cytology such as fine-needle aspiration, brushing and ascites. This study indicated a significant difference in the nuclear morphology of HCC (round shape) and ICC (oval shape) in Papanicolaou-stained cytology specimens. This simple and objective finding is considered to be useful for differential cytodiagnosis of HCC and ICC.

**Abstract:** To investigate useful cytological features for differential diagnosis of hepatocellular carcinoma (HCC) and intrahepatic cholangiocarcinoma (ICC), this study cytologically compared HCC to ICC using image analysis software. Touch smear specimens of surgically resected specimens were obtained from a total of 61 nodules of HCC and 16 of ICC. The results indicated that the major/minor axis ratio of ICC is significantly larger than that of HCC (1.67 ± 0.27 vs. 1.32 ± 0.11, $p < 0.0001$) in Papanicolaou staining. This result means that the nucleus of HCC is close to round and the nucleus of ICC is close to an oval. This significant difference in the major/minor axis ratio between ICC and HCC was consistently observed by the same analyses using clinical samples of cytology (4 cases of HCC and 13 cases of ICC) such a fine-needle aspiration, brushing and ascites (ICC: 1.45 ± 0.13 vs. HCC: 1.18 ± 0.056, $p = 0.004$). We also confirmed that nuclear position center-positioned nucleus ($p < 0.0001$) and granular cytoplasm ($p < 0.0001$) are typical features of HCC tumor cells compared to ICC tumor cells. The research study found a significant difference in the nuclear morphology of HCC (round shape) and ICC (oval shape) in Papanicolaou-stained cytology specimens. This simple and objective finding will be very useful for the differential cytodiagnosis of HCC and ICC.

**Keywords:** hepatocellular carcinoma; cholangiocarcinoma; cytology; touch smear; nuclear atypia

**Citation:** Kitamura, S.; Kai, K.; Nakamura, M.; Tanaka, T.; Ide, T.; Noshiro, H.; Sueoka, E.; Aishima, S. Cytological Comparison between Hepatocellular Carcinoma and Intrahepatic Cholangiocarcinoma by Image Analysis Software Using Touch Smear Samples of Surgically Resected Specimens. *Cancers* **2022**, *14*, 2301. https://doi.org/10.3390/cancers14092301

Academic Editor: Georgios Germanidis

Received: 11 April 2022
Accepted: 4 May 2022
Published: 5 May 2022



## 1. Introduction

In Japan, 94.0% of primary liver cancers are hepatocellular carcinoma (HCC), 4.4% intrahepatic cholangiocarcinoma (ICC), and the remaining small proportion includes combined hepatocellular and cholangiocarcinoma or other rare tumors [1]. Although ICC is generally considered to be a rare tumor, it has been found that the prevalence of ICC is relatively high in limited regions worldwide including Japan and India [2]. It has also been reported that the development of ICC is related to hepatitis B and C virus infections, similar to that of HCC [3,4].

The usefulness of cytology for the investigation of hepatic nodules involving hepatocellular carcinoma (HCC) and intrahepatic cholangiocarcinoma (ICC) is limited at present. The development of imaging modalities, such as multiphasic computed tomography (CT), magnetic resonance imaging (MRI), and positron emission tomography (PET)-CT permits a noninvasive clinical diagnosis of classical HCC or ICC [5–7]. The cases with typical imaging would not require an invasive procedure, such as liver biopsy or fine-needle aspiration cytology (FNAC).

Liver biopsy or FNAC are generally considered useful diagnosis techniques for hepatic nodules without specific imaging features or for difficult cases of imaging diagnosis, such as small nodules (<2 cm) or indistinct nodules in cirrhotic patients [8]. Even in such cases, biopsy specimens are considered to be superior to FNAC samples from the viewpoint of tumor volume assessment and availability for immunohistochemistry. Therefore, FNAC for liver nodule investigation is rarely performed. However, an opportunity to distinguish between HCC and ICC on cytological material is encountered in greatly advanced/unresectable cases, in cases where the patient's general condition is too poor to undergo a liver biopsy, and in cases with ascites or pleural fluid. Considering that the therapeutic strategy is different for HCC and ICC, it is important to distinguish HCC from ICC even with cytology alone.

The cytology of ICC is relatively familiar to cytopathologists compared to HCC because bile juice and brushings, which are obtained by endoscopic retrograde cholangiopancreatography or the biliary drainage route, are frequently submitted for cytological diagnosis [9,10]. Although the cytology of HCC is clinically rare, many studies have reported that the cytology of HCC is mostly obtained by FNAC [11–22]. These FNAC studies focused on cytologic differences between HCC and non-neoplastic hepatocytes or metastatic lesions. However, a definitive cytologic difference between HCC and ICC has not been well defined by researchers and therefore little knowledge has been accumulated regarding the differential diagnosis of HCC and ICC.

The aim of the present study was to investigate the cytological features that can facilitate the differential diagnosis of HCC and ICC, using touch smear samples of resected specimens and morphological analyses using image-analysis software.

## 2. Materials and Methods

### *2.1. Touch Smear Cytology*

Seventy-seven hepatic nodules of 71 patients who underwent surgical resection in Saga University Hospital between 2010 to 2012 and 2020 to 2021 after a clinical diagnosis of HCC or ICC were enrolled in the study. Six patients underwent hepatic resection for 2 HCC nodules during 1 operation. The touch smears of hepatic nodules were obtained from fresh cut surfaces of resected specimens and then subjected to Giemsa and Papanicolaou staining. The details of the hepatic nodules were: HCC, 61 nodules (well differentiated; 8, moderately differentiated; 48, poorly differentiated; 5) and 16 ICC nodules (Table 1). The differentiation of HCC depended on pathological reports. Giemsa-stained touch smears were unavailable for 20 nodules. As a control sample, the touch smears (Papanicolaou staining only) of non-tumorous background liver were obtained from 5 cases. Comprehensive informed consent for the use of resected tissue for research was obtained from all patients, and the study protocol was approved by the Ethics Committee of Saga University Hospital (No. 2018-12-R-11).

**Table 1.** Details of Touch Smear Cytology of Resected Specimens.

| | Papanicolaou | Giemsa |
|---|---|---|
| Hepatocellular carcinoma | 61 | 47 |
| Well differentiated | 8 | 4 |
| Moderately differentiated | 48 | 40 |
| Poorly differentiated | 5 | 3 |
| Intrahepatic cholangiocarcinoma | 16 | 10 |
| Non-tumorous liver tissue | 5 | 0 |

### 2.2. Analysis of Hematoxylin and Eosin-Stained Tissue and Giemsa-Stained Touch Smear Cytology Using Imaging Software

To compare cytological findings, hematoxylin-eosin (HE) stained formalin-fixed paraffin-embedded resected specimens sliced into 4 μm sections were also analyzed in the present study. Three digital images of tumor tissue (×200) of HCC and ICC were analyzed using the imaging analysis software Tissue Studio (Definiens, München, Germany) and data on the major axis, minor axis and the area of the nucleus were automatically calculated (Figure 1a–d). In the same way, three digital images of Giemsa-stained touch smear cytology (×200) of HCC and ICC were also analyzed using Tissue Studio (Figure 2a,b). Each mean value of the major axis, minor axis, major/minor axis ratio and the area of the nuclei were statistically compared between HCC and ICC specimens.

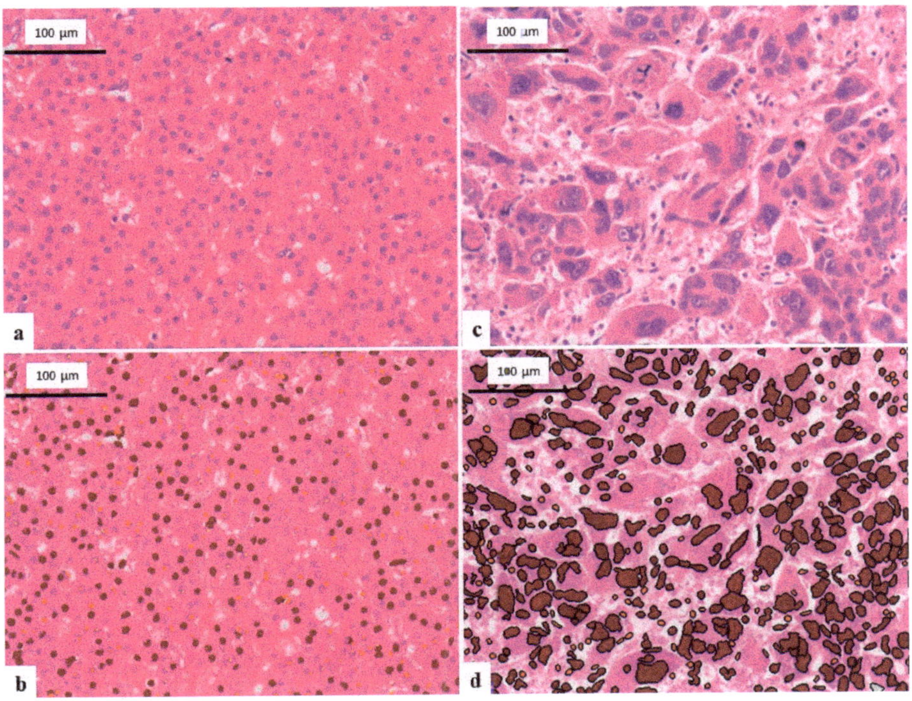


**Figure 1.** Analyzing images of HE-stained tissue section using Tissue Studio. (**a**) Image of well differentiated HCC (×200). (**b**) Analyzing image of Figure 1a. The software appropriately recognizes the nuclei of tumor cells and calculate major axis, minor axis, and area of the nucleus. (**c**) Image of poorly differentiated HCC (×200). The tumor cells are significantly larger than well differentiated HCC. (**d**) Analyzing image of Figure 1c. The software appropriately recognizes the nuclei of tumor cells although the nuclei are markedly pleomorphic.

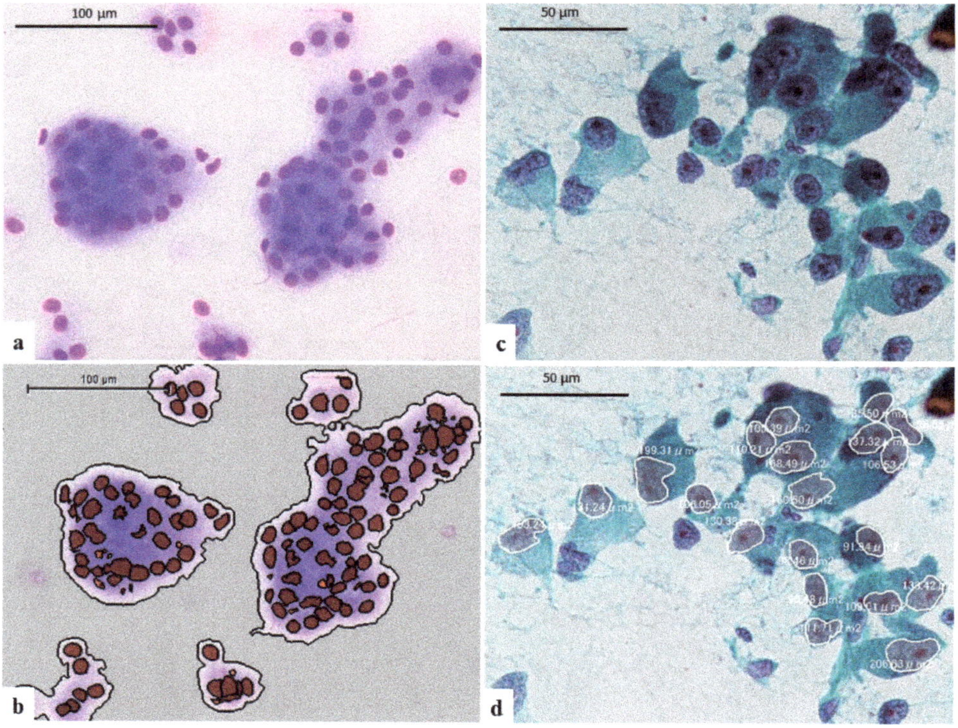


**Figure 2.** (**a**) The image of Giemsa-stained touch smear cytology of well differentiated HCC (×200, same case of Figure 1a). (**b**) Analyzing image of Figure 2a. The software Tissue Studio appropriately recognize the nuclei of tumor cells. (**c**) The image of Papanicolaou-stained touch smear cytology of poorly differentiated HCC (×200, same case of Figure 1c). (**d**) Analyzing image of Figure 2c. The nuclei were manually selected and then of the major axis, minor axis, and area of the nucleus were calculated by attaching software of EXpath III.

### *2.3. Analysis of Papanicolaou-Stained Touch Smear Cytology Specimens Using Imaging Software*

As the software Tissue Studio did not support Papanicolaou staining, analyses of Papanicolaou-stained touch smear cytology were performed using attaching software of EXpath III (INTEC, Toyama, Japan). Three digital images of Papanicolaou-stained touch smear cytology specimens (×200) of HCC and ICC were subjected to the analyses. In addition, Papanicolaou staining of touch smear cytology of non-tumorous hepatocytes was also analyzed. Twenty nuclei in each image were manually selected and then the major axis, minor axis and area of the nucleus were calculated (Figure 2c,d). Each mean value of the major axis, minor axis, major/minor axis ratio and the area of the nucleus were statistically compared for HCC, ICC, and non-tumorous hepatocytes.

### *2.4. Assessment of Cytological Findings of HCC and ICC in Papanicolaou-Stained Touch Smear Cytology Specimens*

The cytological findings of HCC and ICC in Papanicolaou-stained touch smear cytology specimens were assessed by two authors (SK and KK) after discussion following observations under multi-headed microscope images. The following cytological findings were evaluated and categorized into several two titer classifications: nuclear contours (irregular vs. smooth), chromatin pattern (coarse/granular vs. fine), chromatin distribution (homogeneous vs. heterogeneous), nuclear position (center vs. uncentre), number of the nucleolus (single/unclear vs. multiple), cytoplasm (vacuole/foamy vs. granular) and cell boundaries (clear vs. unclear). Each finding was statistically compared between HCC and ICC.

### 2.5. Clinical Materials of Cytology for Validation

As the condition of clinical samples (such as bile juice, brushings, ascites, and fine-needle aspiration may be different from touch smear cytology, we performed the analysis of clinical materials of cytology for validation of results obtained by touch smear cytology of resected samples. A total of 17 samples in which tumor cells of HCC or ICC appeared were found (HCC: four cases, ICC: 13 cases) among 56,383 cytological samples examined at Saga University Hospital between 2014 and 2021. Each mean value of the major axis, minor axis, major/minor axis ratio and the area of the nuclei of tumor cells in Papanicolaou-stained specimens were statistically compared between HCC and ICC.

### 2.6. Statistical Analysis

The statistical analysis was performed using JMP (ver. 15.2 software, AS Institute, Cary, NC, USA). The comparisons of pairs of groups were performed using the Wilcoxon test or Fisher's exact test (two-sided). Values of $p < 0.05$ were considered to be statistically significant findings.

## 3. Results

### 3.1. Comparison of Nuclei among HCC and ICC in HE-Stained Tissue Specimens

The results of comparisons of nuclei between HCC and ICC in each stained section are summarized in Table 2. In the HE-stained tissue specimens, the means and standard deviation (SD) of nuclei in HCC and ICC were evaluated thus: Major axis: HCC: 11.52 ± 2.98 μm vs. ICC: 14.12 ± 2.05 μm ($p$ = 0.0003), Minor axis: HCC: 8.64 ± 1.91 μm vs. ICC: 9.45 ± 1.23 μm ($p$ = 0.031), major/minor axis ratio: HCC: 1.36 ± 0.092 vs. ICC: 1.54 ± 0.083 ($p$ < 0.0001), Nucleus area: HCC: 77.42 ± 38.44 μm$^2$ vs. 93.64 ± 21.89 μm$^2$ ($p$ = 0.0099). The nuclei of ICC were significantly larger and oval-shaped rather than exhibiting a round shape like HCC found in HE-stained tissue specimens.

**Table 2.** Comparison of the Nucleus between HCC and ICC for Each Stain.

| | HE | | | Giemsa | | | Papanicolaou | | |
|---|---|---|---|---|---|---|---|---|---|
| | HCC | ICC | *p* | HCC | ICC | *p* | HCC | ICC | *p* |
| Major axis (mean ± SD, μm) | 11.52 ± 2.98 | 14.12 ± 2.05 | 0.0003 | 16.42 ± 4.47 | 18.46 ± 3.90 | 0.26 | 6.89 ± 2.47 | 8.60 ± 3.06 | 0.057 |
| Minor axis (mean ± SD, μm) | 8.64 ± 1.91 | 9.45 ± 1.23 | 0.031 | 12.04 ± 2.93 | 13.10 ± 2.26 | 0.29 | 5.22 ± 1.79 | 5.22 ± 1.95 | 0.940 |
| Major/Minor axis ratio (mean ± SD) | 1.36 ± 0.092 | 1.54 ± 0.083 | <0.0001 | 1.39 ± 0.10 | 1.44 ± 0.12 | 0.38 | 1.32 ± 0.11 | 1.67 ± 0.27 | <0.0001 |
| Nucleus area (mean ± SD, μm$^2$) | 77.42 ± 38.44 | 93.64 ± 21.89 | 0.0099 | 146.78 ± 68.40 | 176.32 ± 58.20 | 0.14 | 43.80 ± 27.31 | 55.37 ± 35.34 | 0.350 |

HCC: hepatocellular carcinoma, ICC: intrahepatic cholangiocarcinoma, SD: standard deviation.

### 3.2. Comparison of Nuclei among HCC and ICC in Giemsa-Stained Touch Smear Cytology Specimens

In Giemsa-staining, means and SD of major axis, means of minor axis, major axis/minor axis ratio and area of nuclei were: major axis: HCC, 16.42 ± 4.47 μm vs. ICC; 18.46 ± 3.90 μm ($p$ = 0.26), minor axis: HCC, 12.04 ± 2.93 μm vs. ICC 13.10 ± 2.26 μm ($p$ = 0.29); major/minor axis ratio: HCC: 1.39 ± 0.10 vs. ICC; 1.44 ± 0.12 ($p$ = 0.38); Nucleus area: HCC, 146.78 ± 68.40 μm$^2$ vs. 176.32 ± 58.20 μm$^2$ ($p$ = 0.14). No significant difference between HCC and ICC was observed for all variables examined.

### 3.3. Comparison of HCC, CCC and Non-Tumorous Hepatocytes in Papanicolaou-Stained Touch Smear Cytology Specimens

In a comparison between HCC and ICC by Papanicolaou-stained touch smear cytology, means and SD of major axis, minor axis, major axis/minor axis ratio, and area of nuclei of HCC and ICC were: major axis, HCC, 6.89 ± 2.47 µm vs. ICC: 8.60 ± 3.06 µm ($p = 0.057$); Minor axis: HCC, 5.22 ± 1.79 µm vs. ICC, 5.22 ± 1.95 µm ($p = 0.94$); major/minor axis ratio: HCC, 1.32 ± 0.11 vs. ICC, 1.67 ± 0.27 ($p < 0.0001$), Nucleus area: HCC, 43.80 ± 27.31 µm$^2$ vs. 55.37 ± 35.34 µm$^2$ ($p = 0.35$). The nuclei of the ICC were significantly oval rather than round in shape compared to HCC cells, while no significant difference was observed in the nucleus area of Papanicolaou-stained touch smear cytology specimens.

Non-tumorous hepatocytes were also assessed, and the results are as follows: Major axis: 5.78 ± 0.97 µm; Minor axis, 4.77 ± 0.75 µm; major/minor axis ratio: 1.21 ± 0.11; Nucleus area, 36.08 ± 7.37 µm$^2$. In comparison with HCC and ICC, the minor axis in non-tumorous hepatocytes appeared to be the smallest although statistical significance was not reached (vs. HCC: $p = 0.56$, vs. ICC: $p = 0.71$). A significant difference was found for the major/minor axis ratio in comparison to ICC (vs. HCC: $p = 0.037$, vs. ICC: $p = 0.0015$) (Figure 3). This result indicates the major/minor axis ratio of the nucleus (namely the shape of tumor nuclei is close to round or oval shapes) is an important finding for cytological differential diagnosis between HCC and ICC.

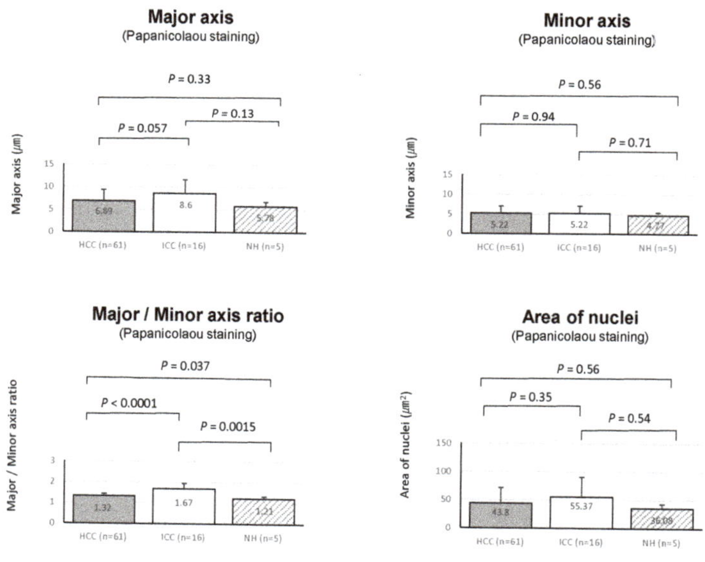


**Figure 3.** Comparison of major axis, minor axis, major/minor axis ratio and the nucleus area of hepatocellular carcinoma (HCC), intrahepatic cholangiocarcinoma (ICC) and non-tumorous hepatocytes (NH).

### 3.4. Comparison of Cytological Findings in Papanicolaou-Stained Touch Smear Cytology Specimens of ICC and HCC

Cytological findings of ICC and HCC in Papanicolaou-stained touch smear cytology specimens are summarized in Table 3. The nuclei of HCC were significantly center-positioned ($p < 0.0001$), having a single nucleolus rather than multiple nucleoli ($p = 0.005$), and a granular cytoplasm ($p < 0.0001$). No significant difference was observed in the chromatin pattern and chromatin distribution between HCC and ICC. Typical cytological images of HCC and ICC in Papanicolaou-stained touch smear cytology specimens are shown in Figure 4.

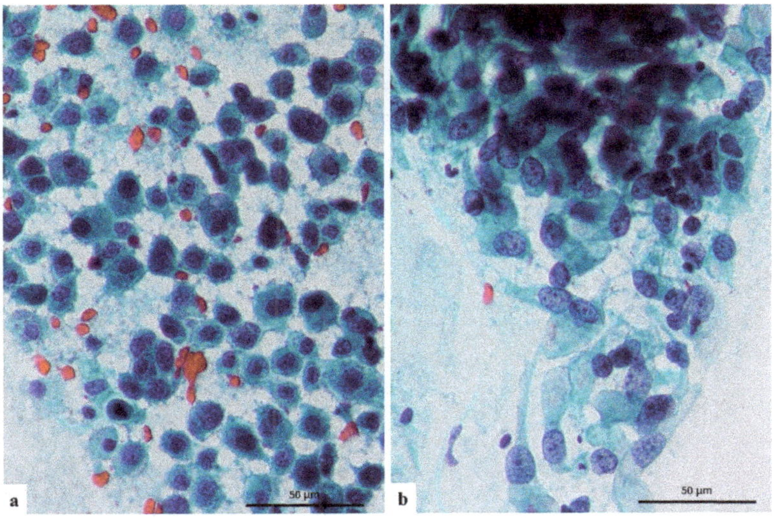


**Figure 4.** Representative cytological figures of hepatocellular carcinoma (HCC). (**a**) and intrahepatic cholangiocarcinoma (ICC). (**b**) In touch smear specimens (Papanicolaou stain, ×400). HCC cells have round and center-positioned nucleus and granular cytoplasm whereas ICC cells have oval and uncenter-positioned nucleus, multiple nucleolus, and foamy cytoplasm.

**Table 3.** Comparison of Cytological Fndings between ICC and HCC.

| | | HCC ($n$ = 61) | CCC ($n$ = 16) | $p$ |
|---|---|---|---|---|
| nuclear contours (%) | irregular | 23 (37.70) | 2 (12.50) | 0.074 |
| | smooth | 38 (62.30) | 14 (87.50) | |
| chromatin pattern (%) | coarse/granular | 34 (55.74) | 6 (37.50) | 0.260 |
| | fine | 27 (44.26) | 10 (62.50) | |
| chromatin distribution (%) | homogeneous | 23 (37.70) | 2 (12.50) | 0.074 |
| | heterogeneous | 38 (62.30) | 14 (87.50) | |
| nuclear position (%) | center | 56 (91.80) | 1 (6.25) | <0.0001 |
| | uncentre | 5 (8.20) | 15 (93.75) | |
| number of nucleolus (%) | single/unclear | 52 (85.25) | 8 (50.00) | 0.005 |
| | multiple | 9 (14.75) | 8 (50.00) | |
| cytoplasm (%) | vacuole/foamy | 8 (13.11) | 14 (87.50) | <0.0001 |
| | granular | 53 (86.89) | 2 (12.50) | |
| cell boundaries (%) | clear | 35 (57.38) | 5 (31.25) | 0.092 |
| | unclear | 26 (42.62) | 11 (68.75) | |

HCC, hepatocellular carcinoma; ICC. intrahepatic cholangiocarcinoma.

### *3.5. Comparison of HCC and CCC in Papanicolaou-Stained Clinical Specimens*

We consider the findings regarding the major/minor axis ratio in Papanicolaou-stained touch smear cytology specimens to be very important as they may be useful in daily cytological practice. As the condition of clinical samples may be different from touch smear cytology specimens, we planned a validation study using Papanicolaou-stained clinical samples. The details of clinical samples are summarized in Table 4. From 2014 to 2021, we found only 4 cytological samples which contained HCC tumor cells. All 4 samples were obtained by FNA (liver: 3, lymph node:1). In addition, 13 clinical samples which contained ICC tumor cells were found during this time period (FNA: 3, brushings: 8, ascites: 2).

**Table 4.** Details of Clinical Specimens.

| Sample Type | | HCC ($n$ = 4) | ICC ($n$ = 13) |
|---|---|---|---|
| | FNA | 4 | 3 |
| | Brushing | 0 | 8 |
| | Ascites | 0 | 2 |
| **Tumor Location** | | | |
| | Primary (liver) | 3 | 11 |
| | Metastasis/Dissemination | 1 (Lymph nodes) | 2 (Ascites) |

HCC: hepatocellular carcinoma, ICC: intrahepatic cholangiocarcinoma, SD: standard deviation, FNA: fine needle aspiration.

The means ± SD of each major axis, minor axis and major axis/minor axis ratio, and the area of nuclei of HCC and ICC were as follows: major axis: HCC, 8.79 ± 1.58 μm vs. ICC, 10.01 ± 1.90 μm ($p$ = 0.308); Minor axis: HCC: 7.67 ± 1.32 μm vs. ICC, 7.00 ± 1.72 μm ($p$ = 0.428), major/minor axis ratio: HCC, 1.18 ± 0.056 vs. ICC: 1.45 ± 0.13 ($p$ = 0.004, Figure 5); Nucleus area: HCC, 72.00 ± 24.87 μm$^2$ vs. 75.49 ± 27.47 μm$^2$ ($p$ = 0.821). These data are summarized in Table 5. Thus, the major/minor axis ratio was significantly different between HCC and ICC, even in clinical samples.

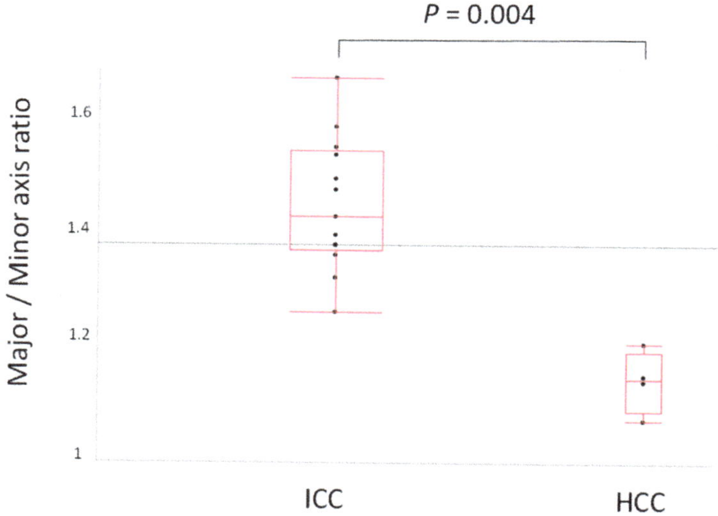


**Figure 5.** Comparison of major minor axis ratio in hepatocellular carcinoma (HCC) and intrahepatic cholangiocarcinoma (ICC) by validation Papanicolaou-stained clinical samples. A significant difference was found in the major/minor axis ratio between HCC and ICC ($p$ = 0.004).

**Table 5.** Comparison of Nuclei of HCC and ICC for Each Stain.

| | HCC ($n$ = 4) | ICC ($n$ = 13) | $p$ |
|---|---|---|---|
| Major axis (mean ± SD, μm) | 8.79 ± 1.58 | 10.01 ± 1.90 | 0.308 |
| Minor axis (mean ± SD, μm) | 7.67 ± 1.32 | 7.00 ± 1.72 | 0.428 |
| Major/minor axis ratio (mean ± SD) | 1.18 ± 0.056 | 1.45 ± 0.13 | 0.004 |
| Area of nuclei (mean ± SD, μm$^2$) | 72.00 ± 24.87 | 75.49 ± 27.47 | 0.821 |

HCC: hepatocellular carcinoma, ICC: intrahepatic cholangiocarcinoma, SD: standard deviation.

## 4. Discussion

The most notable finding of the present study was that the major/minor axis ratio of HCC and ICC was significantly different in Papanicolaou-stained cytological specimens. This means that the nuclei of HCC were close to round shapes and the nuclei of ICC close to oval shapes. This finding may have been previously noticed by other cytopathologists, but it is not well-recognized, presumably because no previous literature documenting this finding with analysis evidence has been published. This result was confirmed not only by the touch smear cytology of resected specimens but also by the validation of clinical cytological materials. Although our study also found a significant difference in the nuclear position, number of nucleoli and the cytoplasm between ICC and HCC, these cytological findings are subjective and depend on the experience of the investigating cytologist. Therefore, we consider that the major/minor axis ratio is very useful for the differential cytodiagnosis of ICC and HCC, because of its objective nature and simpleness. The significant difference in the major/minor axis ratio of ICC and HCC samples was confirmed in HE-stained tissue specimens, but its significance disappeared in Giemsa-stained specimens, likely because of morphological changes due to the dry process involved. The cell morphology is affected by the dry process of Giemsa-staining. Usually, the cell size is significantly enlarged, and cell morphology becomes more circular than Papanicolaou-staining. Therefore, it seems reasonable that the results of morphological analyses by image analyzing software were different between Giemsa-staining and Papanicolaou-staining. The significance of Papanicolaou-staining is important because clinical samples of HCC or ICC are usually evaluated by Papanicolaou-staining. Differential diagnosis of HCC and ICC is sometimes clinically problematic. The definitive diagnosis of HCC or ICC is usually made by pathological evaluation of morphology and immunohistochemical analyses of liver biopsy or resected specimens. Although it is rare, there are situations that require the differential diagnosis of HCC and ICC to be made only from cytological materials, when biopsy or surgical resection specimens could not be safely obtained because of the patient's general physical condition (such as a bleeding tendency or cachexia). In this situation, knowledge of the major/minor axis ratio will greatly help the cytological differential diagnosis of HCC and ICC.

The cytological characteristics of HCC have been well documented by studies of FNAC [11–22]. HCC cells typically have granular cytoplasm but cytological findings such as cellularity, cell dissociation, cell borders, monotony, trabeculae structure, nucleoli, cell size, nucleus/cytoplasmic ratio, nuclear crowding, chromatin distribution, and nuclear pleomorphism vary widely according to the differentiation of HCC [13,16,17]. Regarding ICC, knowledge of cytological characteristics has been accumulated by FNAC and brushing cytology. Typical ICC cells cytologically show adenocarcinoma-like features such as three-dimensional clusters, foamy and/or vacuolated cytoplasm, loss of nuclear polarity and prominent nucleoli, although these features varied according to the stage of tumor differentiation [23–26].

Cytological differences between HCC and ICC have not been well documented by researchers and therefore a paucity of knowledge has been accumulated. We found only one article which tried to distinguish ICC and HCC using cytological findings. Sampatanukul et al. [26] tried to distinguish between ICC and HCC, or metastatic carcinoma using cytological findings of ductular clusters. They concluded that the presence of more than 10 ductular clusters associated with malignant cells was a useful discriminator to separate ICC from metastatic carcinoma but was not useful for discrimination of ICC from HCC. Therefore, the present study is the first to demonstrate that simple cytological findings can discriminate between ICC from HCC, with evidence obtained by image analysis.

## 5. Conclusions

This study evaluated the cytological characteristics of ICC and HCC using image analysis software. The results indicated that the nucleus of HCC is close to a round shape whereas the nucleus of ICC is close to an oval shape. This characteristic was significant in Papanicolaou stained cytological specimens, but this significance disappeared after Giemsa-staining. This simple and objective finding will be very useful for the differential cytodiagnosis of HCC and ICC.

**Author Contributions:** Conceptualization and design, S.K. and K.K.; acquisition of data, S.K., M.N., T.T., T.I. and K.K.; material preparation and cytological assessments, S.K., M.N. and K.K.; analysis and interpretation of data, S.K. and K.K.; statistical analysis, S.K. and K.K.; writing—original draft, S.K. and K.K.; writing—review and editing, T.T., T.I., H.N., E.S. and S.A.; visualization, S.K. and K.K.; funding acquisition, K.K. All authors have read and agreed to the published version of the manuscript.

**Funding:** This study was partly supported by the Japan Society for the Promotion of Science (JSPS) Grants-in-Aid for Scientific Research C (KAKENHI grant numbers JP20K07408).

**Institutional Review Board Statement:** The study protocols were approved by the Ethics Committee of Saga University Hospital (approval number: No. 2018-12-R-11).

**Informed Consent Statement:** Comprehensive informed consent for the use of resected tissue for research was obtained from all patients.

**Data Availability Statement:** The data presented in this study are available on reasonable request from the corresponding author.

**Conflicts of Interest:** The authors declare no conflict of interest.

## References

1. Ikai, I.; Kudo, M.; Arii, S.; Omata, M.; Kojiro, M.; Sakamoto, M.; Takayasu, K.; Hayashi, N.; Makuuchi, M.; Matsuyama, Y.; et al. Report of the 18th follow-up survey of primary liver cancer in Japan. *Hepatol. Res.* **2010**, *40*, 1043–1059. [CrossRef] [PubMed]
2. Matsumoto, K.; Onoyama, T.; Kawata, S.; Takeda, Y.; Harada, K.; Ikebuchi, Y.; Ueki, M.; Miura, N.; Yashima, K.; Koda, M.; et al. Hepatitis B and C virus infection is a risk factor for the development of cholangiocarcinoma. *Intern. Med.* **2014**, *53*, 651–654. [CrossRef]
3. Palmer, W.C.; Patel, T. Are common factors involved in the pathogenesis of primary liver cancers? A meta-analysis of risk factors for intrahepatic cholangiocarcinoma. *J. Hepatol.* **2012**, *57*, 69–76. [CrossRef] [PubMed]
4. Zhou, Y.; Zhao, Y.; Li, B.; Huang, J.; Wu, L.; Xu, D.; Yang, J.; He, J. Hepatitis viruses infection and risk of intrahepatic cholangiocarcinoma: Evidence from a meta-analysis. *BMC Cancer* **2012**, *12*, 289. [CrossRef] [PubMed]
5. Lo, E.C.; Rucker, A.N.; Federle, M.P. Hepatocellular carcinoma and intrahepatic cholangiocarcinoma: Imaging for diagnosis, tumor response to treatment and liver response to radiation. *Semin. Radiat. Oncol.* **2018**, *28*, 267–276. [CrossRef] [PubMed]
6. Jiang, H.Y.; Chen, J.; Xia, C.C.; Cao, L.K.; Duan, T.; Song, B. Noninvasive imaging of hepatocellular carcinoma: From diagnosis to prognosis. *World J. Gastroenterol.* **2018**, *24*, 2348–2362. [CrossRef]
7. Sun, L.; Wu, H.; Guan, Y.S. Positron emission tomography/computer tomography: Challenge to conventional imaging modalities in evaluating primary and metastatic liver malignancies. *World J. Gastroenterol.* **2007**, *13*, 2775–2783. [CrossRef]
8. Roberts, L.R.; Sirlin, C.B.; Zaiem, F.; Almasri, J.; Prokop, L.J.; Heimbach, J.K.; Murad, M.H.; Mohammed, K. Imaging for the diagnosis of hepatocellular carcinoma: A systematic review and meta-analysis. *Hepatology* **2018**, *67*, 401–421. [CrossRef]
9. Victor, D.W.; Sherman, S.; Karakan, T.; Khashab, M.A. Current endoscopic approach to indeterminate biliary strictures. *World J. Gastroenterol.* **2012**, *18*, 6197–6205. [CrossRef]
10. Naito, Y.; Kawahara, A.; Okabe, Y.; Ishida, Y.; Sadashima, E.; Murata, K.; Takase, Y.; Abe, H.; Yamaguchi, T.; Tanigawa, M.; et al. SurePath®® LBC improves the diagnostic accuracy of intrahepatic and hilar cholangiocarcinoma. *Cytopathology* **2018**, *29*, 349–354. [CrossRef]
11. Swamy, M.C.; Arathi, C.; Kodandaswamy, C. Value of ultrasonography-guided fine needle aspiration cytology in the investigative sequence of hepatic lesions with an emphasis on hepatocellular carcinoma. *J. Cytol.* **2011**, *28*, 178–184. [PubMed]
12. Wee, A. Fine-needle aspiration biopsy of hepatocellular carcinoma and related hepatocellular nodular lesions in cirrhosis: Controversies, challenges, and expectations. *Patholog. Res. Int.* **2011**, *2011*, 587936. [CrossRef] [PubMed]
13. Wee, A. Fine needle aspiration biopsy of hepatocellular carcinoma and hepatocellular nodular lesions: Role, controversies and approach to diagnosis. *Cytopathology* **2011**, *22*, 287–305. [CrossRef] [PubMed]
14. Lin, C.C.; Lin, C.J.; Hsu, C.W.; Chen, Y.C.; Chen, W.T.; Lin, S.M. Fine-needle aspiration cytology to distinguish dysplasia from hepatocellular carcinoma with different grades. *J. Gastroenterol. Hepatol.* **2008**, *23*, e146–e152. [CrossRef]
15. Zeppa, P.; Anniciello, A.; Vetrani, A.; Palombini, L. Fine needle aspiration biopsy of hepatic focal fatty change. A report of two cases. *Acta Cytol.* **2002**, *46*, 567–570. [CrossRef] [PubMed]
16. Kulesza, P.; Torbenson, M.; Sheth, S.; Erozan, Y.S.; Ali, S.Z. Cytopathologic grading of hepatocellular carcinoma on fine-needle aspiration. *Cancer* **2004**, *102*, 247–258. [CrossRef]

17. Yang, G.C.; Yang, G.Y.; Tao, L.C. Cytologic features and histologic correlations of microacinar and microtrabecular types of well-differentiated hepatocellular carcinoma in fine-needle aspiration biopsy. *Cancer* **2004**, *102*, 27–33. [CrossRef]
18. de Boer, W.B.; Segal, A.; Frost, F.A.; Sterrett, G.F. Cytodiagnosis of well differentiated hepatocellular carcinoma: Can indeterminate diagnoses be reduced? *Cancer* **1999**, *87*, 270–277. [CrossRef]
19. Wen, C.H.; Lin, C.H.; Tsao, S.C.; Su, Y.C.; Tsai, M.H.; Chai, C.Y. Micronucleus scoring in liver fine needle aspiration cytology. *Cytopathology* **2013**, *24*, 391–395. [CrossRef]
20. Geramizadeh, B.; Asadi, N.; Tabei, S.Z. Cytologic comparison between malignant and regenerative nodules in the background of cirrhosis. *Hepat. Mon.* **2012**, *12*, 448–452. [CrossRef]
21. Kaçar Özkara, S.; Ozöver Tuneli, I. Fine needle aspiration cytopathology of liver masses: 101 cases with cyto-/histopathological analysis. *Acta Cytol.* **2013**, *57*, 332–336. [CrossRef] [PubMed]
22. McGahan, J.P.; Bishop, J.; Webb, J.; Howell, L.; Torok, N.; Lamba, R.; Corwin, M.T. Role of FNA and core biopsy of primary and metastatic liver disease. *Int. J. Hepatol.* **2013**, *2013*, 174103. [CrossRef] [PubMed]
23. Chaudhary, H.B.; Bhanot, P.; Logroño, R. Phenotypic diversity of intrahepatic and extrahepatic cholangiocarcinoma on aspiration cytology and core needle biopsy: Case series and review of the literature. *Cancer* **2005**, *105*, 220–228. [CrossRef] [PubMed]
24. Volmar, K.E.; Vollmer, R.T.; Routbort, M.J.; Creager, A.J. Pancreatic and bile duct brushing cytology in 1000 cases: Review of findings and comparison of preparation methods. *Cancer* **2006**, *108*, 231–238. [CrossRef] [PubMed]
25. Salomao, M.; Gonda, T.A.; Margolskee, E.; Eguia, V.; Remotti, H.; Poneros, J.M.; Sethi, A.; Saqi, A. Strategies for improving diagnostic accuracy of biliary strictures. *Cancer Cytopathol.* **2015**, *123*, 244–252. [CrossRef]
26. Sampatanukul, P.; Leong, A.S.; Kosolbhand, P.; Tangkijvanich, P. Proliferating ductules are a diagnostic discriminator for intrahepatic cholangiocarcinoma in FNA biopsies. *Diagn. Cytopathol.* **2000**, *22*, 359–363. [CrossRef]

*cancers*

MDPI

*Article*

# Determinants of Survival and Post-Progression Outcomes by Sorafenib–Regorafenib Sequencing for Unresectable Hepatocellular Carcinoma

**I-Cheng Lee [1,2], Yee Chao [3], Pei-Chang Lee [1,2], San-Chi Chen [3], Chen-Ta Chi [1,2,4], Chi-Jung Wu [1,2,4], Kuo-Cheng Wu [1,2], Ming-Chih Hou [1,2] and Yi-Hsiang Huang [1,2,4,*]**

1 Division of Gastroenterology and Hepatology, Department of Medicine, Taipei Veterans General Hospital, Taipei 11217, Taiwan; iclee@vghtpe.gov.tw (I.-C.L.); pclee11@vghtpe.gov.tw (P.-C.L.); ctchi2@vghtpe.gov.tw (C.-T.C.); cjwu6@vghtpe.gov.tw (C.-J.W.); kcwu3@vghtpe.gov.tw (K.-C.W.); mchou@vghtpe.gov.tw (M.-C.H.)

2 School of Medicine, National Yang Ming Chiao Tung University, Taipei 11221, Taiwan

3 Cancer Center, Taipei Veterans General Hospital, Taipei 11217, Taiwan; ychao@vghtpe.gov.tw (Y.C.); scchen16@vghtpe.gov.tw (S.-C.C.)

4 Institute of Clinical Medicine, National Yang Ming Chiao Tung University, Taipei 11221, Taiwan

* Correspondence: yhhuang@vghtpe.gov.tw; Tel.: +886-2-28712121 (ext. 7506); Fax: +886-2-28739318

**Simple Summary:** The optimal subsequent treatment and the determinants of survival after sorafenib–regorafenib failure in patients with hepatocellular carcinoma (HCC) remain unclear. The aim of this study was to delineate the determinants of response and survival after regorafenib and evaluate the post-progression outcomes in the era of multiple-line sequential systemic therapy. We retrospectively enrolled 108 patients with unresectable HCC receiving regorafenib after sorafenib failure and reported the predictors of progression-free survival, overall survival, post-progression survival, as well as the next-line treatments after regorafenib failure. We showed that some well-known survival predictors of sorafenib treatment and the response to prior sorafenib also had a prognostic role in patients with HCC undergoing regorafenib treatment. Preserved liver function and subsequent systemic therapy play important roles in survival after regorafenib failure. We conclude that the survival outcomes of regorafenib for HCC have improved in the era of multi-line sequential therapy. Preserved liver function and next-line therapy are important prognostic factors after regorafenib failure.

**Abstract:** The predictors of response and survival in patients with hepatocellular carcinoma (HCC) receiving regorafenib remain unclear. This study aimed to delineate the determinants of response and survival after regorafenib and evaluate post-progression treatment and outcomes. We retrospectively enrolled 108 patients with unresectable HCC receiving regorafenib after sorafenib failure. Progression-free survival (PFS), overall survival (OS), post-progression survival (PPS) and post-progression treatments were evaluated. The median PFS, OS and PPS were 3.1, 13.1 and 10.3 months, respectively. Achieving disease control by prior sorafenib, early AFP reduction and hand-foot skin reaction (HFSR) were associated with significantly better radiologic responses. By multivariate analysis, the time to progression on prior sorafenib, HFSR and early AFP reduction were associated with PFS; ALBI grade, portal vein invasion, HFSR and early AFP reduction were associated with OS. ALBI grade at disease progression, main portal vein invasion, high tumor burden and next-line therapy were associated with PPS. The median PPS was 12 months in patients who received next-line therapy, and the PPS was comparable between patients who received next-line targeted agents and immunotherapy. In conclusion, survival outcomes of regorafenib for HCC have improved in the era of multi-line sequential therapy. Preserved liver function and next-line therapy are important prognostic factors after regorafenib failure.

**Keywords:** hepatocellular carcinoma; sorafenib; regorafenib; progression-free survival; overall survival; post-progression survival

**Citation:** Lee, I.-C.; Chao, Y.; Lee, P.-C.; Chen, S.-C.; Chi, C.-T.; Wu, C.-J.; Wu, K.-C.; Hou, M.-C.; Huang, Y.-H. Determinants of Survival and Post-Progression Outcomes by Sorafenib–Regorafenib Sequencing for Unresectable Hepatocellular Carcinoma. *Cancers* **2022**, *14*, 2014. https://doi.org/10.3390/cancers14082014

Academic Editors: Georgios Germanidis and Ilona Kovalszky

Received: 12 March 2022
Accepted: 14 April 2022
Published: 15 April 2022

# 1. Introduction

Hepatocellular carcinoma (HCC) is the sixth most common cancer in the world and the fourth leading cause of cancer-related mortality [1,2]. Systemic therapy is recommended as the standard of care for patients with HCC at advanced stages or patients with unresectable HCC who are unsuitable for loco-regional therapy (LRT), and it is estimated that about half of patients with HCC may receive systemic therapies at some time point during the course of HCC treatment [3]. For patients with unresectable HCC, the multi-targeted tyrosine kinase inhibitor (TKI) sorafenib has been the standard of treatment since 2008 [4,5], while regorafenib is the first drug approved as the second-line treatment after sorafenib failure for HCC. In the RESORCE trial, regorafenib significantly improved overall survival (OS) and progression-free survival (PFS) compared to the placebo [6]. Currently, the predictors of response and survival under regorafenib treatment for HCC have not been fully clarified. Regorafenib is structurally similar to sorafenib but appears to be more pharmacologically potent than sorafenib [7]. Therefore, regorafenib and sorafenib might share some common predictors of response and survival. Recent studies suggest that response to prior sorafenib treatment is associated with the outcomes of regorafenib treatment [8,9]. Several prognostic predictors in patients with HCC receiving sorafenib, such as the presence of hand-foot skin reaction (HFSR) [10], ALBI grade [11], early AFP response [12], progression pattern [13,14] and the PROSASH-II model [15], may also have prognostic value for regorafenib treatment.

With the advance of systemic therapies for HCC in the past decade, lenvatinib and subsequently the immunotherapy combinations of atezolizumab plus bevacizumab have been approved as first-line systemic therapies for HCC, whereas cabozantinib, ramucirumab, and immune checkpoint inhibitors (ICIs) pembrolizumab and nivolumab plus ipilimumab are also currently available second-line treatment options for HCC [16]. With the increased options for multiple lines of systemic therapies for HCC, the survival of patients with advanced HCC may improve over time. Several real-world studies of regorafenib for HCC reported that the OS might be longer than 12 months [9,17–19], suggesting that the OS of HCC grossly improves under multiple lines of sequential therapy. Nevertheless, the optimal subsequent treatment and the determinants of survival after sorafenib–regorafenib failure remain unclear. The aim of this study was to delineate the determinants of response and survival after regorafenib treatment and evaluate the post-progression outcomes in the era of multiple-line sequential systemic therapy.

# 2. Patients and Methods

## 2.1. Patients

From May 2019 to September 2020, we retrospectively screened 115 patients with unresectable HCC in Taipei Veterans General Hospital who received regorafenib due to sorafenib failure. Patients were enrolled if they had histologically confirmed HCC or clinically confirmed HCC based on magnetic resonance imaging (MRI) or contrast-enhanced computed tomography (CECT) according to the diagnostic criteria of the American Association for the Study of Liver Diseases (AASLD) treatment guidelines [20]; patients with HCC were classified as being in Barcelona Clinic Liver Cancer (BCLC) stage C or in BCLC stage B and not suitable for trans-arterial chemoembolization (TACE) or other LRT. Patients were excluded if they were lost to follow-up within 2 months of treatment ($n$ = 6) or had no measurable lesion when starting regorafenib ($n$ = 1). For each cycle, the standard dose of regorafenib was 160 mg once daily for 3 weeks, followed by 1 week off therapy. Modification of the initial dose of regorafenib was allowed according to the presence of adverse events during prior sorafenib treatment. Regorafenib treatment was stopped when there was confirmation of disease progression by image studies or when patients experienced intolerable toxicity.

This study was approved by the Institutional Review Board in Taipei Veterans General Hospital (IRB number: 2021-04-006BC) and adhered to the guidance of the Declaration of

Helsinki. The Institutional Review Board waived the need for written informed consent due to the retrospective nature of this study.

*2.2. Patient Evaluation*

Demographic profiles, biochemistry data and tumor characteristics at baseline and at the time of disease progression were recorded. The data included age, gender, duration and response to prior sorafenib treatment, prior or concurrent immune checkpoint inhibitors (ICI) therapy, concurrent loco-regional therapy (LRT), tumor size, tumor number, macrovascular invasion, extrahepatic metastasis, serum alpha-fetoprotein (AFP), platelet count, as well as levels of albumin, total bilirubin, creatinine, alanine aminotransferase (ALT), aspartate aminotransferase (AST), hepatitis B surface antigen (HBsAg) and anti-hepatitis C virus antibodies. The ALBI score and grade were calculated as previously described [21]. High tumor burden was defined as the presence of main portal vein thrombosis (Vp4), bile duct invasion or tumor involvement >50% liver volume [22]. The Prediction Of Survival in Advanced Sorafenib-treated HCC (PROSASH)-II model was calculated as previously described [15].

*2.3. Outcome Assessment*

Radiologic responses according to the Response Evaluation Criteria in Solid Tumors version 1.1 (RECIST v1.1) were evaluated every 8–12 weeks during treatment [23]. The objective response rate (ORR) was defined as the percentage of patients with a complete response (CR) or partial response (PR). The disease control rate (DCR) was defined as the percentage of patients with CR, PR or stable disease (SD).

Progression-free survival (PFS) was defined as the time interval between the day of starting regorafenib treatment and the onset of progressive disease (PD). Overall survival (OS) was defined as the time interval between the day of starting treatment and death. Post-progression survival (PPS) was defined as the time interval between the day of PD and death. The tumor progression pattern was classified into intrahepatic or extrahepatic tumor growth (>20% increase in tumor size of the viable target lesions), new intrahepatic lesions, and new extrahepatic lesions (including new vascular invasion and/or metastasis) [13,14]. Early AFP response was defined as greater than a 10% reduction in AFP levels from baseline within 1 month of treatment [24].

*2.4. Statistical Analysis*

All statistical analyses were performed using IBM SPSS Statistics for Windows, Version 22 (IBM, Armonk, NY, USA). Values were expressed as mean $\pm$ SD or as median (range) when appropriate. We used the Mann–Whitney U test to compare continuous variables and the Pearson chi-square analysis to compare categorical variables. We used the Kaplan–Meier method to estimate survival rates and the log-rank test to compare survival curves between patient groups. We used the Cox proportional hazards model to analyze prognostic factors for survival. Variables that achieved statistical significance ($p < 0.05$) or those close to significance ($p < 0.1$) by univariate analysis were subsequently included in the multivariate analysis. Statistical significance was considered as a $p$-value $< 0.05$ determined by two-tailed tests.

## 3. Results

*3.1. Patient Characteristics*

A total of 108 patients receiving regorafenib for unresectable HCC due to sorafenib failure were ultimately enrolled for analysis. The baseline characteristics of the 108 patients are summarized in Table 1. The majority of patients belonged to BCLC stage C (81.5%), Child–Pugh class A (84.3%), and 38 (35.2%) patients presented with a high tumor burden. Regorafenib was given as the second- and third- to fifth-line therapy after sorafenib failure in 88 (81.5%) and 20 (18.5%) patients, respectively. The median duration of prior sorafenib therapy was 3.9 months, and 59.1% and 51% of patients experienced dose reductions and hand-foot skin reactions (HFSR) during sorafenib treatment, respectively. Nineteen patients (17.6%) experienced prior ICI therapy, while sixteen (14.8%) and nineteen (17.6%) patients received concurrent LRT (TACE 14, radiofrequency ablation 2) and ICI therapy (nivolumab 10, pembrolizumab 3, atezolizumab 1, durvalumab 5), respectively. Sixty-two patients (57.4%) experienced dose reduction of regorafenib, and the most frequently reported adverse events were HFSR (29.6%), diarrhea (15.7%) and hypertension (23.1%).

**Table 1.** Characteristics of 108 patients receiving regorafenib therapy.

| Variables | |
|---|---|
| Age (years) | 65.3 ± 12.9 |
| Male gender, *n* (%) | 91 (84.3) |
| HCC etiology: HBV/HCV/HBV + HCV/Non-viral, *n* (%) | 61/17/4/26 (56.5/15.7/3.7/24.1) |
| Lines of regorafenib therapy: 2/3/4/5, *n* (%) | 88/12/6/2 (81.5/11.1/5.6/1.9) |
| Prior immune checkpoint inhibitors therapy, *n* (%) | 19 (17.6) |
| Prior sorafenib duration (months) † | 3.9 (0.5–44) |
|     Dose reduction for sorafenib, *n* (%) | 61 (59.8) |
| Hand-foot skin reaction during sorafenib treatment, *n* (%) | 52 (51) |
| BCLC stage B/C, *n* (%) | 20/88 (18.5/81.5) |
| Portal vein invasion, *n* (%) | 38 (35.2) |
|     Vp4 | 20 (18.5) |
| Extrahepatic metastasis, *n* (%) | 71 (65.7) |
| Tumor size (cm) | 4.65 ± 4.75 |
| Multiple tumors, *n* (%) | 74 (68.5) |
| High tumor burden, *n* (%) † | 38 (35.2) |
| Child–Pugh class A/B, *n* (%) | 91/17 (84.3/15.7) |
| ALBI grade 1/2/3, *n* (%) | 44/63/1 (40.7/58.3/0.9) |
| Bilirubin (mg/dL) | 0.99 ± 1.39 |
| Albumin (g/dL) | 3.74 ± 0.49 |
| ALT (U/L) | 49.5 ± 37.5 |
| AST (U/L) | 67.7 ± 58.6 |
| Creatinine (mg/dL) | 1.07 ± 0.88 |
| Platelet ($10^9$/L) | 154 ± 96 |
| AFP (ng/mL) | 182.4 (1.2-1397041) |
| AFP > 400 ng/mL, *n* (%) | 44 (40.7) |
| Follow-up period (months) | 9.6 (0.3–29.0) |
| Initial dose of regorafenib: 160/120/80/40 mg | 63/2/41/2 (58.3/1.9/38/1.9) |
|     Dose reduction for regorafenib, *n* (%) | 62 (57.4) |
| Adverse events during regorafenib, *n* (%) | |
|     Hand-foot skin reaction | 32 (29.6) |
|     Diarrhea | 17 (15.7) |
|     Hypertension | 25 (23.1) |
| Concurrent loco-regional therapy during regorafenib use, *n* (%) | 16 (14.8) |
|     Transarterial chemoembolization/radiofrequency ablation | 14/2 (13/1.9) |
| Concurrent immune checkpoint inhibitors during regorafenib use, *n* (%) | 19 (17.6) |
|     Nivolumab/Pembrolizumab/Atezolizumab/Durvalumab | 10/3/1/5 (9.3/2.8/0.9/4.6) |
| Disease progression, *n* (%) | 78 (72.2%) |
| Death, *n* (%) | 52 (48.1%) |

† High tumor burden was defined as the presence of main portal vein thrombosis (Vp4), bile duct invasion or tumor involvement >50% liver volume. Sorafenib information was available for 102 (94.4%) patients.

### 3.2. Radiologic Response

Evaluations of the best radiologic response by RECIST v1.1 to regorafenib and to prior sorafenib treatment were available in 103 (95.4%) and 98 (90.7%) of all patients, respectively (Table 2). The ORR and DCR to regorafenib treatment in all patients were 10.7% and 43.7%, respectively. Three patients (2.9%), all in the second-line setting, achieved a complete response. The ORR and DCR to prior sorafenib treatment were 21.4% and 44.9%, respectively. In patients achieving disease control by prior sorafenib treatment, the DCR to regorafenib was significantly higher (59.1% vs. 29.6%, $p$ = 0.006). Patients with HFSR and early AFP responses had significantly better radiologic responses. Patients with early AFP responses also had significantly higher ORR (21.4% vs. 0%, $p = 0.004$) and DCR (64.3% vs. 17.9%, $p < 0.001$). The ORR and DCR in patients who received regorafenib monotherapy were 8.6 and 39.1, respectively (Table S1). There was no significant difference in ORR and DCR between patients who did or did not receive concurrent LRT or ICI therapy (Table S1).

**Table 2.** Best radiologic responses to regorafenib therapy by RECIST v1.1 criteria.

| Radiologic Response † | CR | PR | SD | PD | ORR | DCR |
|---|---|---|---|---|---|---|
| Overall | 3 (2.9%) | 8 (7.8%) | 34 (33%) | 58 (56.3%) | 11 (10.7%) | 45 (43.7%) |
| Line of therapy | | | | | | |
| 2nd line ($n$ = 83) | 3 (3.6%) | 6 (7.2%) | 26 (31.3%) | 48 (57.8%) | 9 (10.8%) | 35 (42.2%) |
| 3rd–5th line ($n$ = 20) | 0 (0) | 2 (10%) | 8 (40%) | 10 (50%) | 2 (10%) | 10 (50%) |
| $p$ value | | | | 0.859 | 1.000 | 0.702 |
| Achieving disease control by prior sorafenib | | | | | | |
| Yes ($n$ = 44) | 1 (2.3%) | 4 (9.1%) | 21 (47.7%) | 18 (40.9%) | 5 (11.4%) | 26 (59.1%) |
| No ($n$ = 54) | 1 (1.9%) | 4 (7.4%) | 11 (20.4%) | 38 (70.4%) | 5 (9.3%) | 16 (29.6%) |
| $p$ value | | | | 0.032 | 0.744 | 0.006 |
| Presence of hand-foot skin reaction | | | | | | |
| Yes ($n$ = 32) | 2 (6.3%) | 3 (9.4%) | 14 (43.8%) | 13 (40.6%) | 5 (15.6) | 19 (59.4) |
| No ($n$ = 71) | 1 (1.4%) | 5 (7.0%) | 20 (28.2%) | 45 (63.4%) | 6 (8.5) | 26 (36.6) |
| $p$ value | | | | 0.032 | 0.310 | 0.052 |
| Early AFP response | | | | | | |
| Yes ($n$ = 28) | 2 (7.1%) | 4 (14.3) | 12 (42.9%) | 10 (35.7%) | 6 (21.4%) | 18 (64.3) |
| No ($n$ = 39) | 0 (0%) | 0 (0%) | 7 (17.9%) | 32 (82.1%) | 0 (0%) | 7 (17.9) |
| $p$ value | | | | <0.001 | 0.004 | <0.001 |

† Evaluations of the best radiologic response to regorafenib and sorafenib treatment were available in 103 (95.4%) and 98 (90.7%) of all patients, respectively. CR, complete response; PR, partial response; SD, stable disease; PD, progressive disease; ORR, objective response rate; DCR, disease control rate.

### 3.3. Factors Associated with Progression-Free Survival (PFS)

During a median follow-up period of 9.3 months, 78 (72.2%) patients developed disease progression with a median PFS of 3.1 months. The median PFSs were 5.6 and 3.0 months, respectively, in patients with BCLC stages B and C ($p = 0.137$, Figure 1A), and was 2.9 and 3.9 months in second-line and later-line settings, respectively ($p = 0.418$, Figure 1B). By multivariate analysis, TTP on prior sorafenib >4 months (hazard ratio (HR) = 0.563, $p = 0.018$, Figure 1C) was the only baseline predictor of PFS, while the presence of HFSR (HR = 0.238, $p < 0.001$, Figure 1D) and early AFP responses (HR = 0.397, $p = 0.003$, Figure 1E) were on-treatment predictors of PFS (Tables 3 and S2).

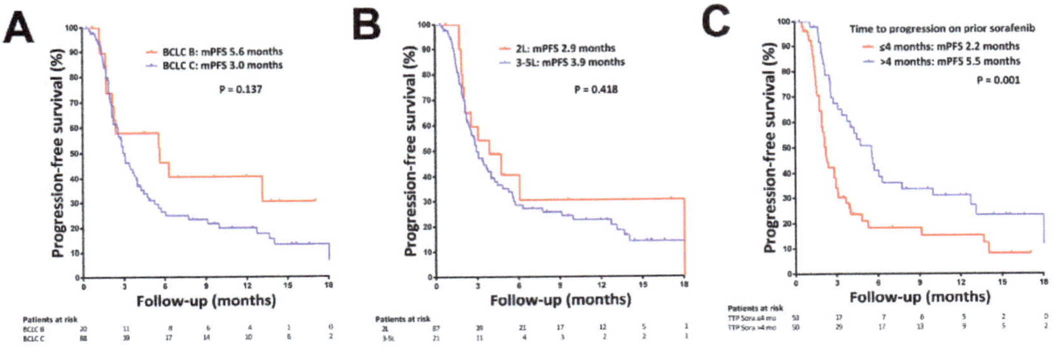


**Figure 1.** *Cont.*

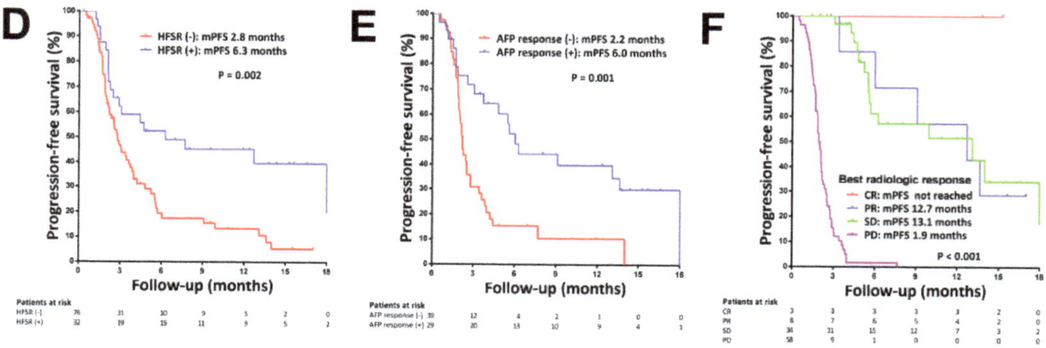


**Figure 1.** Kaplan–Meier curves of progression-free survival (PFS) in patients with HCC receiving regorafenib treatment. (**A**) PFS stratified by BCLC stage. (**B**) PFS stratified by lines of therapy. (**C**) PFS stratified by time-to-progression on prior sorafenib treatment. (**D**) PFS in patients with and without hand-foot skin reaction (HFSR). (**E**) PFS in patients with and without early AFP response. (**F**) PFS stratified by radiologic response by mRECIST criteria.

**Table 3.** Independent factors associated with progression-free survival, overall survival and post-progression survival by multivariate analysis.

| Variables | | Multivariate | |
|---|---|---|---|
| | | **HR (95% CI)** | ***p*** |
| Progression-free survival | | | |
| Baseline factor | | | |
| Time to progression on prior sorafenib (months) | >4/≤4 | 0.485 (0.302–0.781) | 0.003 |
| On-treatment factors | | | |
| Hand-foot skin reaction | Yes/No | 0.238 (0.108–0.525) | <0.001 |
| Early AFP reduction | >10%/≤10% | 0.397 (0.214–0.737) | 0.003 |
| Overall survival | | | |
| Baseline factors | | | |
| ALBI grade | 2-3/1 | 2.758 (1.458–5.216) | 0.002 |
| Portal vein invasion | Yes/No | 3.169 (1.817–5.528) | <0.001 |
| On-treatment factors | | | |
| Hand-foot skin reaction | Yes/No | 0.173 (0.068–0.442) | <0.001 |
| Early AFP reduction | >10%/≤10% | 0.450 (0.215–0.940) | 0.034 |
| Post-progression survival | | | |
| Main portal vein invasion | Yes/No | 5.102 (1.578–16.949) | 0.007 |
| High tumor burden | Yes/No | 9.296 (3.379–25.578) | <0.001 |
| ALBI grade | 1 | 1 | |
| | 2 | 4.499 (1.541–13.137) | 0.006 |
| | 3 | 26.926 (6.638–109.227) | <0.001 |
| Next-line therapy | Yes/No | 0.369 (0.163–0.838) | 0.017 |

None of the three patients achieving CR had disease progression during the observation period, whereas the median PFSs in patients with PR and SD were 12.7 and 13.1 months, respectively (Figure 1F). We validated the PROSASH-II model for predicting RFS after regorafenib treatment, and a significantly poorer RFS was observed in PROSASH-II group 4 ($p$ = 0.001, Figure S1A).

### 3.4. Factors Associated with Overall Survival (OS)

Fifty-two patients (48.1%) died during the observation period, with a median OS of 13.1 months. The median OSs in patients with BCLC stage C and second-line setting were 12 and 14.7 months, respectively (Figure S2A,B). The median OS was significantly better in patients with ALBI grade 1 (not reached vs. 8.5 months for ALBI grades 2–3, $p < 0.001$, Figure 2A) and Child–Pugh class A (14.7 vs. 4.1 months for Child–Pugh class B, $p < 0.001$, Figure S2C). By multivariate analysis, ALBI grades 2–3 (HR = 2.758, $p = 0.002$) and the presence of portal vein invasion (HR = 3.169, $p < 0.001$) were the baseline predictors of OS (Figure 2B). Combining the ALBI grades 2–3 and the presence of portal vein invasion could discriminate patients with high, intermediate and low risk of mortality (Figure 2C). The presence of HFSR (HR = 0.173, $p < 0.001$, Figure 2D) and early AFP response (HR = 0.450, $p = 0.034$, Figure 2E) were on-treatment predictors of OS (Table 3 and Table S3). Combining the risk factors of ALBI grade, portal vein invasion, HFSR and early AFP response could further stratify patients into four mortality risk groups (Figure 2F). The PROSASH-II model could also significantly stratify the OS after regorafenib treatment (median OS in groups 1, 2, 3, 4: not reached, 14.4, 8, 3.8 months, respectively; $p < 0.001$, Figure S1B).

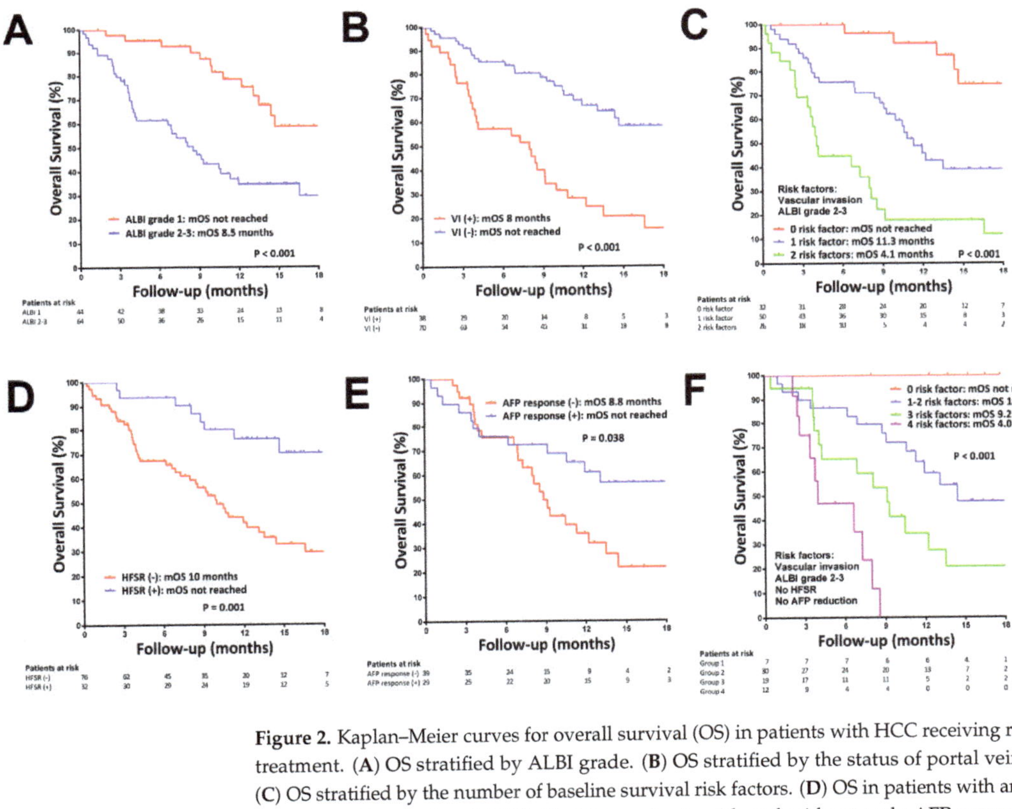


**Figure 2.** Kaplan–Meier curves for overall survival (OS) in patients with HCC receiving regorafenib treatment. (**A**) OS stratified by ALBI grade. (**B**) OS stratified by the status of portal vein invasion. (**C**) OS stratified by the number of baseline survival risk factors. (**D**) OS in patients with and without hand-foot skin reaction (HFSR). (**E**) OS in patients with and without early AFP response. (**F**) OS stratified by the number of baseline and on-treatment survival risk factors.

### 3.5. Factors Associated with Post-Progression Survival (PPS)

Patient characteristics at disease progression and the tumor progression patterns for 78 patients with regorafenib failure are shown in Table 4. Twenty (25.6%) and 25 (32.1%) patients had deterioration of Child–Pugh class and ALBI grade at the time of disease progression, respectively.

**Table 4.** Characteristics at disease progression in 78 patients with regorafenib failure.

| Characteristics | Descriptive Analysis | Median Post-Progression Survival (Months) |
|---|---|---|
| BCLC stage B/C, *n* (%) | 8/78 (10.3/89.7) | |
| Child–Pugh class A/B/C, *n* (%) | 53/19/6 (67.9/24.4/7.7) | |
| Child–Pugh class deterioration, *n* (%) | 20 (25.6) | |
| ALBI grade 1/2/3, *n* (%) | 21/43/14 (26.9/55.1/17.9) | |
| ALBI grade deterioration, *n* (%) | 25 (32.1) | |
| Bilirubin (mg/dL) | 1.84 ± 2.25 | |
| Albumin (g/dL) | 3.43 ± 0.62 | |
| ALT (U/L) | 46.8 ± 49.0 | |
| AST (U/L) | 84.5 ± 119.6 | |
| Creatinine (mg/dL) | 1.11 ± 1.10 | |
| AFP (ng/mL) | 242 (1.39–823.19.9) | |
| AFP > 400 ng/mL, *n* (%) | 34 (43.6) | |
| Tumor progression pattern | | |
| Intrahepatic tumor growth | 39 (50%) | |
| New intrahepatic lesions | 33 (42.3%) | |
| Extrahepatic tumor growth | 26 (33.3%) | |
| New extrahepatic lesions | 24 (30.8%) | |
| Next-line therapy, *n* (%) | 54 (69.2) | |
| Treatment types in 54 patients receiving next-line therapies | | 12.0 |
| Child–Pugh class A at disease progression | 41/53 (77.4%) * | Not reached |
| Child–Pugh class B7 at disease progression | 5/9 (55.6%) * | 4.3 |
| Child–Pugh class B8–9 at disease progression | 7/10 (70%) * | 2.2 |
| Child–Pugh class C at disease progression | 1/6 (16.7%) * | 0.3 |
| ALBI grade 1 at disease progression | 18/21 (85.7%) + | Not reached |
| ALBI grade 2 at disease progression | 30/43 (69.8%) + | 10.3 |
| ALBI grade 3 at disease progression | 6/14 (42.9%) + | 2.5 |
| Tyrosine kinase inhibitor | 29 (53.7%) | Not reached |
| Levnatinib | 22 (40.7%) | Not reached |
| Cabozantinib | 6 (11.1%) | Not reached |
| Ramucirumab | 1 (1.9%) | No death event |
| Immune checkpoint inhibitor-based therapy | 13 (24.1%) | 11.9 |
| Pembrolizumab + Lenvatinib | 10 (18.5%) | 8.9 |
| Atezolizumab + Bevacizumab | 2 (3.7%) | 2.0 and 11.9 |
| Nivolumab | 1 (1.9%) | No death event |
| Transarterial chemoembolization | 7 (13%) | Not reached |
| Chemotherapy (FOLFOX: fluorouracil, leucovorin, oxaliplatin) | 5 (9.3%) | 10.3 |

* $p = 0.009$; + $p = 0.009$.

The median PPS was 10.3 months. The median PPS in patients with ALBI grade 1 was not reached, and was 10.3 and 1.9 months in patients with ALBI grades 2 and 3, respectively ($p < 0.001$, Figure 3A). The median PPS in patients with Child–Pugh class A was not reached, and was 3.7, 2.2 and 0.4 months in patients with Child–Pugh classes B7, B8–9 and C, respectively ($p < 0.001$, Figure 3B). By multivariate analysis, ABLI grade (2 vs. 1: HR = 4.499, $p = 0.006$; 3 vs. 1: HR = 26.926, $p < 0.001$), the presence of main portal vein invasion (HR = 5.102, $p = 0.007$, Figure 3C), a high tumor burden (HR = 9.296, $p < 0.001$, Figure 3D) and receiving next-line therapy (HR = 0.369, $p = 0.017$, Figure 3E) were independent predictors of PPS (Table 3 and Table S4).

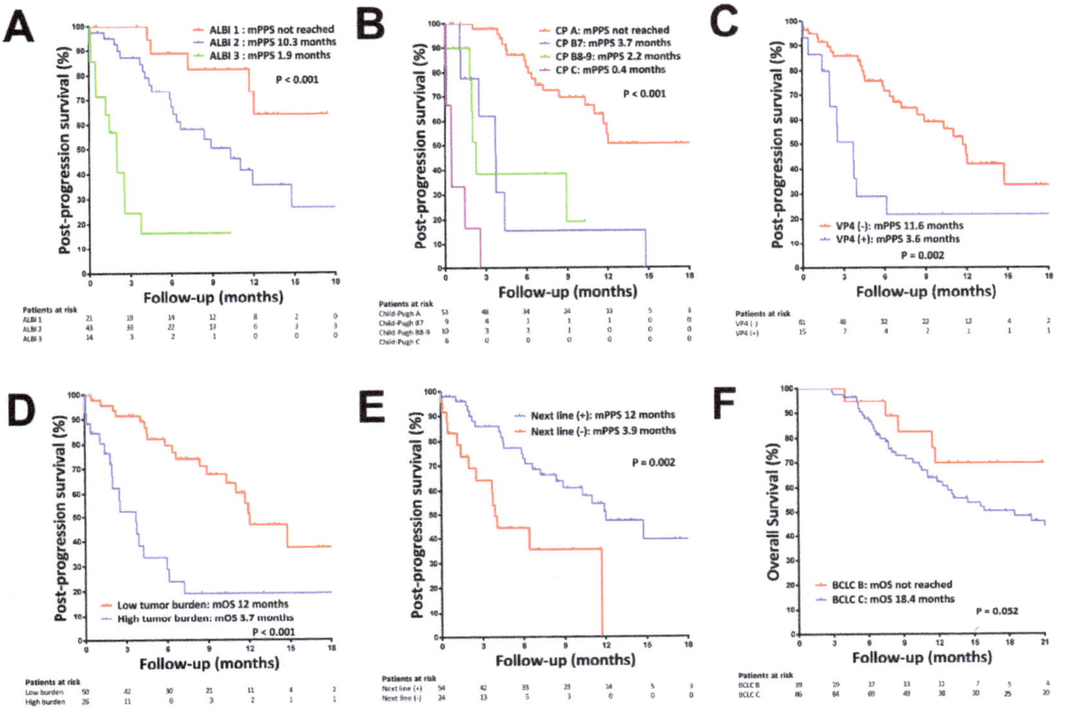

**Figure 3.** Kaplan–Meier curves for post-progression survival (PPS) after regorafenib failure and overall survival (OS) from starting sorafenib treatment. (**A**) PPS stratified by ALBI grade at disease progression. (**B**) PPS stratified by Child–Pugh class at disease progression. (**C**) PPS in patients with and without Vp4 vascular invasion. (**D**) PPS in patients with and without high tumor burden at disease progression. (**E**) PPS in patients who did and did not receive next-line therapy. (**F**) OS from starting sorafenib treatment stratified by BCLC stage.

Fifty-four patients (69.2%) received next-line therapy after disease progression, including twenty-nine (53.7%) patients who received TKI monotherapy (levnatinib 22, cabozantinib 6, ramucirumab 1), thirteen (24.1%) who received ICI-based therapy (pembrolizumab plus lenvatinib 10, atezolizumab plus bevacizumab 2, nivolumab 1), seven (13%) who received TACE and five (9.3%) who received chemotherapy (FOLFOX: fluorouracil, leucovorin, oxaliplatin) (Table 4). The percentages of patients who received next-line therapies were 77.4%, 55.6%, 70% and 16.7% in patients with Child–Pugh classes A, B7, B8–9 and C, respectively ($p = 0.009$), and were 85.7%, 69.8% and 42.9% in patients with ALBI grades 1, 2 and 3, respectively ($p = 0.009$). The median PPS in patients who received next-line therapies was 12.0 months, and the individual median PPS by different next-line therapy is shown in Table 4. There was no significant difference in PPS among patients treated with next-line TKI or ICI-based therapy ($p = 0.446$).

### *3.6. OS since the Start of Prior Sorafenib*

The median OS from the start of sorafenib treatment was 21.2 months. The median OS was not reached in patients classified as BCLC B and was 18.4 months in patients classified as BCLC C ($p = 0.052$, Figure 3F). The median OS was not significantly different in the second-line and the third- to fifth-line settings (21.2 vs.24.4 months $p = 0.982$, Figure S2D).

## 4. Discussion

In this study, we reported the detailed survival outcomes of regorafenib for HCC in the era of multiple-line sequential systemic therapy. The ORR of 10.7% and the PFS of 3.1 months in this study were consistent with the results from RESORCE and recent real-world reports [6,8,9,17,18]. The DCR of 43.7% was lower than that in RESORCE but was similar to the largest real-world report from Korea [9]. The median OS in this study was 14.7 months in patients with Child–Pugh class A, which was longer than the data from RESORCE and previous real-world reports. The median OS of 4.1 months in patients with Child–Pugh B was also similar to the recent Korean report on regorafenib for patients with Child–Pugh B [25]. The median PPS of 10.3 months in our study suggests that post-progression treatment after sorafenib–regorafenib failure may further improve the OS in the era of multiple-line sequential treatment [26].

In our study, the TTP in prior sorafenib treatment was the baseline predictor of PFS under regorafenib treatment, which is consistent with the results of prior reports [8,9]. Although patients with a shorter TTP on prior sorafenib had a poorer tumor response and PFS with regorafenib, an exploratory study from RESORCE showed a consistent TTP benefit over placebo, irrespective of TTP on prior sorafenib, suggesting that shorter TTP on sorafenib does not preclude the survival benefit of regorafenib for HCC [27].

The presence of HFSR and early AFP reduction during regorafenib treatment were on-treatment predictors of radiologic response, PFS and OS. Recent studies showed that HFSR was not only a predictor of survival on sorafenib [10], but also a significant predictor for patients with HCC on regorafenib treatment [9,17]. Early AFP reduction has been shown to be an early predictor of response and survival to sorafenib and ICI therapy [12,24]. Our data showed that early AFP reduction also had a prognostic role for regorafenib treatment.

Compatible with our findings, the ALBI score has been shown to be a predictor of HCC across the diverse BCLC stages, including patients who received sorafenib–regorafenib sequential therapy [11,28–30]. Several studies also reported that the presence of vascular invasion was a poor prognostic factor after sorafenib failure [31–33]. The PROSASH-II model, which comprised albumin, bilirubin, vascular invasion, extrahepatic spread, tumor size and AFP, has been shown to have good discriminative value in predicting the survival of patients with HCC receiving sorafenib treatment [15]. We also confirmed that the PROSASH-II model could discriminate PFS and OS in patients on regorafenib treatment. Based on the independent predictors of OS, we propose simple baseline and on-treatment risk scores that also have good discriminative value for predicting OS after regorafenib treatment. The risk scores could assist physicians with outcome prediction and considering an early switch to next-line treatment for patients with a high risk score.

The predictors of PPS and the impact of post-progression treatment after regorafenib failure remain unclear. In this study, the median PPS was 10.3 months, and 25% and 32% of patients showed a deterioration of Child–Pugh class and ALBI grade, respectively. Liver function reserve is an important determinant of PPS in this study, and patients with liver dysfunction at PD had less chance of receiving next-line therapy. In patients who maintained Child–Pugh A or ALBI grade 1, the median PPS was not reached during the observation period, whereas survival was significantly poorer in patients with liver function deterioration. Although progression patterns may have a prognostic impact after sorafenib failure [13,14], we did not observe a significant correlation between progression pattern and PPS after regorafenib failure, possibly due to the subsequent treatments after regorafenib failure. Next-line systemic therapy was shown to be an independent predictor of PPS after regorafenib failure, and the median PPS was 12 months in patients who were able to receive next-line therapy. The optimal third-line therapy after sorafenib–regorafenib failure remains unclear. Current guidelines and experts' opinions suggest that other options for systemic agents could be applied as multiple-line sequential therapy [16,26,34]. In clinical practice, lenvatinib with or without ICI is the preferred subsequent systemic treatment after regorafenib, followed by cabozantinib. We did not observe a significant difference in PPS among patients treated with next-line TKI or ICI-based therapy. Although lenvatinib has only been evaluated in the first-line setting, recent real-world studies showed that lenvatinib could have survival benefits in the third-line setting after regorafenib failure [17,18,35]. In 2020, the phase Ib study of lenvatinib plus pembrolizumab showed promising results of high ORR and improved OS in the first-line setting [36], and this combination could also be a treatment option after sorafenib–regorafenib failure. Cabozantinib is the only systemic agent that has been investigated in the third-line setting in the CELESTIAL trial, and the survival benefit of cabozantinib is independent of the duration of prior sorafenib treatment [37]. Other treatment options, including ramucirumab, atezolizumab plus bevacizumab, and nivolumab, have also been applied as multiple-line sequential treatment options in real-world practice. In view of the PPS from our data according to different next-line systemic agents, lenvatinib or cabozantinib may be considered following sorafenib–regorafenib failure. In addition, lenvatinib plus ICI in combination with broadening modes of action might also be an option [38].

In the RESORCE trial, the median OSs from starting sorafenib were 26.0 and 21.5 months in the overall cohort and the Asian subgroup, respectively. Other real-world studies from Asia reported an OS of 25.3 to 28.5 months from starting sorafenib [9,17]. In this study, the median OSs from starting sorafenib were 28.3 and 13.1 months in patients with Child–Pugh classes A and B, respectively, and were 35.5 and 13 months in patients with ALBI grades 1 and 2, respectively. Consistent with previous studies, our data underline the crucial role of preserved liver function in the administration of multi-line sequential therapy and improved survival [39].

There are some limitations in this study. First, this is a retrospective study. Unintentional biases might exist in patient enrollment and the evaluation of clinical outcomes. Nevertheless, the National Health Insurance program in Taiwan enforced the strict regulation of clinical and image follow-up for the reimbursement of targeted therapies. Therefore, the majority of patients had regular clinical and image evaluations during sorafenib and regorafenib treatment for further drug reimbursement. Second, this is a single-center study from Taiwan, and the majority of patients had underlying HBV infections. Our findings need to be validated in other ethnicities and in HCC with other etiologies. Third, quality of life is an important issue during the application of systemic therapies for patients with HCC. However, quality of life measurements were not available in this retrospective study. Although TKI-related adverse events have adverse impacts on quality of life, patients with HFSR conferred better PFS and OS in our data.

## 5. Conclusions

In conclusion, the survival outcomes of regorafenib for patients with HCC were consistent with those of the phase III trial result. Survival predictors and responses to sorafenib had a prognostic role in patients with HCC undergoing regorafenib treatment. Subsequent systemic therapy plays an important role in survival after regorafenib failure.

**Supplementary Materials:** The following supporting information can be downloaded at: https://www.mdpi.com/article/10.3390/cancers14082014/s1, Figure S1: Kaplan–Meier curves for progression-free survival (PFS) and overall survival (OS) after regorafenib treatment. Figure S2: Kaplan–Meier curves for overall survival (OS) in patients with HCC receiving regorafenib treatment. Table S1: Best radiologic response to regorafenib therapy by RECIST v1.1 criteria. Table S2: Univariate and multivariate analyses of factors associated with progression-free survival. Table S3: Univariate and multivariate analyses of factors associated with overall survival. Table S4: Univariate and multivariate analyses of factors associated with post-progression survival.

**Author Contributions:** I.-C.L.: data acquisition; analysis and interpretation of data; drafting of the manuscript; statistical analysis. C.-T.C., C.-J.W., P.-C.L., S.-C.C., Y.C., K.-C.W. and M.-C.H.: data acquisition. Y.-H.H. and Y.C.: study concept and design; critical revision of the manuscript for important intellectual content; study supervision. All authors have read and agreed to the published version of the manuscript.

**Funding:** The study was supported by grants from Taipei Veterans General Hospital, Taipei, Taiwan (V110C-094, V110C-144, V111C-114, V111C-107), and the Ministry of Science and Technology, Taiwan (MOST 109-2628-B-075-022, MOST 109-2314-B-010-034-MY3, MOST 110-2314-B-075-052).

**Institutional Review Board Statement:** This study adhered to the guidelines of the Declaration of Helsinki and gained consent from the Institutional Review Board at Taipei Veterans General Hospital (IRB number: 2021-04-006BC).

**Informed Consent Statement:** Due to the retrospective nature of the study, the Institutional Review Board waived the need for written informed consent.

**Data Availability Statement:** The data that support the findings of this study are available from the corresponding author upon reasonable request.

**Acknowledgments:** The authors thank the Clinical Research Core Laboratory, Taipei Veterans General Hospital for providing their facilities to conduct this study.

**Conflicts of Interest:** Y.-H.H. has received research grants from Gilead Sciences and Bristol-Meyers Squibb, and honoraria from Abbvie, Gilead Sciences, Bristol-Meyers Squibb, Ono Pharmaceutical, Merck Sharp & Dohme, Eisai, Eli Lilly, Ipsen and Roche, and has served in an advisory role for Abbvie, Gilead Sciences, Bristol-Meyers Squibb, Ono Pharmaceuticals, Eisai, Eli Lilly, Ipsen, Merck Sharp & Dohme and Roche. I.-C.L. has received honoraria from Gilead Sciences, Bristol-Meyers Squibb, Abbvie, Merck Sharp & Dohme, Bayer, Eisai, Ipsen and Roche, and has served in an advisory role for Gilead Sciences. Other authors declare no conflict of interest.

## References

1. Arnold, M.; Abnet, C.C.; Neale, R.E.; Vignat, J.; Giovannucci, E.L.; McGlynn, K.A.; Bray, F. Global Burden of 5 Major Types of Gastrointestinal Cancer. *Gastroenterology* **2020**, *159*, 335–349.e315. [CrossRef] [PubMed]
2. Llovet, J.M.; Kelley, R.K.; Villanueva, A.; Singal, A.G.; Pikarsky, E.; Roayaie, S.; Lencioni, R.; Koike, K.; Zucman-Rossi, J.; Finn, R.S. Hepatocellular carcinoma. *Nat. Rev. Dis. Primers* **2021**, *7*, 6. [CrossRef] [PubMed]
3. Llovet, J.M.; Montal, R.; Villanueva, A. Randomized trials and endpoints in advanced HCC: Role of PFS as a surrogate of survival. *J. Hepatol.* **2019**, *70*, 1262–1277. [CrossRef] [PubMed]
4. Llovet, J.M.; Ricci, S.; Mazzaferro, V.; Hilgard, P.; Gane, E.; Blanc, J.F.; de Oliveira, A.C.; Santoro, A.; Raoul, J.L.; Forner, A.; et al. Sorafenib in advanced hepatocellular carcinoma. *N. Engl. J. Med.* **2008**, *359*, 378–390. [CrossRef]
5. Cheng, A.L.; Kang, Y.K.; Chen, Z.; Tsao, C.J.; Qin, S.; Kim, J.S.; Luo, R.; Feng, J.; Ye, S.; Yang, T.S.; et al. Efficacy and safety of sorafenib in patients in the Asia-Pacific region with advanced hepatocellular carcinoma: A phase III randomised, double-blind, placebo-controlled trial. *Lancet Oncol.* **2009**, *10*, 25–34. [CrossRef]
6. Bruix, J.; Qin, S.; Merle, P.; Granito, A.; Huang, Y.H.; Bodoky, G.; Pracht, M.; Yokosuka, O.; Rosmorduc, O.; Breder, V.; et al. Regorafenib for patients with hepatocellular carcinoma who progressed on sorafenib treatment (RESORCE): A randomised, double-blind, placebo-controlled, phase 3 trial. *Lancet* **2017**, *389*, 56–66. [CrossRef]
7. Strumberg, D.; Schultheis, B. Regorafenib for cancer. *Expert Opin. Investig. Drugs* **2012**, *21*, 879–889. [CrossRef]
8. Lee, M.J.; Chang, S.W.; Kim, J.H.; Lee, Y.S.; Cho, S.B.; Seo, Y.S.; Yim, H.J.; Hwang, S.Y.; Lee, H.W.; Chang, Y.; et al. Real-world systemic sequential therapy with sorafenib and regorafenib for advanced hepatocellular carcinoma: A multicenter retrospective study in Korea. *Investig. New Drugs* **2021**, *39*, 260–268. [CrossRef]
9. Yoo, C.; Byeon, S.; Bang, Y.; Cheon, J.; Kim, J.W.; Kim, J.H.; Chon, H.J.; Kang, B.; Kang, M.J.; Kim, I.; et al. Regorafenib in previously treated advanced hepatocellular carcinoma: Impact of prior immunotherapy and adverse events. *Liver Int.* **2020**, *40*, 2263–2271. [CrossRef]
10. Diaz-Gonzalez, A.; Sanduzzi-Zamparelli, M.; Sapena, V.; Torres, F.; LLarch, N.; Iserte, G.; Forner, A.; da Fonseca, L.; Rios, J.; Bruix, J.; et al. Systematic review with meta-analysis: The critical role of dermatological events in patients with hepatocellular carcinoma treated with sorafenib. *Aliment Pharmacol. Ther.* **2019**, *49*, 482–491. [CrossRef]
11. Lee, P.C.; Chen, Y.T.; Chao, Y.; Huo, T.I.; Li, C.P.; Su, C.W.; Lee, M.H.; Hou, M.C.; Lee, F.Y.; Lin, H.C.; et al. Validation of the albumin-bilirubin grade-based integrated model as a predictor for sorafenib-failed hepatocellular carcinoma. *Liver Int.* **2018**, *38*, 321–330. [CrossRef] [PubMed]
12. Shao, Y.Y.; Lin, Z.Z.; Hsu, C.; Shen, Y.C.; Hsu, C.H.; Cheng, A.L. Early alpha-fetoprotein response predicts treatment efficacy of antiangiogenic systemic therapy in patients with advanced hepatocellular carcinoma. *Cancer* **2010**, *116*, 4590–4596. [CrossRef] [PubMed]
13. Reig, M.; Rimola, J.; Torres, F.; Darnell, A.; Rodriguez-Lope, C.; Forner, A.; Llarch, N.; Rios, J.; Ayuso, C.; Bruix, J. Postprogression survival of patients with advanced hepatocellular carcinoma: Rationale for second-line trial design. *Hepatology* **2013**, *58*, 2023–2031. [CrossRef] [PubMed]
14. Lee, I.C.; Chen, Y.T.; Chao, Y.; Huo, T.I.; Li, C.P.; Su, C.W.; Lin, H.C.; Lee, F.Y.; Huang, Y.H. Determinants of survival after sorafenib failure in patients with BCLC-C hepatocellular carcinoma in real-world practice. *Medicine* **2015**, *94*, e688. [CrossRef] [PubMed]

15. Labeur, T.A.; Berhane, S.; Edeline, J.; Blanc, J.F.; Bettinger, D.; Meyer, T.; Van Vugt, J.L.A.; Ten Cate, D.W.G.; De Man, R.A.; Eskens, F.; et al. Improved survival prediction and comparison of prognostic models for patients with hepatocellular carcinoma treated with sorafenib. *Liver Int.* **2020**, *40*, 215–228. [CrossRef]
16. Bruix, J.; Chan, S.L.; Galle, P.R.; Rimassa, L.; Sangro, B. Systemic treatment of hepatocellular carcinoma: An EASL position paper. *J. Hepatol.* **2021**, *75*, 960–974. [CrossRef]
17. Wang, W.; Tsuchiya, K.; Kurosaki, M.; Yasui, Y.; Inada, K.; Kirino, S.; Yamashita, K.; Sekiguchi, S.; Hayakawa, Y.; Osawa, L.; et al. Sorafenib-Regorafenib Sequential Therapy in Japanese Patients with Unresectable Hepatocellular Carcinoma-Relative Dose Intensity and Post-Regorafenib Therapies in Real World Practice. *Cancers* **2019**, *11*, 1517. [CrossRef]
18. Ogasawara, S.; Ooka, Y.; Itokawa, N.; Inoue, M.; Okabe, S.; Seki, A.; Haga, Y.; Obu, M.; Atsukawa, M.; Itobayashi, E.; et al. Sequential therapy with sorafenib and regorafenib for advanced hepatocellular carcinoma: A multicenter retrospective study in Japan. *Investig. New Drugs* **2020**, *38*, 172–180. [CrossRef]
19. Rimini, M.; Yoo, C.; Lonardi, S.; Masi, G.; Piscaglia, F.; Kim, H.D.; Rizzato, M.D.; Salani, F.; Ielasi, L.; Forgione, A.; et al. Role of the prognostic nutritional index in predicting survival in advanced hepatocellular carcinoma treated with regorafenib. *Hepatol. Res.* **2021**, *51*, 796–802. [CrossRef]
20. Heimbach, J.K.; Kulik, L.M.; Finn, R.S.; Sirlin, C.B.; Abecassis, M.M.; Roberts, L.R.; Zhu, A.X.; Murad, M.H.; Marrero, J.A. AASLD guidelines for the treatment of hepatocellular carcinoma. *Hepatology* **2018**, *67*, 358–380. [CrossRef]
21. Johnson, P.J.; Berhane, S.; Kagebayashi, C.; Satomura, S.; Teng, M.; Reeves, H.L.; O'Beirne, J.; Fox, R.; Skowronska, A.; Palmer, D.; et al. Assessment of liver function in patients with hepatocellular carcinoma: A new evidence-based approach-the ALBI grade. *J. Clin. Oncol.* **2015**, *33*, 550–558. [CrossRef] [PubMed]
22. Kudo, M.; Finn, R.S.; Qin, S.; Han, K.H.; Ikeda, K.; Piscaglia, F.; Baron, A.; Park, J.W.; Han, G.; Jassem, J.; et al. Lenvatinib versus sorafenib in first-line treatment of patients with unresectable hepatocellular carcinoma: A randomised phase 3 non-inferiority trial. *Lancet* **2018**, *391*, 1163–1173. [CrossRef]
23. Lencioni, R.; Llovet, J.M. Modified RECIST (mRECIST) assessment for hepatocellular carcinoma. *Semin. Liver Dis.* **2010**, *30*, 52–60. [CrossRef] [PubMed]
24. Lee, P.C.; Chao, Y.; Chen, M.H.; Lan, K.H.; Lee, C.J.; Lee, I.C.; Chen, S.C.; Hou, M.C.; Huang, Y.H. Predictors of Response and Survival in Immune Checkpoint Inhibitor-Treated Unresectable Hepatocellular Carcinoma. *Cancers* **2020**, *12*, 182. [CrossRef]
25. Kim, H.D.; Bang, Y.; Lee, M.A.; Kim, J.W.; Kim, J.H.; Chon, H.J.; Kang, B.; Kang, M.J.; Kim, I.; Cheon, J.; et al. Regorafenib in patients with advanced Child-Pugh B hepatocellular carcinoma: A multicentre retrospective study. *Liver Int.* **2020**, *40*, 2544–2552. [CrossRef] [PubMed]
26. Kudo, M. Impact of Multi-Drug Sequential Therapy on Survival in Patients with Unresectable Hepatocellular Carcinoma. *Liver Cancer* **2021**, *10*, 1–9. [CrossRef]
27. Finn, R.S.; Merle, P.; Granito, A.; Huang, Y.H.; Bodoky, G.; Pracht, M.; Yokosuka, O.; Rosmorduc, O.; Gerolami, R.; Caparello, C.; et al. Outcomes of sequential treatment with sorafenib followed by regorafenib for HCC: Additional analyses from the phase III RESORCE trial. *J. Hepatol.* **2018**, *69*, 353–358. [CrossRef]
28. Demirtas, C.O.; D'Alessio, A.; Rimassa, L.; Sharma, R.; Pinato, D.J. ALBI grade: Evidence for an improved model for liver functional estimation in patients with hepatocellular carcinoma. *JHEP Rep.* **2021**, *3*, 100347. [CrossRef]
29. Lee, I.C.; Hung, Y.W.; Liu, C.A.; Lee, R.C.; Su, C.W.; Huo, T.I.; Li, C.P.; Chao, Y.; Lin, H.C.; Hou, M.C.; et al. A new ALBI-based model to predict survival after transarterial chemoembolization for BCLC stage B hepatocellular carcinoma. *Liver Int.* **2019**, *39*, 1704–1712. [CrossRef]
30. Wang, H.W.; Chuang, P.H.; Su, W.P.; Kao, J.T.; Hsu, W.F.; Lin, C.C.; Huang, G.T.; Lin, J.T.; Lai, H.C.; Peng, C.Y. On-Treatment Albumin-Bilirubin Grade: Predictor of Response and Outcome of Sorafenib-Regorafenib Sequential Therapy in Patients with Unresectable Hepatocellular Carcinoma. *Cancers* **2021**, *13*, 3758. [CrossRef]
31. Terashima, T.; Yamashita, T.; Sunagozaka, H.; Arai, K.; Kawaguchi, K.; Kitamura, K.; Yamashita, T.; Sakai, Y.; Mizukoshi, E.; Honda, M.; et al. Analysis of the liver functional reserve of patients with advanced hepatocellular carcinoma undergoing sorafenib treatment: Prospects for regorafenib therapy. *Hepatol. Res.* **2018**, *48*, 956–966. [CrossRef] [PubMed]
32. Kuzuya, T.; Ishigami, M.; Ito, T.; Ishizu, Y.; Honda, T.; Ishikawa, T.; Hirooka, Y.; Fujishiro, M. Clinical characteristics and outcomes of candidates for second-line therapy, including regorafenib and ramucirumab, for advanced hepatocellular carcinoma after sorafenib treatment. *Hepatol. Res.* **2019**, *49*, 1054–1065. [CrossRef] [PubMed]
33. Uchikawa, S.; Kawaoka, T.; Aikata, H.; Kodama, K.; Nishida, Y.; Inagaki, Y.; Hatooka, M.; Morio, K.; Nakahara, T.; Murakami, E.; et al. Clinical outcomes of sorafenib treatment failure for advanced hepatocellular carcinoma and candidates for regorafenib treatment in real-world practice. *Hepatol. Res.* **2018**, *48*, 814–820. [CrossRef] [PubMed]
34. Vogel, A.; Martinelli, E.; ESMO Guidelines Committee. Updated treatment recommendations for hepatocellular carcinoma (HCC) from the ESMO Clinical Practice Guidelines. *Ann. Oncol.* **2021**, *32*, 801–805. [CrossRef]
35. Hiraoka, A.; Kumada, T.; Hatanaka, T.; Tada, T.; Kariyama, K.; Tani, J.; Fukunishi, S.; Atsukawa, M.; Hirooka, M.; Tsuji, K.; et al. Therapeutic efficacy of lenvatinib as third-line treatment after regorafenib for unresectable hepatocellular carcinoma progression. *Hepatol. Res.* **2021**, *51*, 880–889. [CrossRef]
36. Finn, R.S.; Ikeda, M.; Zhu, A.X.; Sung, M.W.; Baron, A.D.; Kudo, M.; Okusaka, T.; Kobayashi, M.; Kumada, H.; Kaneko, S.; et al. Phase Ib Study of Lenvatinib Plus Pembrolizumab in Patients with Unresectable Hepatocellular Carcinoma. *J. Clin. Oncol.* **2020**, *38*, 2960–2970. [CrossRef]

37. Abou-Alfa, G.K.; Meyer, T.; Cheng, A.L.; El-Khoueiry, A.B.; Rimassa, L.; Ryoo, B.Y.; Cicin, I.; Merle, P.; Chen, Y.; Park, J.W.; et al. Cabozantinib in Patients with Advanced and Progressing Hepatocellular Carcinoma. *N. Engl. J. Med.* **2018**, *379*, 54–63. [CrossRef]
38. Wu, C.J.; Lee, P.C.; Hung, Y.W.; Lee, C.J.; Chi, C.T.; Lee, I.C.; Hou, M.C.; Huang, Y.H. Lenvatinib plus pembrolizumab for systemic therapy-naïve and -experienced unresectable hepatocellular carcinoma. *Cancer Immunol. Immunother.* **2022**. *online ahead of print*. [CrossRef]
39. Kirstein, M.M.; Scheiner, B.; Marwede, T.; Wolf, C.; Voigtlander, T.; Semmler, G.; Wacker, F.; Manns, M.P.; Hinrichs, J.B.; Pinter, M.; et al. Sequential systemic treatment in patients with hepatocellular carcinoma. *Aliment Pharmacol. Ther.* **2020**, *52*, 205–212. [CrossRef]

*cancers*

MDPI

*Article*

# Efficacy and Safety of Atezolizumab and Bevacizumab in the Real-World Treatment of Advanced Hepatocellular Carcinoma: Experience from Four Tertiary Centers

**Vera Himmelsbach [1], Matthias Pinter [2,3], Bernhard Scheiner [2,3], Marino Venerito [4], Friedrich Sinner [4], Carolin Zimpel [5,6], Jens U. Marquardt [5,6], Jörg Trojan [1,7], Oliver Waidmann [1,7,†] and Fabian Finkelmeier [1,7,8,*,†]**

1 Department of Gastroenterology, Hepatology and Endocrinology, University Hospital Frankfurt, 60590 Frankfurt, Germany; vera.himmelsbach@kgu.de (V.H.); joerg.trojan@kgu.de (J.T.); oliver.waidmann@kgu.de (O.W.)
2 Division of Gastroenterology and Hepatology, Department of Internal Medicine III, Medical University of Vienna, 1090 Vienna, Austria; matthias.pinter@meduniwien.ac.at (M.P.); bernhard.scheiner@meduniwien.ac.at (B.S.)
3 Liver Cancer (HCC) Study Group Vienna, Medical University of Vienna, 1090 Vienna, Austria
4 Department of Gastroenterology, Hepatology and Infectious Diseases, Otto-Von Guericke University Hospital, 39120 Magdeburg, Germany; m.venerito@med.ovgu.de (M.V.); friedrich.sinner@med.ovgu.de (F.S.)
5 Department of Internal Medicine I, Johannes Gutenberg University, 55131 Mainz, Germany; carolin.czauderna@uksh.de (C.Z.); jens.marquardt@uksh.de (J.U.M.)
6 Department of Medicine I, University Medical Centre Schleswig-Holstein, Campus Lübeck, 23538 Lübeck, Germany
7 University Cancer Center Frankfurt, University Hospital Frankfurt, 60590 Frankfurt, Germany
8 Frankfurt Cancer Institute, Goethe University Frankfurt/Main, 60590 Frankfurt, Germany
* Correspondence: fabian.finkelmeier@kgu.de; Tel.: +49-69-6301-0
† These authors contributed equally to this work.

**Citation:** Himmelsbach, V.; Pinter, M.; Scheiner, B.; Venerito, M.; Sinner, F.; Zimpel, C.; Marquardt, J.U.; Trojan, J.; Waidmann, O.; Finkelmeier, F. Efficacy and Safety of Atezolizumab and Bevacizumab in the Real-World Treatment of Advanced Hepatocellular Carcinoma: Experience from Four Tertiary Centers. *Cancers* **2022**, *14*, 1722. https://doi.org/10.3390/cancers14071722

Academic Editors: Georgios Germanidis and David A. Geller

Received: 7 February 2022
Accepted: 25 March 2022
Published: 28 March 2022



**Simple Summary:** Hepatocellular carcinoma is one of the most common cancers in the world with increasing incidence. In advanced stages, according to the Barcelona Clinic Liver Cancer (BCLC) staging defined by number, size, vessel infiltration status, and patient's performance status, the therapy of choice is systemic therapy. For several years, the tyrosine kinase inhibitor sorafenib was the only therapeutic option. Atezolizumab and bevacizumab are administered as a combination therapy promoting PD-L1 inhibition and anti-VEGF activity, which yields synergistic effects against cancer. The IMBRAVE150 trial investigated the use of this combination therapy versus that of sorafenib and showed an increase in overall patient survival to nearly 20 months. In this work, we investigated the real-world efficacy and safety of this combination in different centers.

**Abstract:** The combination of atezolizumab and bevacizumab (A + B) is the new standard of care for the systemic first-line treatment of hepatocellular carcinoma (HCC). However, up to now there are only few data on the safety and efficacy of A + B in real life. We included patients with advanced HCC treated with A + B as first-line therapy at four cancer centers in Germany and Austria between December 2018 and August 2021. Demographics, overall survival (OS), and adverse events were assessed until 15 September 2021. We included 66 patients. Most patients had compensated cirrhosis (n = 34; 52%), while Child–Pugh class B cirrhosis was observed in 23 patients (35%), and class C cirrhosis in 5 patients (8%). The best responses included a complete response (CR) in 7 patients (11%), a partial response (PR) in 12 patients (18%), stable disease (SD) in 22 patients (33%), and progressive disease in 11 patients (17%). The median progression-free (PFS) survival was 6.5 months, while the median overall survival (OS) was not reached in this cohort (6-month OS: 69%, 12-month OS: 60%, 18-month OS: 58%). Patients with viral hepatitis seemed to have a better prognosis than patients with HCC of non-viral etiology. The real-world PFS and OS were comparable to those of the pivotal IMBRAVE trial, despite including patients with worse liver function in this study. We conclude that A + B is also highly effective in a real-life setting, with manageable toxicity, especially in patients with compensated liver disease. In patients with compromised liver function (Child B and C), the treatment showed low efficacy and, therefore, it should be well considered before administration to these patients.

**Keywords:** hepatocellular carcinoma; atezolizumab; bevacizumab; real world; immunotherapy

## 1. Introduction

Hepatocellular carcinoma (HCC) is the most frequent malignant primary liver cancer and the third leading reason of cancer-related death [1]. Immunotherapy is active and well tolerated in patients with advanced HCC. However, due to formally negative phase 3 trials, the anti-programmed cell death protein 1 (PD-1) antibodies nivolumab and pembrolizumab have not been approved in Europe [2,3]. The IMbrave150 study investigated the combined therapy with the anti-PD-L1 antibody atezolizumab and the vascular endothelial growth factor (VEGF)-targeting antibody bevacizumab compared to treatment with sorafenib in a phase 3 trial. The trial reached its coprimary end points of improving overall survival and progression-free survival and showed a favorable quality of life in the immunotherapy arm [4–6]. Based on these results, atezolizumab and bevacizumab have become the new standard of care for advanced HCC and represent the first immune checkpoint inhibitor-based combination therapy approved for HCC. However, in a real-life setting, many patients at need for anticancer treatment do not fulfill all inclusion criteria of the phase 3 trials, mainly due to impaired liver function [7]. Therefore, data from regular prescription and treatment are urgently needed. Here, we analyzed data from HCC patients treated with atezolizumab/bevacizumab in four referral centers.

## 2. Materials and Methods

### 2.1. Study Design and Selection of Patients

This was a retrospective study of patients treated with atezolizumab and bevacizumab s first-line therapy across four academic hospitals in Germany and Austria.

All patients with confirmed HCC treated with atezolizumab and bevacizumab in the individual centers between December 2018 and August 2021 were included into the analysis. Several patients were treated before the EMA approval, as a result of exceptional approvals by health care insurance companies to cover the costs.

Patients' data including history of the disease, treatment course, laboratory results, radiological data, and follow-up were collected retrospectively from patients' files.

The study was performed in accordance with the 1975 Declaration of Helsinki. The retrospective analysis was approved by the local Ethics Committee (SGI03/18, Amendment 01/19) as well as by the Ethics Committees of the individual centers.

### 2.2. Assessments

Electronic hospital charts were retrospectively analyzed for baseline demographic data and laboratory results.

Radiological response was recorded by computed tomography (CT) or magnetic resonance imaging (MRI) at baseline, 6–12 weeks after treatment initiation, and about every 2–3 months thereafter according to the local guidelines. Tumor response was assessed according to the Response Evaluation Criteria in Solid Tumors (RECIST) V1.1 [8] or modified RECIST [9] (according to centers' preference). Side effects were recorded at every visit and graded according to the Common Terminology Criteria for Adverse Events (CTCAE) version 4.0 [10] or 5.0 [11] according to centers' preference.

### 2.3. Atezolizumab and Bevacizumab

Atezolizumab and bevacizumab are approved by the European Medicines Agency (EMA) and the United States Food and Drug Administration (FDA) for the treatment of patients with HCC who have not yet received systemic treatment for HCC. The recommended doses are 1200 mg for atezolizumab and 15 mg/kg for Bevacizumab every three weeks. Treatment with the two drugs and discontinuation were performed according to the recommendations of the manufacturer and at the discretion of the treating physician.

### 2.4. Statistical Analysis

The present study is as a retrospective cohort study. All patients were followed until death or last contact. The primary end point was overall survival (OS), the secondary end points included progression-free survival (PFS), response rate, occurrence of bleeding complications, and safety.

Data on baseline characteristics, radiological response, and adverse events were summarized using descriptive statistics. Continuous variables are shown as median and full range, and categorical variables are reported as frequencies and percentages. Median duration of therapy was defined as the time from the first administration until the last administration of the drugs. Patients who still received atezolizumab with or without concomitant bevacizumab at data cut-off were censored. Patients with at least one staging imaging assessment were evaluated for radiological response.

Data from patients, who died without radiologically confirmed tumor progression, were censored at the date of the last radiological assessment or death. Progression-free survival (PFS) was defined as the time from the date of the first therapy administration until radiological disease progression or death, whatever occurred first. Patients still alive and without radiologically confirmed progression at the date of the last contact or data cut-off were censored. Overall survival (OS) was defined as the time from the start of the treatment with atezolizumab and bevacizumab until the date of death. Survival curves were calculated with the Kaplan–Meier method and compared by means of the log-rank test. To analyze prognostic parameters uni- and multivariable Cox regression models with forward stepwise likelihood ratio were used. Statistical analyses were performed with SPSS (Version 27.0, IBM, New York, NY, USA) and GraphPad Prism 8.0 (GraphPad Software, La Jolla, CA, USA). Differences between different patient cohorts were determined using the nonparametric Wilcoxon–Mann–Whitney and Kruskal–Wallis tests or Fisher's exact text. For the sub-analysis of multiple comparisons, the Bonferroni correction was used. $p$ values < 0.05 were considered significant.

## 3. Results

### 3.1. Patients

Sixty-six patients from four centers (one Austrian center, and three German centers) were included. Data cut-off for the analysis was 15 September 2021. Fifty-four patients (82%) were male, and the median age was 66 years (range 30–89 years). Additional baseline characteristics are shown in Table 1. Most patients had compensated cirrhosis (n = 34; 52%), while Child–Pugh class B cirrhosis was observed in 23 patients (35%), and class C cirrhosis in 5 patients (8%). For 4 patients, Child–Pugh assessment was not reported.

The median follow-up time was 211 days, with a range of 1 to 995 days from atezolizumab and Bevacizumab treatment initiation. Twenty-seven patients had died at the date of data cut-off.

**Table 1.** Patient characteristics.

| Parameter | Patients |
| --- | --- |
| Epidemiology | |
| Patients, n | 66 |
| Gender, m/f (%) | 54/12 (81.8/18.2) |
| Age, median, range | 65 (30–88) |
| Etiology of liver disease | |
| Alcohol, n (%) | 25 (37.9) |
| Hepatitis C, n (%) | 14 (21.2) |

**Table 1.** *Cont.*

| Parameter | Patients |
| --- | --- |
| Hepatitis B, n (%) | 9 (13.6) |
| NASH/NAFLD [1], n (%) | 18. (27.3) |
| BCLC stage [2] | |
| A, n (%) | 1 (1.5) |
| B, n (%) | 22 (33.3) |
| C, n (%) | 35 (53.0) |
| D, n (%) | 8 (12.1) |
| MVI [3], n (%) | 29 (43.9) |
| EHS [4], n (%) | 18 (27.3) |
| Child–Pugh score | |
| A, n (%) | 35 (53.0) |
| B, n (%) | 23 (34.8) |
| C, n (%) | 5 (7.6) |
| Albumin–Bilirubin (ALBI) grade | |
| 1, n (%) | 14 (21.2) |
| 2, n (%) | 46 (69.7) |
| 3, n (%) | 6 (9.1) |
| MELD [5], median, range | 10 (6–23) |
| Betablocker medication, n (%) | 38 (59.1) |
| Prior Treatment | |
| Resection, n (%) | 9 (13.6) |
| Local ablation *, n (%) | 11 (16.7) |
| Loco-regional (TACE/SIRT) [6], n (%) | 27 (40.9) |
| Laboratory results | |
| BMI [7], median, range | 27.6 (16.9–42.5) |
| ALT [8] (U/L), median, range | 42 (7–1260) |
| AST [9] (U/L), median, range | 64 (10–876) |
| Bilirubin (mg/dL), median, range | 1.5 (0.2–9.4) |
| Albumin (g/dL), median, range | 3.2 (1.8–4.4) |
| INR [10], mean, median, range | 1.27 (0.69–2.99) |
| CRP [11] (mg/dL), median, range | 1.1 (0.15–10.9) |
| AFP [12] (ng/mL), median, range | 17.65 (1–49220) |
| AFP > 400 ng/mL, n (%) | 19 (28.8) |

Abbreviations: [1] NASH, non-alcoholic steatohepatitis; [2] BCLC, Barcelona liver clinic; [3] MVI, Macrovascular invasion; [4] EHS, Extrahepatic spread; [5] MELD, model of end-stage liver disease; [6] TACE/SIRT, transarterial chemoembolization/selective internal radiotherapy; [7] BMI, Body Mass Index; [8] ALT, alanine aminotransferase, [9] AST, aspartate aminotransferase; [10] INR, internationalized ratio; [11] CRP, C-reactive protein; [12] AFP, alpha-Fetoprotein. * including radiofrequency ablation (RFA), microwave ablation (MWA).

### 3.2. *Efficacy*

At last contact, 51 patients (77%) had stopped atezolizumab and bevacizumab treatment. Thirteen patients (20%) were still on treatment, and two patients (3%) were lost to follow-up. The median time of treatment was 110 days (+995 days, 33 month), and the median number of cycles administered to the patients was 21. Overall, 52 patients (79%) had at least one follow-up imaging for the assessment of tumor response. Best responses included complete response (CR) in 7 patients (11%), partial response (PR) in 12 patients (18%), stable disease (SD) in 22 patients (33%), and progressive disease in 11 patients (17%) (Table 2); for 14 patients (21.1%), staging at data analysis was not available, and therefore, they were not evaluable for best response. The median progression-free (PFS) survival was 6.5 months (95% confidence interval (CI) of 4.0–9.1 months) (Figure 1A).

Patients with HCC due to viral hepatitis had a more favorable PFS (median PFS 17.3 months, 95% confidence interval (CI) of 5.6–29 months) than patients without a history of viral hepatitis (median PFS 6.1 months, 95% confidence interval (CI) of 3.1–8.9 months), corresponding to a hazard ratio (HR) of 0.48 with a 95% CI of 0.24–0.99 ($p < 0.05$) (Figure 1B).

Median overall survival (OS) was not reached in this cohort (6-month OS: 69%, 12-month OS: 60%, 18-month OS: 58%) (Figure 2A).

**Table 2.** Radiological response and survival data.

| Parameter | Patients |
|---|---|
| Best documented response | |
| Complete response (CR), n (%) | 7 (11.0) |
| Partial response (PR), n (%) | 12 (18.0) |
| Stable disease (SD), n (%) | 22 (33.0) |
| Progressive disease (PD), n (%) | 11 (17.0) |
| Not evaluable (NA), n (%) | 14 (21.0) |
| Disease control rate (DCR), (%) | 62.0% |
| PFS [1], median (95%CI), month | 6.5 (4.0–9.1) |
| OS [2], median days (95%CI), month | Not reached |

Abbreviations: [1] PFS, progression-free survival; [2] OS, overall survival.

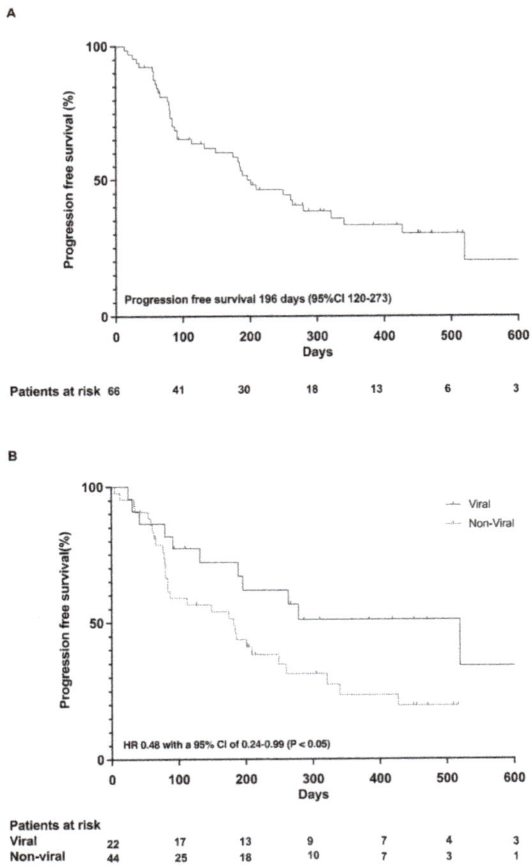


**Figure 1.** Progression free survival (**A**); PFS according to viral and non-viral etiology (**B**).

Patients with compensated liver disease (Child Pugh A) had a much more favorable prognosis than patients with more advanced liver disease (Child Pugh B and C cirrhosis). The 12-month survival rate in patients with Child A cirrhosis was 78%. (Figure 2B).

Patients with viral hepatitis tended to have a more favorable prognosis (median OS not reached) than patients without viral-related HCC (median OS 11.8 months, 95% confidence interval (CI) of 9.4–14.7 months), HR 0.61 with a 95% CI of 0.26–1.45 ($p > 0.2$) (Figure 2C).

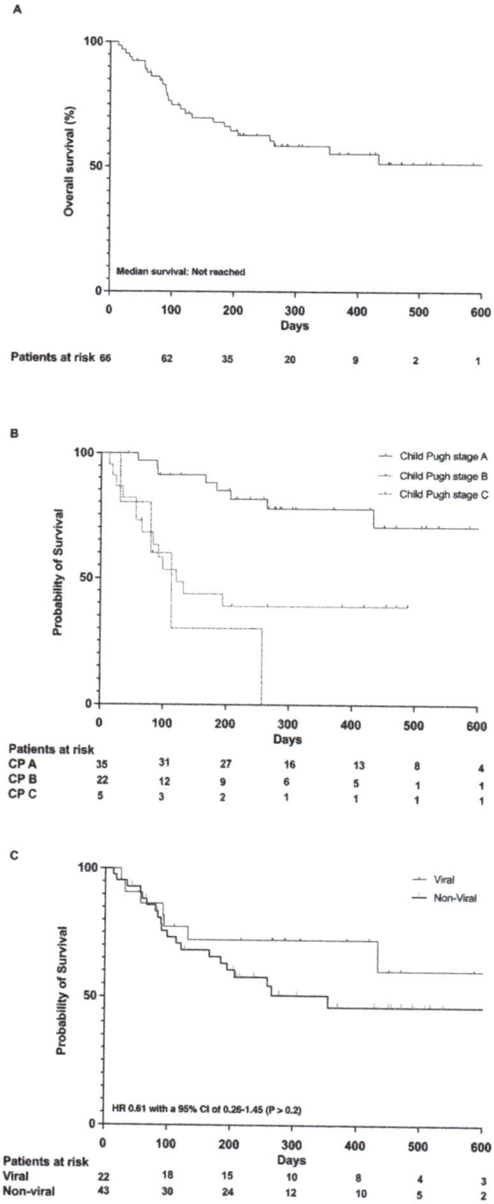


**Figure 2.** Overal survival (**A**); Survival according to Child Pugh stage (**B**); Survival according to viral and non-viral etiology (**C**).

### 3.3. Safety

At the time of data cut-off, 44 of 66 patients (67%) had stopped the treatment with atezolizumab and bevacizumab. In 25 patients (38%), bevacizumab was paused during treatment, and atezolizumab therapy was continued. Fifty-one (77%) of all patients reported at least one adverse event, and 39 patients (59%) experienced a high-grade (grade 3 or higher) event. The most common adverse events were bleeding events in 30.3% of the patients, worsening of renal function in 15.2%, and ascites in 8 patients (12.1%). Seven (10.6%) patients were diagnosed with variceal bleeding after treatment initiation. Of all patients with documented baseline variceal status (n = 55; 83%), 12 patients had grade 1, and 16 patients had varices above grade 1 (13 patients with grade 2, and 3 patients with grade 3 varices). None of the patients underwent prophylactic ligation therapy at baseline in this real-world cohort.

In total, 38 patients were administered a betablocker at the study start (59.1%), whereas 27 (40.9%) were not. Eleven of 38 (28.9%) patients received a selective betablocker, mostly prescribed for past cardiovascular reasons (bisoprolol and nebivolol), while 27 (71.1%) received a non-selective betablocker (19 were administered carvedilol, and 8 patients propranolol) explicitly for variceal bleeding prevention. Eighteen of the 27 patients without betablocker did not present diagnosed varices, while 9 patients with diagnosed varices used a betablocker. There was no statistical difference in betablocker use in patients with variceal bleeding compared to patients without bleeding; however, two patients with a bleeding event did not use betablockers.

Bleeding events were associated with the stage of varices ($p < 0.05$). In a multivariable analysis, the stage of varices was the only significant risk factor for bleeding, whereas betablocker intake, ALBI score, MELD score, and Child–Pugh score where not significantly associated with the risk of bleeding. For a detailed list of adverse events, see Table 3.

**Table 3.** Documented adverse events.

| | Any Grade, (n/%) | ≥Grade 3, (n/%) | Leading to Any Treatment Discontinuation (n/%) | Leading to Death, (n/%) |
|---|---|---|---|---|
| Bleeding events | 20 (30.3) | 18 (27.3) | 11 (16.7) | 3 (4.5) |
| Gastrointestinal bleeding | 14 (21.2) | 14 (21.2) | 6 (9.1) | 1 (1.5) |
| Variceal bleeding | 7 (10.6) | 7 (10.6) | 5 (7.5) | |
| Subarachnoidal hemorrhage | 2 (3.0) | 2 (3.0) | | 2 (3.0) |
| Epistaxis | 4 (6.0) | 2 (3.0) | | |
| Worsening of renal function | 10 (15.2) | 8 (12.1) | | |
| Acute kidney failure | 7 (10.6) | 7 (10.6) | 1 (1.5) | |
| Acute on chronic kidney failure | 1 (1.5) | 1 (1.5) | 1 (1.5) | |
| Ascites | 8 (12.1) | 6 (9.1) | 1 (1.5) | |
| Pruritus | 6 (9.1) | | | |
| Diarrhea | 5 (7.6) | 2 (3.0) | | |
| Rash | 4 (6.1) | | | |
| Fatigue | 4 (6.1) | | | |
| Hyponatremia | 3 (4.5) | 1 (1.5) | | |
| Arterial hypertension | 3 (4.5) | | | |
| Ulcus lower extremities | 3 (4.5) | 1 (1.5) | | |
| Acute on chronic liver failure | 2 (3.0) | 2 (3.0) | 1 (1.5) | |
| Hepatic encephalopathy | 2 (3.0) | 2 (3.0) | | |
| Allergic reaction | 2 (3.0) | 1 (1.5) | 1 (1.5) | |
| Nausea | 2 (3.0) | | | |
| Emesis | 1 (1.5) | | | |
| Cholangitis | 1 (1.5) | 1 (1.5) | 1 (1.5) | |

**Table 3.** *Cont.*

| | Any Grade, (n/%) | ≥Grade 3, (n/%) | Leading to Any Treatment Discontinuation (n/%) | Leading to Death, (n/%) |
|---|---|---|---|---|
| Pyrexia | 1 (1.5) | 1 (1.5) | | |
| Transient ischemic attack | 1 (1.5) | 1 (1.5) | 1 (1.5) | |
| Pulmonary embolism | 1 (1.5) | 1 (1.5) | 1 (1.5) | |
| Flare of autoimmune disease | 1 (1.5) | | | |
| Insomnia | 1 (1.5) | | | |
| Hyperbilirubinemia | 1 (1.5) | | 1 (1.5) | |
| Spontaneous bacterial peritonitis | 1 (1.5) | 1 (1.5) | | |
| Cough | 1 (1.5) | | | |
| Hyperkalemia | 1 (1.5) | | | |
| Hoarseness | 1 (1.5) | | | |
| Vasculitis | 1 (1.5) | | | |
| Anemia | 1 (1.5) | 1 (1.5) | | |
| Proteinuria | 1 (1.5) | | | |
| Edema | 1 (1.5) | | | |
| Worsening of heart failure | 1 (1.5) | | | |
| Stomatitis | 1 (1.5) | | | |
| Nephritis | 1 (1.5) | | | |
| Dry skin | 1 (1.5) | | | |
| Immune checkpoint-inhibitor hepatitis (ICI) | 1 (1.5) | 1 (1.5) | 1 (1.5) | |
| Esophageal candidiasis | 1 (1.5) | 1 (1.5) | 1 (1.5) | |
| Urogenital abscess | 1 (1.5) | 1 (1.5) | 1 (1.5) | |

### 3.4. Factors Associated with Survival

Known predictors of survival in patients with HCC include liver function and AFP levels.

The variables gender, age (≤65 years vs. >65 years), hepatotropic virus infection, BCLC stage, AFP levels (≤400 ng/mL vs. >400 ng/mL), albumin–bilirubin (ALBI) score, Child–Pugh score, extrahepatic spread, and prior therapy were included in a multivariable model. Furthermore, all factors from univariable analysis with a $p$-value < 0.1 were included in the multivariable model. As shown in Table 4, Child–Pugh A cirrhosis and prior local therapy were independently associated with OS.

**Table 4.** Univariate and multivariate analyses of parameters associated with overall survival.

| | Univariate Analysis | | | Multivariate Analysis | | |
|---|---|---|---|---|---|---|
| Parameter | HR | 95% CI | $p$ Value | HR | 95 % CI | $p$ Value |
| Male gender | 1.609 | 0.555–4.662 | 0.381 | | | |
| Age < 65 years | 0.645 | 0.302–1.380 | 0.259 | | | |
| Viral Hepatitis | 0.612 | 0.258–1.449 | 0.264 | | | |
| BCLC AB | 0.632 | 0.276–1.445 | 0.277 | | | |
| AFP < 400 ng/mL | 0.928 | 0.402–2.142 | 0.861 | | | |
| ALBI score 1 | 0.043 | 0.005–0.337 | 0.003 | | | |
| Child Pugh A | 0.152 | 0.045–0.515 | 0.002 | 0.112 | 0.024–0.534 | 0.006 |
| Extrahepatic spread of HCC | 2.182 | 0.996–4.780 | 0.051 | | | |
| Prior local therapy/surgery | 0.450 | 0.205–0.988 | 0.047 | 0.346 | 0.122–0.978 | 0.045 |

Abbreviations: HR, hazard ratio; CI, confidence interval; BCLC stage AB. Barcelona Clinic Liver Cancer stage A and B; AFP, alpha-fetoprotein, ALBI score, albumin–bilirubin score.

## 4. Discussion

Atezolizumab and bevacizumab have become the new standard of care for the first-line treatment of advanced HCC. We report our first experience with atezolizumab and bevacizumab in real-life European patients. Treatment was feasible and effective. The median PFS of 6.5 months was similar to that reported in the pivotal phase 3 trial (6.8 months) [4]. Furthermore, the 12-month survival rate in our patients with compensated liver disease was even better than in the patients in the trial (78% versus 67%). The subgroup of patients with viral hepatitis, namely, hepatitis B or hepatitis C, had a more favorable prognosis than patients without a history of viral hepatitis concerning PFS and OS. Also an exploratory subgroup analysis of the IMbrave150 trial favored immunotherapy for patients with viral hepatitis [12]. The high efficacy of atezolizumab and bevacizumab compared to sorafenib was also supported by the analysis of Chinese patients treated in the IMbrave150 trial [13]. In this group of patients, more than 90% had a history of viral hepatitis infection, mainly hepatitis B; 77% of the patients were alive 12 months after treatment initiation. Treatment response in our real-life cohort was 29%, which is nearly the same as in the phase 3 trial, which reported a treatment response of 27% [4]. In our multivariable analysis, only Child–Pugh stage A cirrhosis and prior local therapy were independently associated with survival. In support of these findings, emerging data are showing less effectiveness of immunotherapy in NASH/NAFLD patients with HCC, most probably due to an altered immune environment [14,15].

Comparable real-world data on atezolizumab/bevacizumab are scarce. Iwamoto and colleagues retrospectively analyzed 61 patients from Japan and found a median PFS of 5.4 month, a disease control rate of 86.3%, and adverse event rates of any grade and higher than grade 3 of 98% and 29.4%, respectively. However, 23 patients (62.7%) in this study underwent at least one line of prior small-molecule treatment, hampering a direct comparison [16].

Another group from Japan published real-world data concerning tumor response and safety for atezolizumab/bevacizumab in 40 patients with Child–Pugh A cirrhosis [17]. Twenty-four patients had a previous treatment experience with molecular agents (TKI). They found an ORR of 22.5% based on mRECIST. Multivariate analysis showed that an AFP ratio <1.0 at 3 weeks (odds ratio 39.2, 95% confidence interval CI 2.37–649.0, $p = 0.0103$) was the only significant factor for predicting an early response.

Hiraoka et al., aimed at the evaluation of early response (6 weeks) to atezolizumab/bevacizumab and included 171 HCC patients from Japan; again, 96 patients were systemically pretreated [18]. In initial imaging examination findings, they described objective response rates for early tumor shrinkage and disease control after 6 weeks (ORR-6W/DCR-6W) of 10.6% and 79.6%, respectively. Hayakawa et al., published a short report describing 52 patients undergoing atezolizumab/bevacizumab treatment (only 23 receiving it as first-line treatment). They found an objective response rate (ORR) and disease control rate (DCR) in all patients of 15.4% and 57.7%, respectively, and suggested AFP response as a predictive marker [19].

Sho et al., investigated 64 patients, 46 of whom 46 (71.9%) did not meet the inclusion criteria of the IMBRAVE 150 trial; 44 of these 46 patients where systemically pretreated. They showed good safety and efficacy for these patients. Interestingly, none of the 15 patients with hepatitis B experienced progressive disease [20].

Liver function is well known to be highly prognostic for survival in patients with HCC [21]. We recently reported that patients with more advanced cirrhosis receiving nivolumab obtain only a marginal benefit from a tumor-specific treatment and have a poor overall prognosis [22,23]. Therefore, immunotherapy seems to be of value only for patients with a relatively well-preserved liver function [24,25]. Interestingly, the 12-month survival in patients with Child B cirrhosis was still 39%, indicating that a subgroup of these patients does benefit from a tumor-specific treatment. These could be patients whose liver function impairment is mainly driven by a large intrahepatic tumor load.

In our cohort, patients with non-viral HCC tended to have a worse outcome compared to patients with viral-related HCC, which, however, was not significant in the uni- or multivariate analysis. Recently, Pfister et al., published highly discussed evidence of lowered effectiveness of immunotherapy in HCC patients than in non-alcoholic steatohepatitis (NASH) patients with HCC due to the presence of special resident-like activated $CD8^+$ cells in patients with NASH [15]. The poorer response of NASH patients is supported by findings of Inada et al., and others [14,26]. A meta-analysis of the two first-line trials that used sorafenib as the comparator (CheckMate 459 trial and IMbrave150 trial) [3,4] showed the same trend (non-viral HCC OS HR = 0.94, but HBV OS HR = 0.65 and HCV OS HR = 0.60).

Patient-reported outcomes of the IMBRAVE 150 trial were published separately [6] and showed benefits in terms of patient-reported quality of life, functioning, and disease symptoms with atezolizumab plus bevacizumab compared with sorafenib. In the pivotal trial, 98% patients with any AE were reported, and 56.5% of them had higher-than-grade 2 events, which corresponds to our data. We documented treatment discontinuation (complete, or stopping, or discontinuation of bevacizumab) due to an AE in 20 patients (30%), which fits the trial data as well. In the trail, 7% of the patients reported bleeding, while our event rate was higher, most probably due to a less strict patient inclusion for treatment in real life and the overall worse liver function in our patients. As recently published, an important adverse event seems to be hypertension in atezolizumab/bevacizumab-treated patients (up to 30% of patients with grade 3) [27], which was reported at a much lower frequency in our cohort, and this may implicate underdiagnosis in real-life practice.

## 5. Conclusions

The combination of atezolizumab and bevacizumab is highly effective for patients with hepatocellular carcinoma in real life. Patients with cirrhosis and hepatocellular carcinoma of viral origin seemed to respond better than those with non-viral HCCs. Variceal bleeding is an important adverse event of this drug combination. In patients with compromised liver function (Child–Pugh B and C cirrhosis), the drug combination showed low efficacy; therefore, treatment for these patients should be well considered.

**Author Contributions:** V.H., M.P., B.S., M.V., F.S., C.Z., J.U.M., J.T., O.W., F.F. collected and analyzed the data. V.H., F.F., O.W. designed the research study and wrote the manuscript. All authors have read and agreed to the published version of the manuscript.

**Funding:** This research received no external funding.

**Institutional Review Board Statement:** The study was conducted according to the guidelines of the Declaration of Helsinki and approved by the Institutional Review Board SGI03/18, Amendment 01/19 Frankfurt/Germany.

**Informed Consent Statement:** Written informed consent by patients was waived due to the retrospective nature of the study.

**Data Availability Statement:** Data sharing not applicable. No new data were created or analyzed in this study.

**Conflicts of Interest:** V.H. reports no conflict of interest. M.P. is an investigator for Bayer, BMS, Lilly, and Roche and received speaker honoraria from Bayer, BMS, Eisai, Lilly, and MSD; he is a consultant for Bayer, BMS, Ipsen, Eisai, Lilly, MSD, and Roche and received travel support from Bayer and BMS. B.S. received travel support from AbbVie, Gilead and Ipsen. M. V. received honoraria from Merck Serono, Bayer Vital, and Sirtex and is a member of the advisory boards of Ipsen, Roche, Bayer, Lilly, Nordic Pharma, BMS, MSD, and Amgen. F.S. has nothing to declare. C.C. received fees and travel support from Eisai, MSD, Ipsen, Falk Foundation. J.U.M. received fees from Roche, Astra, Jansen, Abbvie, Shionogi, leap, Bayer, Ipsen, J.T. Amgen, AstraZeneca, Bayer Healthcare, Bristol Myers-Squibb, Eisai, Ipsen, Merck Serono, Merck Sharp & Dome, Lilly Imclone, Roche Servier (C/A), Ipsen, Roche (RF); O.W. served as a speaker and/or consultant for AstraZeneca, Amgen, Bayer, BMS, Celgene, Eisai, Incyte, Ipsen, Merck Serono, MSD, Novartis, Roche, Servier, and Shire. He received travel support from Abbvie, Bayer, BMS, Gilead, Ipsen, Medac, and Merck. He received financial support for scientific projects from Else Kröner-Fresenius-Stiftung, IPSEN, Medac, Merck Serono, and

Novartis. He is an investigator for Basilea, Incyte, and MSD. F.F. received travel support from Abbvie and Ipsen and speaker fees from AbbbVie.

## References


1. Villanueva, A. Hepatocellular Carcinoma. *N. Engl. J. Med.* **2019**, *380*, 1450–1462. [CrossRef] [PubMed]
2. Finn, R.S.; Ryoo, B.Y.; Merle, P.; Kudo, M.; Bouattour, M.; Lim, H.Y.; Breder, V.; Edeline, J.; Chao, Y.; Ogasawara, S.; et al. Pembrolizumab as Second-Line Therapy in Patients with Advanced Hepatocellular Carcinoma in KEYNOTE-240: A Randomized, Double-Blind, Phase III Trial. *J. Clin. Oncol.* **2020**, *38*, 193–202. [CrossRef] [PubMed]
3. Yau, T.; Park, J.W.; Finn, R.S.; Cheng, A.-L.; Mathurin, P.; Edeline, J.; Kudo, M.; Han, K.-H.; Harding, J.J.; Merle, P.; et al. CheckMate 459: A Randomized, Multi-Center Phase III Study of Nivolumab (NIVO) vs. Sorafenib (SOR) as First-Line (1L) Treatment in Patients (Pts) with Advanced Hepatocellular Carcinoma (AHCC). *Ann. Oncol.* **2019**, *30*, v874–v875. [CrossRef]
4. Finn, R.S.; Qin, S.; Ikeda, M.; Galle, P.R.; Ducreux, M.; Kim, T.Y.; Kudo, M.; Breder, V.; Merle, P.; Kaseb, A.O.; et al. Atezolizumab plus Bevacizumab in Unresectable Hepatocellular Carcinoma. *N. Engl. J. Med.* **2020**, *382*, 1894–1905. [CrossRef] [PubMed]
5. Finn, R.S.; Qin, S.; Ikeda, M.; Galle, P.R.; Ducreux, M.; Kim, T.-Y.; Lim, H.Y.; Kudo, M.; Breder, V.V.; Merle, P.; et al. IMbrave150: Updated Overall Survival (OS) Data from a Global, Randomized, Open-Label Phase III Study of Atezolizumab (Atezo) + Bevacizumab (Bev) versus Sorafenib (Sor) in Patients (Pts) with Unresectable Hepatocellular Carcinoma (HCC). *J. Clin. Oncol.* **2021**, *39*, 267. [CrossRef]
6. Galle, P.R.; Finn, R.S.; Qin, S.; Ikeda, M.; Zhu, A.X.; Kim, T.Y.; Kudo, M.; Breder, V.; Merle, P.; Kaseb, A.; et al. Patient-Reported Outcomes with Atezolizumab plus Bevacizumab versus Sorafenib in Patients with Unresectable Hepatocellular Carcinoma (IMbrave150): An Open-Label, Randomised, Phase 3 Trial. *Lancet Oncol.* **2021**, *22*, 991–1001. [CrossRef]
7. Giannini, E.G.; Aglitti, A.; Borzio, M.; Gambato, M.; Melandro, F.; Morisco, F.; Ponziani, F.R.; Rendina, M. Overview of Immune Checkpoint Inhibitors Therapy Cohort Derived Estimate of Amenability Rate to Immune Checkpoint Inhibitors in Clinical Practice. *Cancers* **2019**, *11*, 1689. [CrossRef] [PubMed]
8. Eisenhauer, E.A.; Therasse, P.; Bogaerts, J.; Schwartz, L.H.; Sargent, D.; Ford, R.; Dancey, J.; Arbuck, S.; Gwyther, S.; Mooney, M.; et al. New Response Evaluation Criteria in Solid Tumours: Revised RECIST Guideline (Version 1.1). *Eur. J. Cancer* **2009**, *45*, 228–247. [CrossRef] [PubMed]
9. Lencioni, R.; Llovet, J. Modified RECIST (MRECIST) Assessment for Hepatocellular Carcinoma. *Semin. Liver Dis.* **2010**, *30*, 052–060. [CrossRef] [PubMed]
10. US Department of Health and Human Services. *Common Terminology Criteria for Adverse Events (CTCAE) Common Terminology Criteria for Adverse Events v4.0 (CTCAE)*; National Institutes of Health, National Cancer Institute: Bethesda, MD, USA, 2009.
11. National Cancer Institute US. *Common Terminology Criteria for Adverse Events (CTCAE).v.5.0*; Cancer Therapy Evaluation Program (CTEP): Rockville, MD, USA, 2017; p. 155.
12. Cheng, A.-L.; Qin, S.; Ikeda, M.; Galle, P.; Ducreux, M.; Zhu, A.; Kim, T.-Y.; Kudo, M.; Breder, V.; Merle, P.; et al. IMbrave150: Efficacy and Safety Results from a Ph III Study Evaluating Atezolizumab (Atezo) + Bevacizumab (Bev) vs Sorafenib (Sor) as First Treatment (Tx) for Patients (Pts) with Unresectable Hepatocellular Carcinoma (HCC). *Ann. Oncol.* **2019**, *30*, ix186–ix187. [CrossRef]
13. Qin, S.; Ren, Z.; Feng, Y.H.; Yau, T.; Wang, B.; Zhao, H.; Bai, Y.; Gu, S.; Li, L.; Hernandez, S.; et al. Atezolizumab plus Bevacizumab versus Sorafenib in the Chinese Subpopulation with Unresectable Hepatocellular Carcinoma: Phase 3 Randomized, Open-Label IMbrave150 Study. *Liver Cancer* **2021**, *10*, 296–308. [CrossRef] [PubMed]
14. Eso, Y.; Taura, K.; Seno, H. Does Immune Checkpoint Inhibitor Exhibit Limited Efficacy against Non-Viral Hepatocellular Carcinoma?: A Review of Clinical Trials. *Hepatol. Res.* **2021**, *52*, 67–74. [CrossRef] [PubMed]
15. Pfister, D.; Núñez, N.G.; Pinyol, R.; Govaere, O.; Pinter, M.; Szydlowska, M.; Gupta, R.; Qiu, M.; Deczkowska, A.; Weiner, A.; et al. NASH Limits Anti-Tumour Surveillance in Immunotherapy-Treated HCC. *Nature* **2021**, *592*, 450–456. [CrossRef] [PubMed]
16. Iwamoto, H.; Shimose, S.; Noda, Y.; Shirono, T.; Niizeki, T.; Nakano, M.; Okamura, S.; Kamachi, N.; Suzuki, H.; Sakai, M.; et al. Initial Experience of Atezolizumab plus Bevacizumab for Unresectable Hepatocellular Carcinoma in Real-World Clinical Practice. *Cancers* **2021**, *13*, 2786. [CrossRef] [PubMed]
17. Ando, Y.; Kawaoka, T.; Kosaka, M.; Shirane, Y.; Johira, Y.; Miura, R.; Murakami, S.; Yano, S.; Amioka, K.; Naruto, K.; et al. Early Tumor Response and Safety of Atezolizumab plus Bevacizumab for Patients with Unresectable Hepatocellular Carcinoma in Real-World Practice. *Cancers* **2021**, *13*, 3958. [CrossRef] [PubMed]
18. Hiraoka, A.; Kumada, T.; Tada, T.; Hirooka, M.; Kariyama, K.; Tani, J.; Atsukawa, M.; Takaguchi, K.; Itobayashi, E.; Fukunishi, S.; et al. Atezolizumab plus Bevacizumab Treatment for Unresectable Hepatocellular Carcinoma: Early Clinical Experience. *Cancer Rep.* **2021**, *5*, e1464. [CrossRef] [PubMed]
19. Hayakawa, Y.; Tsuchiya, K.; Kurosaki, M.; Yasui, Y.; Kaneko, S.; Tanaka, Y.; Ishido, S.; Inada, K.; Kirino, S.; Yamashita, K.; et al. Early Experience of Atezolizumab plus Bevacizumab Therapy in Japanese Patients with Unresectable Hepatocellular Carcinoma in Real-World Practice. *Investig. New Drugs* **2021**, 1–11. [CrossRef] [PubMed]
20. Sho, T.; Suda, G.; Ogawa, K.; Kimura, M.; Kubo, A.; Tokuchi, Y.; Kitagataya, T.; Maehara, O.; Ohnishi, S.; Shigesawa, T.; et al. Early Response and Safety of Atezolizumab plus Bevacizumab for Unresectable Hepatocellular Carcinoma in Patients Who Do Not Meet IMbrave150 Eligibility Criteria. *Hepatol. Res.* **2021**, *51*, 979–989. [CrossRef]
21. Pinter, M.; Trauner, M.; Peck-Radosavljevic, M.; Sieghart, W. Cancer and Liver Cirrhosis: Implications on Prognosis and Management. *ESMO Open* **2016**, *1*, e000042. [CrossRef] [PubMed]

22. Shek, D.; Read, S.A.; Nagrial, A.; Carlino, M.S.; Gao, B.; George, J.; Ahlenstiel, G. Immune-Checkpoint Inhibitors for Advanced Hepatocellular Carcinoma: A Synopsis of Response Rates. *Oncologist* **2021**, *26*, e1216–e1225. [CrossRef] [PubMed]
23. Finkelmeier, F.; Czauderna, C.; Perkhofer, L.; Ettrich, T.J.; Trojan, J.; Weinmann, A.; Marquardt, J.U.; Vermehren, J.; Waidmann, O. Feasibility and Safety of Nivolumab in Advanced Hepatocellular Carcinoma: Real-Life Experience from Three German Centers. *J. Cancer Res. Clin. Oncol.* **2019**, *145*, 253–259. [CrossRef] [PubMed]
24. Pinter, M.; Scheiner, B.; Peck-Radosavljevic, M. Immunotherapy for Advanced Hepatocellular Carcinoma: A Focus on Special Subgroups. *Gut* **2021**, *70*, 204–214. [CrossRef] [PubMed]
25. Scheiner, B.; Kirstein, M.M.; Hucke, F.; Finkelmeier, F.; Schulze, K.; von Felden, J.; Koch, S.; Schwabl, P.; Hinrichs, J.B.; Waneck, F.; et al. Programmed Cell Death Protein-1 (PD-1)-Targeted Immunotherapy in Advanced Hepatocellular Carcinoma: Efficacy and Safety Data from an International Multicentre Real-World Cohort. *Aliment. Pharmacol. Ther.* **2019**, *49*, 1323–1333. [CrossRef] [PubMed]
26. Inada, Y.; Mizukoshi, E.; Seike, T.; Tamai, T.; Iida, N.; Kitahara, M.; Yamashita, T.; Arai, K.; Terashima, T.; Fushimi, K.; et al. Characteristics of Immune Response to Tumor-Associated Antigens and Immune Cell Profile in Patients With Hepatocellular Carcinoma. *Hepatology* **2019**, *69*, 653–665. [CrossRef] [PubMed]
27. Jiang, L.; Tan, X.; Li, J.; Li, Y. Incidence and Risk of Hypertension in Cancer Patients Treated with Atezolizumab and Bevacizumab: A Systematic Review and Meta-Analysis. *Front. Oncol.* **2021**, *11*, 1–10. [CrossRef] [PubMed]

*cancers*

MDPI

*Review*

# Is There Still a Place for Tyrosine Kinase Inhibitors for the Treatment of Hepatocellular Carcinoma at the Time of Immunotherapies? A Focus on Lenvatinib

**Marie Decraecker [1,*], Caroline Toulouse [1] and Jean-Frédéric Blanc [1,2]**

[1] Department of Oncology, Hospital Haut Leveque-CHU Bordeaux, Avenue Magellan, 33604 Pessac, France; ctoulo200e@gmail.com (C.T.); jean-frederic.blanc@chu-bordeaux.fr (J.-F.B.)
[2] INSERM U1053, BaRITOn, University Victor Segalen, 146 Rue Léo Saignat, 33000 Bordeaux, France
* Correspondence: marie.decraecker@gmail.com

**Simple Summary:** The combination of atezolizumab and bevacizumab has changed the therapeutic algorithm of advanced hepatocellular carcinomas. Therefore, the place of tyrosine kinase inhibitors, and among them, lenvatinib, which exhibits promising antitumour activity compared to other TKIs, needs to be redefined. Lenvatinib is still an option in case of contra-indication of immunotherapy or anti-vascular endothelial growth factor (anti-VEGF), but its place can also be discussed in second-line treatment. Otherwise, emerging strategies are currently being studied to assess the efficacy of the combination of lenvatinib with immunotherapy or loco-regional treatment for advanced HCC, as well as the efficacy of lenvatinib alone or in combination at earlier stages of the disease. The aim of this review was to define potential indications for lenvatinib treatment in different clinical situations of hepatocellular carcinoma.

**Abstract:** The systemic treatment of hepatocellular carcinoma is changing rapidly. Three main classes of treatment are now available. Historically, multi-targeted tyrosine kinase inhibitors (TKIs) (sorafenib and lenvatinib as first-line; regorafenib and cabozantinib as second-line) were the first to show an improvement in overall survival (OS). Anti-vascular endothelial growth factor (anti-VEGF) antibodies can be used in first-line (bevacizumab) or second-line (ramucirumab) combination therapy. More recently, immuno-oncology (IO) has profoundly changed therapeutic algorithms, and the combination of atezolizumab-bevacizumab is now the first-line standard of care. Therefore, the place of TKIs needs to be redefined. The objective of this review was to define the place of TKIs in the therapeutic algorithm at the time of IO treatment in first-line therapy, with a special focus on lenvatinib that exhibits one of the higher anti-tumoral activity among TKI in HCC. We will discuss the place of lenvatinib in first line (especially if there is a contra-indication to IO) but also after failure of atezolizumab and bevacizumab. New opportunities for lenvatinib will also be presented, including the use at an earlier stage of the disease and combination with IOs.

**Keywords:** hepatocellular carcinoma; lenvatinib; tyrosine kinase inhibitor

**Citation:** Decraecker, M.; Toulouse, C.; Blanc, J.-F. Is There Still a Place for Tyrosine Kinase Inhibitors for the Treatment of Hepatocellular Carcinoma at the Time of Immunotherapies? A Focus on Lenvatinib. *Cancers* **2021**, *13*, 6310. https://doi.org/10.3390/cancers13246310

Academic Editor: Georgios Germanidis

Received: 14 November 2021
Accepted: 14 December 2021
Published: 16 December 2021



## 1. Introduction

Hepatocellular carcinoma (HCC) is a major public health issue and, as the most common primary liver tumour, its incidence reaches one million new cases per year worldwide [1]. Although screening programs diagnose approximately 40% of HCCs at a curative stage, at least 50% of patients will be diagnosed at an intermediate or advanced stage [2]. The prognosis remains unfavourable at these later stages due to extensive tumour burden, a high frequency of liver dysfunction, and deterioration of health status, which limit access to any treatment, including systemic therapies. HCC is the second leading cause of cancer death worldwide [3–5].

In patients with advanced HCC (Barcelona Clinic Liver Cancer (BCLC) C) or with intermediate-stage (BCLC B) disease not eligible for, or progressing despite, locoregional therapies, systemic therapies are the gold standard of care. Preliminary results of the TARGET-HCC study demonstrated that patients with BCLC stage C were more than twice as likely to receive any systemic therapy compared to all other stages [6].

Sorafenib has been the standard treatment of care since 2007, based on improved overall survival (OS) in randomised controlled trials compared to placebo [7–9]. However, the management of advanced HCC has been modified since 2017 with the development of new and effective systemic treatments that improve both OS and progression free survival (PFS). Lenvatinib has been approved by the United States Food and Drug Administration and the European Medicines Agency after demonstration of the non-inferiority to sorafenib as first-line treatment for patients with advanced or unresectable HCC who have not received prior systemic therapy, based on the results of the phase III REFLECT study. Two other TKIs, regorafenib and cabozantinib, were also approved in second line after sorafenib.

More recently, immune checkpoint inhibitors (ICIs) have shown promising results in the treatment of HCC, and the combination therapy with atezolizumab and bevacizumab is now the first-line standard of care for advanced HCCs from the IMbrave150 study, showing a clear superiority of the combination among sorafenib for OS, PFS, and quality of life.

However, TKIs (lenvatinib and sorafenib) remain an alternative in first-line standard of care in international guidelines (the European Association for the Study of the Liver, American Association for the Study of Liver Diseases, Asian Pacific Association for the Study of the Liver, European Society for Medical Oncology, and National Comprehensive Cancer Network [6,10–13]). However, TKIs are now currently indicated in patients with advanced or unresectable HCC not eligible for atezolizumab plus bevacizumab, with well-preserved liver function (Child-Pugh (CP) class A) and an Eastern Cooperative Oncology Group (ECOG) Performance Status (PS) of 0–2, [14,15]. The place of TKIs in second line after atezolizumab and bevacizumab is not consensually defined due to a lack of data from clinical trials.

This review aims to discuss the place of lenvatinib in HCC at the time of immunotherapy emergence.

## 2. Lenvatinib in the First Line Setting

### 2.1. *Efficacy in Clinical Trials and in the Real Life Compared to Sorafenib*

Lenvatinib is an oral inhibitor of multiple tyrosine kinase receptors, targeting vascular endothelial growth factor receptor (VEGFR1–3), fibroblast growth factor receptor (FGFR1–4), platelet-derived growth factor receptor $\alpha$ (PDGFR $\alpha$), KIT-ligand (stem cell factor receptor), and RET (rearranged during transfection), with a distinct in vitro tyrosine kinase inhibitory profile compared to sorafenib [16–19].

Clinical evidence of the antitumour activity of lenvatinib was demonstrated in preclinical studies, with inhibition of both VEGF- and FGF-driven angiogenesis, and direct antiproliferative activity on liver cancer cells in vitro and in vivo, depending on the FGF-signalling pathway [20–24].

Following positive preliminary data, the REFLECT trial was conducted by Kudo et al. to compare lenvatinib with sorafenib as first-line treatment for unresectable HCC [25,26] (Table 1).

**Table 1.** Summary of first-line validated treatments for unresectable HCC based on the results of the REFLECT and IMbrave150 trials.

| | **Atezolizumab-Bevacizumab** | **Sorafenib** | **Lenvatinib** |
|---|---|---|---|
| | Patients' characteristics at baseline | | |
| OMS | 0/1 | 0/1/2 | 0/1 |
| BCLC B/C, % | 15%/82% | 18%/82% | 22%/78% |
| Age, % | 64 (56–71) | 64.9 ± 11.2 | 63 (20–88) |
| Male Sex, % | 82% | 87% | 85% |
| Non-viral-related HCC aetiology, % | 30% | 52% | 29% |
| Geographic region Asia vs. rest of the world, % | 40%/60% | Unknown | 67%/33% |
| Macrovascular invasion, % | 38% | 36% | 23% |
| Metastatic status, % | 63% | 53% | 61% |
| OS, months | NE | 12.3 (10.4–13.9) //13.2 (10.4–NE) | ↑ 13.6 (12.1–14.9) |
| ORR, % | 27.3 (22.5–32.5) | 9.2 (6.6–11.8)//11.9 (7.4–18.0) | 24.1 (20.2–27.9) |
| PFS, months | 6.8 (5.7–8.3) | 3.7 (3.6–4.6)//4.3 (4.0–5.6) | ↑ 7.4 (6.9–8.8) |
| TTP, months | NE | 3.7 (3.6–5.4) | ↑ 8.9 (7.4–9.2) |
| DCR, % | 73.6% | 55.3% to 60.5% | ↑ 75.5% |
| TEAEs, % | 98.2% | 95% to 98.7% | 94% |
| Hypertension | 29.8% | 24.4% to 30% | ↑ 42% |
| Diarrhoea | 18.8% | 46% to 49.4% | 39% |
| Decreased appetite | 17.6% | 24.4% to 27% | ↑ 34% |
| Decreased weight | 11.2% | 9.6% to 22% | ↑ 31% |
| PPES | 0.9% | 48.8% to 52% | 27% |
| TEAEs grade $\frac{3}{4}$, % | 56.5% | 49% to 55.1% | 57% |
| Hypertension | 15.2% | 14% | ↑ 23% |
| PPES | 0% | 11% | 3% |

OS: Overall Survival; ORR: Objective Response Rate; PFS: Progression-free survival; TTP: Time To Progression; DCR: Disease Control Rate; TEAEs: Treatment-Emergent Adverse Events; PPES: Palmar-plantar erythrodysaesthesia syndrome; NE: Not Evaluated. //: REFLECT vs. IMbrave150. ↑: higher result.

This was a non-inferiority, multicentre, international, open-label, randomised phase 3 trial in 954 patients [14]. Patients were performance status (PS) 0–1, CP-A, and BCLC B or C without previous systemic therapy. Patients were randomised 1:1 to lenvatinib or sorafenib (478 to lenvatinib and 476 to sorafenib), stratified by region (Asia-Pacific or Western), macroscopic portal vein invasion and/or extrahepatic spread (yes or no), PS (0 or 1), and body weight (<60 or ≥60 kg). Baseline patient characteristics were similar between the two groups, except for hepatitis C virus (HCV) aetiology (lower in the lenvatinib group) and alpha-foetoprotein (AFP) baseline levels (higher in the lenvatinib group). The study was positive for all primary and secondary outcomes: the median OS was 13.6 months vs. 12.3 months (hazard ratio (HR): 0.92, 95% CI, 0.79–1.06); the PFS was 7.4 months (6.9–8.8) vs. 3.7 months (3.6–4.6) (HR: 0.66 (0.57–0.77); $p < 0.0001$); the time to progression (TTP) was 8.9 months (7.4–9.2) vs. 3.7 months (3.6–5.4) (HR: 0.63 (0.53–0.73); $p < 0.0001$); the overall response rate (ORR) according to response evaluation criteria in solid tumours (RECIST) was 24.1% (20.2–27.9) vs. 9.2% (6.6–11.8) (Odds ratio (OR) 3.13 (2.15–4.56); $p < 0.0001$), and the disease control rate (DCR) was 75.5% (71.7–79.4) vs. 60.5% (56.1–64.9).

Subsequently, several publications confirmed significant improvements in PFS, TTP, and ORR compared to sorafenib in real life conditions reflecting a greater anti-tumoral activity [27–32].

### 2.2. Safety and Tolerability

Lenvatinib had a manageable tolerability profile in the REFLECT study. Most common treatment-emergent adverse events (TEAEs) were hypertension (42%), diarrhoea (39%), decreased appetite (34%), and decreased weight (31%). The overall safety profile was comparable to other TKIs, with similar rates of grade ≥3 TEAEs. TEAEs led to lenvatinib interruption in 40% of cases, a dose reduction in 37%, and drug withdrawal in 9% [33]. Patients treated with lenvatinib had a higher level of high blood pressure (23% vs. 14% of grades 3–4). The higher response rate and the higher frequency of severe hypertension with lenvatinib may indicate a greater anti-angiogenic effect of this drug. On the other hand, patients treated by lenvatinib exhibited a lower level of palmar-plantar erythrodysaesthesia syndrome (PPES) (3% vs. 11% of grades 3–4). Therefore, in patients with pre-existing skin diseases (outside non-healing ulcers), lenvatinib could be preferred to avoid an additional skin toxicity impacting the quality of life. Similarly, pre-ulcerative diabetic foot might benefit from the absence of the neuropathic-like pain induced by the hand–foot syndrome.

Regarding cost-effectiveness analyses, there was an increase of 0.27 life years (LY), an improvement of 0.23 quality-adjusted LY (QALY), and a decrease in costs for lenvatinib compared with sorafenib [34–36]. The AE treatment costs were very small and accounted for <1% of the total cost, suggesting that lenvatinib could represent a new long-awaited alternative option to sorafenib for the first-line systemic treatment of patients with unresectable HCC.

### 2.3. Predictive Biomarkers for Response to Lenvatinib

Preclinical studies demonstrated that inhibition of FGF19 signalling may play a role in the anti-tumour effects of TKIs against HCC. FGF19 is expressed in approximately one-third of HCC tissue samples and is associated with tumour aggressiveness, represented by a poorly differentiated tumour and an unfavourable prognosis [37].

In a post hoc analysis of the REFLECT study of Finn et al., lenvatinib (and not sorafenib) was associated with an increase in FGF19 and FGF23 levels at four weeks (FGF19: 55.2% vs. 18.3%, $p = 0.0140$; FGF23: 48.4% vs. 16.4%; $p = 0.0022$, respectively), suggesting efficient inhibition of the FGF signalling pathway [38]. In the lenvatinib arm, patients with a complete or partial response had a greater increase in FGF19 and FGF23 from baseline vs. non-responders (FGF19: 55.2% vs. 18.3%, $p = 0.0140$; FGF23: 48.4% vs. 16.4%; $p = 0.0022$).

Otherwise, early changes in serum FGF19 and Ang-2 (an angiogenesis regulator that plays a role through TEK tyrosine kinase and endothelium receptor levels during lenvatinib treatment) might predict clinical response and PFS. In a recent study of 74 patients (BCLC stages B and C), including patients previously treated with sorafenib or regorafenib, with a median follow-up of 157 days, significantly increased FGF19 levels and decreased Ang-2 levels were seen in lenvatinib responders compared with non-responders (ratio of FGF19 level at 4 weeks/baseline in responders vs. non-responders: 2.09 vs. 1.32, respectively, $p = 0.0004$; ratio at 8 weeks: 2.19 vs. 1.40, $p = 0.0015$) [39,40]. In multivariate analysis, the combination of serum FGF19 and Ang-2 was the most independent predictive factor for lenvatinib response (OR: 9.143; $p = 0.0012$) and PFS (HR: 0.171; $p = 0.0240$). The ability of FGF19 to predict an early lenvatinib response had a receiver operating characteristic (ROC) curve area of 0.726 at the optimal cut-off value of 1.51 for the FGF19 ratio vs. baseline, and with 68.6% specificity and sensitivity in discriminating the responder group from the non-responder group. Similarly, patients who experienced a greater decrease in Ang-2 levels were observed in the responder group compared with the non-responder group at 2 weeks (Ang-2 level ratio at 2 weeks vs. baseline: 0.709 vs. 0.893, $p = 0.0041$), 4 weeks (Ang-2 ratio: 0.584 vs. 0.810, $p = 0.0002$), and 8 weeks (Ang-2 ratio: 0.500 vs. 0.804, $p < 0.0001$).

## 3. Could Lenvatinib Compete with Atezolizumab Plus Bevacizumab?

Following the results of the recent IMbrave150 trial, the combination of atezolizumab and bevacizumab has become the first-line standard of care [41] (Table 1).

This open-label phase 3 study compared the combination of atezolizumab/bevacizumab with sorafenib in patients with advanced unresectable and never treated HCC. The HR for mortality was 0.58 (95% CI, 0.42–0.79; $p = 0.001$) in favour of atezolizumab/bevacizumab. The PFS was 6.8 months (95% CI, 5.7–8.3) for the atezolizumab/bevacizumab group vs. 4.4 months (95% CI, 4.0–5.6) for the sorafenib group. The ORR was 27.3% vs. 11.9%, with a 5.5% complete response in the atezolizumab/bevacizumab group. Arterial hypertension was the most common Grade 3 or 4 adverse reaction in the atezolizumab/bevacizumab group (15.2% of patients).

This shift in first-line therapies led us to reconsider the place of lenvatinib in the sequential management of patients. To date, all randomised trials compare new first line therapies with sorafenib as the control arm, making it difficult to demonstrate the superiority of a specific drug.

Several recent reviews including meta-analysis aimed to compare first-line therapies of advanced HCC, most of them featuring sorafenib as the comparator [42–45]. Lenvatinib was associated with the greatest ORR benefit (OR, 3.34, 95% CI 2.17–5.14) in one study, whereas atezolizumab plus bevacizumab was superior to all other therapies including lenvatinib in others.

A cost-utility analysis was conducted in Canada, using a partitioned survival analysis, comparing atezolizumab and bevacizumab (from a de novo network meta-analysis based on patients data and clinical input from REFLECT, extrapolated using parametric survival models), and lenvatinib and sorafenib [46]. Lenvatinib was associated with cost savings of CAD 4640 and CAD 120,095 and an improvement in quality of life of 0.15 and −0.28 vs. sorafenib, and atezolizumab and bevacizumab, respectively.

However, we still need more large cohort observational studies or randomised controlled trials to confirm the difference in efficacy and safety between lenvatinib and atezolizumab–bevacizumab combination.

Lenvatinib is first line in cases of contraindication to immunotherapy plus anti-angiogenic combinations or in special populations.

Sorafenib and lenvatinib remain two possible first-line drugs in patients with a contraindication to atezolizumab and/or bevacizumab, e.g., patients with active dysimmunity disease, cardiovascular comorbidities such as uncontrolled arterial hypertension, recent myocardial stroke, recent surgery or lack of wound healing, or marked portal hypertension (including significant oesophageal or cardio-tuberositary varices).

If atezolizumab–bevacizumab is not suitable for the patient, the choice of TKI treatment should consider clinical, radiological, and biological features: (i) tumour characteristics (number of tumours, vascular invasion, extrahepatic spread, and AFP level), (ii) underlying liver disease (CP score and portal hypertension), and (iii) general status (ECOG, comorbidities and symptoms associated with the disease). Thus, the distinct safety profile of TKI could be taken into account in the choice of TKI according to comorbidities as arterial hypertension or vascular diseases for example.

The higher response rate and the improved TTP with lenvatinib compared to other TKI reflecting a high anti-tumoral activity could also be a selection criterion especially in symptomatic patients with a high tumour burden and threatening lesions. In patients with major liver involvement, a significant response is likely to prevent or to delay liver failure due to tumoral infiltration. Similarly, a symptomatic lesion (e.g., a painful bone metastasis) will benefit from a combination of TKIs and locoregional antalgic treatments. Less frequently in HCC, a deep response in selected patients can also lead to the consideration of curative treatments (e.g., surgical resection or ablation) through downstaging [28].

First-line treatment in some special situations:

### 3.1. Child-Pugh B Patients

Like most pivotal HCC phase III studies, patients with a CP-B score were not included in the REFLECT or the IMbrave150 study.

However, the benefit of TKI in CP-B patients remains highly uncertain. Thus, results of the use of sorafenib in CP-B patients are poor according to large observational studies [47]. Moreover, in a recent multicentric prospective randomized trial reporting the role for sorafenib versus best supportive care (BSC) in CP-B patients with HCC [48], median TTP and OS were similar in the sorafenib and BSC arms; nevertheless, there was a possible benefit in the CP-B7 sub-group.

Real-world studies have attempted to describe the profile of efficacy and safety of lenvatinib within this fragile population [29–32]. CP-B patients (19%, $n = 10$ in Wang's study, $n = 18$, 19.6% in Cheon's study) had a similar survival compared to CP-A [29–31,49]. The ORR ($p = 0.8293$) and DCR ($p = 0.7965$) were not statistically different according to REFLECT inclusion criteria, for example, in Sho's study. However, in a recent multicentre retrospective study, the OS at 12 months was significantly different between CP 5–7 (59.2%) and CP 8 patients (34.8%, $p = 0.003$) [50].

The data on the atezolizumab–bevacizumab combination in CP-B patients are also too scarce to rule on the ratio benefit/risk of this treatment in this weak population.

These results confirm the importance of considering hepatic function before introducing a treatment and the 2018 guidance AASLD noticed that systemic treatment could be administered in "well-selected patients with CP-B". Systemic therapies, among them lenvatinib, could therefore represent an alternative to palliative care after discussion in the case of a multidisciplinary consultation meeting.

### 3.2. Liver Transplantation

Liver transplantation is one of the curative treatments for HCC and a classic exclusion criterion of most pivotal phase III studies. Despite the optimisation of selection criteria, post-transplant HCC recurrence remains a major cause of death, but there is no standard of care for these poor prognosis diseases. While immunotherapies are today contra-indicated after liver transplantation, TKIs appear to be the treatment of choice. Some small retrospective cohort studies have reported heterogenous effects of sorafenib on post-recurrence survival, but there are very little data concerning the use of lenvatinib in this indication [51–53]. A recent case report of a patient with a five-year recurrence of HCC after liver transplantation who received lenvatinib as first-line systemic therapy reported no severe AEs, no liver dysfunction, and stable blood tacrolimus levels during the entire follow-up period.

In the pre-transplantation setting, the use of systemic treatment can be considered, particularly because of the expected long time on the waiting list. There are few data about the safety of immunotherapy in this situation. A recent review of three liver transplant recipients treated by immunotherapy before liver transplantation showed that two patients developed an acute rejection [54]. The authors also performed a retrospective cohort study with seven transplant receipts previously treated by PD-1 inhibitors of their centre (camrelizumab or pembrolizumab combined with lenvatinib). An acute rejection occurred in 14.3% of patients. Moreover, to avoid transplant rejection, immunosuppressive treatments should be prescribed at an optimal dose, and an addition of corticoids may be necessary, which could increase the risk of tumour recurrence. Overall, due to the unknown duration effects of immunotherapy, and because anti-VEGF should be stopped within six weeks before liver transplant, whereas its date is uncertain, the use of a TKI is considered in clinical practice. The prolonged time to progression with lenvatinib compared to sorafenib could be a strong argument for using lenvatinib in this situation, but this hypothesis requires further studies.

### 3.3. Severe Portal Hypertension

In case of advanced portal hypertension with an increased hemorrhagic risk, the use of anti-angiogenesis agents could be limited. For patients who cannot receive beta blockers, the protocol of ligation of esophageal varices may be long before eradication. If the risk of bleeding is increased by bevacizumab in the IMBrave study, there is no clear signal for an increase in bleeding related to portal hypertension in TKIs studies. In preclinical studies, sorafenib had beneficial effects on portocollateral circulation in cirrhotic animals [55]. Similarly, Hidaka demonstrated that sorafenib could reduce the portal venous flow in a prospective cohort study with 25 CP-A patients with advanced HCC [56]. The congestion index (portal venous area (PVA)/portal venous flow velocity (PVV)), which reflects the pathophysiological haemodynamics of portal venous system, significantly decreased ($3.9 \pm 1.7$ vs. $3.0 \pm 1.4$, $p = 0.042$), but there were no significant differences in the portal venous flow velocity (PVV; cm/s). Sorafenib could therefore be used in case of severe portal hypertension.

There are few studies regarding the safety of lenvatinib in case of advanced portal hypertension. Hidaka conducted analyses of the portal venous flow in 28 patients with advanced HCC treated with lenvatinib [57]. There was, in this study, 15% CP-B patients. The congestion index significantly worsened ($0.037 \pm 0.025$ vs. $0.043 \pm 0.024$, $p = 0.045$), but there were no significant differences ($p = 0.39$) in the portal venous area ($p = 0.665$). Finally, in the REFLECT trial, as well as in a recent prospective multicenter study of 93 patients treated with lenvatinib, the efficacy and the safety of lenvatinib do not seem to be impacted by the level of portal hypertension.

Overall, there is no strong argument for choosing sorafenib rather than lenvatinib in case of severe portal hypertension, and lenvatinib could be administered in case of advanced portal hypertension (without recent bleeding).

### 3.4. Etiology of the Liver Disease

The efficacy of lenvatinib does not appear to be influenced by the etiology of the liver disease. Hiraoka et al. recently conducted a multicentre retrospective study with 103 patients with NAFLD and 427 patients with AFLD or viral -related HCC treated by lenvatinib [58] without significant difference in overall survival (OS) (20.5 vs. 16.9 months, $p = 0.057$) between viral and NAFLD group.

On the other hand, in the IMbrave150 study, non-viral HCC etiologies (i.e., nonalcoholic fatty liver disease and alcohol) seem to be associated with a lower response rate and a lower survival with immunotherapies when compared to viral etiologies [59].

A recent meta-analysis of eight first-line high-quality phase three randomised clinical trials in advanced HCC (2002–2020) was published, reporting the relationship between aetiology and outcomes after systemic therapies with either TKI, anti-angiogenic, or ICI therapy [60]. Of these, five trials studied TKI/anti-VEGF (REACH, REACH-2, METIV-HCC, CELESTIAL, and JET-HCC; total of 2083 patients), and three studied immunotherapies (CheckMate-459, Journal Pre-proof 14, IMbrave150 and KEYNOTE-240; with a total of 1656 patients). The pooled HR for OS of patients with viral-related HCC treated with ICI was 0.64 (95% CI 0.5–0.83), compared with those in the standard of care group. The pooled HR for OS in patients with non-viral-related HCC treated with ICI was 0.92 (95% CI 0.77–1.11). The difference in efficacy between viral and non-viral-related HCC treated with ICI was significant ($p$ of heterogeneity = 0.0259), and the effect of ICI was remarkably similar in HBV- and HCV-related HCC (HR 0.64 (95% CI 0.49–0.83) vs. HR 0.68 (95% CI 0.47–0.98), respectively).

Since not all patients benefit from immunotherapy, the role of the aetiology of the underlying liver disease deserves to be investigated in further studies. However, even if viral-related HCCs could therefore benefit more from immunotherapy, there are currently not enough data to support the use of TKIs rather than ICI in first-line setting in non-viral-related HCCs.

## 4. Lenvatinib as Second-Line Treatment

In patients eligible for second-line therapy, after progression on atezolizumab/ bevacizumab, treatment options include TKIs (sorafenib, lenvatinib, regorafenib, and cabozantinib), ramucirumab, and IO (pembrolizumab), according to local approvals.

The development of IO as a gold standard at first line has opened new perspectives of the use of TKIs and among them lenvatinib as a second line therapy. In vitro studies of PD-1 inhibitor demonstrated that anti-PD-1 antibodies can remain bound to CD8+ T cells for more than 20 weeks [61]. The introduction of a TKI, and among them, lenvatinib, could act synergistically with anti-PD-1 antibodies even after the interruption of the immunotherapy. Aoki et al. reported encouraging results of lenvatinib when used after failure of PD-1/PD-L1 antibodies [62]. The ORR was 55.6%, the DCR was 86.1%, PFS was 10 months, and OS was 15.8 months. The OS since initiation of ICI therapy was 29.8 months, which is much longer than that conferred by lenvatinib alone as first-line therapy [63]. Yamauchi et al. conducted a study of 40 patients with HCC and reported that lenvatinib achieved a high response rate (81%) in tumours with a high expression of FGFR4 [64]. In addition, treatment with lenvatinib resulted in longer PFS in patients with a high FGFR4 expression than in those without FGFR4 expression (5.5 vs. 2.7 months, respectively), indicating that lenvatinib shows higher antitumour activity against tumours with high FGFR4 expression. However, there is a positive correlation between β-catenin mutations and FGFR4 expression, and its expression is higher in the population of tumours with WNT/β-catenin-activating mutations, which are found in approximately 20–30% of all HCCs [37,62,64,65].

Thus, even in patients who did not respond well to previous treatment with atezolizumab plus bevacizumab due to β-catenin activating mutations, subsequent treatment with lenvatinib would still provide better results due to its potent inhibitory effect on FGFR4.

A multinational, multicentre, and retrospective study reported clinical outcomes of patients who received subsequent systemic therapies after progression on atezolizumab-bevacizumab [66]. Of the 49 patients, 19 received lenvatinib. The ORR and DCR were 6.1 and 63.3%, respectively, across all patients. With a median follow-up duration of 11.0 months, PFS and OS were 3.4 months (95 CI 1.8–4.9) and 14.7 months (95% CI 8.1–21.2), respectively. Median PFS with lenvatinib was significantly longer than with sorafenib (6.1 vs. 2.5 months; $p = 0.004$), although there was no significant difference in median OS (16.6 vs. 11.2 months; $p = 0.347$). Patients treated with sorafenib had significantly more hand–foot syndromes than those treated with lenvatinib (69.0 vs. 26.3%, $p = 0.004$), while patients with lenvatinib seemed to have more fatigue and hypertension than those with sorafenib (fatigue; 42.1 vs. 17.2%, $p = 0.058$, and hypertension; 42.1 vs. 17.2%, $p = 0.058$).

In addition, a retrospective study has recently investigated the potential use of lenvatinib (based on real-life experience and in vitro assessment) as second-line for patients intolerant to sorafenib, and as third-line for patients resistant to regorafenib [67]. The results suggest that lenvatinib is active and safe as a second/third-line treatment for unresectable HCC. Another study in a few patients treated with at least three different systemic therapies also reported the efficacy of lenvatinib as later treatment, with a tolerable toxicity profile [68].

Cabozantinib has demonstrated an improved OS and PFS in the phase 3 CELESTIAL study and is now validated for patients progressive after sorafenib [69]. Only retrospective data are available about the use of cabozantinib after ICI in HCC. In the recent multinational multicentre retrospective study of 49 patients who received subsequent systemic therapy after progression on atezolizumab-bevacizumab, only one received cabozantinib as second line [66]. There is not enough evidence in the literature to choose from the four available TKIs after failure of atezolizumab and bevacizumab.

## 5. Emerging Strategies

### *5.1. Lenvatinib and Pembrolizumab*

Lenvatinib combined with immunotherapy has demonstrated promising antitumour activity with a tolerable safety profile in preclinical and clinical studies.

Regarding the mechanism of action of pembrolizumab and lenvatinib combination therapy, a preclinical study including in vitro and in vivo studies showed that suppression of tumour-associated macrophages, regulatory T cells, and other constituents of the tumour-suppressive microenvironment resulted in decreases in TGF-β and IL-10, the downregulation of PD-1 and Tim3, and the upregulation of ICOS and OX40, thereby inducing tumour immunity through IL-12 [70].

The combination of lenvatinib and pembrolizumab was recently approved as a second-line therapy for advanced endometrial carcinoma after the failure of systemic therapy [71]. For patients with advanced HCC, a phase 1b trial has recently shown promising results in terms of antitumoral activity, with a median OS of 22 months (95% CI, 20.4 months–not estimable), and an acceptable toxicity profile [72]. In this study, the ORR was 46.0% with mRECIST. Including a complete response in 11 patients, the median duration of response was 8.6 months, and median PFS was 9.3 months.

A phase three study is currently underway to confirm these promising results as first-line therapy [73]. For patients with intermediate HCC, eligible for locoregional treatment, a phase three trial (LEAP-012) is investigating lenvatinib and pembrolizumab vs. placebo in combination with transarterial chemoembolisation (TACE) [74]. Recruitment for this study began in April 2020 (Supplementary Data Table S1).

### *5.2. Lenvatinib in the Intermediate Stage*

TACE is one of the most widely used palliative treatments in the world. However, repeated chemoembolisation sessions can lead to impaired liver functions, limiting access to subsequent systemic treatments, and cohort studies have shown that only <20% of patients treated with chemoembolisation will be able to receive systemic treatment [75]. In addition, patients with intermediate-stage HCC constitute a very heterogeneous group, and some systems for subclassification (based in particular on CP score, beyond Milan, and up-to-seven criteria [76,77]) have been proposed to identify patients who will not benefit from TACE [78,79]. The arrival of new effective systemic treatments has led to consideration of the optimal timing for the switch from loco-regional to systemic therapies including lenvatinib as an interesting alternative to TACE as first-line treatment.

A proof-of-concept retrospective study with a propensity matching score showed that lenvatinib could be more beneficial than TACE in HCC beyond up-to-seven criteria [80]. The lenvatinib group had a significantly higher ORR (73.3% vs. 33.3%; $p < 0.001$), a longer median PFS (16.0 vs. 3.0 months; $p < 0.001$), and a longer OS (37.9 vs. 21.3 months; HR: 0.48). Liver function, based on the albumin–bilirubin (ALBI) score, was well preserved in the lenvatinib group compared to the TACE group at the end of treatment. The preservation of liver function with lenvatinib allows full dose administration over a long period and can indirectly explain the high response rate with this treatment, optimising the access to a second-line therapy. In this same study, more than half the patients previously treated with lenvatinib were subsequently treated by TACE [81].

Patients who receive early lenvatinib administration may have a better prognosis than those who receive TACE [82]. A cohort study with 208 patients who were considered candidates for repeated TACE showed that cumulative survival rates were higher in patients treated with lenvatinib vs. patients treated with TACE (the 6-, 12-, 18-, and 24-month cumulative survival rates were 96.0, 90.4, 65.7, and 65.7%, respectively, vs. 94.1, 78.5, 65.3, and 48.4%, respectively, $p < 0.001$). The median survival time in the lenvatinib group was not available (95% CI, 17.1–not available), while that in the TACE group was 22.5 (95% CI, 20.9–26.7) months.

Further studies are necessary to confirm these encouraging results in patients with intermediate stage HCC, in whom TACE alone is not helpful (Supplementary Data Table S1).

*5.3. Lenvatinib in an Adjuvant Setting*

Despite a high recurrence rate after curative surgery (estimated at 70% at 5 years), there are currently no validated adjuvant therapies for patients with HCC [83,84]. The phase three, randomised, double-blind, placebo-controlled STORM trial studied sorafenib as an adjuvant treatment after surgical resection or local ablation of HCC [85]. There was no statistical difference in median recurrence-free survival between the sorafenib group vs. the placebo group (33·3 months vs. 33·7 months, respectively; $p = 0{\cdot}26$), and sorafenib could not, therefore, be recommended as an adjuvant setting.

Another recent preliminary study demonstrated a potential benefit of adjuvant lenvatinib on disease-free survival and recurrence in high-risk patients with HBV-related HCC following liver transplantation [86]. A phase two study is underway to assess the efficacy and safety of adjuvant lenvatinib for patients after radical resection of HCC with a high risk of tumour recurrence [87]. In addition, the interim analysis of the LANCE study suggests that lenvatinib combined with TACE would be efficient and safe [88]. More studies attesting a benefit in recurrence-free survival of lenvatinib are necessary (Table S1).

## 6. Conclusions

The rapid implementation of new therapeutic options, including immunotherapies and combination therapies, has dramatically modified the treatment landscape of HCC, requiring a reassessment of sequential therapeutic strategy (Figure 1).

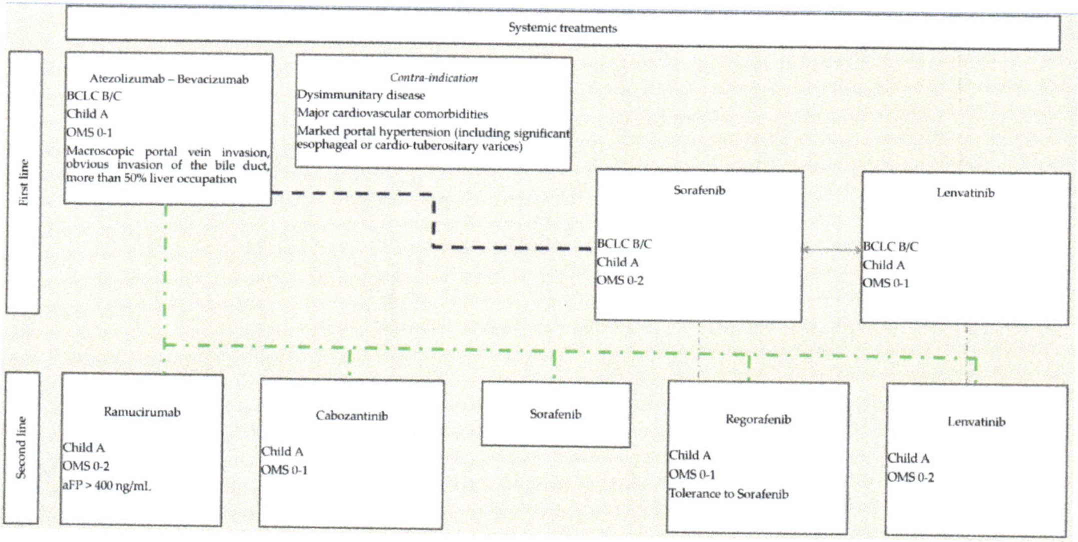


**Figure 1.** Proposed therapeutic algorithm for the use of systemic treatments in advanced, unresectable HCC.

As the combination of bevacizumab and atezolizumab is now the first-line standard of care, the place of TKIs in monotherapy is moving to subsequent lines of treatment. Lenvatinib displays strong antitumoral activity, with the highest response rate among TKIs, and therefore, its early use in sequential therapy should be considered. Future developments could include the use of lenvatinib at an earlier stage of the disease, at stage BCLC B (in association with or vs. TACE), and first line in stage BCLC C HCC in association with immunotherapies, but further studies are required with atezolizumab and bevacizumab as a group control.

**Supplementary Materials:** The following are available online at https://www.mdpi.com/article/10.3390/cancers13246310/s1, Table S1. Clinical trials of lenvatinib for HCC.

**Author Contributions:** Conceptualization, M.D. and J.-F.B.; methodology, M.D. and J.-F.B.; validation, M.D. and J.-F.B.; writing—original draft preparation, M.D., C.T. and J.-F.B.; writing—review and editing, M.D. and J.-F.B.; visualization, M.D. and J.-F.B.; supervision, J.-F.B. All authors have read and agreed to the published version of the manuscript.

**Funding:** This research received no external funding.

**Conflicts of Interest:** Marie Decraecker and Caroline declare no conflict of interest. Jean-Frédéric Blanc: Bayer, ESAI, MSD, IPSEN, Roche, BMS, Astra-Zeneca. The funders had no role in the design of the study; in the collection, analyses, or interpretation of data; in the writing of the manuscript, or in the decision to publish the results.

## Abbreviations

| | |
|---|---|
| AE | adverse events |
| AFP | alpha-foetoprotein |
| Anti-VEGFR | anti-vascular endothelial growth factor |
| BCLC | Barcelona Clinic Liver Cancer |
| BSC | best supportive care |
| CP | Child–Pugh |
| DCR | disease control rate |
| ECOG | Eastern Cooperative Oncology Group |
| FGFR | fibroblast growth factor receptor |
| HCC | hepatocellular carcinoma |
| HBV | hepatitis B virus |
| HCV | hepatitis C virus |
| HR | hazard ratio |
| ICIs | immune checkpoint inhibitors |
| IO | immuno-oncology |
| IT | immunotherapy |
| NE | not evaluated |
| OR | odds ratio |
| ORR | overall response rate |
| OS | overall survival |
| PFS | progression-free survival |
| PDGFR | platelet-derived growth factor receptor |
| PPES | palmar–plantar erythrodysaesthesia syndrome |
| PS | performance status |
| PVA | portal venous area |
| PVV | portal venous flow velocity |
| QALY | quality-adjusted life years |
| TACE | transarterial chemoembolisation |
| TEAEs | treatment-emergent adverse events |
| TTP | time to progression |
| TKIs | tyrosine kinase inhibitors |
| RECIST | response evaluation criteria in solid tumours |

## References

1. Kulik, L.; El-Serag, H.B. Epidemiology and Management of Hepatocellular Carcinoma. *Gastroenterology* **2019**, *156*, 477–491.e1. [CrossRef]
2. Marrero, J.A.; Kulik, L.M.; Sirlin, C.B.; Zhu, A.X.; Finn, R.S.; Abecassis, M.M.; Roberts, L.R.; Heimbach, J.K. Diagnosis, Staging, and Management of Hepatocellular Carcinoma: 2018 Practice Guidance by the American Association for the Study of Liver Diseases. *Hepatology* **2018**, *68*, 723–750. [CrossRef]
3. Khalaf, N.; Ying, J.; Mittal, S.; Temple, S.; Kanwal, F.; Davila, J.; El-Serag, H.B. Natural History of Untreated Hepatocellular Carcinoma in a US Cohort and the Role of Cancer Surveillance. *Clin. Gastroenterol. Hepatol.* **2017**, *15*, 273–281.e1. [CrossRef] [PubMed]
4. Huang, Y.; Wallace, M.C.; Adams, L.A.; Macquillan, G.; Garas, G.; Ferguson, J.; Samuelson, S.; Tibballs, J.; Jeffrey, G.P. Rate of Nonsurveillance and Advanced Hepatocellular Carcinoma at Diagnosis in Chronic Liver Disease. *J. Clin. Gastroenterol.* **2018**, *52*, 551–556. [CrossRef]

5. Giannini, E.G.; Farinati, F.; Ciccarese, F.; Pecorelli, A.; Rapaccini, G.L.; Di Marco, M.; Benvegnù, L.; Caturelli, E.; Zoli, M.; Borzio, F.; et al. Prognosis of Untreated Hepatocellular Carcinoma. *Hepatology* **2015**, *61*, 184–190. [CrossRef] [PubMed]
6. Vogel, A.; Cervantes, A.; Chau, I.; Daniele, B.; Llovet, J.M.; Meyer, T.; Nault, J.-C.; Neumann, U.; Ricke, J.; Sangro, B.; et al. Hepatocellular Carcinoma: ESMO Clinical Practice Guidelines for Diagnosis, Treatment and Follow-Up. *Ann. Oncol.* **2018**, *29*, iv238–iv255. [CrossRef]
7. Llovet, J.M.; Ricci, S.; Mazzaferro, V.; Hilgard, P.; Gane, E.; Blanc, J.-F.; de Oliveira, A.C.; Santoro, A.; Raoul, J.-L.; Forner, A.; et al. Sorafenib in Advanced Hepatocellular Carcinoma. *N. Engl. J. Med.* **2008**, *359*, 378–390. [CrossRef]
8. Keating, G.M. Sorafenib: A Review in Hepatocellular Carcinoma. *Target. Oncol.* **2017**, *12*, 243–253. [CrossRef] [PubMed]
9. Bruix, J.; Cheng, A.-L.; Meinhardt, G.; Nakajima, K.; De Sanctis, Y.; Llovet, J. Prognostic Factors and Predictors of Sorafenib Benefit in Patients with Hepatocellular Carcinoma: Analysis of Two Phase III Studies. *J. Hepatol.* **2017**, *67*, 999–1008. [CrossRef] [PubMed]
10. European Association for the Study of the Liver (EASL); European Association for the Study of Diabetes (EASD); European Association for the Study of Obesity (EASO). EASL-EASD-EASO Clinical Practice Guidelines for the Management of Non-Alcoholic Fatty Liver Disease. *J. Hepatol.* **2016**, *64*, 1388–1402. [CrossRef] [PubMed]
11. Heimbach, J.K.; Kulik, L.M.; Finn, R.S.; Sirlin, C.B.; Abecassis, M.M.; Roberts, L.R.; Zhu, A.X.; Murad, M.H.; Marrero, J.A. AASLD Guidelines for the Treatment of Hepatocellular Carcinoma. *Hepatology* **2018**, *67*, 358–380. [CrossRef]
12. Omata, M.; Cheng, A.-L.; Kokudo, N.; Kudo, M.; Lee, J.M.; Jia, J.; Tateishi, R.; Han, K.-H.; Chawla, Y.K.; Shiina, S.; et al. Asia-Pacific Clinical Practice Guidelines on the Management of Hepatocellular Carcinoma: A 2017 Update. *Hepatol. Int.* **2017**, *11*, 317–370. [CrossRef] [PubMed]
13. Benson, A.B.; D'Angelica, M.I.; Abbott, D.E.; Abrams, T.A.; Alberts, S.R.; Saenz, D.A.; Are, C.; Brown, D.B.; Chang, D.T.; Covey, A.M.; et al. NCCN Guidelines Insights: Hepatobiliary Cancers, Version 1.2017. *J. Natl. Compr. Cancer Netw.* **2017**, *15*, 563–573. [CrossRef] [PubMed]
14. Kudo, M.; Finn, R.S.; Qin, S.; Han, K.-H.; Ikeda, K.; Piscaglia, F.; Baron, A.; Park, J.-W.; Han, G.; Jassem, J.; et al. Lenvatinib versus Sorafenib in First-Line Treatment of Patients with Unresectable Hepatocellular Carcinoma: A Randomised Phase 3 Non-Inferiority Trial. *Lancet* **2018**, *391*, 1163–1173. [CrossRef]
15. Park, J.-W.; Chen, M.; Colombo, M.; Roberts, L.R.; Schwartz, M.; Chen, P.-J.; Kudo, M.; Johnson, P.; Wagner, S.; Orsini, L.S.; et al. Global Patterns of Hepatocellular Carcinoma Management from Diagnosis to Death: The BRIDGE Study. *Liver Int.* **2015**, *35*, 2155–2166. [CrossRef]
16. Matsui, J.; Yamamoto, Y.; Funahashi, Y.; Tsuruoka, A.; Watanabe, T.; Wakabayashi, T.; Uenaka, T.; Asada, M. E7080, a Novel Inhibitor That Targets Multiple Kinases, Has Potent Antitumor Activities against Stem Cell Factor Producing Human Small Cell Lung Cancer H146, Based on Angiogenesis Inhibition. *Int. J. Cancer* **2008**, *122*, 664–671. [CrossRef] [PubMed]
17. Matsui, J.; Funahashi, Y.; Uenaka, T.; Watanabe, T.; Tsuruoka, A.; Asada, M. Multi-Kinase Inhibitor E7080 Suppresses Lymph Node and Lung Metastases of Human Mammary Breast Tumor MDA-MB-231 via Inhibition of Vascular Endothelial Growth Factor-Receptor (VEGF-R) 2 and VEGF-R3 Kinase. *Clin. Cancer Res.* **2008**, *14*, 5459–5465. [CrossRef]
18. Tohyama, O.; Matsui, J.; Kodama, K.; Hata-Sugi, N.; Kimura, T.; Okamoto, K.; Minoshima, Y.; Iwata, M.; Funahashi, Y. Antitumor Activity of Lenvatinib (E7080): An Angiogenesis Inhibitor That Targets Multiple Receptor Tyrosine Kinases in Preclinical Human Thyroid Cancer Models. *J. Thyroid Res.* **2014**, *2014*, 638747. [CrossRef]
19. Yamamoto, Y.; Matsui, J.; Matsushima, T.; Obaishi, H.; Miyazaki, K.; Nakamura, K.; Tohyama, O.; Semba, T.; Yamaguchi, A.; Hoshi, S.S.; et al. Lenvatinib, an Angiogenesis Inhibitor Targeting VEGFR/FGFR, Shows Broad Antitumor Activity in Human Tumor Xenograft Models Associated with Microvessel Density and Pericyte Coverage. *Vasc. Cell* **2014**, *6*, 18. [CrossRef] [PubMed]
20. Matsuki, M.; Hoshi, T.; Yamamoto, Y.; Ikemori-Kawada, M.; Minoshima, Y.; Funahashi, Y.; Matsui, J. Lenvatinib Inhibits Angiogenesis and Tumor Fibroblast Growth Factor Signaling Pathways in Human Hepatocellular Carcinoma Models. *Cancer Med.* **2018**, *7*, 2641–2653. [CrossRef]
21. Ichikawa, K.; Watanabe Miyano, S.; Minoshima, Y.; Matsui, J.; Funahashi, Y. Activated FGF2 Signaling Pathway in Tumor Vasculature Is Essential for Acquired Resistance to Anti-VEGF Therapy. *Sci. Rep.* **2020**, *10*, 2939. [CrossRef]
22. Okamoto, K.; Kodama, K.; Takase, K.; Sugi, N.H.; Yamamoto, Y.; Iwata, M.; Tsuruoka, A. Antitumor Activities of the Targeted Multi-Tyrosine Kinase Inhibitor Lenvatinib (E7080) against RET Gene Fusion-Driven Tumor Models. *Cancer Lett.* **2013**, *340*, 97–103. [CrossRef] [PubMed]
23. Ogasawara, S.; Mihara, Y.; Kondo, R.; Kusano, H.; Akiba, J.; Yano, H. Antiproliferative Effect of Lenvatinib on Human Liver Cancer Cell Lines In Vitro and In Vivo. *Anticancer Res.* **2019**, *39*, 5973–5982. [CrossRef] [PubMed]
24. Hoshi, T.; Watanabe Miyano, S.; Watanabe, H.; Sonobe, R.M.K.; Seki, Y.; Ohta, E.; Nomoto, K.; Matsui, J.; Funahashi, Y. Lenvatinib Induces Death of Human Hepatocellular Carcinoma Cells Harboring an Activated FGF Signaling Pathway through Inhibition of FGFR-MAPK Cascades. *Biochem. Biophys. Res. Commun.* **2019**, *513*, 1–7. [CrossRef] [PubMed]
25. Ikeda, K.; Kudo, M.; Kawazoe, S.; Osaki, Y.; Ikeda, M.; Okusaka, T.; Tamai, T.; Suzuki, T.; Hisai, T.; Hayato, S.; et al. Phase 2 Study of Lenvatinib in Patients with Advanced Hepatocellular Carcinoma. *J. Gastroenterol.* **2017**, *52*, 512–519. [CrossRef]
26. Ikeda, M.; Okusaka, T.; Mitsunaga, S.; Ueno, H.; Tamai, T.; Suzuki, T.; Hayato, S.; Kadowaki, T.; Okita, K.; Kumada, H. Safety and Pharmacokinetics of Lenvatinib in Patients with Advanced Hepatocellular Carcinoma. *Clin. Cancer Res.* **2016**, *22*, 1385–1394. [CrossRef]

27. Kuzuya, T.; Ishigami, M.; Ito, T.; Ishizu, Y.; Honda, T.; Fujishiro, M. FRI-484-The Early Clinical Response at 2 Weeks of Lenvatinib Therapy for Patients with Advanced HCC. *J. Hepatol.* **2019**, *70*, e611. [CrossRef]
28. Kudo, M. Lenvatinib May Drastically Change the Treatment Landscape of Hepatocellular Carcinoma. *Liver Cancer* **2018**, *7*, 1–19. [CrossRef]
29. Wang, D.-X.; Yang, X.; Lin, J.-Z.; Bai, Y.; Long, J.-Y.; Yang, X.-B.; Seery, S.; Zhao, H.-T. Efficacy and Safety of Lenvatinib for Patients with Advanced Hepatocellular Carcinoma: A Retrospective, Real-World Study Conducted in China. *World J. Gastroenterol.* **2020**, *26*, 4465–4478. [CrossRef]
30. Sho, T.; Suda, G.; Ogawa, K.; Kimura, M.; Shimazaki, T.; Maehara, O.; Shigesawa, T.; Suzuki, K.; Nakamura, A.; Ohara, M.; et al. Early Response and Safety of Lenvatinib for Patients with Advanced Hepatocellular Carcinoma in a Real-world Setting. *JGH Open* **2019**, *4*, 54–60. [CrossRef]
31. Cheon, J.; Chon, H.J.; Bang, Y.; Park, N.H.; Shin, J.W.; Kim, K.M.; Lee, H.C.; Lee, J.; Yoo, C.; Ryoo, B.-Y. Real-World Efficacy and Safety of Lenvatinib in Korean Patients with Advanced Hepatocellular Carcinoma: A Multicenter Retrospective Analysis. *Liver Cancer* **2020**, *9*, 613–624. [CrossRef]
32. Hiraoka, A.; Kumada, T.; Kariyama, K.; Takaguchi, K.; Atsukawa, M.; Itobayashi, E.; Tsuji, K.; Tajiri, K.; Hirooka, M.; Shimada, N.; et al. Clinical Features of Lenvatinib for Unresectable Hepatocellular Carcinoma in Real-world Conditions: Multicenter Analysis. *Cancer Med.* **2018**, *8*, 137–146. [CrossRef]
33. Sung, M.W.; Finn, R.S.; Qin, S.; Han, K.-H.; Ikeda, K.; Cheng, A.-L.; Kudo, M.; Tateishi, R.; Ikeda, M.; Breder, V.; et al. Association between Overall Survival and Adverse Events with Lenvatinib Treatment in Patients with Hepatocellular Carcinoma (REFLECT). *J. Clin. Oncol.* **2019**, *37*, 317. [CrossRef]
34. Kobayashi, M.; Kudo, M.; Izumi, N.; Kaneko, S.; Azuma, M.; Copher, R.; Meier, G.; Pan, J.; Ishii, M.; Ikeda, S. Cost-Effectiveness Analysis of Lenvatinib Treatment for Patients with Unresectable Hepatocellular Carcinoma (UHCC) Compared with Sorafenib in Japan. *J. Gastroenterol.* **2019**, *54*, 558–570. [CrossRef]
35. Ikeda, S.; Kudo, M.; Izumi, N.; Kobayashi, M.; Azuma, M.; Meier, G.; Pan, J.; Ishii, M.; Kaneko, S. Cost-Effectiveness of Lenvatinib in the Treatment of Patients with Unresectable Hepatocellular Carcinomas in Japan: An Analysis Using Data from Japanese Patients in the REFLECT Trial. *Value Health Reg. Issues* **2021**, *24*, 82–89. [CrossRef]
36. Kim, J.J.; McFarlane, T.; Tully, S.; Wong, W.W.L. Lenvatinib Versus Sorafenib as First-Line Treatment of Unresectable Hepatocellular Carcinoma: A Cost-Utility Analysis. *Oncologisty* **2020**, *25*, e512–e519. [CrossRef]
37. Kanzaki, H.; Chiba, T.; Ao, J.; Koroki, K.; Kanayama, K.; Maruta, S.; Maeda, T.; Kusakabe, Y.; Kobayashi, K.; Kanogawa, N.; et al. The Impact of FGF19/FGFR4 Signaling Inhibition in Antitumor Activity of Multi-Kinase Inhibitors in Hepatocellular Carcinoma. *Sci. Rep.* **2021**, *11*, 5303. [CrossRef]
38. Finn, R.S.; Kudo, M.; Cheng, A.-L.; Wyrwicz, L.; Ngan, R.; Blanc, J.F.; Baron, A.D.; Vogel, A.; Ikeda, M.; Piscaglia, F.; et al. Final Analysis of Serum Biomarkers in Patients (Pts) from the Phase III Study of Lenvatinib (LEN) vs. Sorafenib (SOR) in Unresectable Hepatocellular Carcinoma (UHCC) [REFLECT]. *Ann. Oncol.* **2018**, *29*, viii17–viii18. [CrossRef]
39. Finn, R.S.; Kudo, M.; Cheng, A.-L.; Wyrwicz, L.; Ngan, R.K.-C.; Blanc, J.-F.; Baron, A.D.; Vogel, A.; Ikeda, M.; Piscaglia, F.; et al. Pharmacodynamic Biomarkers Predictive of Survival Benefit with Lenvatinib in Unresectable Hepatocellular Carcinoma: From the Phase 3 REFLECT Study. *Clin. Cancer Res.* **2021**, *27*, 4848–4858. [CrossRef]
40. Chuma, M.; Uojima, H.; Numata, K.; Hidaka, H.; Toyoda, H.; Hiraoka, A.; Tada, T.; Hirose, S.; Atsukawa, M.; Itokawa, N.; et al. Early Changes in Circulating FGF19 and Ang-2 Levels as Possible Predictive Biomarkers of Clinical Response to Lenvatinib Therapy in Hepatocellular Carcinoma. *Cancers* **2020**, *12*, 293. [CrossRef]
41. Finn, R.S.; Qin, S.; Ikeda, M.; Galle, P.R.; Ducreux, M.; Kim, T.-Y.; Kudo, M.; Breder, V.; Merle, P.; Kaseb, A.O.; et al. Atezolizumab plus Bevacizumab in Unresectable Hepatocellular Carcinoma. *N. Engl. J. Med.* **2020**, *382*, 1894–1905. [CrossRef]
42. Park, R.; Lopes da Silva, L.; Nissaisorakarn, V.; Riano, I.; Williamson, S.; Sun, W.; Saeed, A. Comparison of Efficacy of Systemic Therapies in Advanced Hepatocellular Carcinoma: Updated Systematic Review and Frequentist Network Meta-Analysis of Randomized Controlled Trials. *J. Hepatocell. Carcinoma* **2021**, *8*, 145–154. [CrossRef]
43. Sonbol, M.B.; Riaz, I.B.; Naqvi, S.A.A.; Almquist, D.R.; Mina, S.; Almasri, J.; Shah, S.; Almader-Douglas, D.; Uson Junior, P.L.S.; Mahipal, A.; et al. Systemic Therapy and Sequencing Options in Advanced Hepatocellular Carcinoma: A Systematic Review and Network Meta-Analysis. *JAMA Oncol.* **2020**, *6*, e204930. [CrossRef]
44. Trueman, D.; Liu, Y.; Lucero, M.; Meier, G. The Comparative Efficacy of Atezolizumab and Bevacizumab versus Lenvatinib in Patients with Unresectable Hepatocellular Carcinoma (UHCC). *J. Clin. Oncol.* **2021**, *39*, e16151. [CrossRef]
45. Casadei-Gardini, A.; Tada, T.; Shimose, S.; Kumada, T.; Niizeki, T.; Cascinu, S.; Cucchetti, A. Is Atezolizumab Plus Bevacizumab for Unresectable Hepatocellular Carcinoma Superior Even to Lenvatinib? A Matching-Adjusted Indirect Comparison. *Target. Oncol.* **2021**, *16*, 249–254. [CrossRef]
46. Trueman, D.; Liu, Y.; Geadah, M.; Hon, N.; Sabapathy, S.; Kamboj, L.; Li, H.; Lucero, M.; Meier, G. The Cost Effectiveness of Lenvatinib versus Atezolizumab and Bevacizumab or Sorafenib in Patients with Unresectable Hepatocellular Carcinoma (UHCC) in Canada. *J. Clin. Oncol.* **2021**, *39*, 4098. [CrossRef]
47. Marrero, J.A.; Kudo, M.; Venook, A.P.; Ye, S.-L.; Bronowicki, J.-P.; Chen, X.-P.; Dagher, L.; Furuse, J.; Geschwind, J.-F.H.; de Guevara, L.L.; et al. Observational Registry of Sorafenib Use in Clinical Practice across Child-Pugh Subgroups: The GIDEON Study. *J. Hepatol.* **2016**, *65*, 1140–1147. [CrossRef]

48. Blanc, J.-F.; Khemissa, F.; Bronowicki, J.-P.; Monterymard, C.; Perarnau, J.-M.; Bourgeois, V.; Obled, S.; Abdelghani, M.B.; Mabile-Archambeaud, I.; Faroux, R.; et al. Phase 2 Trial Comparing Sorafenib, Pravastatin, Their Combination or Supportive Care in HCC with Child-Pugh B Cirrhosis. *Hepatol. Int.* **2021**, *15*, 93–104. [CrossRef]
49. Ogushi, K.; Chuma, M.; Uojima, H.; Hidaka, H.; Numata, K.; Kobayashi, S.; Hirose, S.; Hattori, N.; Fujikawa, T.; Nakazawa, T.; et al. Safety and Efficacy of Lenvatinib Treatment in Child-Pugh A and B Patients with Unresectable Hepatocellular Carcinoma in Clinical Practice: A Multicenter Analysis. *Clin. Exp. Gastroenterol.* **2020**, *13*, 385–396. [CrossRef]
50. Hiraoka, A.; Kumada, T.; Atsukawa, M.; Hirooka, M.; Tsuji, K.; Ishikawa, T.; Takaguchi, K.; Kariyama, K.; Itobayashi, E.; Tajiri, K.; et al. Prognostic Factor of Lenvatinib for Unresectable Hepatocellular Carcinoma in Real-World Conditions-Multicenter Analysis. *Cancer Med.* **2019**, *8*, 3719–3728. [CrossRef]
51. Mancuso, A.; Mazzola, A.; Cabibbo, G.; Perricone, G.; Enea, M.; Galvano, A.; Zavaglia, C.; Belli, L.; Cammà, C. Survival of Patients Treated with Sorafenib for Hepatocellular Carcinoma Recurrence after Liver Transplantation: A Systematic Review and Meta-Analysis. *Dig. Liver Dis.* **2015**, *47*, 324–330. [CrossRef]
52. Piñero, F.; Thompson, M.; Marín, J.I.; Silva, M. Lenvatinib as First-Line Therapy for Recurrent Hepatocellular Carcinoma after Liver Transplantation: Is the Current Evidence Applicable to These Patients? *World J. Transplant.* **2020**, *10*, 297–306. [CrossRef]
53. Eilard, M.S.; Andersson, M.; Naredi, P.; Geronymakis, C.; Lindnér, P.; Cahlin, C.; Bennet, W.; Rizell, M. A Prospective Clinical Trial on Sorafenib Treatment of Hepatocellular Carcinoma before Liver Transplantation. *BMC Cancer* **2019**, *19*, 568. [CrossRef]
54. Qiao, Z.; Zhang, Z.; Lv, Z.; Tong, H.; Xi, Z.; Wu, H.; Chen, X.; Xia, L.; Feng, H.; Zhang, J.; et al. Neoadjuvant Programmed Cell Death 1 (PD-1) Inhibitor Treatment in Patients with Hepatocellular Carcinoma Before Liver Transplant: A Cohort Study and Literature Review. *Front. Immunol.* **2021**, *12*, 653437. [CrossRef]
55. Mejias, M.; Garcia-Pras, E.; Tiani, C.; Miquel, R.; Bosch, J.; Fernandez, M. Beneficial Effects of Sorafenib on Splanchnic, Intrahepatic, and Portocollateral Circulations in Portal Hypertensive and Cirrhotic Rats. *Hepatology* **2009**, *49*, 1245–1256. [CrossRef]
56. Hidaka, H.; Nakazawa, T.; Kaneko, T.; Minamino, T.; Takada, J.; Tanaka, Y.; Okuwaki, Y.; Watanabe, M.; Shibuya, A.; Koizumi, W. Portal Hemodynamic Effects of Sorafenib in Patients with Advanced Hepatocellular Carcinoma: A Prospective Cohort Study. *J. Gastroenterol.* **2012**, *47*, 1030–1035. [CrossRef] [PubMed]
57. Hidaka, H.; Uojima, H.; Nakazawa, T.; Shao, X.; Hara, Y.; Iwasaki, S.; Wada, N.; Kubota, K.; Tanaka, Y.; Shibuya, A.; et al. Portal Hemodynamic Effects of Lenvatinib in Patients with Advanced Hepatocellular Carcinoma: A Prospective Cohort Study. *Hepatol. Res.* **2020**, *50*, 1083–1090. [CrossRef] [PubMed]
58. Hiraoka, A.; Kumada, T.; Tada, T.; Tani, J.; Kariyama, K.; Fukunishi, S.; Atsukawa, M.; Hirooka, M.; Tsuji, K.; Ishikawa, T.; et al. Efficacy of Lenvatinib for Unresectable Hepatocellular Carcinoma Based on Background Liver Disease Etiology: Multi-Center Retrospective Study. *Sci. Rep.* **2021**, *11*, 16663. [CrossRef]
59. Briggs, A.; Daniele, B.; Dick, K.; Evans, T.R.J.; Galle, P.R.; Hubner, R.A.; Lopez, C.; Siebert, U.; Tremblay, G. Covariate-Adjusted Analysis of the Phase 3 REFLECT Study of Lenvatinib versus Sorafenib in the Treatment of Unresectable Hepatocellular Carcinoma. *Br. J. Cancer* **2020**, *122*, 1754–1759. [CrossRef]
60. Haber, P.K.; Puigvehí, M.; Castet, F.; Lourdusamy, V.; Montal, R.; Tabrizian, P.; Buckstein, M.; Kim, E.; Villanueva, A.; Schwartz, M.; et al. Evidence-Based Management of HCC: Systematic Review and Meta-Analysis of Randomized Controlled Trials (2002–2020). *Gastroenterology* **2021**, *161*, 879–898. [CrossRef]
61. Osa, A.; Uenami, T.; Koyama, S.; Fujimoto, K.; Okuzaki, D.; Takimoto, T.; Hirata, H.; Yano, Y.; Yokota, S.; Kinehara, Y.; et al. Clinical Implications of Monitoring Nivolumab Immunokinetics in Non–Small Cell Lung Cancer Patients. *JCI Insight* **2018**, *3*, e59125. [CrossRef] [PubMed]
62. Aoki, T.; Kudo, M.; Ueshima, K.; Morita, M.; Chishina, H.; Takita, M.; Hagiwara, S.; Ida, H.; Minami, Y.; Tsurusaki, M.; et al. Exploratory Analysis of Lenvatinib Therapy in Patients with Unresectable Hepatocellular Carcinoma Who Have Failed Prior PD-1/PD-L1 Checkpoint Blockade. *Cancers* **2020**, *12*, 3048. [CrossRef] [PubMed]
63. Kudo, M. Sequential Therapy for Hepatocellular Carcinoma after Failure of Atezolizumab plus Bevacizumab Combination Therapy. *Liver Cancer* **2021**, *10*, 85–93. [CrossRef]
64. Yamauchi, M.; Ono, A.; Ishikawa, A.; Kodama, K.; Uchikawa, S.; Hatooka, H.; Zhang, P.; Teraoka, Y.; Morio, K.; Fujino, H.; et al. Tumor Fibroblast Growth Factor Receptor 4 Level Predicts the Efficacy of Lenvatinib in Patients With Advanced Hepatocellular Carcinoma. *Clin. Transl. Gastroenterol.* **2020**, *11*, e00179. [CrossRef]
65. Yi, C.; Chen, L.; Lin, Z.; Liu, L.; Shao, W.; Zhang, R.; Lin, J.; Zhang, J.; Zhu, W.; Jia, H.; et al. Lenvatinib Targets FGF Receptor 4 to Enhance Antitumor Immune Response of Anti-Programmed Cell Death-1 in HCC. *Hepatology* **2021**, *74*, 2544–2560. [CrossRef]
66. Yoo, C.; Kim, J.H.; Ryu, M.-H.; Park, S.R.; Lee, D.; Kim, K.M.; Shim, J.H.; Lim, Y.-S.; Lee, H.C.; Lee, J.; et al. Clinical Outcomes with Multikinase Inhibitors after Progression on First-Line Atezolizumab plus Bevacizumab in Patients with Advanced Hepatocellular Carcinoma: A Multinational Multicenter Retrospective Study. *Liver Cancer* **2021**, *10*, 107–114. [CrossRef]
67. Tomonari, T.; Sato, Y.; Tanaka, H.; Tanaka, T.; Fujino, Y.; Mitsui, Y.; Hirao, A.; Taniguchi, T.; Okamoto, K.; Sogabe, M.; et al. Potential Use of Lenvatinib for Patients with Unresectable Hepatocellular Carcinoma Including after Treatment with Sorafenib: Real-World Evidence and in Vitro Assessment via Protein Phosphorylation Array. *Oncotarget* **2020**, *11*, 2531–2542. [CrossRef]
68. Jefremow, A.; Wiesmüller, M.; Rouse, R.A.; Dietrich, P.; Kremer, A.E.; Waldner, M.J.; Neurath, M.F.; Siebler, J. Safety and Efficacy of Lenvatinib in HCC beyond Second-Line Treatment. *J. Clin. Oncol.* **2020**, *38*, e16592. [CrossRef]

69. Abou-Alfa, G.K.; Meyer, T.; Cheng, A.-L.; El-Khoueiry, A.B.; Rimassa, L.; Ryoo, B.-Y.; Cicin, I.; Merle, P.; Chen, Y.H.; Park, J.-W.; et al. Cabozantinib in Patients with Advanced and Progressing Hepatocellular Carcinoma. *N. Engl. J. Med.* **2018**, *379*, 54–63. [CrossRef]
70. Kato, Y.; Bao, X.; Macgrath, S.; Tabata, K.; Hori, Y.; Tachino, S.; Matijevici, M.; Funahashi, Y.; Matsui, J. Lenvatinib Mesilate (LEN) Enhanced Antitumor Activity of a PD-1 Blockade Agent by Potentiating Th1 Immune Response. *Ann. Oncol.* **2016**, *27*, vi1. [CrossRef]
71. Makker, V.; Taylor, M.H.; Aghajanian, C.; Oaknin, A.; Mier, J.; Cohn, A.L.; Romeo, M.; Bratos, R.; Brose, M.S.; DiSimone, C.; et al. Lenvatinib Plus Pembrolizumab in Patients with Advanced Endometrial Cancer. *J. Clin. Oncol.* **2020**, *38*, 2981–2992. [CrossRef] [PubMed]
72. Finn, R.S.; Ikeda, M.; Zhu, A.X.; Sung, M.W.; Baron, A.D.; Kudo, M.; Okusaka, T.; Kobayashi, M.; Kumada, H.; Kaneko, S.; et al. Phase Ib Study of Lenvatinib Plus Pembrolizumab in Patients with Unresectable Hepatocellular Carcinoma. *J. Clin. Oncol.* **2020**, *38*, 2960–2970. [CrossRef] [PubMed]
73. Llovet, J.M.; Kudo, M.; Cheng, A.-L.; Finn, R.S.; Galle, P.R.; Kaneko, S.; Meyer, T.; Qin, S.; Dutcus, C.E.; Chen, E.; et al. Lenvatinib (Len) plus Pembrolizumab (Pembro) for the First-Line Treatment of Patients (Pts) with Advanced Hepatocellular Carcinoma (HCC): Phase 3 LEAP-002 Study. *J. Clin. Oncol.* **2019**, *37*, TPS4152. [CrossRef]
74. Ogasawara, S.; Llovet, J.; El-Khoueiry, A.; Vogel, A.; Madoff, D.; Finn, R.; Ren, Z.; Modi, K.; Li, J.; Siegel, A.; et al. P-107 LEAP-012: A Randomized, Double-Blind, Phase 3 Study of Pembrolizumab plus Lenvatinib in Combination with Transarterial Chemoembolization (TACE) in Patients with Intermediate-Stage Hepatocellular Carcinoma Not Amenable to Curative Treatment. *Ann. Oncol.* **2020**, *31*, S124–S125. [CrossRef]
75. Raoul, J.-L.; Decaens, T.; Burak, K.; Koskinas, J.; Villadsen, G.E.; Heurgue-Berlot, A.; Bayh, I.; Cheng, A.-L.; Kudo, M.; Lee, H.C.; et al. Practice Patterns and Deterioration of Liver Function after Transarterial Chemoembolization (TACE) in Hepatocellular Carcinoma (HCC): Final Analysis of OPTIMIS in Europe and Canada. *Ann. Oncol.* **2018**, *29*, viii240. [CrossRef]
76. Lingiah, V.A.; Niazi, M.; Olivo, R.; Paterno, F.; Guarrera, J.V.; Pyrsopoulos, N.T. Liver Transplantation Beyond Milan Criteria. *J. Clin. Transl. Hepatol.* **2020**, *8*, 69–75. [CrossRef]
77. Lei, J.-Y.; Wang, W.-T.; Yan, L.-N. Up-to-Seven Criteria for Hepatocellular Carcinoma Liver Transplantation: A Single Center Analysis. *World J. Gastroenterol. WJG* **2013**, *19*, 6077–6083. [CrossRef] [PubMed]
78. Bolondi, L.; Burroughs, A.; Dufour, J.-F.; Galle, P.R.; Mazzaferro, V.; Piscaglia, F.; Raoul, J.L.; Sangro, B. Heterogeneity of Patients with Intermediate (BCLC B) Hepatocellular Carcinoma: Proposal for a Subclassification to Facilitate Treatment Decisions. *Semin. Liver Dis.* **2012**, *32*, 348–359. [CrossRef]
79. Yamakado, K.; Hirota, S. Sub-Classification of Intermediate-Stage (Barcelona Clinic Liver Cancer Stage-B) Hepatocellular Carcinomas. *World J. Gastroenterol. WJG* **2015**, *21*, 10604–10608. [CrossRef]
80. Kudo, M.; Ueshima, K.; Chan, S.; Minami, T.; Chishina, H.; Aoki, T.; Takita, M.; Hagiwara, S.; Minami, Y.; Ida, H.; et al. Lenvatinib as an Initial Treatment in Patients with Intermediate-Stage Hepatocellular Carcinoma Beyond Up-To-Seven Criteria and Child–Pugh A Liver Function: A Proof-Of-Concept Study. *Cancers* **2019**, *11*, 1084. [CrossRef]
81. Kudo, M. A New Treatment Option for Intermediate-Stage Hepatocellular Carcinoma with High Tumor Burden: Initial Lenvatinib Therapy with Subsequent Selective TACE. *Liver Cancer* **2019**, *8*, 299–311. [CrossRef] [PubMed]
82. Tada, T.; Kumada, T.; Hiraoka, A.; Michitaka, K.; Atsukawa, M.; Hirooka, M.; Tsuji, K.; Ishikawa, T.; Takaguchi, K.; Kariyama, K.; et al. Impact of Early Lenvatinib Administration on Survival in Patients with Intermediate-Stage Hepatocellular Carcinoma: A Multicenter, Inverse Probability Weighting Analysis. *Oncology* **2021**, *99*, 518–527. [CrossRef]
83. Bruix, J.; Sherman, M. Management of Hepatocellular Carcinoma. *Hepatology* **2005**, *42*, 1208–1236. [CrossRef]
84. Bruix, J.; Sherman, M. Management of Hepatocellular Carcinoma: An Update. *Hepatology* **2011**, *53*, 1020–1022. [CrossRef]
85. Bruix, J.; Takayama, T.; Mazzaferro, V.; Chau, G.-Y.; Yang, J.; Kudo, M.; Cai, J.; Poon, R.T.; Han, K.-H.; Tak, W.Y.; et al. Adjuvant Sorafenib for Hepatocellular Carcinoma after Resection or Ablation (STORM): A Phase 3, Randomised, Double-Blind, Placebo-Controlled Trial. *Lancet Oncol.* **2015**, *16*, 1344–1354. [CrossRef]
86. Han, B.; Ding, H.; Zhao, S.; Zhang, Y.; Wang, J.; Zhang, Y.; Gu, J. Potential Role of Adjuvant Lenvatinib in Improving Disease-Free Survival for Patients With High-Risk Hepatitis B Virus-Related Hepatocellular Carcinoma Following Liver Transplantation: A Retrospective, Case Control Study. *Front. Oncol.* **2020**, *10*, 562103. [CrossRef]
87. Zhou, J. Safety and Efficacy of Lenvatinib as an Adjuvant Therapy in Patients with Hepatocellular Carcinoma Following Radical Resection: A Single-Arm and Open-Label Prospective Study. 2020. Available online: Clinicaltrials.gov (accessed on 14 November 2021).
88. Chen, J.; Lu, L.; Wen, T.-F.; Huang, Z.-Y.; Zhang, T.; Zeng, Y.-Y.; Li, X.-C.; Xiang, B.-D.; Lu, C.; Xu, X.; et al. Adjuvant Lenvatinib in Combination with TACE for Hepatocellular Carcinoma Patients with High Risk of Postoperative Relapse (LANCE): Interim Results from a Muticenter Prospective Cohort Study. *J. Clin. Oncol.* **2020**, *38*, 4580. [CrossRef]

MDPI
St. Alban-Anlage 66
4052 Basel
Switzerland
www.mdpi.com

*Cancers* Editorial Office
E-mail: cancers@mdpi.com
www.mdpi.com/journal/cancers